Methods of Geometry

Methods of Geometry

James T. Smith
San Francisco State University

A Wiley-Interscience Publication
JOHN WILEY & SONS, INC.
New York • Chichester • Weinheim • Brisbane • Singapore • Toronto

This book is printed on acid-free paper. ♾

Published simultaneously in Canada.

For ordering and customer service, call 1-800-CALL WILEY.

Library of Congress Cataloging in Publication Data:

Smith, James T., 1939–
Methods of geometry / James T. Smith.
p. cm.
"A Wiley-Interscience publication."
Includes bibliographical references.
ISBN 0-471-25183-6 (alk. paper)
1. Geometry. I. Title.
QA445.S57 2000
516—dc21 98-10306

10 9 8 7 6 5 4 3 2 1

This book is dedicated to the place where it was conceived, in the Wind River Mountains. The nearest official landmark is indicated by a cross on a topographical map. That refers to a little brass plate embedded in a rock about the size of your desk. The plate reads

U S Geological Survey
for information write the Director
Washington, DC
Cooperation with the State
Elevation
Above △ sea
10342 feet
GWM 8
1938
Benchmark

The map, based on the 1937–1938 survey, locates the benchmark at 42°54.08′ N, 109°21.40′ W. It's in the Jim Bridger Wilderness, in Wyoming. That world is like geometry: tranquil, but grand; lonely, but populated by wonders. Entering it is an adventure. You'll experience fascination, challenge, and apprehension. And when you become comfortable there, you'll feel great joy.

Contents

Preface

This book is a text for an upper-division general geometry course offered by many universities. It builds on knowledge gained in school and lower-division university courses, to introduce advanced geometry, emphasizing transformations and symmetry groups in two and three dimensions. These theories underly many fields of advanced mathematics, and are essential for application in science, engineering, and graphics. The book analyzes the content of the lower-level courses from an advanced standpoint, to lend prospective and in-service teachers insight into the structure of their subject as well as excitement about its beautiful methods and where they lead.

Mathematical preparation for this course should include vector and matrix algebra and the solution of square and nonsquare linear systems. Appendix C provides a streamlined summary of that material, with most of the nonobvious proofs. The book often uses simple concepts related to equivalence relations, described briefly in appendix A. Concepts related to transformations, their compositions and inverses, and transformation groups are developed in detail in section 6.1, but this book applies them only to isometries and similarities. The least upper bound principle is discussed in Appendix B and applied once in section 3.14, which is about π.

This should probably not be a student's first upper-division course with an algebraic flavor. All topics mentioned in the previous paragraph are indeed covered here, but if this is the first exposure to too many of them, there won't be enough time nor energy for geometry.

Chapter 4 consists entirely of nonroutine exercises related to chapter 3, on elementary Euclidean geometry. These are not the sort that you'd find in a school text. Their goal is to show what you can do with school mathematics beyond what's covered there routinely. You'll also find large exercise sets in the other chapters, to enrich and extend the theory presented in the text, and provide experience with the related computations. It's intended that results stated in some exercises be used in solving later ones.

Often one aim of an exercise is to make students ask questions. They may need help to determine just what the problem is, what constitutes a solution, and where to look for a strategy. Instructors should welcome questions; that's what we're here for. The author generally prefers to see several attempts at most exercises, each one more detailed or progressing farther than its predecessor. In the end students will have portfolios of *correct* solutions, an accomplishment they can build on later in applying the techniques of this book.

A complete solution manual is available to instructors. You can obtain one from John Wiley & Sons, Inc., Mathematics Editorial Office, if you properly identify your academic affiliation. Please use the manual only for class preparation. The solutions are copyrighted, and their further dissemination is not authorized.

When pursuing geometry—or any other subject—it's often instructive for students to study a related area using some of the text's methods, then write a paper on it in a style modeled on the text. That gives practice in independent research, organization, and writing. One strategy for topic selection is to choose a subject that doesn't depend much on what's done in the course after the first weeks. That way, students aren't forced to postpone research and outlining until it's nearly too late. The first two chapters of this book raise a number of questions that could lead to excellent papers, questions to which it gives no specific attention later. Some of those are listed in sections 1.5 and 2.10. You're invited to add to the list.

There's more than enough material in this text for a semester's course. There's *always* more than enough material! A major goal is to prepare students to go further on their own. Chapters 1 and 2 are intended for independent reading. The author tries to present the chapter 3 précis of elementary geometry in about three weeks, and assigns selected chapter 4 exercises at the onset. Some of these may occupy students for the rest of the semester. Alternatively, chapter 4 exercises could be assigned at the beginning, but chapter 3 covered only at the very end. From chapter 5 an instructor is expected to pick favorites. Chapter 6 is essential. Only a little of chapter 7 is required for chapter 8, and that could be covered in a couple of lectures. From chapter 8, pick favorites again. *Warning*: The author has found it only barely within his capacity to present the classification of frieze groups (section 8.2) in lecture. Attempting to classify all the wallpaper groups (8.3) in lecture is unfeasible. A summary or limited selection is possible, however.

The author acknowledges a tremendous debt to those who taught him geometry. His most vivid recollections are of teachers in Springfield, Ohio: Charles Sewell, Dietrich Fischer, and William Armstrong for middle-school shop, art, and algebra; William Waln and Hugh Barber in high-school geometry and trigonometry. In later work, he learned from professors Richard Brauer at Harvard, H. N. Gupta at Stanford and the University of Saskatchewan at Regina, and Friedrich Bachmann in Kiel.

The author drafted and typeset this book entirely in *WordPerfect*. All illustrations not otherwise credited he produced with the DOS *X*(*Plore*) software, by David Meredith (1993). Text was added to the illustrations with *WordPerfect*. The author thanks Prof. Meredith for taking his needs into account when developing that product. Both *WordPerfect* and *X*(*Plore*) work because of the generality in their design.

The author also appreciates the assistance of San Francisco State graduate student Claus Schubert, who read and criticized most of the text and many of the exercises. Finally, editor Stephen Quigley provided support and maximum flexibility with publishers.

[1]

[1] The author's signature is the work of Scott Kim (see Kim 1981). Turn it upside down! Kim's techniques are ingenious. He fashioned the signature at a public book-signing in about one minute. The scratch work is his.

About the author

James T. Smith, Professor of Mathematics at San Francisco State University (SFSU), holds degrees in that subject from Harvard, SFSU, Stanford University, and the University of Saskatchewan. His mathematical research lies mostly in the foundations of geometry, with emphasis on multidimensional transformational geometry. His computer experience began with the UNIVAC I. During the early 1960s, Smith helped the Navy develop mainframe and hybrid digital-analog simulators for radiological phenomena. He's taught geometry and computing techniques since 1969, and was instrumental in setting up the SFSU Computer Science program. During the early 1980s, he worked in the IBM PC software industry. Smith has written seven research papers in geometry and six books on software development.

Chapter 1

Introduction

Recently this author presented to prospective students a five-minute sketch of a geometry course, one of a sequence by various instructors. After the preceding sketch a student asked, "Where will we see applications of that?" So the author began, *Geometry is everywhere*!

Its very name stems from Greek words for *earth* and *measure*. How do you know how far from home you've traveled? The subject evidently originated—long before classical Greece—with problems of measuring land, and problems of constructing things. It extends to space and the stars. Instead of making difficult physical measurements on the ground, we now use the Global Positioning System (GPS).[1] There are satellites in orbit whose positions we know exactly at all times. We determine electronically the distances to four of them, then use programs based on advanced geometry to analyze that data and compute our position with extreme accuracy. The programs use trigonometry developed by ancient Greek mathematicians to solve astronomy problems.[2]

Some of us are enthralled by *pure* mathematics, too. From time to time even this author wants to ignore the world and think, just for peace and enjoyment. Often, geometry has been included as part of pure mathematics. Does it really depend on our physical experience? Ironically, pursuit of pure mathematics enhances the efficiency of our intellectual tools. A theory developed independent of an application can be applied wherever appropriate, often in areas far removed from the problem that stimulated the original study.

This introductory chapter continues with a sampling of experiences that have excited this author about geometry. Perhaps they'll stimulate you, too. They'll suggest ways you might focus to gain insight about the geometry

---

[1] See Kaplan 1996, or—better—search for current GPS information on the Internet.

[2] See van der Waerden 1963, chapter VIII.

of your surroundings. The organization of advanced geometry is outlined in section 1.2. That's followed by an overview of the book and some suggestions for general reading in geometry.

If you're using this as a course text, you may be searching for a paper topic. One strategy for topic selection is to choose a subject that doesn't depend much on what's done in the course after the first weeks. That way, you're not forced to postpone your research and outlining until it's nearly too late. The first two chapters raise a number of questions that could lead to excellent papers, questions to which the book gives no specific attention later. Some of those are listed in sections 1.5 and 2.10.

1.1 Episodes

Concepts

Geometric designs in Southwest American art
Geometric designs in kitchenware
Perspective drawing
Horizon line and vanishing points
Desargues' theorem
Geometry is everywhere!

Art of the Mimbres people

Driving home from an East Coast mathematics conference one winter Sunday afternoon, ready for a break from the highway, the author noticed a small crowd gathering at the public library in Deming, a county seat in southern New Mexico. An opportunity? They were assembling for a lecture on Mimbres pottery. A *geometric* opportunity! The author browsed through the adjoining museum, noting the fine examples of craftsmanship and art recovered from ancient village sites near the Mimbres River, which runs through this small town. Some of the displays had been organized by J. J. Brody, of the University of New Mexico, the scheduled speaker. Prof. Brody showed and interpreted stunning work from about 1000 years ago. The water bowl in figure 1.1.1 depicts curlews and eels the artist might have seen at the river, arranged with twofold symmetry. If you rotate the bowl 180° about its axis, it appears the same. This bowl is about five inches deep and ten in diameter. The artist had to distort the images on its curved inner surface so they would look right when viewed from above! Figures 6.0.1 and the more abstract 8.0.1 and 8.5.2b are more abstract examples of Mimbres designs. From early times to the present, the many cultures of the American

Figure 1.1.1
Mimbres bowl[3]

Figure 1.1.2
Potato mashers

Southwest have produced wonderful geometric art.[4] A journey through that region is a succession of illustrations of many ideas in this book, particularly its study of symmetry in chapter 8.

Mashers

Two years later, after a summer conference, the author was shopping in an antique store in Atlanta, Georgia. Look! Potato mashers! He created a stir, rearranging the display to compare them. They're different! You can see that, in figure 1.1.2. One masher is bilaterally symmetric; its left and right halves are mirror images. It would display the twofold, translational and reflectional symmetry of a sine wave were it continued indefinitely. Another is distorted to break the reflectional symmetry.[5] The wire pattern of the third has been doubled—its upper and lower parts are mirror images. Why did the designers employ these symmetry types? Are they purely decorative, like those on the Mimbres kitchen bowl? Or do these designs enhance the tools' usefulness? Investigate some antique stores. You should be able

---

3 Brody, Scott, and LeBlanc 1983, figure 114.

4 The masterpieces of the contemporary Acoma artists of the Chino and Lewis families (see Dillingham 1994) incorporate some of the most intricate geometric designs this author has ever seen. You'll find more examples in fine art shops in the Southwest.

5 Dan Wheeler provided this masher.

to find at least two other types of symmetry employed in potato masher designs. Analyze other tools the same way. Learn about symmetry in design. And watch the proprietors' reactions when you reveal that you're studying *geometry*!

A mural in Napier

Several years ago, in Napier, New Zealand,[6] the author, a tourist, somewhat academic in appearance, very much so in demeanor, was driving along the main street, downtown. Suddenly he pulled aside, leaped out with camera, strode into the street at risk of life and limb, gestured vigorously to stop traffic in both directions, and snapped a photo of a scene alongside the street! What wonder could have stimulated that wild reaction? All downtown Napier is geometrically exciting. It was destroyed by a great earthquake in 1931, and rebuilt during the next few years. In what style? What was then current? Of course: Art Deco, with its parallels and streamlines and dynamic curves.[7] Did a striking facade seize the academic's attention? No, but the scene surely stemmed from an artist's absorption with that style. The snapshot is figure 1.1.3. It shows a mural on a fence that shields a construction project: a little plaza between street and harbor. The mural depicted what the viewer would see when construction was complete.

Was it accurate? That's not an entirely fair question, since the mural didn't *need* to be accurate, but rather suggestive and exciting. This was the same situation artists met, or created, during the Renaissance in Italy. Earlier painters had designed for psychological effect by choosing the position and size of their figures; the most important were most central and largest. They made little attempt to produce on flat surfaces pictures corresponding closely to what we see in three dimensions. To fifteenth-century Italian artists, the effect produced by realistic depiction of three-dimensional scenes became important. They wanted us to see—or *believe* we're seeing—some aspects of the world the way they *really are*.

You need mathematics to draw a three-dimensional scene realistically on a flat surface. The Italians and their German student Dürer studied the technique—called *perspective* drawing—deeply. They mastered it, and soon began producing magnificent examples, competing with each other in virtuosity. One idea, described in texts of those times, was to regard

- the drawing surface as a window in a vertical plane ε between observer and scene (see figure 1.1.4),
- each point P of the scene connected with the observer's eye O,

[6] Napier is named for a nineteenth-century British military figure, a descendant of the Scottish mathematician John Napier (1550–1617), who invented logarithms.

[7] See Duncan 1988.

Figure 1.1.3
Mural in Napier

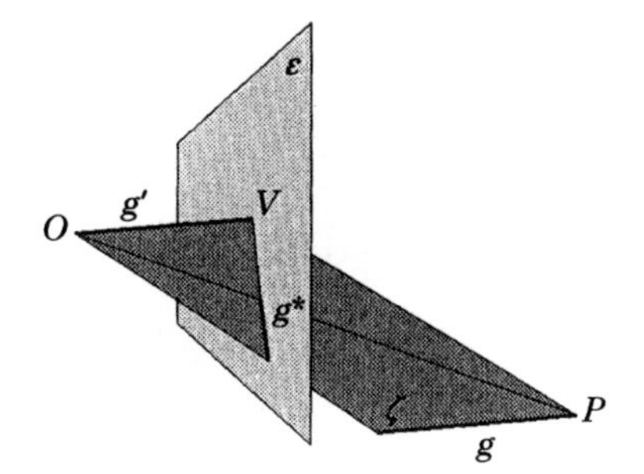

Figure 1.1.4 Creating a perspective image

- and the intersection $\vec{OP} \cap \varepsilon$ as the image of point P.[8]

The image is part of the drawing if it falls in the window. Consider the image of a line g in the scene. It's formed by the intersections with ε of all lines through O and points on g. Should O happen to fall on g, the image would be a single point on ε. Otherwise, it would be the intersection $g^* = \varepsilon \cap \zeta$ of ε with the plane ζ determined by O and g. That image g^* is a line on ε.

Was the Napier mural constructed according to this principle? Look at it. The lines separating the tiles on the ground appear as lines in the mural. But on the ground, you could draw diagonal lines g through a sequence of opposite vertices of the rectangular tiles. Those lines don't pass through O, because they're on the ground; so their images should be *lines*. Is that so? Test figure 1.1.3 with a straightedge. No! The images of the diagonals aren't straight! That's what caused the excitement that stopped traffic in downtown Napier! It's precisely the same phenomenon you'll witness if you inspect many European paintings created during the century or so after the introduction of perspective drawing. Other artists wanted to produce the effect of perspective, but didn't know exactly how, and produced unrealistic—although probably not traffic-stopping—curved diagonals.

Now continue studying figure 1.1.4. Let g' denote the parallel to g through O. Suppose $g \not\parallel \varepsilon$ and let V be the intersection $\varepsilon \cap g'$ of ε with g'. Then V is on the image g^* because g' lies in ζ. Consider any other line g_1 in the scene, not through O, with $g_1 \parallel g$, as in figure 1.1.5. Use it to construct plane ζ_1, line $g_1^* = \varepsilon \cap \zeta_1$, line $g_1' \parallel g_1$ through O, and point $V_1 = \varepsilon \cap g_1$ analogously, so that V_1 lies on the image g_1^* of g_1. Then $g_1' = g'$ because $g_1 \parallel g$, so $V_1 = V$. That is, *if* $g \not\parallel \varepsilon$, *then the images*

---

[8] The term *perspective* reflects this concept; it stems from the Latin verb *specio* and prefix *per-*, which mean *see* and *through*.

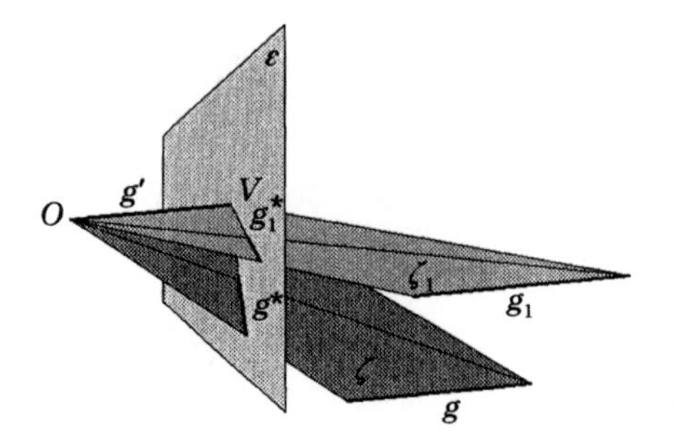

Figure 1.1.5
Images of parallel lines

of all lines in the scene parallel to g *pass through the single point* V, which is called their *vanishing point.* On the other hand, *if* $g \parallel \varepsilon$ *then* $g \parallel g^*$, *and the images of all lines in the scene parallel to* g *are themselves parallel.*

In the Napier mural, the lines g lay on the ground, a plane $\gamma \perp \varepsilon$. But the discussion leading to the notion of vanishing point made no use of that; it holds for the set of all parallels to *any* line g. What's special about lines g on the ground? It's more appropriate to consider lines g *on or parallel to* γ. For all those, $g' \parallel \gamma$; these g' all lie in the plane $\delta \parallel \varepsilon$ through O. Their vanishing points lie in the intersection $h = \delta \cap \varepsilon$, a line in ε parallel to γ at the same distance from the ground as O. That is, the vanishing points of all lines parallel to the ground γ lie in a line h, called the *horizon*, the same distance above ground as O.

Artists don't really erect windows and run cords through them from their noses to points in scenes they're drawing. Can you produce a perspective drawing using only points on the drawing plane? For example, a true perspective drawing of the tiled plaza? Figure 1.1.6 shows a perspective drawing $PQRS$ of *one* rectangular tile $\mathcal{T}$ on the ground. The edges of $\mathcal{T}$ that correspond to lines $\overleftrightarrow{PQ}$ and $\overleftrightarrow{RS}$ are parallel to the drawing plane, and the other edge lines intersect at point Y on the horizon h. Figure 1.1.7 shows how to draw the image $RSTU$ of a tile adjacent to $PQRS$:

1. draw $\overleftrightarrow{PR}$,
2. locate $Z = h \cap \overleftrightarrow{PR}$,
3. draw $\overleftrightarrow{SZ}$,
4. locate $U = \overleftrightarrow{QY} \cap \overleftrightarrow{SZ}$,
5. draw $\overleftrightarrow{QS}$,
6. locate $X = h \cap \overleftrightarrow{QS}$,
7. draw $\overleftrightarrow{RX}$,
8. locate $T = \overleftrightarrow{PY} \cap \overleftrightarrow{RX}$.

It *looks* like $\overleftrightarrow{TU} \parallel h$, so that $\overleftrightarrow{TU}$ is the image of another line on the ground parallel to the drawing plane. Can we *prove* that? Another way to phrase the same question is to consider the following alternatives to steps 5 to 8:

5.′ draw $g \parallel h$ through U,
6.′ locate $T' = g \cap \overleftrightarrow{PY}$.

Condition $\overleftrightarrow{TU} \parallel h$ is equivalent to equation $T = T'$, which says that the two methods produce the same result.

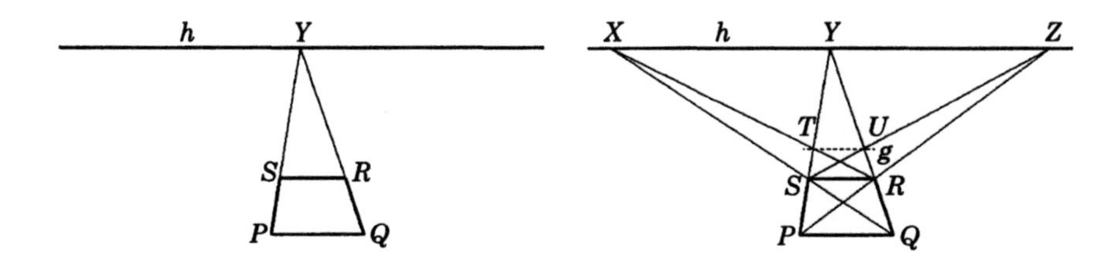

Figure 1.1.6 Perspective image of one tile

Figure 1.1.7 Perspective image of two tiles

There are many related questions. You're invited to adapt this construction to draw adjacent tile images left and right of $PQRS$, then images of its diagonal neighbors above them. You could proceed from either of those back to the image of the tile immediately above $PQRS$. Do you *necessarily* get the same image $RSTU$? Should you answer all these consistency questions affirmatively, another big one would remain: Is such a consistent plane diagram always a perspective image of *something*? That is, can you construct a corresponding three-dimensional scene? You could imitate mathematicians' work over the centuries, using elementary geometry to settle questions such as these. You'd find it frustrating; the arguments aren't difficult individually, but there are so many intertwined problems, and no organizing system is obvious. As an introduction to some of the methods in chapter 5—elementary geometry beyond that covered in school—the following paragraphs prove that $\overleftrightarrow{TU} \parallel h$ in figure 1.1.7. You may find the proof roundabout and unnecessarily complicated. There's certainly a simpler proof of this particular result. But it's only one of a host of related questions. Mathematicians sought a general technique that would apply to the whole class of problems.

Desargues' theorem

By the late nineteenth century mathematicians realized that a theorem, proved by Gérard Desargues two centuries before, played a central role in this study. Several forms of his result are derived in section 5.3. One of them, the parallel case of theorem 5.3.1, is used here:

> Consider points $A = A$, $B = B$, and $C = C$ on distinct lines l, m, and n, where $l \parallel n$, as in figure 1.1.8. Suppose $\overleftrightarrow{AB}$ intersects $A\ B$ at N, $\overleftrightarrow{BC}$ intersects $B\ C$ at L, and $\overleftrightarrow{CA}$ intersects $C\ A$ at M. Then $l \parallel m$ if and only if L, M, and N fall on a single line.

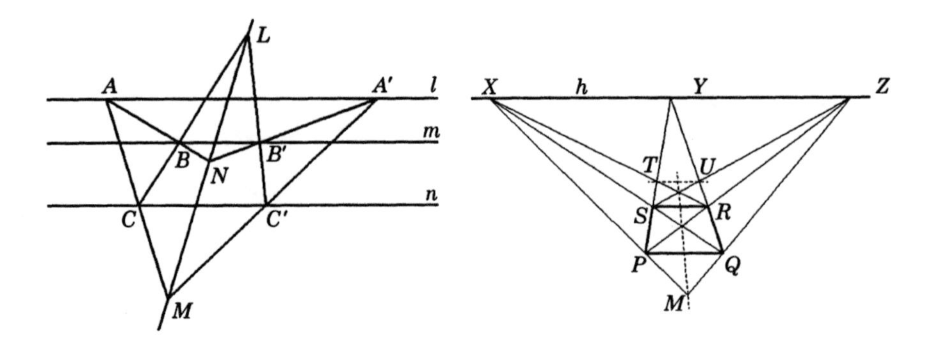

Figure 1.1.8
A theorem of Desargues

Figure 1.1.9
Proving $\overleftrightarrow{TU} \parallel h$

It's stated with different notation, so you need to work out the correspondence between points in the two figures:

$$\begin{array}{rlrl} & \textit{Figure 1.1.8} & & \textit{Figure 1.1.9} \\ & N & & \overleftrightarrow{RX} \cap \overleftrightarrow{SZ} \\ A, B, C;\ A', B', C'; & L & X, R, P;\ Z, S, Q; & \overleftrightarrow{PR} \cap \overleftrightarrow{QS} \\ & M & & \overleftrightarrow{PX} \cap \overleftrightarrow{QZ}. \end{array}$$

According to the *only if* part of Desargues' theorem, the indicated points in figure 1.1.9—the three intersections listed vertically at right above—do fall on one dotted line. Now apply the theorem again, to a different pair of triangles:

$$\begin{array}{rlrl} & L & & Y \\ A, B, C;\ A', B', C'; & N & X, S, P;\ Z, R, Q; & \overleftrightarrow{PR} \cap \overleftrightarrow{QS} \\ & M & & \overleftrightarrow{PX} \cap \overleftrightarrow{QZ}. \end{array}$$

Thus, according to Desargues' theorem, Y falls on the dotted line, too. Finally, apply the *if* part of the theorem to a third pair of triangles:

$$\begin{array}{rlrl} & L & & Y \\ A, B, C;\ A', B', C'; & N & X, T, P;\ Z, U, Q; & \overleftrightarrow{RX} \cap \overleftrightarrow{SZ} \\ & M & & \overleftrightarrow{PX} \cap \overleftrightarrow{QZ}. \end{array}$$

In this situation, lines l, m, and n in figure 1.1.8 correspond to h, $\overleftrightarrow{TU}$, and $\overleftrightarrow{PQ}$. Since the three points listed vertically at right above fall on the dotted line, Desargues' theorem implies $h \parallel \overleftrightarrow{TU}$, which was to be proved.[9]

[9] This proof should really verify that the intersections mentioned all exist. If any could fail to—that is, if certain lines could be parallel—additional arguments would be necessary.

Desargues' theorem has entered this argument three times. Its various forms appear whenever you consider perspective questions. In exercise 5.11.13 you're invited to apply another form of the theorem to draw an analogous picture with an image plane that's vertical, but neither parallel nor perpendicular to any tile edge. The most general form of Desargues' theorem is used as an axiom in *projective geometry*, the theory developed by Desargues and others over the next 250 years to handle questions in perspective drawing.[10]

Geometry is everywhere!

These vignettes may suggest that you'll find geometry in unexpected places. But that's not the right idea. You should *expect* it. Look about you. Enjoy what you see! Study the background, theory, and applications in this book, its exercises and projects. Working through these should convince you that *geometry is everywhere*!

1.2 Advanced geometry

Concepts

AMS *Mathematics subject classification*
Algebraic and differential geometry; topology
Convex and discrete geometry; polyhedra
Linear incidence geometry and ordered geometries; finite geometry
Metric geometry and transformational geometry
Geometric constructions
Extremum problems
Hyperbolic and elliptic non-Euclidean geometries
Projective geometry and homogeneous coordinates
Computational geometry

This book is an introduction to a huge field in mathematics. The book itself covers only a tiny part of that. The field is so large that experts may not agree on whether some current research is in fact geometry. Geometry researchers usually cannot understand current work in other parts of the field. In fact, many of us have never considered an organized, thorough overview of geometry. This book can't provide that. But this section will attempt a rough sketch, so you can place the book in context and get some idea of directions you might follow later.

---

[10] For historical information, consult Cromwell 1997, chapter 3; Field 1997; and Pedoe 1976.

Consider the American Mathematical Society's *Mathematical subject classification* (1991–). You'll find about sixty main classes. Each has several subclasses, and some of those have a pageful of subsubclasses. Standard surveys of geometry (consult the references in section 1.4 for tastes of most areas) mention subjects in six classes:

14. *Algebraic geometry* (with 15 research subclasses)
51. *Geometry* (14 subclasses)
52. *Convex and discrete geometry* (3 subclasses)
53. *Differential geometry* (4 subclasses)
55. *Algebraic topology* (8 subclasses)
57. *Manifolds and cell complexes* (7 subclasses)

The *algebraic geometry* class is listed nearer other algebra classes than the rest of geometry. It's devoted to the analytic geometry of polynomial equations and systems of such equations, excluding methods that make essential use of calculus. Investigations where derivatives play major roles—for example, the study of arc length and curvature—are classified as *differential geometry*. In some parts of geometry the size of objects plays a major role; in others, size is secondary to shape. For example, if you study photographs of a scene, size isn't very relevant; you seldom distinguish between corresponding features of different enlargements of the same negative. In *topology*, some aspects of shape are irrelevant. Topologists regard a sphere as essentially unchanged if you stretch or squeeze it like a balloon, as long as you don't pinch it together or tear it. Topologists study properties like connectivity: How many cuts of what sort are necessary to disconnect a rubber pretzel? The answer would change if you pinched it together or tore it first, but not if you merely stretched or squeezed it. Topology problems have led to so much mathematics that two of these classes, *algebraic topology* and *manifolds and cell complexes*, are devoted to them. The former is distinguished by its emphasis on group-theoretic methods.

This text contains hardly any material from the classes mentioned in the previous paragraph. The situation changes with class 52, *convex and discrete geometry*. It has three subclasses:

- 52A. *General convexity*
- 52B. *Polytopes and polyhedra*
- 52C. *Discrete geometry*

Subclass 52A doesn't overlap this book much, although convex sets play major roles in chapters 3 and 8. In section 8.4, though, you'll study in detail many polyhedra—certain surfaces bounded by finitely many polygonal regions. Related current research—particularly in higher dimensions—could belong to 52B. The third subclass, *discrete geometry*, is the place for questions about how tiles fit together to fill a plane area, how cells analogous to those in a honeycomb fit together to fill space, etc. One long-unsolved problem in this class is whether we can pack an extremely large number of balls of the same size more efficiently than the arrangement suggested by figure 8.0.7.

The central class 51, *geometry*, is appropriate for most of the material in this book. Questions arising from detailed study of the axioms in chapter 3, and related axiom systems, fall in several subclasses of class 51:

51A. Linear incidence geometry (related to the section 3.1 incidence axioms)
51E. *Finite geometry and special incidence structures*
51F. *Metric geometry* (related to the notions of congruent figures, perpendicular lines, etc.)
51G. *Ordered geometries* (related to the notion of order of points on a line)
51M. *Real and complex geometry*

Questions such as whether any of the incidence axioms can be derived from others lead to the study of finite systems of points, lines, and planes that satisfy some axioms but not others: subclass 51E.[11] This is a surprisingly large area of geometry. Much of it you'd regard as pure mathematics, but it is applied in the design of computer codes and statistical experiments.

Chapters 6 and 7 present the theory of isometries and similarities in two and three dimensions. These are correspondences between points of a plane, or between points of space, that preserve distances or multiply distances by a fixed ratio. If you slide a photograph across a table, the correspondence between the original location of each feature and its new location is an isometry. If you compare that photograph with a different enlargement, the correspondence between the locations of a single feature in the two prints is a similarity. Because you can define all relevant properties of these correspondences in terms of distance, their theory falls entirely in subclass 51F, *metric geometry*. These correspondences are special kinds of mathematical transformations, so this theory is also part of *transformational geometry*.

Subclass 51M, *real and complex geometry*, is a catchall. Here are some of its subsubclasses:

51M04. *Elementary problems in Euclidean geometries*
51M10. *Hyperbolic and elliptic geometries (general) and generalizations*
51M15. *Geometric constructions*
51M16. *Inequalities and extremum problems*

Most of the material in chapters 4 and 5 belongs to 51M04. Subsubclass 51M15 is devoted to the process of constructing geometric figures with limited tools. Many of us learned to use ruler and compass for a large class of such problems. Identifying figures that *cannot* be constructed that way requires

---

[11] See exercises 4.1.2 to 4.1.6.

advanced mathematics in this area. Subsubclass 51M16 includes such famous extremum problems as determining which smooth closed curves with a given length contain the largest area. You guessed it—circles! But that's not easy to prove. Recently, researchers have intensely studied compound soap bubbles. For example, how do four of these with a given total volume abut each other about a common point so that their total area is minimized?

51M10 contains *non-Euclidean* geometry: the study of geometric structures that fail to satisfy the Euclidean parallel axiom. According to Euclid's axiom, given a line g and a point P not on g, there's exactly one parallel to g through P. It can be shown that if Euclid's axiom failed but all the others held, then either there would *always be more than one* parallel to g through P, or else there would *never be any* parallel to g through P. These are called the *hyperbolic* and *elliptic* cases.

An important area has almost escaped notice: *projective geometry.* It's part of subclass 51A, *Linear incidence geometry.* In projective geometry, two lines in the same plane are always regarded as intersecting, as they might appear in a perspective drawing. This theory was first developed in the seventeenth century to handle that application. The modern version of the application belongs to *computer graphics.* The corresponding part of analytic geometry uses *homogeneous* coordinates; points in the plane correspond to coordinate *triples* $<x,y,z>$, and two such triples correspond to the same point if one is a multiple of the other. For example, $<1,2,3>$ and $<2,4,6>$ correspond to the same point. During the early nineteenth century, mathematicians found that algebraic geometry is simplified by using homogeneous coordinates. During the late nineteenth century, others found that the two non-Euclidean geometries mentioned in the previous paragraph, as well as Euclidean geometry, can be regarded as the studies of special sets of points in projective geometry, defined by three special cases of a single condition. Projective geometry plays such a fundamental role in graphics, algebraic geometry, and non-Euclidean geometry, that it should be the next geometry course you study after this book.

One further area of geometry is part of subsubclass 68Q20, *nonnumerical algorithms,* in class 68, *computer science.* This research, currently of great technological interest, is concerned with the efficiency of the computations involved in all the areas already mentioned. In earlier years, many of them required so much time that putting them to use in practical computer programs was entirely unfeasible. Therefore no great effort was made to refine the algorithms. But fast computer hardware is now so cheap that these computations are possible.[12] Sophisticated geometric algorithms underlie even the most popular graphics software. The new area of *computational geometry* is devoted to the development and refinement of these algorithms.

---

[12] The author changed computers three times as this book was written. His current machine is about five hundred times faster than the first one, but they cost about the same.

1.3 This book

Concepts

- Overview
- Two- and three-dimensional geometry
- Vector and matrix algebra for analytic geometry
- Equivalence relations and least upper bounds
- Familiar concepts and notation
- Numbering results, exercises, and figures
- Symbol for terminating proofs
- Exercises and projects
- Bibliography
- Historical and etymological remarks

Overview

This book builds on mathematical knowledge you've gained in school and lower-division university courses, to introduce advanced geometry. Chapters 1 and 2 provide background material, setting the subject in its intellectual context. The next two chapters review in considerable detail the content of school geometry courses from a more advanced viewpoint using sophisticated organization and more detailed arguments. They feature a wealth of non-routine exercises to help you develop your techniques. This type of geometry doesn't stop there. Chapter 5, on triangle and circle geometry, extends these methods to open up several fields, including a few aspects of analytic geometry and trigonometry used in more advanced courses but rarely covered adequately in introductory ones. A new approach starts with chapters 6 and 7: the geometry of motions and similarities, often called *transformational* geometry. This theory melds concepts of motion, originating from synthetic techniques of school geometry, with analytic geometry phrased in terms of vector and matrix algebra. It's used to analyze dynamic systems, whose objects may change position and size. On the other hand, it's used to keep track of the description of an object relative to a changing coordinate system. When you study a transformation of a system, you often gain information by concentrating on aspects it does *not* change. An object's *symmetry* properties are those unchanged by motions that transform it. For example, sliding an infinite checkerboard two squares horizontally or vertically doesn't change it: The checkerboard displays two-dimensional translational symmetry. Rotating a milk carton 180° about its vertical axis doesn't change it: The carton displays twofold rotational symmetry. Chapter 8 is a thorough introduction to symmetry theory and some of its applications, particularly in art and design.

Three-dimensional geometry receives the same emphasis in this book as plane geometry. That should fill gaps in many students' backgrounds.

Although the solid geometry course disappeared from schools about the same time the author was graduated in 1957, contemporary applications in biology, chemistry, design, and engineering—for example, the GPS mentioned earlier—are *essentially* three-dimensional. The axiom system presented in chapter 3 is three-dimensional, and spherical trigonometry is included in chapters 4 and 5. Chapter 7 is devoted to the theory of motions in space, and section 8.4 applies some of its concepts to study polyhedra and their symmetries.

Appendices

Analytic geometry is best pursued using vector and matrix algebra and the theory of systems of linear equations. For three and higher dimensions, they're virtually indispensable. Most students encounter this algebra first in school, then scattered among lower division calculus and physics courses. Appendix C presents an outline of the topics from this area that you'll need. It's all in one place, presented with consistent notation. A few additional concepts related to vector products are considered in detail in section 5.6.

Appendices A and B briefly cover two more topics used here but not normally covered in geometry courses: equivalence relations and the least upper bound principle.

Familiar concepts and notation

This text assumes that you're familiar with logical and set-theoretic concepts and notation commonly used in school and lower-division mathematics. In particular, it uses symbols &, $\Rightarrow$, and $\leftrightarrow$ for *and*, *implies*, and *if and only if*. Moreover, notation like

the function $x \to x^2$ stands for *the function* f *such that* $f(x) = x^2$ *for all appropriate* x;

a function $g : A \to B$ stands for *a function* g *with domain* A *and range included in* B.

The abstraction operator $\{:\}$ is used to build sets such as $\{x : x \geq 0\}$, and the empty set is denoted by $\phi = \{x : x \neq x\}$. Finally, the membership and subset relations $\in$ and $\subseteq$ are used frequently, as are the union and intersection operators $\cup$ and $\cap$.

Many figures in the text imitate the style used in elementary geometry books, with tick marks like + indicating equal segments, curved ticks like ∡ indicating equal angles, and symbols like ∟ for right angles.

Numbering of results, exercises, and figures; proofs

Items in the text are numbered by chapter and section. For example, *Lemma 3.5.6* and *Theorem 3.5.7* (*Triangle inequality*) are the sixth and seventh numbered results in section 3.5. The former is a lemma, or auxiliary theorem; the latter, a noted result that will be referred to later by name. Within section 3.5, they might be referred to simply as lemma 6 and theorem 7. Proofs of such results are usually preceded by the word *Proof* and terminated by the symbol ♦. Sometimes, however, proofs are part of the discussion leading to the statement of a theorem. Exercises and figures are numbered similarly. Items in chapter introductions are numbered as though they were in section 0.

Exercises and projects

Exercises are gathered in special sections. Chapter 4 consists entirely of nonroutine exercises related to chapter 3, *Elementary Euclidean geometry*. These are not the sort of exercises you'd find in a school text. They're much more substantial, and often involve several topics. The final sections of the other chapters contain all the exercises for those chapters. Each of these sections is clearly subdivided into exercises on various subjects. When you start a chapter, check out the organization of the corresponding exercises. As you read through the chapter, refer to the appropriate group of exercises for ones that will help you consolidate your knowledge. There were two reasons for not distributing the exercises into the text material. First, keeping them separate makes the text flow more gracefully. Second, in some instances the organization of the exercises is different from the text's. Sometimes, an exercise appropriate for one topic necessarily involves others, and placing it with the material on just one of those would be misleading.

It's intended that results stated in some exercises be used in solving later ones.

Often the intent of an exercise is to make you ask questions. You may need help to determine just what the problem is, what constitutes a solution, and where to look for a strategy. Don't be reluctant to ask. That's what instructors are for. This author generally prefers to see several attempts at most exercises, each one more detailed or progressing farther than its predecessor. If you can arrange that, you'll have in the end a portfolio of accomplishments that you can use later in applying the techniques of this book.

When pursuing geometry—or any other subject—it's often instructive to study a related area using some of the text's methods, then write a paper on it in a style modeled on the text. This gives practice in independent research, organization, and writing. One strategy for topic selection is to choose a subject that doesn't depend much on what's done in the course after the first weeks. That way, you're not forced to postpone your research and outlining until it's nearly too late. The first two chapters of this book raise a number of questions that could lead to excellent papers, questions to which it gives no specific attention later. Some of those are listed in sections 1.5 and 2.10.

Bibliography

The bibliography contains references to all material cited in the text.[13] For example, the reference "Altschiller-Court [1935] 1964" leads to the second item in the bibliography. That particular edition is dated 1964, but the original one was published in 1935. When the author's identity is clear, a reference may consist simply of the date(s) in parentheses. Bibliographic entries present ISBN numbers and Library of Congress (LC) catalog numbers when they're available. Many include brief descriptive comments.

Historical and etymological remarks

Relevant historical comments are included throughout the text, with biographical sketches of some of the mathematicians responsible for this material. Don't regard these as authoritative; the author is no historian. Generally, no references are given for biographical material. Your best single source for that is the *Dictionary of scientific biography*.[14] The author gleaned much information as well from published biographies and histories; and from obituaries, commemorative biographical sketches, and historical articles in journals.

Original meanings of mathematical terms sometimes help you understand the mathematics, and they're almost always entertaining. This text includes many notes on the etymology of terms that occur here but are uncommon outside geometry. Your best source for more information of that kind is Schwartzman's *Words of mathematics* (1994).[15]

1.4 Reading about geometry

Concepts
Survey literature

You may want to spend some time reading about geometry in general— diverse parts of the outline in section 1.2—while you work through the foundations and elementary topics in chapters 2 and 3. That would be appropriate, for example, if you were searching for a topic for independent, parallel study. You may find some of the following references accessible and interesting:

[13] And all items cited in the *Instructor's solution manual*.

[14] Gillispie 1970.

[15] Schwartzman is thorough; for example, he includes entries for about 80% of the nouns that occur from the beginning of section 5.1 through the proof of theorem 5.1.2.

Beck, Bleicher, and Crowe 1969 . . . . *Excursions into mathematics*
Coxeter 1969 . . . . *Introduction to geometry*
Eves 1963–1965 . . . . *A survey of geometry*
Forder [1950] 1962 . . . . *Geometry: An introduction*
Hilbert and Cohn-Vosson [1932] 1952 . . . . *Geometry and the imagination*
Klein [1908] 1939 . . . . *Elementary mathematics from an advanced standpoint: Geometry*
Melzak 1983 . . . . *Invitation to geometry*
Russell [1897] 1956 . . . . *An essay on the foundations of geometry*
Tuller 1967 . . . . *A modern introduction to geometries*

Some references in section 1.1 and in the introduction to chapter 8, although specialized, may be suitable as well, since they present a variety of aspects of geometry at an introductory level.

1.5 Projects

Concepts

Elementary geometry in school and university
Conics and quadrics
Geometric knowledge and physical experience
Hyperbolic and elliptic area theory
Perspective drawing
Orthogonal projection
Art Deco

When pursuing geometry—or any other subject—it's often instructive to study a related area using some of your text's methods, then write a paper on it in a style modeled on the text. This gives practice in independent research, organization, and writing. One strategy for topic selection is to choose a subject that doesn't depend much on what's done in the course after the first weeks. That way, you're not forced to postpone your research and outlining until it's nearly too late. This chapter has raised a number of questions that could lead to excellent papers, questions to which it gives no specific attention later. For some of these topics, you can follow references cited elsewhere in this text. But for others, no references are offered. Search your library and ask faculty members for advice.

Project 1. How is geometry taught to children? The author doesn't remember anything specifically about geometry in his schooling until middle-school shop and art. (Girls didn't take shop!) How is it done now? You might want to compare an approach familiar to you with a different one.

Project 2. How is geometry taught in secondary schools? Has secondary-school geometry changed over the years? Compare your experience with a different one.

Project 3. What geometry is presented in elementary mathematics, science, and engineering courses in university? How has this changed over the years?

Project 4. Some years ago, conic sections and quadric surfaces received more attention in analytic geometry and calculus courses than they do now. Present a more complete treatment than that in your calculus text, using to best advantage some of the methods in this book.

Project 5. Elite education in nineteenth-century England relied on Euclidean geometry to an extreme degree. Describe this practice and its effects.[16]

Project 6. Is geometric knowledge dependent on physical experience? (There's an enormous literature on this subject.)

Project 7. Develop the theory of area in hyperbolic or elliptic geometry.

Project 8. Investigate the development and/or use of perspective techniques in fine or practical art.

Project 9. Use a software package to demonstrate, in perspective drawing, the effect of changing the relationships of eye, drawing plane, and important features of the scene.

Project 10. How do orthographic, isometric, trimetric, and oblique projections differ from perspective drawing? What are their advantages and disadvantages?

Project 11. Investigate the use of geometric design in the Art Deco style.

Project 12. Investigate the life and work of David Eugene Smith. The editor of several items in the bibliography, Smith wrote the solid geometry text[17] that the present author studied in secondary school.

---

[16] See Richards 1988 and Russell [1951] 1967.

[17] Smith 1924. David Eugene Smith is not related to the present author.

Chapter

2

Foundations

This book presents geometry as a branch of *applied* mathematics: the mathematical theory of certain properties of space. That classical viewpoint contrasts with a more recent view of geometry as a branch of *pure* mathematics, studied for its own sake, independent of concrete applications. This chapter is a brief historical introduction to foundational studies in geometry—the investigation of its most basic concepts and principles.

Pure mathematics began to dominate our discipline about 1900, largely because of its efficiency and flexibility. That era also saw the first establishment of truly solid foundations for geometry. This chapter displays the connection. It places in context the axiomatic structure of Euclidean geometry presented in chapter 3. Its conclusion, section 2.9, shows how you can regard a suitably formulated applied mathematical theory also as pure mathematics. The difference lies not in what mathematics you do, but in why you do it and how you describe it. Thus you can actually read chapter 3 with an applied or a pure mathematics interpretation.

Much of chapter 3 should be familiar to you from elementary mathematics courses. It assumes a knowledge of the arithmetic and algebra of real numbers, then develops from axioms the traditional synthetic techniques of solid geometry, and justifies the use of coordinates. The care for detail and rigor, however, reflects the standards of advanced mathematics. The theorems derived in chapter 3 are used extensively in the rest of the book to study a great variety of geometric phenomena.

The present chapter, on foundations, shows what a geometric axiom system must do, and why. It discusses the extent to which traditional Euclidean geometry met those needs, and how it evolved into a more adequate framework. It emphasizes why the chapter 3 axiomatization was chosen from several alternatives, and explains the need for detail and rigor.

Each major area of mathematics includes a subdiscipline devoted to foundations. Most students consider this first, even if briefly and to little depth. But the foundation of a mathematical discipline rarely develops first historically. Mathematical problem solving combines the study of examples, imaginative speculation, precise and extensive deduction or calculation, and checking tentative results against examples. Mathematicians know—but rarely admit publicly—that the trial and error aspect of mathematics is mostly error. Usually, a mathematical subject evolves for years before it's clear of false starts, misconceptions, and inefficient reasoning. Only then do the critical foundational questions become apparent, and ripe for mathematicians to attack and settle. For geometry this process took not years, not centuries, but millennia.

Foundations of geometry is a broad field with specialized subdivisions, but only the background for the one axiom system in chapter 3 is relevant to this book. You can't use chapter 2 as a general introduction to foundations of geometry, because other topics are mentioned at most fleetingly. Read it once, before chapter 3, to see what lies ahead, so that you can recognize landmarks of that theory as you pass them. Studying chapter 3 in detail, you'll probably lose sight of its overall purpose. So when you've finished, read this foundational chapter again, to review what you've accomplished.

2.1 Geometry as applied mathematics

Concepts

- Mathematical modeling
- An example in astronomy
- Testing for accuracy, not correctness
- Testing a geometric model
- Conventionalism

Applied mathematics is characterized by its use of *models*. The process circulates, as shown in figure 2.1.1. A *model* is a collection of precisely formulated concepts and principles that describe certain aspects of the subject under study. (A "principle" is a statement involving the concepts.) The *theory* consists of all true statements, or *theorems*, about the model. Normally, a subject area specialist—a scientist—*formulates* the concepts and principles that make up the model. Specialists and mathematicians together *deduce* interesting theorems. Theorems don't refer to the subject matter directly; they stem from the model alone. Scientists then *verify* some theorems against empirical data from the subject area. *Agreement* of theorems with data confirms the model's accuracy. You can use further theorems about a model that has been judged suitably accurate, to explain or predict events

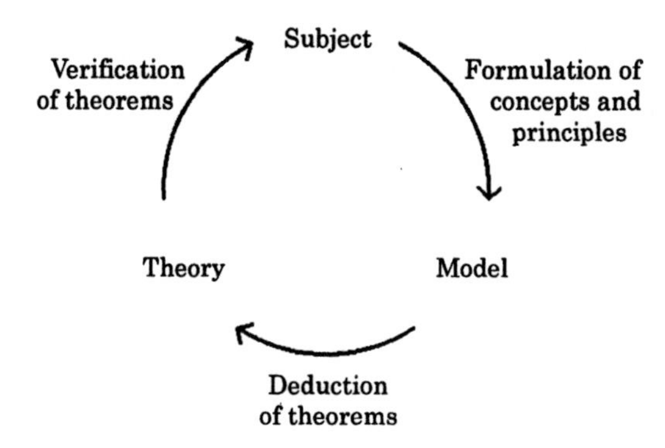

Figure 2.1.1 Mathematical modeling

concerning the subject area. If testing reveals *disagreement*, then either the subject matter was improperly delineated (the tests applied outside the appropriate area) or the model is improperly formulated. Reassessment is called for, and the process recycles.

For example, astronomers may observe a comet's approach and work with mathematicians to formulate some equations as a model for its motion. They may compare locations predicted by the model with observations on later orbits. These may seem reasonable, but some adjustment of the equations may yield greater accuracy. On the other hand, the comet may once approach the sun so closely that it partially breaks up and the model no longer holds. Or perturbations from planets not considered in the model may add up over several orbits and cause the predictions to fail. The model should be presented with an explicit statement of the conditions under which it ought to be valid.

Often at first, the subject matter of an applied mathematics investigation is only vaguely specified. After a model has been tested and judged acceptable, the subject can be more precisely delineated by including just those facets that the model most successfully describes. The comet model is a simple example—it might be restricted to apply only to a certain number of orbits.

Subtle questions arise when geometry itself is regarded as a model of a physical system. The subject area—certain properties of space—is adjusted to include only those properties that the model describes with sufficient accuracy. Ordinarily, not much effort is devoted to describing the verification stage, though. After all, many scientific and engineering achievements over the centuries attest to the accuracy of geometric models.

Notice that success isn't judged by a model's *correctness*, but by its *accuracy*, relative to the task for which it will be used. This criterion is particularly important for geometric models. For centuries, scientists studied essentially only one model, Euclidean geometry. Models whose theorems contradicted Euclid's—even only slightly—would have been regarded as incorrect, hence useless. After nineteenth-century mathematicians began to consider non-Euclidean models, philosophers of science wrestled mightily with the question of which model was correct. That was perhaps unfortunate; it would have been emotionally less complicated, and likely more productive, to concentrate on the *accuracy* of one's predictions, than to challenge the correctness of others' views.

However, attempts to judge the accuracy of geometric models by testing theorems against empirical data have led to difficult conceptual problems in physics. The measuring techniques are themselves intertwined with the geometric theories. For example, Euclidean geometry includes the theorem that the sum S of the angles in a triangle is always exactly 180°, while the corresponding non-Euclidean theorem is $S < 180°$, with the difference $180° - S$ increasing in proportion to the triangle's area. This seems a reasonable theorem to test. In the early 1800s, two of the founders of non-Euclidean geometry, Carl Friedrich Gauss and Nikolai Ivanovich Lobachevski, both performed such measurements inconclusively. The $180° - S$ values that they measured for some very large triangles were small enough to fall within the expected errors of their instruments.[1] Better equipment might not help much, because in the non-Euclidean case, the sum might differ significantly from 180° only for triangles of vast astronomical dimensions. Validity of the required measurements would depend on the assumption that transportation over enormous distances (and times) wouldn't affect any critical properties of measuring devices. That is, you could also explain why an angle sum S seemed to measure less than 180° by assuming that the familiar physical theory of your measuring devices is inaccurate in that case, not your geometric model. You might judge it necessary to correct your measurements in much the same way that surveyors used to adjust theirs when extreme heat or cold affected their equipment. The belief that a change in physical assumptions can compensate for inaccuracies in a geometric model is called *conventionalism*. For an engaging introductory account, consult Rudolf Carnap's *Philosophical foundations of physics* (1966).

---

[1] See Trudeau 1987, 147–150; Miller 1972; and Daniels 1975.

2.2 Need for rigor

Concepts
How much rigor is necessary?
A sophism
Analysis: Assumptions weren't checked; the diagram is wrong.
Opposite sides of a line
Inside and outside of a segment or triangle
Need for rigorous proofs

Before the geometric modeling process is described in detail, it may help to consider a question you'll encounter in proving geometric theorems. How much rigor or formality is needed? This has probably troubled you already in an elementary geometry course. At that level it's hard to present and justify an answer. Rigor is certainly required to avoid errors in complicated investigations, but you need to start with simple exercises to learn the proof techniques. Moreover, it's even hard to distinguish what's simple from what's complicated. As a result, elementary geometry courses are often characterized by their demand for seemingly excessive formality and detail in proofs of results that are entirely obvious. Students may remember geometry as classes "where we had to write proofs line by line in two columns" instead of experiences studying significant and beautiful shapes in art and nature and learning practical techniques for drawing and design.

This section presents a *sophism*—an example of an incorrect argument—to show why geometric investigations demand rigor even for apparently simple situations. It should give you an idea of the type of proof that the axiomatic framework developed in chapter 3 must accommodate. You'll see that quite simple geometric constructions can be very tricky to analyze. That will justify the use of formal techniques even at the beginning of a critical study of elementary geometry.

Consider triangle ΔABC in figure 2.2.1. With a suitable scale,

$$AB = 1.03 \qquad m\angle B = 50.9° \qquad BC = 1.00\,.$$[2]

The problem is to compute CA from these data. You may recognize this as a routine trigonometry problem that you can solve with the law of cosines (derived in section 5.5):

$$CA = \sqrt{(AB)^2 + (BC)^2 - 2(AB)(BC)\cos 50.9°} \approx 0.873\,.$$

On the other hand, you may attempt to solve it using a seemingly more elementary technique, as follows. As in figure 2.2.1, find the midpoint A'

[2] $m\angle B$ means *measure of angle* B.

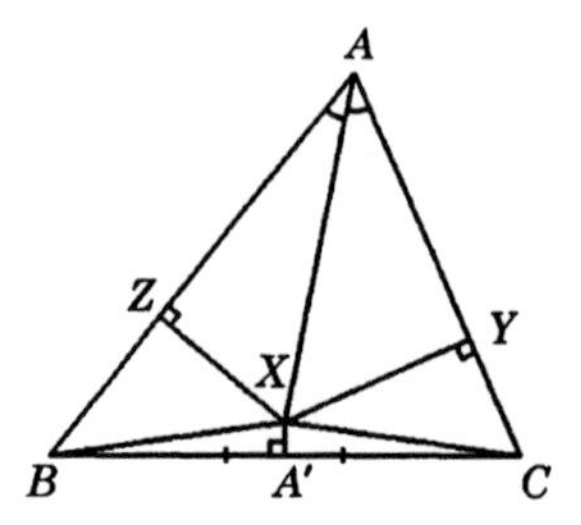

Figure 2.2.1
An incorrect diagram

of segment $\overline{BC}$, the intersection X of the bisector of $\angle A$ with the perpendicular bisector of $\overline{BC}$, then the feet Y and Z of the perpendiculars from X to $\overleftrightarrow{CA}$ and $\overleftrightarrow{AB}$. First, notice right triangles ΔAYX and ΔAZX They have the same hypotenuse AX and equal[3] angles at A, so they're congruent—corresponding pairs of legs are equal. That is, $ZA = YA$ and $ZX = YX$. Second, $\Delta BA'X$ and $\Delta CA'X$ are right triangles with a common leg $A'X$ and equal legs $A'B = A'C$. Thus, they're also congruent, so their hypotenuses are equal: $BX = CX$. Third, right triangles ΔBZX and ΔCYX are congruent because they have equal hypotenuses $BX = CX$ and a pair of equal legs $ZX = YX$. Therefore, their remaining legs are equal: $BZ = CY$. It seems to follow that $CA = CY + YA = BZ + ZA = AB$, which *contradicts* the trigonometry in the previous paragraph.[4]

It's easy to see that the *second* computation is wrong, because it concludes that ΔABC is isosceles without using any of the measurements AB, $m\angle B$, or BC. It seems to show that *every triangle is isosceles*, which is absurd! But where's the error?

If you review the reasoning with congruences, which is discussed in detail in sections 3.5 to 3.8, you'll find no error. The fault lies instead in the failure of the assumptions that underlie equations $CA = CY + YA$ and $BZ + ZA = AB$. They're valid only when Y lies between C and A and Z between A and B. Although these relationships are shown in figure 2.2.1, they depend on point X, which is located wrong. The line $\overleftrightarrow{AX}$ there is *not* the true bisector. In fact, the contradiction really proves that in a nonisosceles triangle ΔABC, the constructed points X, Y, and Z cannot all fall within the triangle and on its legs as shown.

In order to highlight the consequences of an error, and make it reasonable to resolve in a text, a simple example was chosen. What would happen if you committed a similar fault during a much more complicated analysis? The contradiction would probably not be clear—you'd just get a wrong answer.

---

[3] For readability, this paragraph uses *equal* to mean *having equal measure.*

[4] This example probably appeared first in 1892. See Ball and Coxeter [1892] 1987, 80.

How can you avoid pitfalls like that? It doesn't help just to insist on accurate diagrams, because geometry should enable you to make conclusions about configurations that you can't measure directly nor draw conveniently. What's necessary is to state clearly all the assumptions required by each step, and to verify that each one is valid before accepting the conclusion.

Before you use a diagram to guide geometric reasoning, you must ensure that it really represents the situation you want to study. In this example, you could prove the triangle congruences without referring to any diagram—your argument would be hard to understand but nevertheless correct. However, before proceeding with the step *It follows that* $CA = CY + YA = BZ + ZA = AB$, you'd need to check the validity of the underlying assumptions. A diagram certainly helps with that, and you must be sure it's accurate. You could prove this preliminary result:

> If $CA < AB$ in ΔABC, then the bisector of $\angle A$ meets the perpendicular bisector of segment $\overline{BC}$ at a point X on the opposite side of line $\overleftrightarrow{BC}$ from A. Moreover, the feet Y and Z of the perpendiculars from X to lines $\overleftrightarrow{CA}$ and $\overleftrightarrow{AB}$ lie outside and inside the triangle legs, respectively.

That is, the appropriate diagram is figure 2.2.2. The proof details are considered later, in exercise 4.5.6, after suitable tools have been introduced. Using the new diagram as a guide, you'd then modify the last step of the argument to read *It follows that* $CA = YA - CY = ZA - BZ = \ldots$. But you couldn't fill in the last gap, the dots. In fact, this argument with congruences doesn't directly yield any numerical value for CA. You *must* use the law of cosines or some equivalent trigonometry.

To facilitate correct arguments, any general framework for studying geometry must provide means for carrying out detailed proofs of results like the description of figure 2.2.2, as well as the more straightforward arguments with congruences and trigonometry. Until the mid-1900s, no elementary geometry textbooks provided that capability. Each troublesome example had to be considered separately, often with ad hoc methods that went beyond those explicitly described. The axiomatic system presented in chapter 3

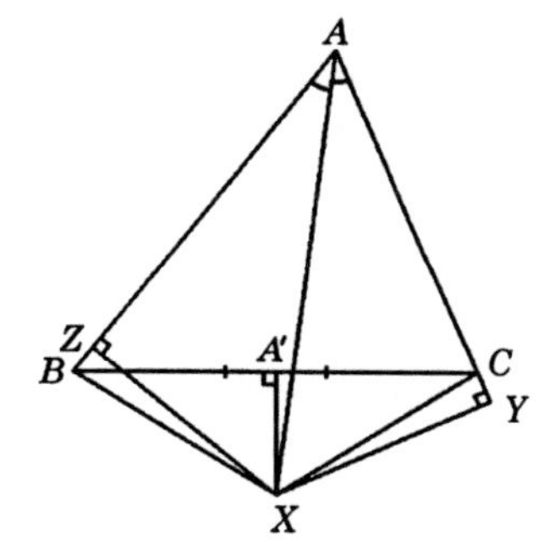

Figure 2.2.2
The correct diagram

fills the need, but at a price. For example, to permit fully rigorous proofs, the notions of opposite sides of a line, and of the inside and outside of a line segment, are discussed in detail *before* the traditional congruence techniques. Thus, some of the most delicate parts of elementary geometry are presented at the very beginning. Chapter 3 is appropriate for students at your level to use as a framework for precise reasoning about complicated problems, but it can't reasonably serve as an *introduction* to geometry.

2.3 The axiomatic method

Concepts

Undefined concepts and definitions
Unproved axioms and proved theorems
Euclidean geometry
The Declaration of Independence
Controversy over the role of undefined concepts

An effective model building technique is the *axiomatic method*, described by Aristotle around 330 B.C. in the text *Posterior analytics*.[5] He emphasized the need for *precise definition* of the concepts used in a model and its theory, and the *rigorous proof* of theorems about the model. Complicated concepts should be defined, and difficult theorems proved, from simpler ones. Imagine a careful exposition of a mathematical model. Among its concepts are some that are *specific* to the model; they're not part of the underlying pure mathematics or logic. You can't define some of these specific concepts— for instance, the first one mentioned—in terms of the others without circularity. Thus you must designate certain specific concepts as *undefined* in the model, and define all the others from those. Similarly, among the theorems about the model are some that are specific to it. You can't deduce some of these specific theorems—for instance, the first one mentioned—from the others without circularity. Thus you must designate certain specific theorems as *unproved* in the model, and prove all the others from those. The unproved theorems are called *axioms*.[6] Axioms should be self-evident truths not based on simpler ones. The undefined concepts, axioms and definitions constitute the model.

[5] Aristotle 1975, chapters A2–A3, A10, and the associated Synopsis and Notes. *Posterior* means simply *second book of*.

[6] Some authors write *postulate* instead of *axiom*, and some use both terms, with slightly different meanings. This book never makes such distinctions, and always uses the word *axiom*.

ARISTOTLE was born in 384 B.C. at Stagira, in Greece, the son of a physician. From age 17 to 37 he studied at Plato's Academy in Athens. After Plato's death Aristotle spent twelve years moving about Greece. During 343–342 B.C. he was tutor to the teenage prince of Macedonia, the future Alexander the Great. At age 49 he founded his own school in Athens. He retired in 322 B.C. and died that same year. Aristotle's works, mostly lecture notes published during his second period in Athens, cover the spectrum of human inquiry. They guided western philosophy for centuries. His formulation of the axiomatic method and his influence on other Greek thinkers, especially Euclid, are particularly important for mathematics.

A statement involving the concepts of the model is regarded as a theorem only after it has been rigorously proved from the axioms.

This framework for organizing science was first applied extensively in Euclidean geometry, as described in section 2.4 and practiced in chapter 3. Its *general* effectiveness is clear, however, and Aristotle felt that *all* scholarly inquiries should be reported this way. The method now pervades western culture. One of its most vivid occurrences outside mathematical disciplines lies in the *Declaration of Independence in Congress, July 4, 1776*:

Axioms	We hold these truths to be self-evident, that all men are created equal, that they are endowed by their Creator with certain unalienable Rights, that among these are Life, Liberty and the pursuit of Happiness.—That to secure these rights, Governments are instituted among Men, deriving their just powers from the consent of the governed,
Theorem	—That whenever any Form of Government becomes destructive of these ends, it is the Right of the People to alter or to abolish it ...
Axioms	(Many destructive acts of the King are enumerated.)
Theorem	The history of the present King of Great Britain is a history of repeated injuries and usurpations, all having in direct object the establishment of an absolute Tyranny over these States....
Theorem	We, therefore, ... declare, That these United Colonies are ... Absolved from all Allegiance to the British Crown...

The *Declaration*'s authors, principally Thomas Jefferson, were familiar with the axiomatic method and with Euclid, and assumed that those whom they addressed would be, too.[7]

---

[7] Gary Wills (1978) and I. Bernard Cohen (1995) discuss the wording of the Declaration and Jefferson's intellectual *milieu* at length. Wills devotes an entire chapter to the use of the term *self-evident*. Cohen's first two chapters describe the fundamental position of Newtonian science and its Euclidean basis in the education and political thought of that era. Florian Cajori (1890, chapter 1) showed that most college graduates of Jefferson's time were familiar with Euclid.

The axiomatic method provides general guidelines for organizing a scientific investigation. Once an acceptable axiomatic model has been constructed, the system of undefined concepts, definitions, and proofs is useful in reporting the results, and can serve later as an effective framework for learning. Axiomatic models are sometimes adapted for instruction by moving into the list of axioms some theorems with particularly difficult proofs. Students *start* from those theorems instead of struggling with their proofs. On the other hand, polishing an axiom system for archival publication sometimes involves exactly the opposite step. Investigators may discover that some axiom can be proved from the others. The proof may contain interesting and useful arguments. It's proper, then, to record that by relabeling the axiom as a theorem, and publishing the proof. This also makes the model simpler and perhaps easier to understand.

Attributing this current description of the axiomatic method entirely to Aristotle is historically questionable. His work is not a polished text, but rough lecture notes, and it's often obscure. In particular, he may not have seen as clearly as we do now why some particular concepts should be left undefined. Authors using the method sometimes started with circular or nonsensical definitions. Euclid's are criticized in section 2.4. Jefferson gave no definitions. Until the nineteenth century an axiomatic model was not ordinarily applied to any subject area beyond that for which it was originally devised. To facilitate this practice, which is now common, mathematicians learned to leave the most basic definitions for later specification. Then they realized that those definitions couldn't be used in proving theorems, hence were superfluous. This observation also eliminated the need for philosophical discussion of the nature of such definitions.

The general status of undefined concepts in axiomatic models—particularly in geometry—remained muddled through the centuries. By the end of the nineteenth, geometric models had been developed that *could* have been presented entirely axiomatically in the sense just described. But their authors didn't quite do that, and controversy over the role of undefined concepts continued into the twentieth century. As discussed later in section 2.9, the turning point finally occurred around 1900.[8] From then on, the full axiomatic framework was used commonly and routinely.

---

[8] Blaise Pascal ([1658] 1948) described the axiomatic method concisely, but in essentially its full form, explicitly referring to undefined concepts. This work, published years after his death, was studied by generations of European students and scholars. But it seems to have had little influence in mathematics until the 1890s. Why 340 years passed before Pascal's outline was fully implemented is an intriguing question.

2.4 Euclid's *Elements*

Concepts

Evolution and influence of the *Elements*
Its lack of undefined notions
Euclid's axioms
Their insufficiency
Straightedge and compass constructions
Incommensurable segments
Segment calculus
Hilbert's and Birkhoff's rigorous foundations
What was the subject matter of the *Elements*?

The earliest surviving exposition of an axiomatic model is Euclid's *Elements* ([1908] 1956). Written about 300 B.C., perhaps sixty years after Aristotle's *Posterior analytics*, this textbook organized a large part of the geometry of that era into a single system. A masterful compendium, it proved so effective that it dominated mathematics instruction for two millennia. Euclid's thoroughness has been celebrated by the immense wealth of science, engineering, art and design based on the *Elements*. The oldest surviving edition dates from about A.D. 400, although some ancient commentaries discuss earlier editions in detail. During its first seven centuries, the text evidently suffered revision by editors and scribes, so it's not completely reliable as a record of the state of Greek mathematics in Euclid's time. Over later centuries, though, editions of the *Elements* varied only slightly. They provided the core—often nearly the entirety—of an educated westerner's experience with mathematics. Only about 1800 did mathematicians feel it appropriate to write new geometry textbooks, and until about 1950 even those differed little from Euclid's original presentation.

The *Elements* has some major faults, which persisted for centuries in elementary geometry texts. They lie particularly in the definitions, axioms, and early theorems, at the very beginning of the book.

If you should edit it according to Aristotle's guidelines, you'd probably include *point*, *line*, *length*, and *angle equality*, for example, among the undefined concepts. Euclid's intent isn't clear, since the *Elements* purports to define *all* geometric concepts used. Some of these "definitions" are vague, and it's not clear what's being defined in terms of what: for example,

1. A *point* is that which has no part.
2. A *line* is breadthless length.
3. The extremities of a line are points.
4. A *straight line* is a line which lies evenly with the points on itself.[9]

[9] Euclid [1908] 1956, vol. 1, 153.

Euclid uses the term *line* for what we call *line segment.* How do these descriptions distinguish a circular arc, for example, from a straight line segment?

Euclid postulates only five axioms:

1. To draw a straight line from any point to any point.
2. To produce a finite straight line continuously in a straight line.
3. To describe a circle with any centre and distance.
4. That all right angles are equal to one another.
5. That, if a straight line falling on two straight lines make the interior angles on the same side less than two right angles, the two straight lines, if produced indefinitely, meet on that side on which are the angles less than the two right angles.[10]

With suitable interpretation, these become clear. Axiom 1 says that any given points P and Q lie in some segment $\overline{PQ}$. Axiom 2 says that any segment $\overline{PQ}$ can be extended, if necessary, to produce a segment $\overline{PQ'}$ whose length exceeds any given distance. To interpret axiom 3, replace *to describe* by *there is*. Axiom 4 may seem strange until you see how Euclid defines a right angle: If Q is on the segment $\overline{PS}$, R is not on line $\overleftrightarrow{PQ}$, and $m\angle PQR = m\angle RQS$, then these two angles are right. No numerical angle measure is involved. Axiom 4 then says, if $m\angle PQR = m\angle RQS$ and $m\angle P'Q'R' = m\angle R'Q'S'$ as in figure 2.4.1, then $m\angle PQR = m\angle P'Q'R'$. Axiom 5 is the famous parallel axiom, figure 2.4.2.

Euclid's axioms, although clearly stated, are *inadequate*. They guarantee existence only of figures constructible with the classical Greek instruments: an unmarked straightedge and a compass that might collapse if lifted off the drawing surface. You draw and extend segments with the

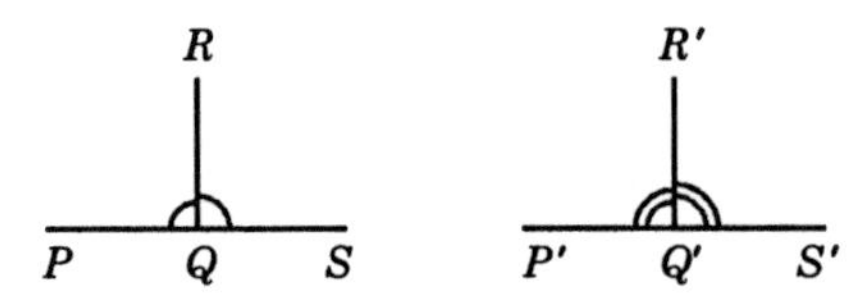

Figure 2.4.1 Euclid's axiom 4

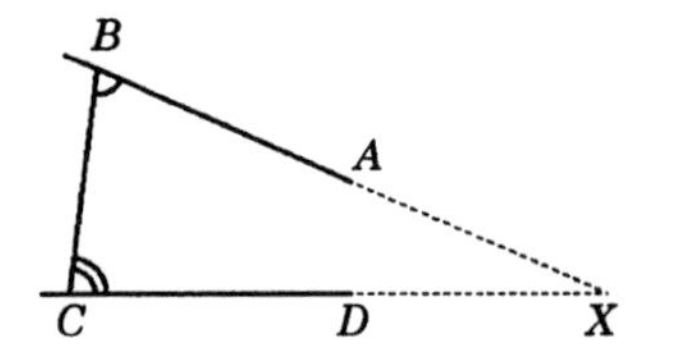

Figure 2.4.2 Euclid's axiom 5, the parallel axiom:

If $m\angle ABC + m\angle BCD <$ *two right angles, then lines* $\overleftrightarrow{AB}$ *and* $\overleftrightarrow{CD}$ *meet at a point* X.

---

[10] Ibid. 154–155.

straightedge. With the compass you can draw a circle with a given center O and radius OP. These tools are idealized conceptually to permit arbitrary magnitude and accuracy. But even then, later mathematics has shown, they won't suffice for drawing *all* line segments and circles—for example, a line segment whose length is π times that of a given unit segment. Also lacking are axioms required to describe the order of points on a line and to guarantee intersection of various lines and circles. Euclid's proof of the existence of an equilateral triangle ΔOPQ with a given edge $\overline{OP}$, his very first theorem, is defective. It fails to establish the intersection Q of the two circles in the familiar construction of figure 2.4.3. Further, Euclid uses in some early proofs arguments by *superposition*—placing one figure atop another—a technique that isn't even mentioned in his definitions or axioms. Finally, many of Euclid's proofs can be criticized because they don't cover all cases. In these ways, the sophism of section 2.2 closely resembles a typical proof in the *Elements*!

Nineteenth-century mathematical progress, especially the discovery of non-Euclidean geometry and investigations into the nature of the real number system, placed heavy strain on Euclid's approach. Throughout that century, mathematicians criticized the *Elements*. Gradually they identified and repaired its defects. In 1899—see section 2.8—David Hilbert published the first completely rigorous axiomatic presentation of Euclidean geometry.

A major triumph in the *Elements* is its *calculus of lengths* of line segments. Now we use real numbers to represent lengths, and manipulate them with algebra and decimal arithmetic. But those techniques weren't developed until about fifteen centuries after Euclid! He calculated with the segments directly, not with numerical lengths. Euclid could do arithmetic only with integers and their ratios. He couldn't base segment computations on integer arithmetic alone, else he would have been limited to commensurable cases. (Segments A and B are *commensurable* if for some integers m and n, you can make m copies of A, placed end to end, coincide with n similarly arranged copies of B.) This restriction was unacceptable because—for example—the diagonal D and side S of a unit square are not commensurable.

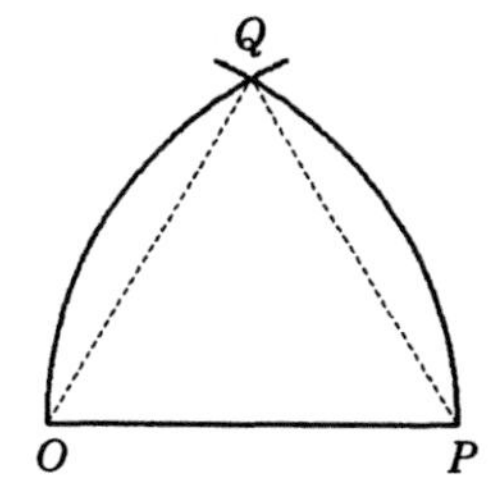

Figure 2.4.3
Euclid's equilateral triangle construction

Little is known of EUCLID's life except that he was teaching and writing mathematics around 300 B.C. in the university at Alexandria. The Macedonian King, Alexander the Great, had founded that Egyptian port city in 332 B.C. after wresting from King Darius of Persia the entire Eastern Mediterranean region between Greece and Egypt. After Alexander's death in 323 B.C., the local commander, who had been one of Alexander's top generals, prevailed in a power struggle and eventually became King Ptolemy I, the root of a long Egyptian dynasty. Ptolemy founded the university.

Euclid's great work, the *Elements*, a compendium of theorems in elementary geometry organized as an axiomatic theory, has survived only in editions and commentaries from later centuries. For two millennia it served as the foundation for virtually all advanced mathematical studies. Two other elementary geometry books by Euclid survive, and three on astronomy, optics, and music. One further elementary geometry text and two advanced ones are known only through sparse references in subsequent literature.

In modern language, the length $\sqrt{2}$ of D is not a ratio of integers.[11] Euclid's method was essentially this: To compute with incommensurable segments like D, *approximate* them by segments commensurable with S, use integer arithmetic to compute with the approximations, then be sure the resulting error can be made arbitrarily small.

That's the hardest mathematics in the *Elements*, a major obstacle to its use in teaching geometry. Until recently, texts could simplify or avoid this segment calculus only by departing significantly from the rigor demanded by the axiomatic method. Finally, in 1932, George David Birkhoff—see section 2.8—formulated an axiomatic model of geometry, similar to those of Euclid and Hilbert, that avoids this obstacle without loss of rigor. Birkhoff explicitly built the real number system into the geometric model. His framework has evolved into the axiomatic development of geometry presented in chapter 3.

Mathematicians and historians continue to study the *Elements* and other ancient mathematics to determine just how Euclid regarded the axiomatic foundation of his work. In view of the great skill he showed with the really difficult mathematical parts, why are his initial definitions so muddled, and why did he seem unaware that his axioms were so inadequate? There's a spectrum of interpretations. Many scholars now agree that the aspects of space constituting Euclid's subject matter included not *all* figures consisting of points, line segments, etc., but only those that can be constructed with the classical instruments.[12] In the paper "Did Euclid's *Elements*, Book I, develop geometry axiomatically?" Abraham Seidenberg (1974) suggested

---

[11] If $\sqrt{2} = m/n$ then $2n^2 = m^2$, and m would be *even*—that is, $m = 2p$ for some p. Thus $n^2 = \frac{1}{2}m^2 = 2p^2$, so n would be even, too. That would contradict the possibility of writing m/n with lowest terms.

[12] See Trudeau 1987, chapter 2; Mueller 1981, chapter 1; and the sources cited in the latter.

that excising most of the axioms and vague definitions from the *Elements* would leave a coherent and masterly, but clearly nonaxiomatic, text about the classical geometric constructions. Could the objectionable material have been inserted by later editors? More recently, Lucio Russo (1998) noted its similarity to text by Hellenistic mathematicians, particularly Hero, who worked in Alexandria during the first century A.D., about 300 years after Euclid. They wrote detailed *descriptions* of Euclid's axioms, evidently not intended as definitions of the terms mentioned, but as clarifications. Russo suggests that Euclid, like today's mathematicians, left the initial concepts undefined, and Hellenistic scientists understood why. But by A.D. 400, when Greek science had lost its vitality, editors awkwardly and inaccurately interpolated the descriptions into the *Elements*, ignoring the resulting logical problems.

Is it reasonable that Euclid should have been so preoccupied with the artificially restricted straightedge and compass construction methods? Isn't one of the purposes of geometry to provide a theory on which you can base science and engineering, which certainly must transcend those methods? Seidenberg considered those questions deeply, particularly in the light of rigidly stylized mathematical prescriptions for religious rituals discovered in texts surviving from cultures preceding the Greek. He concluded that

> the elements of geometry as found in the ancient civilizations, in Greece, Babylonia, Egypt, India, and China, are a derivative of a system of ritual practices as disclosed in the *Sulvasutras*. They also suggest that the ritualists knew some deductive mathematics. Did they, then, also have the axiomatic method? Of course not, but they were concerned with exact thought. The ritual in general was to be carried out exactly ... the wrath of the gods followed the [slightest error]. Why the ritualists should want so much to be right is hard to say, unless it is that they were concerned with symbolic action and that there is not much point to symbolic action unless it is right. (Seidenberg 1974, 292)

2.5 Coordinate geometry

Concepts

Thinking about numbers, independently of the objects they measure
Development of decimal arithmetic and algebra
Solving cubic and quartic equations
Negative numbers
Descartes and coordinate geometry

It's 1637. Nearly two thousand years have passed, and Euclid's *Elements* is still the core of European mathematics. Several major new ideas and techniques have taken root, though, over that time. People can now think of numbers independently of the objects they measure. Requirements of commerce have forced the gradual introduction of decimal notation and arithmetic, which are now routine. Computation is much more efficient than it was in Euclid's time.

The same period has seen the slow introduction of algebraic methods, and the refinement of algebraic notation. Algebra is now commonly applied to various types of geometry problems, many of which arise in science or engineering. During the 1500s mathematicians achieved a *tour de force*, using algebra, complex numbers, and trigonometry flamboyantly to solve formidable cubic and quartic equations. Nevertheless, in 1637 algebra is not yet a familiar skill. It will soon become the medium for a great explosion of mathematical theory and applications in science. But most scientists aren't yet comfortable enough with algebra to rely on it heavily as the basis for research. They're surprisingly inept, for example, with negative numbers. You don't need algebra yet to run the family business, so the logic of the balance sheet hasn't permeated the language that will become the medium for science.

In isolated cases during the preceding decades, applications of algebra to geometry have foreshadowed coordinate methods, but no general guidelines are yet established. War and chaos reign in central Europe. Churches and governments suppress independent thought everywhere. There isn't any mathematical mainstream yet to direct any flow of energy.

This year, René Descartes, a French philosopher of independent means living in Holland, has published *La géométrie* ([1637] 1954), a deep and impressive account of *coordinate geometry*. It presents a new type of geometric model, very different from Euclid's, based on the concept of number.

René DESCARTES was born in France near Tours, in 1596. His father was a lawyer, a minor noble. His early schooling was with the Jesuits at La Flèche, who recognized his distinction in mathematics. In 1616 he earned a law degree from the university at Poitiers, but did not enter the profession. Instead, he traveled in Holland and Germany, witnessing the beginning of the Thirty Years' War, then to Italy and back to France. During that time he was apparently preoccupied with questions of philosophy, mathematics, and science. After 1628, Descartes lived in Holland, moving from city to city to find tolerance for his views. There he published a series of works on mathematics, physical science, and general philosophy which had tremendous influence on western thought. He was involved in controversy with other European thinkers throughout his life. His work of greatest importance for geometry is the introduction of the coordinate method. Descartes left Holland in 1649 to take an appointment at the Swedish court. He died there in 1650.

In modern language, a coordinate model regards points as triples of numbers, or coordinates, and lines as sets of points whose coordinates are generated by linear parametric equations. It's clear now that coordinates let us translate difficult geometry problems about points, lines, curves, etc., into problems about numbers, then apply efficient algebraic tools. But Descartes' immediate aim is almost the exact opposite. While he does describe the translation explicitly, he's mostly concerned with *geometric* analysis of difficult *algebra* problems: solutions of cubic, quartic, and higher degree equations. He wants to use the familiar, inefficient geometric methods to study new problems.

La géométrie is not a beginner's introduction to coordinate geometry, but rather a demonstration of its use in solving a complicated problem. It was published as an appendix to a controversial and widely circulated philosophical treatise that analyzed the basis of all understanding: *Discours de la méthode pour bien conduire sa raison, et chercher la verité dans les sciences*. Descartes' work was translated, clarified, and publicized during the middle 1600s. With that of many other mathematicians who were concerned with the same problems at the same time, it laid the foundation for the application of algebra—and later, calculus—to geometry and science. By the time of Newton's work in the late 1600s, mathematical mainstreams had formed, algebra had become routine, and coordinate geometry had reached the form we know today.

2.6 Foundation problem

Concepts

- Are the Euclidean and coordinate models equivalent?
- What are numbers?
- Is every number the length of some segment?
- Defining the real number system
- Rephrasing and augmenting Euclid's axioms

As mentioned in section 2.4, Euclid didn't use numbers per se as lengths of line segments. He calculated with segments directly using integer arithmetic, and handled incommensurable segments with approximation techniques. But coordinate geometry was *based* on the number system as it had evolved by Descartes' time. Euclidean geometry and coordinate geometry provide quite different but complementary models of the same subject. We know from experience that some theorems are easier to derive in Euclid's system, others with coordinate methods. Are there any that can be derived in one but not the other? This is the *foundation problem*: whether the two models are equivalent. That is, do they have the same theorems?

It's easy to see that Euclid's axioms are true if you interpret points as triples of coordinates and lines as point sets defined by linear equations. It follows that the theorems of Euclidean geometry are all true statements in coordinate geometry. What about the reverse? Can every true statement about the coordinate model be derived from Euclid's axioms?

To settle this remaining half of the foundation problem you must justify the use of numbers in geometry, based on Euclid's axioms. What are numbers? In Descartes' time numbers could be described only vaguely, as conceptual objects manipulated by algebra and decimal arithmetic. Can you justify associating them with the points on a line, so that you can measure the length of a line segment as with a ruler? And so that you can identify points with triples of numbers and define lines by linear equations?

Even if you successfully introduce numbers into Euclidean geometry, assign numerical lengths to all segments, and find linear equations for all lines, part of the foundation problem remains unsettled. Does *every* number represent the length of some segment? Answering that requires a solid understanding of both notions: number and segment. During the nineteenth century, mathematicians partially answered the question, in the negative. With the Greek instruments—unmarked straightedge and compass—you can't construct any segment of length π, for example. (See section 3.14.) At the time of those investigations, however, mathematicians didn't believe that Euclidean geometry ought to be restricted to constructible figures only, because they regarded the *Elements* as a foundation for science, and science didn't restrict itself that way. The foundation problem was veiled in the same fog that clouded the axiomatic basis of Euclidean geometry.

The centuries after Descartes saw an enormous expansion of mathematical theory and applications, particularly involving differential and integral calculus. The use of limit operations became more and more refined, and their analysis involved delicate properties of numbers. An example was the demonstration that no constructible segment has length π. Puzzling questions arose about the number system. Mathematics needed a definitive analysis of numbers, their relationships, and our methods for computing with them. Various mathematicians worked on that, and by about 1900 had provided several alternative definitions of the real number system that we use today. Moreover, they invented a method to show that any two definitions—for example, one based on the geometry of points on a line and one on decimal expansions—really yield the same arithmetic and algebra. (See section 2.9.)

A precise definition of the real number system was what they needed to clarify and solve the foundation problem. Euclid's axioms did not in themselves justify identifying the set of points on a line with the real number system, just as they were inadequate even for rigorous proofs of the theorems in the *Elements*. Therefore the foundation problem reduced to this:

rephrase and augment Euclid's axioms to justify identifying the real number system with the set of points on some line and to justify using triples of real coordinates and linear equations to represent and reason with points and lines in general.

2.7 Parallel axiom

Concepts

The parallel axiom is more complicated than the others.
Mathematicians tried to prove it as a theorem, but failed.
Is it independent of the other axioms?
An answer requires a complete list of Euclid's unstated assumptions.
That's equivalent to the foundation problem.
Hilbert finally established the independence in 1899.
But non-Euclidean geometry had strongly suggested it much earlier.

Another train of thought also led mathematicians to the foundation problem: the possibility of non-Euclidean geometry. While that question is central to a general study of the foundations of geometry, it's only a sidelight to this book. Therefore only the core ideas are mentioned here.

Euclid postulated explicitly only the five axioms listed in section 2.4. He also assumed others that were apparently so simple he didn't notice their use, and left them unstated. He seemed to avoid axiom 5, the parallel axiom, except when it was absolutely necessary. He referred to it first in the proof of his twenty-seventh theorem! Euclid didn't explain his strategy, but later mathematicians speculated that the parallel axiom was more complicated than the others. Its word count alone justifies such a conclusion. You can also argue that axioms 1 to 4 involve portions of space that are clearly delimited once you know the given points and distances. But that's not true of the parallel axiom. Given points A to D as in figure 2.4.2, with $m\angle ABC + m\angle BCD < 180°$, you can't locate the intersection X of lines $\overleftrightarrow{AB}$ and $\overleftrightarrow{CD}$ without a complicated trigonometric calculation. The closer the angle sum to $180°$, the farther X lies from the given points.

Consequently, later mathematicians tried to find a proof of the parallel axiom based on axioms 1 to 4. They failed repeatedly. In each case, they left unjustified some steps in their proofs, and the missing arguments required axiom 5. Mathematicians began to suspect that the parallel axiom was *independent*—that it could *not* be proved from the other axioms. But what *are* the other axioms? Proving rigorously even the simplest theorems requires assumptions that Euclid had left unstated. A rigorous published *proof* of the parallel axiom would settle the matter—you could *read* which assumptions

David HILBERT was born in 1862 near Königsberg, in Prussia. His father and grandfather were judges. His father was very conservative in thought and behavior; his mother was the greater intellectual stimulus. Hilbert decided on mathematics during his gymnasium[13] studies, and entered the university at Königsberg, where he earned the doctorate under Lindemann in 1884. (Lindemann had just proved in 1882 that no segment of length π was constructible with the Greek instruments.) Hilbert's early research was in the area now called algebraic geometry, which is concerned with polynomial equations for curves and surfaces. By 1892 he had solved one of the most important problems in the subject by a new, very original and powerful method. Hilbert was appointed Professor at Göttingen in 1895. During the succeeding years he worked with Felix Klein to make that university the mathematics center of the world.

In his next research, Hilbert applied new algebraic techniques to great effect in the theory of numbers. Then he turned to the foundations of geometry. Hilbert published in 1899 the first completely rigorous axiomatic development of Euclidean geometry. Soon after, he and his students extended this work to include various non-Euclidean geometries as well. In Paris, at the 1900 International Congress of Mathematicians, Hilbert posed a list of twenty-three problems that he thought would guide mathematics during the twentieth century. To a large extent his prediction has been accurate. During the early 1900s, he did research in analysis, recasting the theory of integral equations into its current form. Hilbert later considered questions in theoretical physics, and returned to problems in the foundations of mathematics. In the last area, his leadership directly inspired many important discoveries by other researchers in mathematical logic.

Hilbert was the predominant figure in mathematics for the first third of the twentieth century. He directed the doctoral research of sixty-nine mathematicians, many of whom became leaders in their areas. He was one of the last mathematicians to master and make important contributions to *many* diverse parts of the discipline. Hilbert's mathematical activity had largely ceased by 1932, but he lingered on, sad and alone near the empty university, through the war. He died in Göttingen in 1943.

were used. But even a *suggestion* that the parallel axiom can *not* be proved from some assumptions requires a list of those assumptions.

Before mathematicians could argue convincingly for the independence of the parallel axiom they would have to list all the geometric assumptions that Euclid had omitted from the axioms. Then they could ask a clear question: Is the parallel axiom provable from his other axioms and the tacit assumptions, or not? But they couldn't consider *all* the geometry that will

[13] In many countries, the terms *gymnasium* and *lycée* designate secondary schools devoted to preparation for university education. They include courses offered in the first years of American universities.

ever be invented, searching for tacit assumptions. How should they know when the list is complete?

A solution to the foundation problem would suffice. Once mathematicians had become confident that they could derive any desired geometric theorem—even if awkwardly—by coordinate methods, they realized that a list of all the geometric assumptions required to justify those techniques would yield a fully rigorous axiomatic model of Euclidean geometry. David Hilbert achieved that with his 1899 monograph, *Foundations of geometry* (1899).

The independence of the parallel axiom had been strongly suggested years earlier. About 1830, János Bolyai and Nikolai Lobachevski had derived a large body of theorems, called *non-Euclidean* geometry, by replacing the parallel axiom with its negation. Included, for example, was the theorem that the sum of the angles in a triangle is always *less than* 180°. Of course these theorems contradicted the familiar Euclidean geometry, but they seemed consistent among themselves. This work remained unfamiliar for about twenty years, but was then studied intensively. Mathematicians discovered about 1870 that non-Euclidean geometry describes faithfully some well defined mathematical structures that had been overlooked until then. But if the parallel axiom were deducible from the others, these structures would satisfy that axiom, not its negation.[14]

In *Foundations of geometry*, Hilbert filled in the last steps of this argument. He gave a complete list of axioms, and rigorously demonstrated the independence of the parallel axiom. Two years later, he showed how to derive all non-Euclidean theorems from his axioms, with the parallel axiom negated.[15]

2.8 Firm foundations

Concepts

What must a firm foundation provide?
Hilbert and the Italian school
Geometric construction of real numbers
Can a rigorous foundation be used for teaching?
Birkhoff, School Mathematics Study Group (SMSG), and chapter 3

By the late nineteenth century the need to establish a firm foundation for geometry was clear. Euclid's axiom system should be adapted and augmented as necessary, to

[14] Bonola [1911] 1955.

[15] Hilbert's 1902 paper on non-Euclidean geometry is translated and included as Appendix 3 in later editions of Hilbert 1899.

- clarify the initial definitions,
- justify all the steps in the proofs,
- justify the use of real numbers as coordinates, and
- facilitate study of the parallel axiom.

In order to gain acceptance, this work should fit in with other mathematical fields under development. In particular, any geometric construction of the real number system should be recognizable as equivalent to those used as a foundation for differential and integral calculus.

Two attacks on the foundation problem appeared simultaneously: the work of David Hilbert in Göttingen, and that of an Italian group including Giuseppe Peano, Alessandro Padoa, and Mario Pieri. These researchers had all studied earlier foundational work by German mathematicians and philosophers. Hilbert proceeded in isolation, but his book *Foundations of geometry* (1899) was the more complete and received the most attention. Hilbert's main concepts differed little from Euclid's, his mathematics was close to the mainstream, and his presentation was complete and polished. The Italians explored new territory and invented new techniques and language, but produced no really complete report. Pieri was particularly interested in finding out which concepts he could designate as undefined. In a major work published the same year (Pieri 1899), he defined all other concepts in terms of *point* and *motion*. At this time, Hilbert was busy developing Göttingen into the center that would dominate the world of mathematics until about 1930. The Italian group faded away after a decade. But around 1900, as described in section 2.9, their influence led to significant redirection and progress in foundations of mathematics.

Like Euclid, Hilbert didn't mention numbers in his axioms. He introduced x and y axes with origin O and unit points U and U'. Considering the x coordinates *as though* they were numbers, he derived with great labor all properties required for one of the accepted definitions of the real number system. For example, figure 2.8.1 shows how to construct the product $C = A \cdot B$ of two x coordinates A and B. Hilbert had to derive many algebraic rules like $A \cdot B = B \cdot A$, whose geometric counterparts are not easy to see. Those results made it possible to regard the x coordinates as *actually being* numbers. Then he proceeded to develop coordinate geometry, much as any introductory text would.

Hilbert's approach is unsuitable for beginning students because it's too difficult to verify the required properties of the real number system.[16] It took nearly seventy years' more research, hindered by two world wars and the depression, to develop a foundation for Euclidean geometry appropriate for elementary texts. George David Birkhoff published an equivalent axiom

[16] The author has found only one beginner's text based on Hilbert's axiomatization: the apparently unsuccessful Halsted 1907.

Figure 2.8.1 The product C of x coordinates A and B:

If $OU = OU' = 1$, $OB = OB'$, *and* $\overleftrightarrow{U'A} \parallel \overleftrightarrow{B'C}$, *then* $A/1 = OA/OU' = OC/OB' = OC/OB = C/B$, *hence* $A \cdot B = C$.

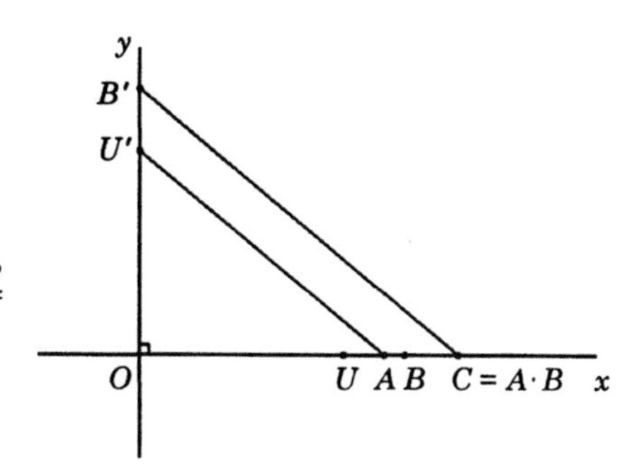

system (1932) that avoids the obstacle of constructing the real numbers. Collaborating with education professor Ralph Beatley, he produced an elementary text, *Basic geometry* (Birkhoff and Beatley 1941). Unlike Euclid and Hilbert, he introduced the real number system explicitly in his axioms for measuring lengths, angles, areas, and volumes. Thus Birkhoff avoided the difficult proofs required to derive these same statements as theorems from simpler axioms.

Birkhoff's axioms received little notice until the 1950s, when his method was proposed by the School Mathematics Study Group (SMSG) as a way to reform elementary geometry instruction in the United States.[17] Their

The SCHOOL MATHEMATICS STUDY GROUP (SMSG) was formed in 1958, when popular opinion held that the United States was not producing enough scientists and engineers to maintain its world stature. A principal cause of the deficiency, it was felt, was the poor quality of the secondary-school college-preparatory mathematics curriculum. SMSG was encouraged by the American Mathematical Society and financed by the National Science Foundation. Its organization consisted of a Director, Edward G. Begle, of Yale and later Stanford, a varying staff, and several committees, including an advisory committee of mathematicians from industry, government, universities, and secondary schools. During several consecutive summers from 1958 on, SMSG writing groups met on various campuses to devise new secondary and (later) elementary school mathematics curricula, and to create appropriate experimental texts. Among these texts was one for the tenth grade: elementary Euclidean geometry using the axiom system of George David Birkhoff. The text was planned by a committee, and a draft was written by Edwin E. Moise and revised by the committee. It eventually became the text *Geometry*, by Moise and Floyd L. Downs, Jr. SMSG also produced an alternative geometry text that used coordinate methods more extensively. The School Mathematics Study Group was disbanded in 1974.

[17] See Wooton 1965.

George David BIRKHOFF was born near Holland, Michigan, in 1884. His father was a physician. He received his higher education at Harvard and at the University of Chicago, where he earned the doctorate under E. H. Moore in 1907. Birkhoff was one of the first distinguished American mathematicians to receive his training entirely in this country. From 1914 to his death in 1944 he was Professor at Harvard. Birkhoff's mathematical research lay mostly in the area of analysis and its applications to physics, though his interests ranged widely and led to books on geometry and aesthetics. He was the acknowledged leader of the American mathematical community, and also served Harvard as Dean of the Faculty. Birkhoff's son Garrett (1911–1996) followed in his footsteps, serving as Professor at Harvard 1936–1981. Also a mathematician with broad interests, Garrett Birkhoff helped develop lattice theory, which is used as a framework for multidimensional geometry.

program met general acceptance, although efforts are under way during the 1990s to supplant it. Chapter 3 of this text is a detailed outline of the SMSG approach. It follows closely the first commercial text that adhered to their guidelines: *Geometry*, by SMSG authors Edwin E. Moise and Floyd L. Downs, Jr. (1964). Moise also published an advanced text ([1963] 1990) that follows this approach.

2.9 Geometry as pure mathematics

Concepts

The axiomatic method in foundations of geometry
Confusion over definitions
Pieri's and Hilbert's use and views of undefined concepts
Hilbert's dispute with Frege
Implicit definitions
The 1900 congresses and Padoa's paper
Pure, abstract mathematics
Russell's reaction
Huntington's axiomatization of the real number system
Modularization of mathematics

This book presents geometry as a branch of applied mathematics, through use of a model. A geometric model is a collection of precisely formulated concepts and principles that describe certain aspects of space. The theorems—true statements about the model—can explain observed properties of space and predict new ones. To dispel confusion and ensure against error, we follow the axiomatic method. An axiomatic model consists of undefined concepts, definitions, and unproved principles called axioms. We define all the required concepts in terms of simpler ones, except the most basic concepts, which we leave undefined. We accept a statement as a theorem only after proving it rigorously from the axioms.

During the 1890s the axiomatic method was clarified to accommodate a rigorous solution of the foundation problem. This encouraged a shift in mathematicians' view of their discipline: the emergence of pure, abstract mathematics, independent of any application. Its efficiency and flexibility soon made it the dominant approach to higher mathematics. Pure mathematicians don't really *do* mathematics differently, though. The distinction lies in *why they do it* and *how they describe it.*

In this section, you'll observe the shift at a critical moment, through the words of some of its major figures. You'll see how their accounts of what they were doing differed at first from the substance of their mathematics, then converged to a streamlined, effective presentation of the axiomatic method of pure mathematics.

Although philosophers as early as Aristotle recognized the need for undefined concepts in an axiomatic presentation, scientists seemed uncomfortable with them. Euclid's *Elements,* for example, purports to define every concept used. It's easy for us now to criticize that as bad exposition betraying muddled thinking. Through the centuries, however, philosophers tried to elucidate the definitional status of the concepts of an axiomatic theory, while preserving Euclid's traditional style. They created a mess of distinctions between different kinds of concepts and definitions. As technology enabled more precise observations, and mathematical progress brought experience with a great variety of theories, it became hard to justify those sophisticated distinctions.

David Hilbert and Mario Pieri presented the first really firm foundations for Euclidean geometry independently in 1899. They did leave their initial concepts undefined. Pieri (1899, appendix C, 1) *says* exactly what he's doing:

> [This] ... system of geometry ... originates ... with two concepts not taken from any other deductive science and about which one supposes to know nothing from the beginning: these are points and motion. All the other notions to which we are able to make reference concern pure logic ... or else they derive their origin from only these two ideas, combined with ... definitions.
>
> For the goals of the purely deductive method, it is beneficial to preserve the major indetermination possible for the content of the primitive ideas, [because that would never enter into the discussion except] by means of the logical relations expressed in the ... primitive propositions. For that reason, we are not obliged to connect with these ... terms ... even a specification: it is sufficient to the understanding of the entire system ... that it introduces itself generically around these ideas in the ... primitive propositions.

But Hilbert (1899, 4–6) simply *does* it, with little comment:

> Explanation. We consider three distinct systems of things. Those of the first system we call *points* ... the things of the second system, . *lines* ... those of the third system we call *planes* ...
>
> We consider the points, lines, and planes as having certain mutual relations, and indicate these with words like "lie on", "between", ... The exact and complete description of these relations follows from the *axioms of geometry* ...
>
> The axioms of Group I establish a *connection* between the things just explained ...
>
> Axiom I 1. Two distinct points always determine a line ...
>
> Axiom I 2. Any two distinct points of a line determine this line. ...
>
> Axiom I 7. On any line there are at least two points. In any plane there are at least three points that don't lie on any line. There are at least four points that don't lie in any plane. ...
>
> The axioms of Group II define the concept "between", and make possible, based on this concept, the *ordering* of the points on a line. ...
>
> Axiom II 1. If A, B, and C are points on a line and B lies between points A and C, then B is also between C and A.[18]

Hilbert and Pieri disagreed about their ultimate purpose. In another paper (Pieri 1901, 368) written about the same time, Pieri stated that

> this science, in its more advanced parts as well as the more modest ones, is affirming and consolidating itself more and more *as the study of a certain class of logical relations*; it is freeing itself little by little of the bonds that still tie it (however feebly) to *intuition*, and it is consequently

Mario PIERI was born in Lucca in 1860. His father was a lawyer. He studied at the universities in Bologna and Pisa, earning the doctorate in 1884 with a dissertation in algebraic geometry. From 1888 to 1900, Pieri taught projective and descriptive geometry in Turin, at the Military Academy and the University. There he worked with a group of mathematicians led by Giuseppe Peano, researching and reformulating the foundations of various areas of mathematics. Pieri continued this work as he moved on in 1900 and 1908 to professorships at Catania and Parma. Pieri provided one of the first completely rigorous foundations of Euclidean geometry, based on the concepts *point* and *motion*, and explored the use of several other combinations of basic concepts. He applied the same axiomatic methods to projective geometry, and was able to clarify the relationships between Euclidean, projective, and complex projective geometry. Pieri died of cancer in Lucca in 1913.

[18] There are seven axioms in Group I, five in Group II, and eight further axioms.

> vesting itself with the form and qualities of an ideal science, *purely deductive and abstract*, like arithmetic.

But Hilbert wrote in his short introduction (1899, 1),

> For a logically correct construction of geometry—just like arithmetic—only a few simple fundamental facts are required. These are called *axioms* of geometry. Selecting these axioms and exploring their interrelationships ... amounts to the logical analysis of our spatial intuition.[19]

Later, Hilbert demonstrated his axioms' logical correctness. They're all true when he interprets points, lines, and planes in the sense of coordinate geometry, hence any contradiction would lead to an inconsistency in arithmetic.

Those were confusing words from Hilbert, the more renowned scientist. What was he doing? In December 1899, the eminent philosopher Gottlob Frege wrote to Hilbert (Frege 1980, chapter 4). Shouldn't axioms "express fundamental facts of our [spatial] intuition?" How can a statement like Axiom II 1 *define* betweenness? Properly formulated definitions shouldn't assert anything. They just simplify the language, and can't lead to contradiction. How can points be intuitive things that we perceive in space, and at the same time be triples of coordinates?

Hilbert's response seems to indicate that the nature of concepts was not his major concern, that he wrote *Foundations of geometry* to provide a framework for understanding important geometrical problems like the proposition "that the parallel axiom is not a consequence of the other axioms." Nevertheless, he countered Frege's criticism by claiming that "the whole structure of axioms yields a complete definition" at once for the entire system of concepts not explicitly defined. This idea, called *implicit* definition, was popular at the time. For example, Pieri used it to moderate his extreme position against *any* specification for undefined concepts: "This demand for intuitive clarity shouldn't bother anybody, if one considers that the primitive concepts ... can be given ... by means of implicit definitions" (Pieri 1901, 387). The implicit definition idea faded from use, because it requires us to accept, before *any* analysis, that certain systems of undefined concepts can be interpreted consistently in essentially one way only.

Continuing, Hilbert pointed out "that every theory is only a scaffolding ... of concepts together with their necessary relations to one another, and that the basic elements can be thought of in any way one likes." But he contentiously overstated his case: "if the ... axioms ... do not contradict one another ... then they are true and the things defined by the axioms exist." Here he has stopped communicating. The two scientists were

---

[19] Hilbert started his first page with a short quotation from the philosopher Immanuel Kant ([1781] 1965) that used the same word *Anschauung* for *intuition*. In other writings, Hilbert supported and elaborated Kant's principle that mathematics is about certain objects, our knowledge of which is due to our intuition, rather than to experience or logic. See Peckhaus 1994, section 4.

proceeding from different bases. To Frege, geometry was mathematics *applied* to an existing subject, space. But Hilbert was trying to say that *the theory itself, without the application,* provides sufficient rationale for his study: He was presenting *pure* mathematics with general applicability. For about a decade, he and some other German and Italian mathematicians[20] had proceeded from that standpoint, without fanfare. Hilbert had no need for Frege's analysis of concepts, but he couldn't describe his position convincingly.[21]

A week later, exasperated with the inadequate answer, Frege fired back these axioms:

> E x p l a n a t i o n. We consider objects, that we call Gods.
>
> Axiom 1. All gods are omnipotent.
>
> Axiom 2. All gods are omnipresent.
>
> Axiom 3. There is at least one God.

According to Hilbert's argument, if the existence of an omnipresent omnipotent being entailed no contradiction, then God *must* exist! Frege incorporated this savage parody into a widely circulated book review published three years later.[22]

This kind of controversy and discussion continued for several years. But with the arrival of the twentieth century it became obsolete. In August 1900, philosophers and mathematicians from the whole world assembled at congresses in Paris to bid adieu to the old century and greet the new. Alessandro Padoa, a member of the Italian delegation led by Peano, presented the same paper at both meetings. He included an overview of the axiomatic method as it had evolved in his group, and was apparently the first mathematician

---

[20] Notably Gino Fano, who presented an axiomatization of projective geometry (1892) in almost the same style as Hilbert 1899. His axioms described higher-dimensional projective subspaces, objects that were clearly nonintuitive—in fact entirely unfamiliar to most mathematicians. Toepell (1986) suggests that Hilbert was little influenced by the Italians. But Fano did spend 1892–1893 in Göttingen!

[21] Steiner (1964) describes the dispute and presents Frege's position in detail.

[22] Frege is tacitly, but sarcastically, comparing Hilbert's response with St. Anselm's *ontological* argument for the existence of God. According to that argument (Hick 1967) *God* is synonymous with *a perfect* (*omnipotent omnipresent*) *being.* We *can conceive* of a perfect being. Could we also conceive that no such being existed? *No,* because otherwise our concept of perfection would include the possibility of its own nonexistence—it wouldn't be perfect. Since the nonexistence of a perfect being is inconceivable, God *must* exist. Kant refuted this argument in chapter 3, section 4, of the work Hilbert had cited in his introduction (see footnote 19). Kant claimed that *existence* is a property not of *beings,* but rather of *noun phrases*; it's inappropriate to ask whether a *being* exists, but reasonable to ask whether a noun phrase such as *a perfect being* describes anything real. With that interpretation, St. Anselm's argument makes no sense. Hilbert had studied at the same gymnasium and university as Kant, and had supported Kant's philosophy of arithmetic in his secondary Ph.D. subject examination and in lectures during the 1890s; see Reid's Hilbert biography (1970, 17) and Toepell's account (1986, 1.7.2) of the genesis of Hilbert 1899.

to get all the ideas concerning defined and undefined concepts completely straight:

> ... during the period of *elaboration* of any deductive theory we choose the *ideas* to be represented by the undefined symbols and the *facts* to be stated by the unproved propositions; but, when we begin to *formulate* the theory, we can imagine that the undefined symbols are *completely devoid of meaning* and that the unproved propositions (instead of stating ... *relations* between the ideas ...) are simply *conditions* imposed upon the undefined symbols.
>
> Then, the *system* of *ideas* that we have initially chosen is simply *one interpretation* of the ... *undefined symbols*; but ... this interpretation can be ignored by the reader, who is free to replace it in his mind by *another interpretation* that satisfies the conditions ...
>
> *Logical* questions thus become completely independent of *empirical* or *psychological* questions ... and every question concerning the *simplicity of ideas* and the *obviousness of facts* disappears ...
>
> It may be that there are several ... *interpretations* of the ... *undefined* symbols that verify the [conditions] .... The system of undefined symbols can then be regarded as the *abstraction* obtained from all these interpretations, and the ... *theory* can be regarded as the *abstraction* obtained from the *specialized theories* that result when ... the system of undefined symbols is successively replaced by each of the interpretations ....
>
> Thus, by means of just one argument that proves a proposition of the [abstract] theory we prove implicitly a proposition in each of the specialized theories.[23] (Padoa 1901, 120–121)

Padoa was describing lucidly what Hilbert was doing: pure, abstract mathematics, independent of any specialized application.

Also attending was Bertrand Russell, who would soon become a major figure in philosophy, particularly in foundations of mathematics. He recalled the event and the Italians fifty years later in his autobiography:

> The Congress was a turning point in my intellectual life, because I there met Peano ... In discussions ... I observed that he was always more precise than anyone else, and that he invariably got the better of any argument ... I decided that this must be owing to his mathematical logic ... By the end of August I had become completely familiar with all the work of his school ... The time was one of intellectual intoxication ... Suddenly, in the space of a few weeks, I discovered what appeared to be definitive answers to the problems which had baffled me for years ... I was introducing a new mathematical technique, by which regions formerly abandoned to the vaguenesses of philosophers were conquered for the precision of exact formulae. Intellectually, the month of September 1900 was the highest point of my life. (Russell 1951, 232–233)

---

[23] Padoa explicitly compares his position with Blaise Pascal's much earlier work (Pascal [1658] 1948). See section 2.3, note 8.

Gottlob FREGE was born in Wismar in 1848. His father was a schoolmaster. Frege was schooled there, and from 1869 studied at the universities in Jena, then Göttingen, where he earned the doctorate in 1873 with a dissertation in geometry. He took a position in Jena, attaining full professorship in 1896. His first major publication, the 1879 *Begriffsschrift*, presented a formula language for expressing mathematical reasoning. There Frege introduced the notion of universal and existential quantifiers, and developed the underlying structure of modern first order logic. Despite the formal nature of the language, the underlying concepts lacked really precise definition in the modern sense—Frege was actually opposed to that! Frege based on the *Begriffsschrift* several later works in logic and a major project, the two volume 1893–1903 work *Grundgesetze der Arithmetic*, in which he developed the arithmetic of natural numbers based solely on logic and simple set theory. Bertrand Russell discovered a fundamental contradiction—now known by his name—in a prepublication version of the second volume. Frege was unable to fix it; this crisis led to the later development of type theory and axiomatic set theory by Russell and others. Frege never accepted the modern abstract approach in mathematics, which often makes use of his work in logic. After 1903 his work passed from center stage; he died in 1925.

These meetings formed a watershed in philosophy of science. Soon after, the full axiomatic framework described by the Italian school was used commonly and routinely.[24]

Padoa's proposed change in mathematical practice was small, but profound. It involved not so much the conduct of research, but rather the presentation of results. We should "imagine that the undefined symbols are ... devoid of meaning and that the unproved propositions ... are simply conditions." But, he continues, we can use an *intended* interpretation as commentary to guide us, as long as it doesn't enter into the deductions of theorems. The benefit of such an abstract approach is that we can apply the theory to *any* interpretation that satisfies the axioms. The axiomatic development of Euclidean geometry in chapter 3 follows Padoa's guidelines.

By removing all emphasis on intended applications, the abstract method encourages mathematicians to devise axiom systems that apply to broad classes of interpretations. This has led to a tremendous increase in the intellectual efficiency of our discipline—the *modularization* of mathematics. Pure mathematicians, uninterested in any particular application, can work out in detail the consequences of small, manageable axiom systems. During the first half of the twentieth century advanced mathematics—particularly abstract algebra and analysis—was reorganized along these lines. A major application might draw on several of these "prefabricated" theories to study

[24] For another facet of the Italians' activity at the congress, see note 11 in the introduction to chapter 8.

particular aspects of a complicated model. In later chapters, this book will use that technique to apply results from the theory of real numbers and from group theory.

One of the first successes of abstract mathematics was Edward V. Huntington's axiomatization of the real number system (1902). Casting earlier mathematicians' work—particularly that of Richard Dedekind—into an abstract framework, he characterized the real number system as merely an interpretation of his model. From the model he derived all the standard theorems of arithmetic, algebra, and analysis. They could now be applied with confidence to any interpretation—for example, the decimal expansions used in arithmetic or the x coordinates used in geometry. What about the vagueness suggested by the indefinite article in the phrase "*an* interpretation of his model"? Huntington showed that his axioms are *categorical*— two interpretations may differ in details of their construction, but not in any property of their arithmetic or algebra. His work filled the last gap in the foundations of Euclidean geometry. It has been used by every approach to the foundation problem since then.

Any properly constructed axiomatic model can be adapted easily to the abstract form. You can regard the axiomatic presentation in chapter 3 either as applied mathematics—the development of a model for space—or as pure mathematics, independent of any application. And casting Hilbert's *Foundations of geometry* into abstract form requires changing only a few sentences of his expository prose, not his mathematics.

To mathematicians, Hilbert's inadequate explanation of his philosophy was no obstacle. They read his mathematics, understood what he was *doing*, and adapted his methods as their own. The geometer Hans Freudenthal explained (1962, 619) why most mathematicians identify the birth of abstract mathematics with Hilbert's book: "This thoroughly and profoundly elaborated piece of axiomatic workmanship was infinitely more persuasive than programmatic and philosophical speculations on space and axioms ever could be."

Alessandro PADOA was born in Venice in 1868. He attended the technical university at Padua, then the university at Turin, where he earned the doctorate in mathematics in 1895. He taught in secondary schools until 1909, when he took a position with the technical university in Genoa. Padoa was an effective contributor to and expositor of the Peano school's techniques in axiomatics and logic. The Italian academy of science honored him in 1934 for his statesmanly activity with secondary teachers' organizations and conferences. Padoa died in 1937.

2.10 Exercises and projects

Concepts

Scholarship skills
Decimal arithmetic
Algebra
Paradigmatic shift and cultural evolution

This section lists exercises and projects that involve the material covered in chapter 2, and extend it in several directions. Because this material is mainly historical and philosophical, there are few exercises here like those in most mathematics texts. Instead, there are suggestions for research papers. Some could be short reports on specific, narrow questions. For others, your work could continue alongside your study of the rest of this text, and result in a major term paper.

These projects are intended to help you acquire skills you'll need as a scholar:

library research,	deciding when you have an answer,
critical reading,	organizing a report,
asking questions,	detailed technical writing.

This book's bibliography, particularly the items referred to in this chapter, provides a start for these topics. Many listed sources themselves contain references that will lead you further. You may also want to consult *Mathematical reviews* and the journal *Historia mathematica*, which report current literature and work in progress.

Project 1. Find an interesting application in an applied mathematics, science, or engineering book, and one in the popular press. Describe all facets of the modeling processes used. Are they reported accurately and understandably?

Project 2. Report on Gauss' experiment concerning the angle sum of a triangle. How are such measurements done now with lasers? What accuracy is possible?

Project 3. This project is for those who've studied advanced physics. Describe the major features of conventionalism, and its application to relativity theory. The idea was introduced by Henri Poincaré around 1900 and developed further by Hans Reichenbach. How did they differ? What questions are still debated? You'll find advanced discussions in the Reichenbach memorial volume *Synthese* 34 (1977).

Project 4. Analyze one of the other sophisms in Ball and Coxeter [1892] 1987.

Project 5. Describe a nonmathematical work that's based on the axiomatic method. How closely does it adhere to the section 2.3 guidelines?

Project 6. How is Euclid's *Elements* ([1908] 1956) organized? Examine a later edition: for example, Playfair [1795] 1860. How do they differ?

Project 7. Show how to add, subtract, multiply, divide, and compute square roots using ruler and compass. That is, given segments with lengths a and b, construct segments with lengths $a + b$, $a - b$, ab, a/b, and $\sqrt{a}$.

Project 8. Describe some additional historical work along the lines mentioned at the end of section 2.4.

Project 9. Find and interpret pictures of ritual artifacts such as those Seidenberg mentions.

Project 10. Describe the introduction of decimal notation and the decimal algorithms for computing $a + b$, $a - b$, ab, a/b, and $\sqrt{a}$. Suppose you need a result accurate to n decimal places. Can you specify how many decimals of a and b you need to know?

Project 11. Describe the introduction of algebra to European mathematics.

Project 12. Report on the solution of the general cubic equation. What was its connection with Descartes' book?

Project 13. How does the position taken by Frege in his letter to Hilbert stand up to the problem that gave rise to conventionalism? Was Frege aware of the difficulty?

Project 14. What else was going on in Paris in summer 1900 that relates to the substance of this chapter? Was mathematics ahead of or behind the trends?

Project 15. Discuss this chapter in the light of the characteristics of paradigmatic shift and cultural evolution developed by Thomas Kuhn in *The structure of scientific revolutions* ([1962] 1970) and Raymond L. Wilder in *Evolution of mathematical concepts* (1968).

This chapter should suggest many more questions that could lead to similar projects, just as fruitful as those listed. In most cases, you should consult a mentor who is familiar with the area. It's easy to miss standard information sources. Before you dive in too deep, you should find out what background is required, and how closely such a project may be related to the subjects of this book. If you can find appropriate material, and have the background to relate it to geometry, you can gain great satisfaction from such a study.

Chapter

3

Elementary Euclidean geometry

This chapter is an axiomatic development of Euclidean geometry. It doesn't aim to *introduce* you to this material, for you should have studied much of it already in elementary classes. Rather, most of the chapter is a review, and it provides a common basis for readers with different backgrounds. It also shows in action the axiomatic method, as described in sections 2.3 and 2.9.

This book presents geometry as *applied* mathematics. The undefined concepts in this chapter are meant to refer to certain features of space; the axioms describe some of their fundamental properties. But the precise interpretation of the undefined concepts is left unspecified. Most of us acquire similar spatial experience as children. We regard these features, and the interpreted axioms, as intuitively clear. The axioms' validity is confirmed by engineering successes based on Euclidean geometry. Intuitive clarity persists, however, only until higher mathematics and physics presents us with feasible alternatives. At that point, mathematicians evade the issue. We're not concerned with the difficult problems of interpreting the undefined concepts and verifying the axioms. We leave those to philosophers of science.

Thus, this chapter holds together even if you *never* interpret the undefined concepts nor confirm the truth of the axioms. That is, you can regard it as a chapter of *pure* mathematics. That's up to you.

Euclidean geometry can be derived from several quite different axiom systems. This chapter is based on the SMSG axiomatization,[1] whose genesis was outlined in section 2.8.

This approach avoids the difficult mathematics employed by earlier axiomatizations to construct the real number system on a geometric basis.

---

[1] See Moise and Downs 1964, and Moise [1963] 1990.

The axioms themselves involve numbers as well as geometric concepts. Thus, a minimal familiarity with the real number system is required from the start to understand this material. You'll need to reason and manipulate with sets and functions, too. The necessary techniques are presented in detail in any standard intermediate algebra text and reviewed in most college algebra and calculus texts. For the proof of theorem 3.3.3 and the definition of arc length in section 3.14, some facts about equivalence relations and least upper bounds of sets of real numbers are required. If necessary, you can consult appendices A and B for details of those concepts.

You'll really find here just an outline of an axiomatic development. Routine proofs are omitted. Most others are only sketched. That's appropriate for a review. By supplying missing details, you familiarize yourself with material that you may at first recall only vaguely. The style is informal, like that of most introductory university mathematics texts. Use this style when you supply the missing arguments. But also convince yourself that greater formalization is *possible*—whatever is necessary to justify even the simplest steps in every proof.

Chapter 3 includes relatively few figures. A rigorous axiomatic development of geometry must proceed by logical steps solely from the undefined concepts and axioms. It must avoid any real *dependence* on figures. The sophism in section 2.2 showed why! Nevertheless, figures are essential guides to any study of geometry. Professional mathematicians generally construct more figures to follow material like this than beginners do. You should draw a figure for every case of every definition, every theorem, and every step in every proof.

Edwin E. MOISE was born in New Orleans in 1918. He attended Louisiana State and Tulane universities. After naval service in World War II, he studied at the University of Texas, which awarded him the Ph.D. in 1947 for work with R. L. Moore in topology. From 1947 to 1960 Moise rose through the ranks to full professor at Michigan, then moved to Harvard as Professor of Education. He worked with the School Mathematics Study Group (SMSG), writing its high school geometry text. Moise published that with coauthor Floyd L. Downs Jr. in 1964, accompanied by a now classic text to train teachers to use the SMSG approach. He moved again in 1971 to Queens College in New York, from which he retired in 1987. Moise continued work in topology throughout his career, and wrote several other texts, both about elementary subjects of interest to teachers and about advanced topology. Moise served in many professional leadership roles, including presidency of the Mathematical Association of America during 1967–1968. He died in 1998.

Since this chapter is not an introduction, few routine exercises are appropriate. The most instructive exercises relate several elementary geometry topics. Therefore, a selection of mostly nonroutine exercises in elementary Euclidean geometry is presented in chapter 4.

The *undefined concepts* enter into this axiomatic theory gradually. There are seven, and they first appear in the sections noted here:

point, line, plane	3.1	area	3.8
distance	3.2	volume	3.10
angle measure	3.4		

The *axioms* are presented in several groups:

incidence axioms	3.1	congruence axiom	3.5
ruler axiom	3.2	parallel axiom	3.7
Pasch's axiom	3.3	area axioms	3.8
protractor axioms	3.4	volume axioms	3.10

3.1 Incidence geometry

Concepts

- Points, lines, and planes
- Incidence
- Collinearity, coplanarity, and concurrence
- Incidence axioms
- Incidence geometry

Point, *line*, and *plane* are the first undefined concepts. Lines and planes are regarded as *point sets*—subsets of the set of all points, which is called *space*. Letters A to Z are used to denote points, g to l for lines, and α to ε for planes. The words *in*, *on*, and *through* describe the membership and subset relations $\in$ and $\subseteq$ among points and point sets; *incident with* means *on or through*. For example, the phrases *P is on g, g passes through P*, and *g and P are incident* mean $P \in g$. And *g is on α, α passes through g*, and *α and g are incident* mean $g \subseteq \alpha$. A point set is called *collinear* if all its members fall on one line; a family of points and point sets is *coplanar* if all its members lie in one plane. A family of point sets is called *concurrent* if all its members pass through a single fixed point.

If you wish, you can interpret these terms intuitively, as we learned when children. You can then assemble the corresponding images into figures that will be invaluable for steering you through the theory of this chapter. But the intuitive representation will never be used essentially in the logical development of geometry. That will be based solely on the axioms.

The first of these are the *incidence axioms* I1 to I5 below; they deal exclusively with the concepts just mentioned. Hilbert first stated them, in a slightly different form, in 1899.[2]

Axiom I1. Any two distinct points O and P lie on a unique line $\overleftrightarrow{OP}$. Each line passes through at least two distinct points.

Axiom I2. Any three noncollinear points O, P, and Q lie in a unique plane OPQ. Each plane passes through at least three noncollinear points.

Axiom I3. If two distinct points O and P lie in a plane ε, then the line $\overleftrightarrow{OP}$ lies entirely in ε.

Axiom I4. If two planes pass through a point O, then they pass through another point $P \neq O$.

Axiom I5. There exist noncoplanar points O, P, Q, and R such that O, P, and Q are noncollinear and $O \neq P$.

This book's pages illustrate axioms I3 and I4: they intersect not in just a single point, but along its whole spine. Axiom I5 ensures that geometry is three-dimensional. To axiomatize plane geometry, you could omit I4 and modify I5 to read, *There's exactly one plane.*

Listed next are some immediate consequences of the incidence axioms; they're used in the proofs of most later theorems. The language of theorem 1 has been simplified by disregarding the distinction between a point and the set consisting of that point alone; this usage will be continued. You can supply simple proofs for the first two theorems. Theorem 4 is more complicated, so its proof is sketched here. It's convenient to organize the first part of the proof as a *lemma*: an auxiliary theorem not particularly interesting in itself.

Theorem 1. The intersection of two distinct intersecting lines is a point. The intersection of a line and an intersecting but nonincident plane is a point. The intersection of two distinct intersecting planes is a line.

Theorem 2. Through a point and a nonincident line passes a unique plane. Through two distinct intersecting lines passes a unique plane.

Lemma 3. Points O to R of axiom I5 are distinct, and no three are collinear.

---

[2] Hilbert 1899, chapter 1. Actually, Hilbert used the axioms in lectures five years earlier; see Toepell 1986, section 2.4.

Proof. Suppose O, P, and R lay on a line g; since O, P, and Q are noncollinear, Q couldn't lie on g. By theorem 2, g and Q would lie in a plane, hence O, P, Q, and R would be coplanar—contradiction! Thus O, P, and R must be noncollinear. Similarly, triples O, Q, R and P, Q, R are noncollinear.

If $O = Q$ then O, P, and Q would lie on the line $\overleftrightarrow{OP}$ —contradiction! Thus $O \neq Q$. Similarly, $O \neq R$, $P \neq Q$, and $P \neq R$. If $Q = R$, then O, P, Q, and R would lie in the plane OPQ —contradiction! Thus $Q \neq R$. ♦[3]

Theorem 4. Through any point S pass at least three noncoplanar lines and three distinct planes. Through any line g pass at least two distinct planes.

Proof. S can't lie in all four of the planes determined by points O to R of lemma 3, for then S would lie on

$$(OPQ \cap OPR) \cap (OPQ \cap OQR) = \overleftrightarrow{OP} \cap \overleftrightarrow{OQ},$$

hence $S = O$, which isn't on PQR. Without loss of generality, you can assume S isn't on OPQ. Then S lies on $\overleftrightarrow{OS}$, $\overleftrightarrow{PS}$, and $\overleftrightarrow{QS}$. If these lines lay in a plane ε, then O, P, and Q would fall on $\varepsilon \cap OPQ$, which is a line by theorem 1—contradiction! Thus S lies in these three noncoplanar lines and in three distinct planes OPS, OQS, and PQS.

To show that g lies in at least two distinct planes, select any point S on g. By the previous paragraph, S lies on three noncoplanar lines. Select two lines h and k from these three so that g, h, and k are noncoplanar. Pairs g,h and g,k determine distinct planes through g. ♦

We could continue developing *incidence geometry*—the consequences of the incidence axioms alone. But there's only such result, Desargues' theorem, that's appropriate for this book. It's proved in section 5.3.

3.2 Ruler axiom and its consequences

Concepts

- Distance
- Scale and coordinates on a line
- Ruler axiom and ruler placement theorem
- Betweenness
- Segments and rays
- Convex sets

---

[3] The symbol ♦ is used throughout the book to mark the end of a proof.

The next undefined concept is *distance*. Actually, it's partly specified; it's a function that assigns to each pair P, Q of points a nonnegative real number PQ. Further details of the function are left undefined.

A *scale* for a line g is a one-to-one function c from the point set g onto the set of all real numbers, such that for any points P and Q on g,

$$|c(P) - c(Q)| = PQ.$$

The value $c(P)$ is called the c *coordinate* of P. The following axiom was introduced by George David Birkhoff (1932).

Ruler axiom. Each line has a scale.

You may think of a scale as the correspondence between points and numbers that results when you place a ruler with its edge along g; and you can interpret PQ as the distance you'd measure by subtracting the numbers opposite P and Q. However, this book will make no essential use of that intuitive idea. The idea does suggest that a line has many scales, all obtained from a given one by sliding the ruler or flipping it end over end. The idea underlies the proof of theorem 3 and is explored further in exercise 4.1.7. First, here's a simple theorem. You supply the proof.

Theorem 1. $PQ = QP$ for all points P and Q. Moreover, $PQ = 0$ if and only if $P = Q$.

One use of a scale is to describe the order of points on a line by referring to the order of their coordinates. Theorem 2 shows that this is independent of the scale used. Its proof depends on the following property of real numbers x, y, and z:

$$|x - z| = |x - y| + |y - z| \text{ if and only if } x \le y \le z \text{ or } x \ge y \ge z.$$

When these conditions hold, y is said to lie *between* x and z.

Theorem 2. Let c and d be scales for a line g through points X, Y, and Z. If $c(Y)$ lies between $c(X)$ and $c(Z)$, then $d(Y)$ lies between $d(X)$ and $d(Z)$.

The next result is a much stronger version of the ruler axiom. It shows what conditions determine a scale uniquely.

Theorem 3 (Ruler placement theorem). Let O and P be distinct points. Then there's a unique scale c for $\overleftrightarrow{OP}$ such that $c(O) = 0$ and $c(P) > 0$.

Proof. By the ruler axiom there's a scale a for $g = \vec{OP}$. Define a function b by setting $b(X) = a(X) - a(O)$ for each point X on g. Then b is also a scale for g and $0 = b(O) \neq b(P)$. In case $b(P) > 0$, let $c = b$. In case $b(P) < 0$, define c by setting $c(X) = -b(X)$ for each point X on g. In either case, c is a scale for g and $0 = c(O) < c(P)$. Thus a scale c as described *exists*. To prove its *uniqueness*, suppose d is a scale for g such that $d(O) = 0$ and $d(P) > 0$. It must be shown that $c(X) = d(X)$ for each point X on g. First,

$$|c(X)| = |c(X) - c(O)| = XO = |d(X) - d(O)| = |d(X)|. \qquad (*)$$

If $c(X) \geq 0$, then $c(X)$ lies between $c(O)$ and $c(P)$, or $c(P)$ lies between $c(O)$ and $c(X)$; by theorem 2, $d(X)$ lies between $d(O)$ and $d(P)$, or $d(P)$ lies between $d(O)$ and $d(X)$. Thus $c(X) \geq 0$ implies $d(X) \geq 0$, and $c(X) = d(X)$ follows from equations (*). By a similar argument, $c(X) < 0$ implies $d(X) < 0$, hence $c(X) = d(X)$. ♦

A point Y is said to lie *between* points X and Z, written X-Y-Z, if all three fall on a line g and $c(Y)$ lies between $c(X)$ and $c(Z)$ for some, hence any, scale c for g. Apparently, the concept of betweenness for points on a line has the same properties as the analogous concept for real numbers. Since the latter was defined using *inclusive* $\leq$ and $\geq$ inequalities, the conditions X-X-Z, X-Z-Z, and X-X-X are true for any points X and Z.

Any points X and Z determine a *segment*

$$\overline{XZ} = \{ Y : X\text{-}Y\text{-}Z \}.$$

Its *length* is XZ. It contains its *ends* X and Z; the rest of the segment is called its *interior*. A segment $\overline{XX}$ consists of the point X alone; its length is zero and its interior empty. A symbol or word that denotes a segment is often used, as well, to denote its length. To determine the meaning, you must examine the context.

Two distinct points O and P determine a *ray*

$$\vec{OP} = \{ Q \neq O : O\text{-}Q\text{-}P \text{ or } O\text{-}P\text{-}Q \}.$$

The ray's *origin* is the point O; that doesn't belong to the ray itself. (See figure 3.2.1.)

Theorem 4. Two rays $\vec{OP}$ and $\vec{NQ}$ are equal if and only if $N = O$ and O-P-Q or O-Q-P. If Q-O-P, the line $\vec{OP}$ consists of O and the two *opposite* rays $\vec{OP}$ and $\vec{OQ}$.

A point set is called *convex* if it includes segment $\overline{XZ}$ whenever it contains X and Z. Figure 3.2.2 shows example convex and nonconvex sets. Convex

Figure 3.2.1
Segment $\overline{XZ}$ and ray $\vec{OP}$

Figure 3.2.2
Convex and nonconvex sets

sets are involved in many applications of geometry. Theorem 5 lists some of their properties that are needed in this chapter.

Theorem 5. The empty set and all sets consisting of single points are convex. All segments, rays and lines are convex. All planes and the set of all points are convex. The intersection of any family of convex point sets is convex.

3.3 Pasch's axiom and the separation theorems

Concepts
Triangles, vertices, and edges
Pasch's axiom
Line, plane, and space separation theorems
Sides of a point in a line, of a line in a plane, and of a plane

This section is concerned with the *sides* of a point within a line, of a line within a plane, and of a plane in space. Euclid virtually ignored these notions in the *Elements*. Section 2.2 showed that carelessness with them can lead to serious errors. Because betweenness considerations enter into so many other topics in elementary geometry, they must be appear early in any axiomatic development. Unfortunately, they're often delicate; this section must consider many details and so may appear tedious.

Pasch's axiom connects incidence geometry with the ordering of points on a line. It lets you apply the betweenness concept in two- and three-dimensional situations, to describe when points lie on the same or opposite sides of a line or a plane. Basically, the axiom indicates how a line can intersect a triangle.

What *is* a triangle? Is it three points, or three edges? Does it contain interior points? There are several ways to define the concept. This seems the most effective: A *triangle* ΔXYZ is an ordered triple of noncollinear points X, Y, and Z. These are called the triangle's *vertices*. Segments

$\overline{YZ}$, $\overline{ZX}$, and $\overline{XY}$ are called the *edges* of ΔXYZ *opposite* X, Y, and Z.

The vertices of ΔXYZ are *ordered.* This is not the same triangle as ΔYXZ, because these triangles have different first and second vertices. You can make six different triangles from the same set of vertices. That complication is not useful now, but it streamlines considerably the discussion of triangle congruence in section 3.5.

Pasch's axiom. If a line in the plane of a triangle contains an interior point of one edge, it must intersect another edge.[4] (See figure 3.3.1.)

Theorem 1. No line can contain points interior to all three edges of a triangle.

Proof. (See figure 3.3.2.) Suppose collinear points O, P, and Q were interior to edges $\overline{YZ}$, $\overline{XZ}$, and $\overline{XY}$ of ΔXYZ. One of them must lie between the other two; without loss of generality you can suppose O-P-Q. Then the line $g = \overleftrightarrow{XZ}$ would contain point P interior to edge $\overline{OQ}$ of ΔYOQ. By Pasch's axiom, g would intersect another edge, $\overline{YO}$ or $\overline{YQ}$. But the *lines* that include those edges intersect g at Z and X, which lie *outside* the edges themselves—contradiction! ♦

Because the betweenness relation for points on a line has the same properties as the corresponding relation for real numbers, it's easy to handle the notion of the *sides* of a point within a line. Theorem 2, the line separation theorem, gives the definition and main properties of this concept. Its proof doesn't require Pasch's axiom. The theorem is phrased just like theorems 3 and 4, the analogous plane and space separation theorems. This formulation

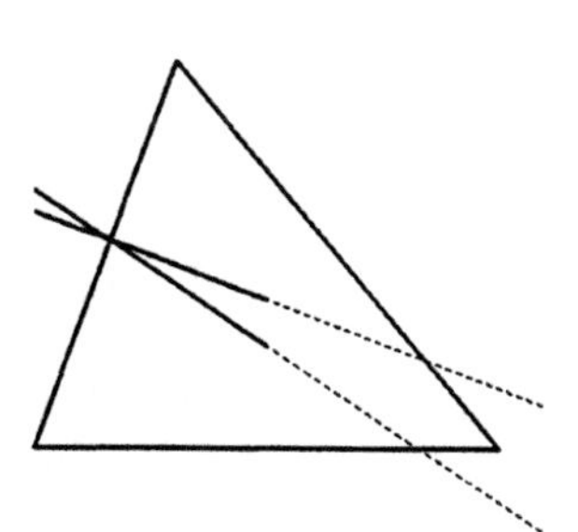

Figure 3.3.1
Pasch's axiom

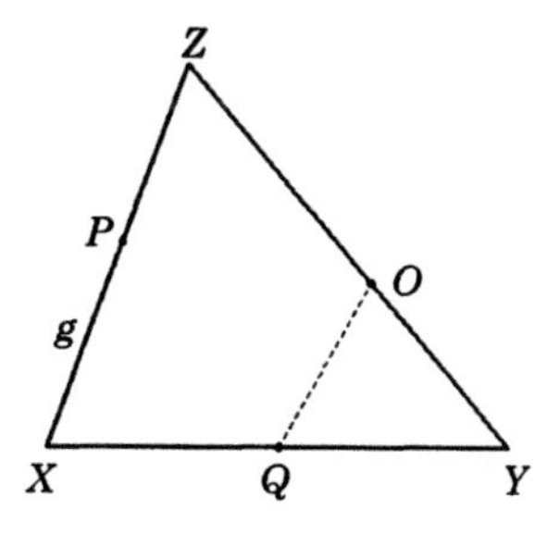

Figure 3.3.2
Proof of theorem 1

[4] First stated by Moritz Pasch ([1882] 1926, 20).

was devised by SMSG about 1959.[5] The proofs of the latter two theorems are too involved for beginning students, so SMSG actually included these results among their axioms and proved Pasch's axiom as a theorem.[6]

***Theorem 2* (*Line separation theorem*).** Let O be a point on a line g. Then g is the union of three disjoint convex sets: O itself and two opposite rays, called the *sides* of O in g. Two points X and Y on g lie on the same side of O in g if and only if segment $\overline{XY}$ doesn't contain O. If P and Q are points distinct from O such that P-O-Q, then $\vec{OP}$ and $\vec{OQ}$ are the sides of O in g.

***Theorem 3* (*Plane separation theorem*).** Let g be a line in a plane ε. Then ε is the union of three disjoint convex point sets: g itself and two *sides* of g in ε. Two points X and Y in ε lie on the same side of g in ε if and only if segment $\overline{XY}$ doesn't intersect g. If P is any point in ε not on g, then the two sides of ε are the sets

$$\{Q \in \varepsilon : \overline{PQ} \cap g = \phi\}, \qquad \{Q \in \varepsilon : Q \notin g \ \& \ \overline{PQ} \cap g \neq \phi\}.$$

Proof. (To understand this proof, you must be familiar with the notion of an equivalence relation. See appendix A for details of that theory.) Define a binary relation E on ε by setting

$$X\,E\,Y \leftrightarrow X,Y \in g \text{ or } \overline{XY} \cap g = \phi$$

for all X and Y on ε. Clearly, E is reflexive and symmetric. It's transitive as well, for if $X\,E\,Y$ and $Y\,E\,Z$, then one of the following cases holds:

1. Any—hence all—of X, Y, and Z lie in g.
2. None of X, Y, and Z lies in g. Then g can't intersect $\overline{XY}$ or $\overline{YZ}$. By Pasch's axiom, it can't intersect $\overline{XZ}$, hence $X\,E\,Z$.

Therefore E is an equivalence relation on ε. The equivalence class of any point O on g is $\{Q : O\,E\,Q\} = g$ itself. That of P is

$$\{Q : P\,E\,Q\} = \{Q \in \varepsilon : \overline{PQ} \cap g = \phi\}.$$

Let X and Y be points on ε lying in neither of these classes. Then g contains points interior to edges $\overline{PX}$ and $\overline{PY}$ of ΔPXY. It follows that $X\,E\,Y$; otherwise, g would contain points interior to all three edges of that triangle, contrary to theorem 1. Therefore, there's only one equivalence class besides g and $\{Q : P\,E\,Q\}$, namely

$$\{Q : X\,E\,Q\} = \{Q \in \varepsilon : \overline{PQ} \cap g \neq \phi\}.$$

Any equivalence class is convex: If $X\,E\,Z$ and X-Y-Z, then $\overline{XZ}$ lies in g or is disjoint from g; the same must hold for $\overline{XY}$, so $X\,E\,Y$ also. ♦

---

[5] See Wooton 1965.

[6] See Moise and Downs 1964, chapter 3.

Moritz PASCH was born in 1843 in Breslau, Prussia, to a merchant family. He studied chemistry and mathematics at the university there, earning the doctorate in 1865. From 1870 on he worked at the University of Giessen, attaining full professorship in 1875, and serving as dean and rector from 1883 to 1894. Pasch's early research was in the area of algebraic geometry, but he's best known for his work in the foundations of geometry. His 1882 book, *Vorlesungen über neuere Geometrie*, was the first rigorous axiomatic presentation of geometry, developing the Euclidean theory via projective geometry. Particularly noteworthy is his treatment of betweenness and separation properties. Pasch retired in 1911 and died at age 87 in 1930.

***Theorem 4* (*Space separation theorem*).** Let ε be a plane. Then the set of all points is the union of three disjoint convex point sets: ε itself and its two *sides*. Two points X and Y lie on the same side of ε if and only if segment $\overline{XY}$ doesn't intersect ε. If P is any point not on ε, then the two sides of ε are the sets

$$\{Q : \overline{PQ} \cap \varepsilon = \phi\}, \qquad \{Q : Q \notin \varepsilon \ \& \ \overline{PQ} \cap \varepsilon \neq \phi\}.$$

Proof. Define a binary relation E on the set of all points by setting

$$X E Y \Leftrightarrow X, Y \in \varepsilon \text{ or } \overline{XY} \cap \varepsilon = \phi$$

for all X and Y. Clearly, E is reflexive and symmetric. It's transitive as well, for if $X E Y$ and $Y E Z$, then one of the following cases holds:

1. Any—hence all—of X, Y, and Z lie in ε.
2. None of X, Y, and Z lies in ε but all lie in a plane δ whose intersection with ε is a line g. Then X, Y, and Z lie on the same side of g in δ, hence $\overline{XZ} \cap \varepsilon = \overline{XZ} \cap g = \phi$.
3. The only plane through X, Y, and Z is disjoint from ε, hence $\overline{XZ} \cap g = \phi$.

Therefore E is an equivalence relation. The equivalence class of any point O on ε is $\{Q : O E Q\} = \varepsilon$ itself. That of P is

$$\{Q : P E Q\} = \{Q : \overline{PQ} \cap \varepsilon = \phi\}.$$

Let X and Y be points lying in neither of these classes. Then P, X, and Y fall in a plane δ whose intersection with ε is a line g; and P is on the side of g in δ opposite X and Y. Thus X and Y are on the same side of g in δ, and $Q E X$ because $\overline{QX} \cap \varepsilon = \overline{QX} \cap g = \phi$. Therefore, there's only one equivalence class besides g and $\{Q : P E Q\}$, namely

$$\{Q : X E Q\} = \{Q : Q \notin \varepsilon \ \& \ \overline{PQ} \cap \varepsilon \neq \phi\}.$$

You can supply the proof that the two sides of ε are convex. ♦

The notions of the sides of a line in a plane and of a plane in space are used later in this chapter to define the interior of a triangle and of various other figures.

3.4 Angles and the protractor axioms

Concepts
- Angles
- Interior of an angle and of a triangle
- Crossbar theorem
- Degree measure
- Protractor axioms
- Linear and vertical pairs of angles

Section 3.3 dealt with triangles without ever considering angles. This section defines that concept, and uses the sides of lines in a plane to define the interior of an angle and clarify that notion. You're invited to supply many of the figures for this section, and the simple proofs for most of the theorems. Theorems 4 and 5 are more complicated. The proofs sketched here are due to Moise.[7] After this preparation, the section introduces angle measurement via new axioms.

Theorem 1. Consider a point O on a line g, a point P not on g, and let ε be the plane through g and P. Then the ray $\vec{OP}$ lies entirely on one side of g in ε.

An *angle* is a point set of the form

$$\angle POQ = \vec{OP} \cup O \cup \vec{OQ},$$

where O, P, and Q are noncollinear points; O is called its *vertex*. Often, when no confusion results, it's written simply $\angle O$. This convention is particularly common with triangles: the angles of ΔXYZ are $\angle X = \angle ZXY$, $\angle Y = \angle XYZ$, and $\angle Z = \angle YZX$. By theorem 1 you can define the *interior* of $\angle POQ$ to be the set of all points in OPQ that lie on the same side of $\vec{OP}$ as the ray $\vec{OQ}$ and on the same side of $\vec{OQ}$ as $\vec{OP}$. (See figure 3.4.1.)

Theorem 2. The interior of $\angle POQ$ is convex. All interior points of segment $\overline{PQ}$ lie in the interior of $\angle POQ$. If R is a point in the interior of $\angle POQ$, then the ray $\vec{OR}$ lies entirely in the interior of $\angle POQ$.

---

[7] Moise [1963] 1990, section 4.3.

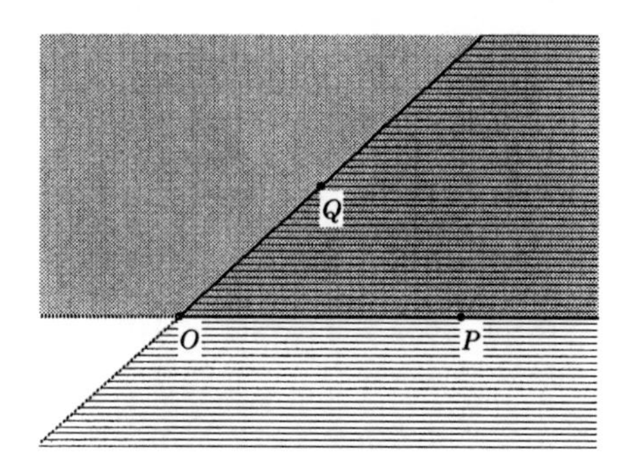

Figure 3.4.1
Interior of $\angle POQ$
(intersection of shaded and scored half planes)

Theorem 3. Suppose Q and R are points on the same side of a line $\overleftrightarrow{OP}$ in a plane ε. Then exactly one of the following conditions holds:

1. Q lies in the interior of $\angle POR$,
2. $\overrightarrow{OQ} = \overrightarrow{OR}$,
3. R lies in the interior of $\angle POQ$.

Theorem 4. Let O be an interior point of edge $\overline{SQ}$ of ΔSPQ and let R be a point in plane SPQ on the same side of $\overleftrightarrow{SQ}$ as P. Then the ray $\overrightarrow{OR}$ intersects $\overline{SP}$ or $\overline{PQ}$.

Proof. The line $\overleftrightarrow{OR}$ intersects $\overline{SP}$ or $\overline{PQ}$; otherwise S, P, and Q would all lie on the same side of $\overleftrightarrow{OR}$ and $\overline{QS}$ couldn't intersect $\overleftrightarrow{OR}$. Let T be an intersection of $\overleftrightarrow{OR}$ with $\overline{SP}$ or $\overline{PQ}$. If T weren't in $\overrightarrow{OR}$, then T would lie on the side of $\overleftrightarrow{SQ}$ opposite R, hence opposite P, so T couldn't lie in $\overline{SP}$ or $\overline{PQ}$ by theorem 1. Thus T lies in $\overrightarrow{OR}$. ♦

Theorem 5 (Crossbar theorem). Let R be a point in the interior of $\angle POQ$. Then $\overrightarrow{OR}$ intersects $\overline{PQ}$. (See figure 3.4.2.)

Proof. Select a point $S \neq O$ such that S-O-Q. Then S, P, and Q are noncollinear, O lies in $\overline{SQ}$, and R is on the same side of $\overleftrightarrow{SQ}$ as P. By theorem 4, $\overrightarrow{OR}$ intersects $\overline{SP}$ or $\overline{PQ}$. By theorem 1, $\overrightarrow{OR}$ can't intersect $\overline{SP}$. ♦

Angle measurement, like length, is another partly undefined concept. It's a function that assigns to each angle $\angle O$ a real number $m\angle O$, called its *measure*, such that $0 < m\angle O < 180$. Further details of the function are undefined. There are two common systems of angle measurement; the one described here is called *degree* measure. To distinguish it from the other system (radian measure), we write the symbol $^\circ$ after the measure—for example, $m\angle O = 90^\circ$, to be read "ninety *degrees*." Some properties of angle measurement are postulated by the *protractor axioms* P1 to P3 below. They

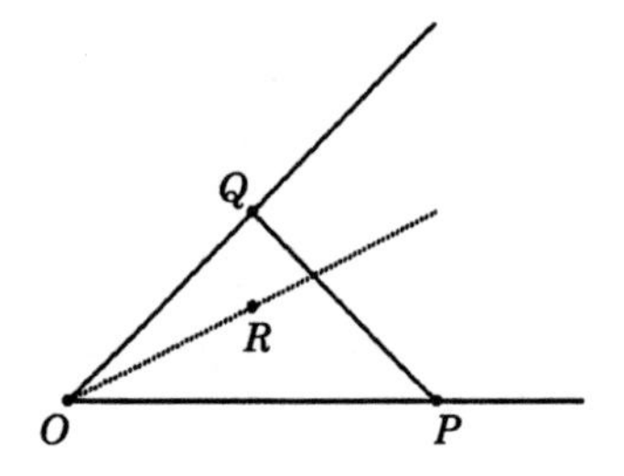

Figure 3.4.2
Crossbar theorem

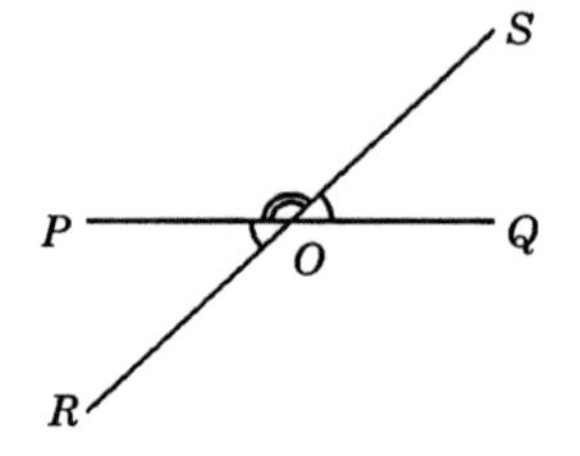

Figure 3.4.3 Linear and vertical pairs of angles

were first stated, in a slightly different form, by George David Birkhoff (1932).

Axiom P1. Let O and P be distinct points in a plane ε and $0 < x < 180$. Then on each side of $\overleftrightarrow{OP}$ in ε there's a point Q such that $m\angle POQ = x°$.

Axiom P2. If Q is a point in the interior of $\angle POR$, then $m\angle POQ + m\angle QOR = m\angle POR$.

Axiom P2 and theorem 3 yield the uniqueness of ray $\overrightarrow{OQ}$ in axiom P1, as follows.

Theorem 6. In axiom P1, if R is a point on the same side of $\overleftrightarrow{OP}$ in ε as Q and $m\angle POQ = m\angle POR$, then $\overrightarrow{OQ} = \overrightarrow{OR}$.

Proof. Otherwise $m\angle QOR = 0°$. ♦

You may think of angle measurement as the correspondence between rays $\overrightarrow{OQ}$ and numbers produced by placing a protractor with its base along line $\overleftrightarrow{OP}$ and centered at O. The protractor can lie in ε on either side of the line. However, this book will make no essential use of that intuitive idea.

Let g and h be distinct lines intersecting at a point O. Let P, Q and R, S be pairs of points on g and h distinct from O such that P-O-Q and R-O-S. Then $\angle POR$ and $\angle POS$ are said to form a *linear* pair of angles, and $\angle POR$ and $\angle QOS$, a *vertical* pair. (See figure 3.4.3.)

Axiom P3. The sum of the measures of the angles of a linear pair is $180°$.

Theorem 7 (Vertical angle theorem). Angles of a vertical pair have the same measure.

3.5 Congruence

Concepts

Congruent segments and angles
Congruent triangles
SAS congruence axiom
ASA, SSS, and SAA congruence theorems
There's no SSA congruence theorem!
Isosceles triangle theorem
Equilateral triangles, scalene triangles
Exterior angle theorem
Triangle inequality
Hinge theorem

Using one new axiom, this section develops the theory of congruence—the precise concept corresponding to the informal notion "same size and shape." Congruence is indicated by the symbol $\cong$. Two segments are said to be *congruent* if they have the same length; two angles, if they have the same measure. Two triangles ΔXYZ and $\Delta X'Y'Z'$ are called *congruent* if their corresponding edges and angles are congruent:

$$XY = X'Y' \qquad YZ = Y'Z' \qquad ZX = Z'X'$$
$$\angle Z \cong \angle Z' \qquad \angle X \cong \angle X' \qquad \angle Y \cong \angle Y'.$$

Segment congruence, angle congruence, and triangle congruence are all equivalence relations.

Euclid recognized the fundamental nature of the following axiom, but presented it as a theorem, with a faulty proof. Hilbert first stated it as an axiom (1899, section 6).

SAS congruence axiom. Consider triangles ΔXYZ and $\Delta X'Y'Z'$. If $XY = X'Y'$, $\angle Y \cong \angle Y'$, and $YZ = Y'Z'$, then $\Delta XYZ \cong \Delta X'Y'Z'$.

SAS means *side-angle-side*: If two triangles have two pairs of congruent edges (sides), and the angles *between* them are also congruent, then the triangles themselves are congruent. If you're given three equations in SAS pattern among those that define triangle congruence, then you can infer the other three. Theorems 1, 3, and 5 allow you to start with other sets of three equations: two pairs of angles and one of edges (*ASA* or *SAA*), or three pairs of edges (*SSS*). To avoid the difficult proofs of these theorems,

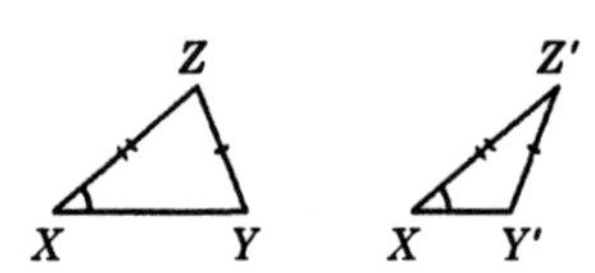

Figure 3.5.1
There's no SSA theorem!

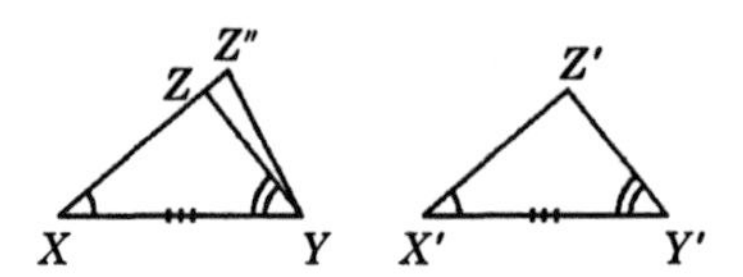

Figure 3.5.2
Proof of theorem 1

SMSG included the ASA and SSS results among their axioms, and postponed SAA until they could use the parallel axiom to prove it easily.[8]

Figure 3.5.1 shows *there's no SSA congruence theorem!* That is, triangles ΔXYZ and $\Delta X'Y'Z'$ may be incongruent, yet have two pairs of edges and one of angles congruent in the SSA pattern. See exercise 4.6.2 for further details. Our experience with incongruent similar triangles, which have congruent angles and edges with proportional but unequal lengths, shows *there's no AAA congruence theorem, either.*

***Theorem 1* (*ASA congruence theorem*).** Consider triangles ΔXYZ and $\Delta X'Y'Z'$. If $\angle X \cong \angle X'$, $XY = X'Y'$, and $\angle Y \cong \angle Y'$, then $\Delta XYZ \cong \Delta X'Y'Z'$.

Proof. (See figure 3.5.2.) Find the point Z'' on $\overrightarrow{XZ}$ such that $XZ'' = X'Z'$. By the SAS congruence axiom, $\Delta XYZ'' \cong \Delta X'Y'Z'$, so $\angle XYZ'' \cong \angle Y' \cong \angle Y$. Therefore, $\overrightarrow{YZ} = \overrightarrow{YZ''}$, hence $Z = Z''$, and finally $XZ = XZ'' = X'Z'$. By the SAS axiom, $\Delta XYZ \cong \Delta X'Y'Z'$. ♦

***Corollary 2* (*Isosceles triangle theorem*).** Two angles of a triangle are congruent if and only if the opposite edges are.

Proof. Consider ΔXYZ. If $\angle X \cong \angle Y$, then $\Delta XYZ \cong \Delta YXZ$, hence $YZ = XZ$; and conversely. ♦

A triangle with two congruent edges and two congruent angles is called *isosceles*. A triangle with three congruent edges is called *equilateral*. Clearly, a triangle is equilateral if and only if its angles are all congruent. A triangle that has no congruent edges nor angles is called *scalene*.

***Theorem 3* (*SSS congruence theorem*).** Consider triangles ΔXYZ and $\Delta X'Y'Z'$. If $XY = X'Y'$, $YZ = Y'Z'$, and $ZX = Z'X'$, then $\Delta XYZ \cong \Delta X'Y'Z'$.

---

8 Moise and Downs 1964, chapter 5. This section's proofs are modeled after Moise [1963] 1990, section 6.2 and chapter 7.

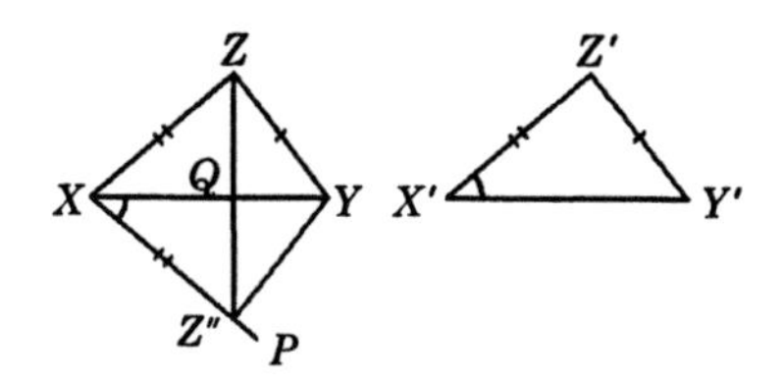

Figure 3.5.3
Proof of theorem 3

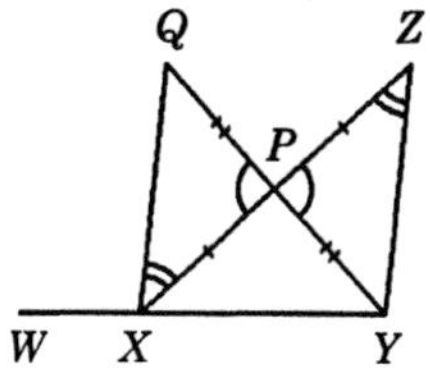

Figure 3.5.4
Proof of theorem 4

Proof. (See figure 3.5.3.) It's enough to show $\angle Z \cong \angle Z'$. First, find a point P in plane XYZ on the side of $\overleftrightarrow{XY}$ opposite Z so that $\angle YXP \cong \angle X'$. Next, find Z'' on $\overrightarrow{XP}$ so that $XZ'' = X'Z' = XZ$. By the SAS axiom, $\Delta XYZ'' \cong \Delta X'Y'Z'$, hence $YZ'' = Y'Z' = YZ$. Segment $\overline{ZZ''}$ intersects $\overleftrightarrow{XY}$ at a point Q. There are now five cases:

1. Q-X-Y & $Q \neq X$,
2. $Q = X$,
3. $X \neq Q$ & X-Q-Y & $Q \neq Y$,
4. $Q = Y$,
5. $Y \neq Q$ & X-Y-Q.

The proof will be completed only for case 3, shown in figure 3.5.3. The other cases are left to you. By the isosceles triangle theorem, $\angle XZQ \cong \angle XZ''Q$ and $\angle QZY \cong \angle QZ''Y$. Since Q lies in the interiors of $\angle XZY$ and $\angle XZ''Y$, axiom P2 yields $\angle Z \cong \angle YZ''X \cong \angle Z'$. The result then follows from the SAS axiom. ♦

The following exterior angle theorem is a weak version of corollary 3.7.11, which is proved after the parallel axiom is introduced. Often, you need only this weak version. Including it here allows presentation of several major results—for example, the SAA congruence theorem, triangle inequality, and hinge theorem—before that axiom.

Theorem 4 (Exterior angle theorem). Consider ΔXYZ and a point $W \neq X$ such that W-X-Y. Then $m\angle Y, m\angle Z < m\angle WXZ$.

Proof. (See figure 3.5.4.) Find the points P and Q such that

$$X\text{-}P\text{-}Z \qquad XP = PZ$$
$$Y\text{-}P\text{-}Q \qquad YP = PQ.$$

Note that Q lies in the interior of $\angle WXZ$. By the vertical angle theorem and the SAS axiom, $\Delta ZPY \cong \Delta XPQ$, hence

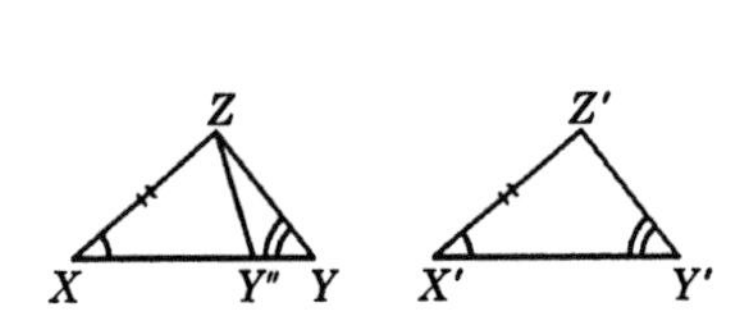

Figure 3.5.5
Proof of theorem 5

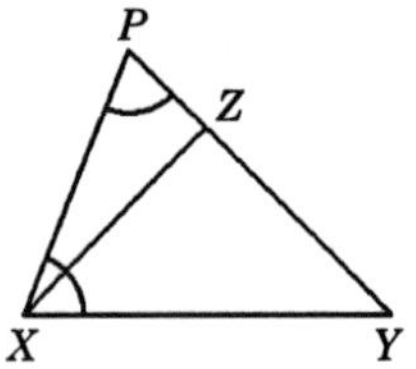

Figure 3.5.6
Proof of lemma 6

$$m\angle Z = m\angle ZXQ = m\angle WXZ - m\angle WXQ < m\angle WXZ.$$

The proof that $m\angle Y < m\angle WXZ$ is left to you. ♦

***Theorem 5* (*SAA congruence*).** Consider triangles ΔXYZ and $\Delta X'Y'Z'$. If $\angle X \cong \angle X'$, $\angle Y \cong \angle Y'$, and $ZX = Z'X'$, then $\Delta XYZ \cong \Delta X'Y'Z'$.

Proof. Find the point Y'' on $\overrightarrow{XY}$ such that $XY'' = X'Y'$. Suppose $X\text{-}Y''\text{-}Y$ and $Y'' \neq Y$ as shown in figure 3.5.5; then $\Delta XY''Z \cong \Delta X'Y'Z'$ by the SAS axiom, so $\angle XY''Z \cong \angle Y' \cong \angle Y$, contrary to the exterior angle theorem. Thus $X\text{-}Y''\text{-}Y$ is impossible. You can show similarly that $X\text{-}Y\text{-}Y''$ & $Y'' \neq Y$ is impossible. Therefore, $Y'' = Y$, and the result follows from the SAS axiom. ♦

Lemma 6 is a weak preliminary version of theorem 8.

***Lemma 6*.** In ΔXYZ, $YZ < XY$ if and only if $m\angle X < m\angle Z$.

Proof. (See figure 3.5.6.) Suppose $YZ < XY$. Find the point P on $\overrightarrow{YZ}$ such that $YP = XY$. By the isosceles triangle and exterior angle theorems,

$$m\angle YXZ < m\angle YXP = m\angle YPX < m\angle YZX.$$

Conversely, suppose $m\angle X < m\angle Z$. If $YZ > XY$, then $m\angle Z < m\angle X$ by the preceding argument, contradiction! If $YZ = XY$, then $m\angle Z = m\angle X$ by the isosceles triangle theorem, contradiction! Thus $YZ < XY$. ♦

***Theorem 7* (*Triangle inequality*).** If points X, Y, and Z are not collinear, then $XY + YZ > XZ$.

Proof. Find the point P such that $Z\text{-}Y\text{-}P$ and $YP = YX$. Then apply the isosceles triangle theorem and lemma 6 to ΔXZP. ♦

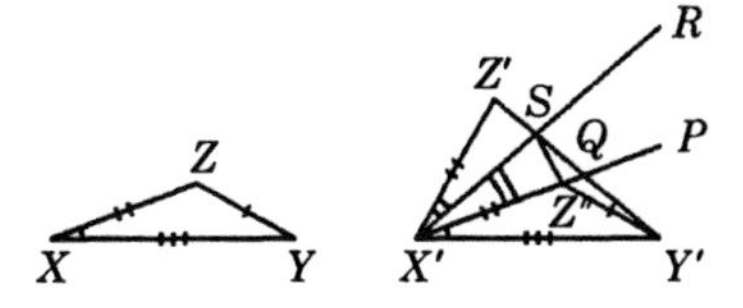

Figure 3.5.7
Proof of theorem 8

Theorem 8 (Hinge theorem). In ΔXYZ and $\Delta X'Y'Z'$, suppose $XY = X'Y'$ and $XZ = X'Z'$. Then $m\angle X < m\angle X'$ if and only if $YZ < Y'Z'$.

Proof. (See figure 3.5.7.) Suppose $m\angle X < m\angle X'$. Find a point P in the interior of $\angle Y'X'Z'$ so that $\angle Y'X'P \cong \angle X$. By the crossbar theorem, $\overrightarrow{X'P}$ intersects $\overline{Y'Z'}$ at a point Q. Find the point Z'' on $\overrightarrow{X'P}$ such that $X'Z'' = XZ$. By the SAS axiom, $\Delta XYZ \cong \Delta X'Y'Z''$, hence $YZ = Y'Z''$. Find a point R in the interior of $\angle Z''X'Z'$ so that $\angle Z''X'R \cong \angle RX'Z'$. By the crossbar theorem $\overrightarrow{X'R}$ intersects $\overline{QZ'}$ at a point S. By the SAS axiom $\Delta Z''X'S \cong \Delta Z'X'S$, hence $SZ' = SZ''$. Finally,

$$YZ = Y'Z'' < Y'S + SZ'' = Y'S + SZ' = Y'Z'.$$

(This follows from the triangle inequality unless Y', S, and Z'' are collinear; in that case, it's trivial.)

You can provide the proof that $YZ < Y'Z'$ implies $m\angle X < m\angle X'$. ♦

3.6 Perpendicularity

Concepts

- Right, acute, and obtuse angles
- Perpendicular lines and planes
- Feet of perpendiculars
- Distance from a point to a line or a plane
- Perpendicular bisectors
- Dihedral angles

This section presents many frequently used and closely related definitions and theorems about perpendicular lines and planes. It introduces the notion of the dihedral angles formed by intersecting planes, then uses properties of perpendicular lines to show how to measure these angles. No new axioms are necessary. The proofs are generally straightforward, and most are left to you for completion. Do that! And provide the figures, too. These proofs are excellent exercises, for they use most of the geometry developed earlier

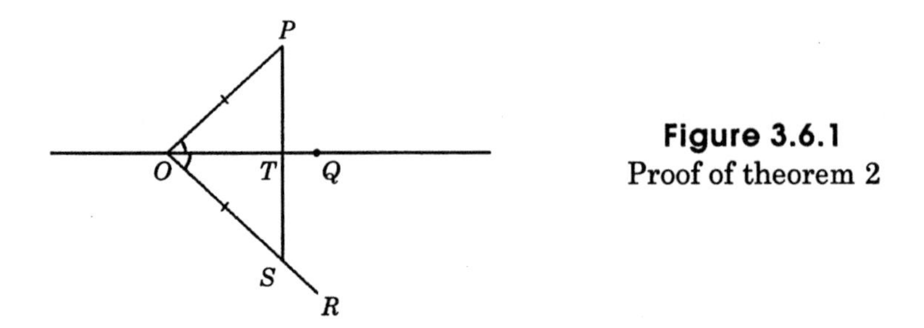

Figure 3.6.1
Proof of theorem 2

in this chapter. The order of the theorems in this section is crucial. If you change it, you'll probably find that the proofs become much more difficult.

An angle whose measure is $90°$ is called a *right* angle. Angles measuring less than that are called *acute*; those greater are *obtuse*. Two angles whose measures total $90°$ are called *complementary*. Two distinct intersecting lines g and h form four angles. If one of these is right, then all are, and the lines are said to be *perpendicular*, written $g \perp h$. This adjective is often applied to a segment or ray if it would be appropriate to apply it to the corresponding line. For example, a segment could be perpendicular to a line, ray, or another segment. The word *perpendicular* is sometimes used as a noun to mean *perpendicular line*.

Theorem 1. Let P be a point on a line g in a plane ε. Then there's a unique perpendicular to g in ε through P.

Theorem 2. Let P be a point not on a line g. Then there's a unique perpendicular to g through P.

Proof. (See figure 3.6.1.) Let ε be the plane through P and g. Select any distinct points O and Q on g, then a point R in ε on the side of g opposite P so that $\angle QOR \cong \angle QOP$. Now find the point S on $\overrightarrow{OR}$ such that $OS = OP$. Segment $\overline{PS}$ intersects g at a point T. If $O = T$, then $\angle QOR$ and $\angle QOP$ form a congruent linear pair. If $O \neq T$, then $\Delta TOP \cong \Delta TOS$, hence $\angle OTP$ and $\angle OTS$ are a congruent linear pair. In either case, $\overleftrightarrow{PS} \perp g$. Uniqueness follows from the exterior angle theorem. ♦

Point T in the proof of theorem 2 is called the *foot* of the perpendicular from P to g.

Theorem 3. The foot of the perpendicular to a line g through a point P not on g is the unique point on g whose distance from P is minimum.

The minimum distance in theorem 3 is called the *distance from* P *to* g. The distance from a point P to a line through P is defined to be zero.

Lemma 4. Let P,Q and X,Y be pairs of distinct points. If $PX = XQ$ and $PY = YQ$, then $PZ = ZQ$ for every point Z on $\overleftrightarrow{XY}$.

Theorem 5. Let g, h, and k be lines through a point O such that $g \perp h,k$ and $h \neq k$. Let ε be the plane through h and k, and l be any line through O. Then $g \perp l$ if and only if l lies in ε.

Proof. Suppose l lies in ε. Select any distinct points P and Q on g such that $PO = OQ$ and find points X and Y on h and k lying on different sides of l in ε. The segment $\overline{XY}$ intersects l at a point $Z \neq O$. The SAS axiom and lemma 4 yield $PX = XQ$, $PY = YQ$, and $PZ = ZQ$, so $g \perp l$ follows from the SSS congruence theorem.

Conversely, suppose $g \perp l$. Let δ be the plane through g and l. If l weren't in ε, then $\delta \cap \varepsilon$ would be a line $m \neq l$. By the previous paragraph, m would be another perpendicular to g in δ through O, which would contradict theorem 1. Thus l must lie in ε. ♦

A line g and a plane ε are said to be *perpendicular*, written $g \perp \varepsilon$, if $g \cap \varepsilon$ is a point O and g is perpendicular to every line through O in ε.

Theorem 6. If P is a point and g a line, then P lies in a unique plane perpendicular to g.

Proof. First, suppose P is on g. Select any two distinct planes γ and δ through g. By theorem 1 there exist perpendiculars h and k to g through P in γ and δ; clearly, $h \neq k$. Let ε be the plane through h and k, so that $g \perp \varepsilon$ by theorem 5. If g were perpendicular to a plane $\varepsilon' \neq \varepsilon$ through P, then lines $\delta \cap \varepsilon$ and $\delta \cap \varepsilon'$ would be distinct perpendiculars to g through P in δ, which would contradict theorem 1! Thus ε is unique.

You can provide the proof for the case that P is not on g. ♦

A point X is said to be *equidistant* from points P and Q if $PX = XQ$. The *midpoint* of segment $\overline{PQ}$ is the point X in $\overline{PQ}$ equidistant from P and Q. If $P \neq Q$ then the *perpendicular bisector* of $\overline{PQ}$ is the plane through X perpendicular to $\overleftrightarrow{PQ}$.

Theorem 7 (Perpendicular bisector theorem). For any distinct points P and Q, the perpendicular bisector of $\overline{PQ}$ is the set of all points equidistant from P and Q.

Theorem 8. Any two lines g and h perpendicular to a plane ε are coplanar.

Proof. Let $P = g \cap \varepsilon$ and $Q = h \cap \varepsilon$. If $P = Q$ the result is trivial. Assume $P \neq Q$ and let X be the midpoint of $\overline{PQ}$. Let k be the perpendicular to

$\vec{PQ}$ through X in ε, and select any distinct points Y and Z on k equidistant from X. It's easy to show that both g and h lie in the perpendicular bisector of $\overline{YZ}$. ♦

Theorem 9. If P is a point and ε a plane, then P lies on a unique line g perpendicular to ε.

Proof. First, suppose P is on ε. Select any two distinct lines h and k through P in ε. By theorem 6 there exist planes α and β through P perpendicular to h and k. Then $g = \alpha \cap \beta$ is a line through P perpendicular to ε. The proof that g is unique is left to you.

Now suppose P is not on ε. Select any point Q in ε. By the previous paragraph there's a line $h \perp \varepsilon$ through Q. If P is on h, let $g = h$. Suppose P is not on h. Let δ be the plane through P and h, and $k = \delta \cap \varepsilon$. By theorem 1 there's a perpendicular g' to k through P; let O be its foot. By the previous paragraph there's a line $g \perp \varepsilon$ through O. By theorem 8, g and h are coplanar; since $O \neq Q$, g must lie in δ. Thus $g = g'$ because $g, g' \perp k$. In either case, P lies on a line $g \perp \varepsilon$. You can prove that g is unique. ♦

The point $g \cap \varepsilon$ in Theorem 9 is called the *foot* of the perpendicular to ε through P.

Theorem 10. The foot of the perpendicular to a plane ε through a point P is the unique point in ε whose distance from P is minimum.

The minimum distance in theorem 10 is called the *distance from* P *to* ε.

Let δ and ε be planes whose intersection is a line g. Let δ' be one of the sides of g in δ and ε' be one of the sides of g in ε. The point set $\delta' \cup g \cup \varepsilon'$ is called a *dihedral angle*; g is its *edge*. Note that δ' lies entirely on one side of ε, and ε' entirely on one side of δ. The *interior* of the dihedral angle is the set of all points that lie on the same side of ε as δ' and on the same side of δ as ε'; it's apparently convex. Analogs of theorems 2 to 4 of section 3.4 hold for dihedral angles. You should formulate and prove them.

Let γ be a plane perpendicular to the edge g of a dihedral angle $\delta' \cup g \cup \varepsilon'$, intersecting g at a point O. Then $\gamma \cap \delta' = \vec{OP}$ and $\gamma \cap \varepsilon' = \vec{OQ}$ for some points P and Q noncollinear with O. The angle $\angle POQ$ is said to *belong* to the dihedral angle. (See figure 3.6.2.) The proof of the following theorem is due to H. G. Forder.[9]

[9] Forder [1928] 1958, chapter IV, paragraph 84.

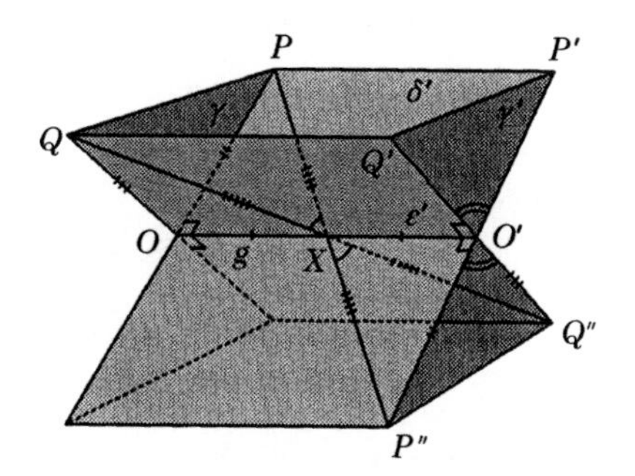

Figure 3.6.2
Proof of theorem 11

Theorem 11 (Dihedral angle theorem). All angles belonging to dihedral angle $\delta' \cup g \cup \varepsilon'$ are congruent.

Proof. Let γ, O, P, and Q be as just described for figure 3.6.2. Consider a plane $\gamma' \perp g$ distinct from γ and find points O', P', and Q' such that $\gamma' \cap g = O'$ and

$$\gamma' \cap \delta' = \overrightarrow{O'P'}$$
$$\gamma' \cap \varepsilon' = \overrightarrow{O'Q'}.$$

Let X be the midpoint of $\overline{OO'}$ and find points P'' and Q'' such that

$$P'\text{-}O'\text{-}P'' \qquad OP = O'P''$$
$$Q'\text{-}O'\text{-}Q'' \qquad OQ = O'Q''.$$

By the SAS axiom, $\Delta OXP \cong \Delta O'XP''$ and $\Delta OXQ \cong \Delta O'XQ''$, hence

$$XP = XP'' \qquad \angle OXP \cong \angle O'XP''$$
$$XQ = XQ'' \qquad \angle OXQ \cong \angle O'XQ''.$$

These angle congruences imply that the triples P, X, P'' and Q, X, Q'' are collinear, hence $\Delta XPQ \cong \Delta XP''Q''$ by the vertical angle theorem and the SAS axiom. Thus $PQ = P''Q''$, hence $\Delta POQ \cong \Delta P''O'Q''$ by the SSS theorem. Therefore $\angle POQ \cong \angle P''O'Q'' \cong \angle P'O'Q'$. ♦

The *measure* of a dihedral angle is defined to be the common measure of all the angles that belong to it. Two distinct intersecting planes δ and ε form four dihedral angles. If one of them has measure $90°$, then all do, and the planes are said to be *perpendicular*, written $\delta \perp \varepsilon$.

Theorem 12. Let δ and ε be planes whose intersection is a line g. Let h be a line in δ perpendicular to g. Then $h \perp \varepsilon$ if and only if $\delta \perp \varepsilon$.

3.7 Parallel axiom and related theorems

Concepts

Absolute geometry
Parallel lines and planes, and the distance between them
Alternate interior angles
Parallel axiom—Euclid's and Playfair's versions
Transitivity theorems for parallels
Sum of the measures of the angles of a triangle
Strong exterior angle theorem
Invariance of betweenness under parallel projection
Convex quadrilaterals: trapezoids, parallelograms, rhombi, rectangles, and squares
Right triangles, their hypotenuses and legs; $30°-60°$ right triangles
Oblique triangles

All results in chapter 3 through corollary 3 of this section belong to *absolute* geometry: They're independent of the parallel axiom. To prove theorem 4, the book uses the axiom for the first time, and thus specializes to Euclidean geometry. The theorems of absolute geometry could also have served as an introduction to *non-Euclidean* geometry. (That theory starts like Euclidean geometry, but postulates the *denial* of the parallel axiom.) This section continues with theorems specifically about parallel lines, then presents a catalog of frequently used properties of various types of quadrilaterals. The proofs of these results are all straightforward. Most of them, *and the corresponding figures*, are left to you as exercises.

Coplanar disjoint lines g and h are said to be *parallel*, written $g \parallel h$. This adjective is often applied to a segment or ray if it would be appropriate to apply it to the corresponding line. For example, a segment could be parallel to a line, ray, or another segment. The word *parallel* is sometimes used as a noun to mean *parallel line.*

Theorem 1 (Alternate interior angles theorem, part 1). Let O, P, Q, and R be points in a plane ε, as shown in figure 3.7.1, with $O \neq P$ and with Q and R on different sides of $\overleftrightarrow{OP}$. If $\angle QOP \cong \angle RPO$, then $\overleftrightarrow{OQ} \parallel \overleftrightarrow{PR}$.

Proof. The lines' intersection would contradict the weak exterior angle theorem. ♦

Corollary 2. Two coplanar perpendiculars to the same line are parallel. Two lines perpendicular to the same plane are parallel.

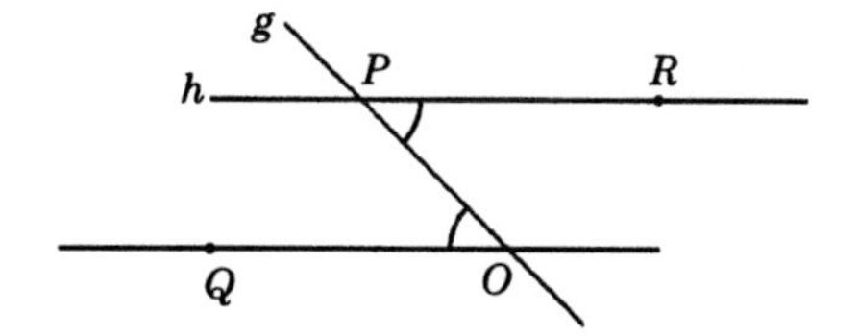

Figure 3.7.1
Alternate interior angles theorem

Corollary 3. If P is a point not on a line g, then there exists *at least* one parallel to g through P.

John Playfair[10] introduced the following version of the *parallel axiom* in 1795. It's a simplification of Euclid's original version, which was quoted in section 2.4. Euclid's axiom, slightly modified, follows as theorem 4.

Parallel axiom. If P is a point not on a line g, then there exists *at most* one parallel to g through P.

Theorem 4 (Euclid's parallel axiom). Let O, P, Q, and R be points in a plane ε with $O \neq P$ and with Q and R on different sides of $\overleftrightarrow{OP}$. If $m\angle QOP > m\angle RPO$, then $\overrightarrow{OQ}$ intersects $\overrightarrow{PR}$ on the same side of $\overleftrightarrow{OP}$ as R.

Proof. Construct R' on the same side of $\overleftrightarrow{OP}$ as R, so that $m\angle OPR' = m\angle QOP$. Then R lies in the interior of $\angle OPR'$. By theorem 1, $\overleftrightarrow{OQ} \parallel \overleftrightarrow{PR'}$, so $\overrightarrow{OQ}$ and $\overrightarrow{PR}$ must intersect as claimed. ♦

Theorem 5 (Alternate interior angles theorem, part 2). Let O be a point in a line g in a plane ε, let Q and R be points in ε on different sides of g, and let h be a line in ε through R. If $\overleftrightarrow{OQ} \parallel h$, then h intersects g at a point $P \neq O$, and $\angle QOP \cong \angle RPO$. (See figure 3.7.1.)

Corollary 6. Let g, h, and k be coplanar lines such that $g \perp h$ and $h \neq k$. Then $h \parallel k$ if and only if $g \perp k$.

Corollary 7. Let ε be a plane and h and k be lines such that $h \perp \varepsilon$ and $h \neq k$. Then $h \parallel k$ if and only if $k \perp \varepsilon$.

Corollary 8 (Transitivity theorem for parallel lines). Let g, h, and k be distinct lines. If $g \parallel k$ and $h \parallel k$, then $g \parallel h$.

Proof. Construct a plane perpendicular to k, then apply corollaries 7 and 2. ♦

---

[10] Playfair [1795] 1860, 11.

John PLAYFAIR was born near Dundee, Scotland, in 1748. He was educated for the ministry at the University of St. Andrews, ordained, and succeeded his father as a parish priest. But he soon resumed scientific study at the University of Edinburgh, and remained there for the rest of his life. Playfair became Professor of Mathematics at Edinburgh in 1785. His popular 1795 book *Elements of geometry*, an edition of Euclid, included his well known improved version of the parallel axiom. Playfair was editor of the *Transactions of the Royal Society of Edinburgh*, and his interests included several other sciences. He is best known for the pioneering work in geology he published in the 1802 treatise, *Illustrations of the Huttonian theory of the earth*. Playfair became Professor of Natural Philosophy at Edinburgh in 1805, and remained active until his death in 1819.

***Corollary 9* (*Triangle sum theorem*).** The sum of the measures of the angles of a triangle is 180°.

Proof. Through a vertex construct a parallel to the opposite edge, and apply theorem 5. ♦

***Corollary 10*.** An equilateral triangle has three 60° angles. An isosceles right triangle has two 45° angles.

Part of the triangle sum theorem belongs to absolute geometry: it's possible to prove, without using the parallel axiom, that the sum of the measures of the angles of a triangle is $\leq 180°$.[11] The argument, first published by Girolamo Saccheri in 1733, is not given here, because it uses specialized methods and, of course, is unnecessary for the development of *Euclidean* geometry.

The triangle sum theorem affords a much simpler proof of the SAA congruence theorem than the one given in section 3.5. Consider two triangles with a pair of congruent edges and two pairs of congruent angles in the SAA pattern. By the triangle sum theorem, the third pair of angles are congruent as well, so the triangles are congruent by the ASA congruence theorem. The triangle sum theorem also yields a stronger form of the exterior angle theorem than the one originally presented in section 3.5:

***Corollary 11* (*Strong exterior angle theorem*).** Consider ΔXYZ and a point $W \neq X$ such that W-X-Y. Then $m\angle WXZ = m\angle Y + m\angle Z$.

Two planes δ and ε are said to be *parallel*, written $\delta \parallel \varepsilon$, if they're disjoint. A line g and a plane ε are said to be *parallel*, written $g \parallel \varepsilon$, if they're disjoint.

[11] Moise [1963] 1990, section 10.4.

Theorem 12. Let g be a line, δ and ε be distinct planes and $g \perp \delta$. Then $\delta \parallel \varepsilon$ if and only if $g \perp \varepsilon$.

Corollary 13 (Transitivity theorem for parallel planes). Let α, β, and γ be distinct planes. If $\alpha \parallel \gamma$ and $\beta \parallel \gamma$, then $\alpha \parallel \beta$.

Theorem 14 (Invariance of betweenness under parallel projection). Let parallel lines g, h, and j intersect a line k at points X, Y, and Z, and a line k' at points X', Y', and Z'. If X-Y-Z, then X'-Y'-Z'.

Proof. All these lines lie in a plane ε. The points X and X' lie on the same side of h in ε, as do Z and Z'. Since X and Z lie on different sides of h, so must X' and Z'. ♦

By theorem 14, it makes sense to speak of the points in a plane ε *between* two parallel lines in ε, and of the points *between* two parallel planes.

Consider four coplanar points P, Q, R, and S, no three of which are collinear, such that segments $\overline{PQ}$, $\overline{QR}$, $\overline{RS}$, and $\overline{SP}$ intersect only at their ends, if at all. The ordered quadruple $PQRS$ is called a *quadrilateral*; the four points are its *vertices*, and the four segments, its *edges*. Its *angles* are

$$\angle P = \angle SPQ \qquad \angle S = \angle RSP$$
$$\angle Q = \angle PQR \qquad \angle R = \angle QRS.$$

Disjoint edges are said to be *opposite*. Two vertices are *opposite* if they're not ends of the same edge; this terminology is also applied to the corresponding angles. Segments $\overline{PR}$ and $\overline{QS}$ are called the *diagonals* of $PQRS$.[12]

Theorem 15. The following conditions on a quadrilateral in a plane ε are equivalent:

1. each edge lies entirely on one side of the line in ε that includes the opposite edge,
2. each vertex lies in the interior of the opposite angle,
3. the diagonals intersect.

A quadrilateral that satisfies the theorem 15 conditions is called a *convex quadrilateral*. (This usage is inconsistent with the earlier definition of *convex*; it does *not* mean that the quadrilateral is a convex point set.)

[12] *Diagonal* is composed of the Greek prefix *dia-* meaning *across* and root *-gon* meaning *angle*.

A *trapezoid* is a quadrilateral two of whose edges lie in parallel lines.[13] If the other two are congruent, it's called *isosceles*.

Theorem 16. A trapezoid is a convex quadrilateral, hence its diagonals intersect.

Proof. Consider a quadrilateral $PQRS$ such that $\overleftrightarrow{PQ} \parallel \overleftrightarrow{RS}$. In verifying condition 1 of theorem 15, only the following step is tricky. Suppose S and P lay on opposite sides of $\overleftrightarrow{QR}$. Then $\overline{SP}$ would intersect $\overleftrightarrow{QR}$ at a point X. Construct line g parallel to lines $\overleftrightarrow{PQ}$ and $\overleftrightarrow{RS}$ through X. Then P-X-S would imply Q-X-R by theorem 14, hence $\overline{SP}$ and $\overline{QR}$ would intersect at a point other than a vertex—contradiction! Therefore $\overline{SP}$ must lie entirely on one side of $\overleftrightarrow{QR}$. ♦

A *parallelogram* is a quadrilateral $PQRS$ such that $\overleftrightarrow{PQ} \parallel \overleftrightarrow{RS}$ and $\overleftrightarrow{QR} \parallel \overleftrightarrow{SP}$. Clearly, a parallelogram is a trapezoid, hence its diagonals must intersect.

Theorem 17. For any noncollinear points P, Q, and R, there's a unique point S such that $PQRS$ is a parallelogram.

Theorem 18 (Parallelogram theorem). The following conditions on a quadrilateral $PQRS$ are equivalent:

1. $PQRS$ is a parallelogram,
2. $PQ = RS$ and $QR = SP$,
3. $\angle P \cong \angle R$ and $\angle Q \cong \angle S$,
4. $\overleftrightarrow{PQ} \parallel \overleftrightarrow{RS}$ and $PQ = RS$,
5. the diagonals of $PQRS$ intersect at their common midpoint.

A *rhombus* is a parallelogram with all edges congruent.[14]

Theorem 19. The diagonals of a parallelogram lie in perpendicular lines if and only if it's a rhombus.

A *rectangle* is a parallelogram with all angles congruent. A *square* is a rectangle with all edges congruent—a rectangular rhombus.

Theorem 20. All angles of a rectangle are right. Any parallelogram with one right angle is a rectangle.

---

[13] *Trapezoid* stems from the Greek word for *table*, which is derived in turn from *tetra-* and *ped-* meaning *four* and *foot*, and a suffix meaning *shaped like*.

[14] Schwartzman (1994, 189) reports that *rhombus* stems from the Greek word for an object of that shape used in religious rituals.

Theorem 21. If g and h are parallel lines and P and P' are points in g, then the distances from P and P' to h are equal. Similar results hold for a parallel line and plane, and for parallel planes.

The distances described in theorem 21 are called the *distances between* the parallel lines or planes.

A triangle with a right angle is called a *right* triangle; the edge opposite the right angle is called its *hypotenuse* and the other edges, its *legs*. A triangle that's not right is called *oblique*.

Theorem 22. The midpoint of the hypotenuse of a right triangle is equidistant from the vertices.

Theorem 23. Consider ΔOPQ with right angle $\angle O$. Then $OP = \frac{1}{2} PQ$ if and only if $m\angle P = 60°$ and $m\angle Q = 30°$.

3.8 Area and Pythagoras' theorem

Concepts

- Triangular and polygonal regions
- Interior and boundary points, edges, and vertices
- Area axioms
- Bases and altitudes of trapezoids and triangles
- Areas of squares, rectangles, triangles, trapezoids, and parallelograms
- Pythagoras' theorem
- Existence of a triangle with given sides
- Alternative area theories based on similarity or integration

An area theory must answer two big questions:

- To what figures does it assign areas?
- How does it compute their areas?

For the first question, in elementary geometry, the answer is fairly clear: figures built from triangles, with their interior points. These include all polygonal regions. Later, it may seem unfortunate that figures bounded by curves are excluded. But a theory general enough to include a large variety of curved figures is beyond elementary geometry. Section 3.14 does present an ad hoc theory that considers some figures bounded by circular arcs. The first task of the present section is to define precisely the polygonal regions that will be assigned areas. The computations are then based on some new area axioms that closely describe the process we use naturally to determine areas of complicated figures from known results for simpler ones. One of the most important results in mathematics—Pythagoras' theorem—fits gracefully into this theory, even though you might not ordinarily consider it related to area. The section concludes by sketching some alternative, more difficult approaches to area theory.

Consider three noncollinear points X, Y, and Z. The set of all points that lie between points on edges of ΔXYZ is called the *triangular region* determined by ΔXYZ.

Theorem 1. If points X, Y, and Z are noncollinear, then the following sets coincide:

1. the triangular region determined by ΔXYZ,
2. the union of the edges of ΔXYZ with the intersection of the interiors of any two of its angles,
3. the intersection of all convex point sets containing X, Y, and Z.

This point set is convex.

Proof. Show first that sets (1) and (2) coincide. Then show that (2) is convex, and finally that (1) and (3) coincide. ♦

Theorem 2. If $PQRS$ is a convex quadrilateral, then the following point sets coincide:

1. the union of the triangular regions determined by ΔPQR and ΔRSP,
2. the union of the triangular regions determined by ΔQRS and ΔSPQ,
3. the union of the edges of $PQRS$ with the intersection of the interiors of any two opposite angles,
4. the intersection of all convex point sets containing P, Q, R, and S.

Theorem 3. If $PQRS$ is a nonconvex quadrilateral, then exactly one pair of opposite vertices lie on opposite sides of the line through the other two.

The *region Σ determined by a quadrilateral* $PQRS$ is defined to be the point set described four ways by theorem 2 if $PQRS$ is convex; otherwise, if Q and S lie on opposite sides of $\overleftrightarrow{PR}$ in the plane PQR, then Σ is defined to be the union of the triangular regions determined by ΔPQR and ΔRSP.

Theorem 4. The region determined by a quadrilateral $PQRS$ is convex if and only if $PQRS$ is a convex quadrilateral.

This book will often use the same terms to describe a triangle or quadrilateral and the associated region.

In general, a *polygonal region*[15] is defined to be the union of a finite set of triangular regions, any two of which intersect, if at all, in a point or segment. Clearly, all triangular regions are polygonal regions; and the region determined by a quadrilateral is a polygonal region in the general sense. Figure 3.8.1 shows some point sets that aren't polygonal regions. Figure 3.8.2 shows a point set that is a polygonal region, although its construction, indicated by solid outlines, doesn't reveal that. One way to reconstruct it from properly intersecting triangular regions is to dissect the left-hand triangle as suggested by the dotted lines.[16]

It's evident which points of the polygonal region displayed in figure 3.8.2 should be regarded as *interior*: those interior to the left- or right-hand triangle. But it's surprisingly hard to define that notion for polygonal regions in general. In exercises 4.7.9 and 4.7.10 you'll describe some examples to show that the criterion just applied to this example doesn't work in general. Then you'll work out the details of a general definition of the concepts of interior and boundary points of a polygonal region Σ. Those are needed for the discussion of polyhedra in section 8.4. In particular, you can distinguish whether an edge or vertex of a constituent triangular region is a boundary point of Σ; if so, it's called an *edge* or *vertex* of Σ.

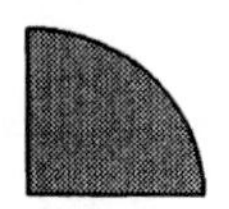
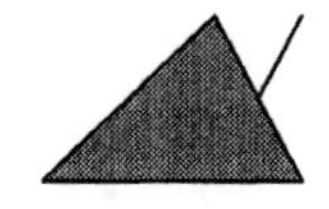
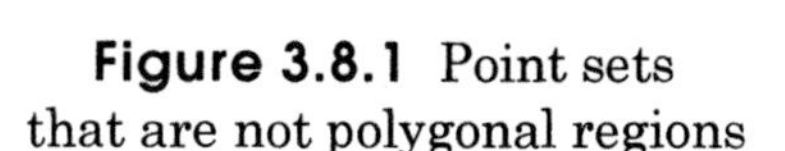

Figure 3.8.1 Point sets that are not polygonal regions

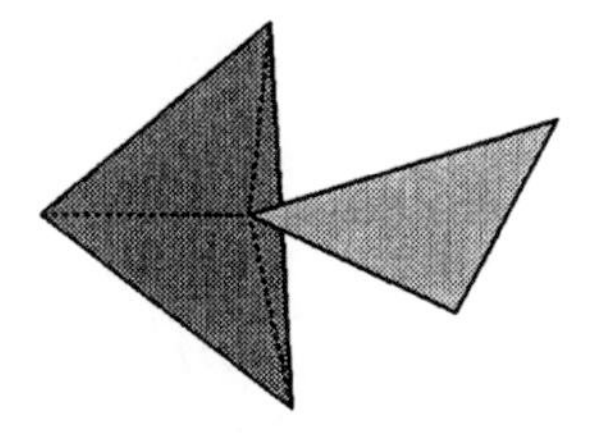

Figure 3.8.2 Questionable region

[15] *Polygon* stems from the Greek *poly-* and *gonia* for *many* and *angle.*

[16] For some applications it's best to define *polygonal region* so that two constituent triangular regions must intersect, if at all, at a vertex of each or along an entire edge of each. To reconstruct the figure 3.8.2 region according to that definition, you could also dissect its right-hand triangle into three smaller triangles. With the more stringent intersection requirement, theorem 5 would require a rather complicated proof. That condition would make results in *some* applications of this theory easier. Since those aren't included in this book, however, the simpler definition is used.

The first result in this section about polygonal regions is required for stating area axiom A2. You can write its proof in a sentence or two.

Theorem 5. If the intersection of polygonal regions Σ and T is empty or the union of a finite set of points and segments, then $\Sigma \cup T$ is a polygonal region.

Area is another partly undefined concept, like length and angle measure. It's a function that assigns to each polygonal region Σ a positive real number $\mathrm{a}\Sigma$ called the *area of* Σ. Further details of the function are undefined. Some properties of area are postulated by the following *area axioms* A1 to A3. They were first stated in 1941, in a slightly different form, by George David Birkhoff and Ralph Beatley.[17]

Axiom A1. Regions determined by congruent triangles have the same area.

Axiom A2. If the intersection of polygonal regions Σ and T is empty or the union of a finite set of points and segments, then $\mathrm{a}(\Sigma \cup T) = \mathrm{a}\Sigma + \mathrm{a}T$.

Axiom A3. The area of a square is the square of one of its edges.

Often it's convenient to call two parallel edges of a trapezoid its *bases*, and the distance between the lines including these edges, the corresponding *altitude*.

Theorem 6. The area of a rectangle is the product of a base b by the corresponding altitude a.

Proof. (See figure 3.8.3.) Axioms A1 and A2 yield $\mathrm{a}PQRS = \mathrm{a}RUVW$ because $\Delta PQS \cong \Delta VUW$ and $\Delta QRS \cong \Delta URW$. By axioms A2 and A3,

$$\begin{aligned} \mathrm{a}PQRS &= \tfrac{1}{2}(\mathrm{a}PTVX - \mathrm{a}QTUR - \mathrm{a}RWXS) \\ &= \tfrac{1}{2}((a+b)^2 - a^2 - b^2) = ab. \blacklozenge \end{aligned}$$

Corollary 7. The area of a right triangle is half the product of its legs.

Often it's convenient to call one edge of a triangle its *base*. (If the triangle is isosceles but not equilateral, and *nothing else* is stated, then *base* means the edge that's not congruent to the others.) If V is the vertex opposite the base and F the foot of the perpendicular from V to the base line, then segment $\overline{FV}$ is called the *altitude* on that base. Analyses of triangles often split into cases depending on whether F falls between the base vertices.

[17] Birkhoff and Beatley [1941] 1959, chapter 7.

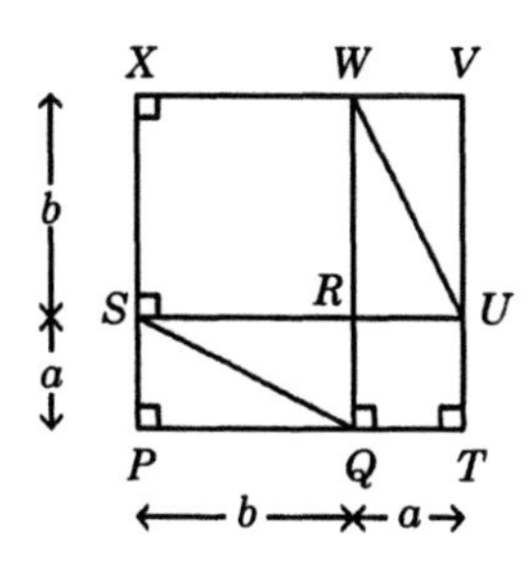

Figure 3.8.3
Proof of theorem 6

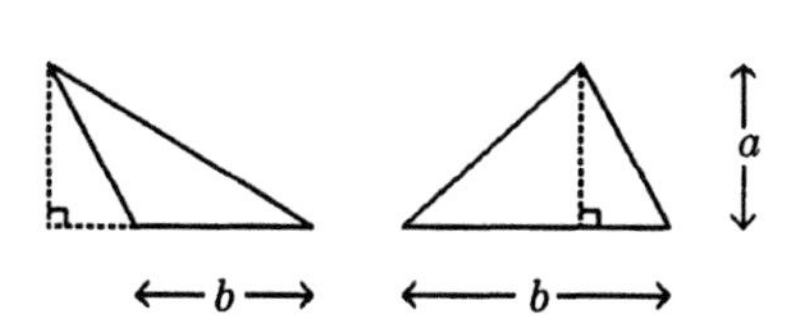

Figure 3.8.4
Proof of theorem 9

Theorem 8. If the foot of an altitude of a triangle falls outside the base, then the nearer base angle is obtuse. If both base angles are acute, the foot is on the base.

Theorem 9. The area of a triangle is half the product of a base b by the corresponding altitude a.

Proof. Consult figure 3.8.4. There are two cases. ♦

Corollary 10. The area of a trapezoid is half the product of an altitude by the sum of the corresponding bases.

Proof. Draw a diagonal. ♦

Corollary 11. The area of a parallelogram is the product of a base by the corresponding altitude.

Pythagoras' theorem that the sum of the squares of the legs of a right triangle equals the square of its hypotenuse is often called the most important in mathematics. While that may be mere grandiloquence, the result does prove fundamental for many different subjects. Set in modern algebraic language, it may seem to be about *lengths*, hence inappropriate for inclusion in this section. But it was originally viewed as a theorem about the *areas* of squares erected on the triangle's edges. Therefore, it's appropriate to present a proof based on area. Section 3.9 will give another proof, based on similarity theory.

Theorem 12 (Pythagoras' theorem). The square of the hypotenuse of a right triangle is the sum of the squares of its legs.

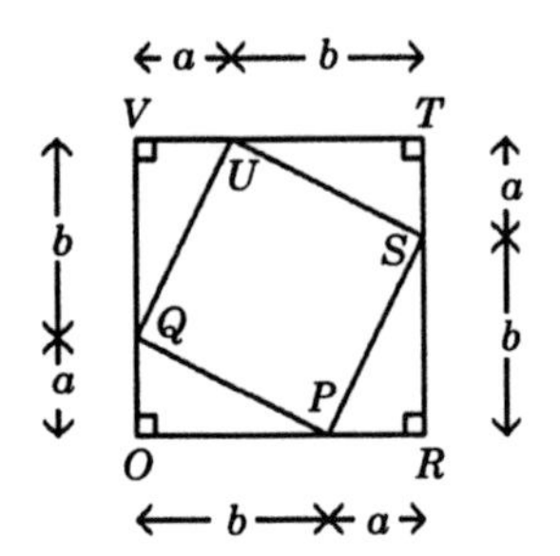

Figure 3.8.5
Proof of
Pythagoras' theorem

Proof. Consult figure 3.8.5. Start with ΔOPQ where $\angle O$ is a right angle. Construct R, S, T, U, and V as shown. Verify that V, Q, and O are collinear and $VQ = b$. Note that $\Delta OPQ \cong \Delta RSP \cong \Delta TUS \cong \Delta VQU$ and that $ORTV$ and $PSUQ$ are squares. Then

$$(PQ)^2 = \text{a}PSUQ = \text{a}ORTV - 4\text{a}\Delta OPQ$$
$$= (a+b)^2 - 4(\tfrac{1}{2}ab) = a^2 + b^2 \;. \blacklozenge$$

The following result is a converse of Pythagoras' theorem: A triangle is right if its edges satisfy Pythagoras' equation. For example, a triangle with edges of length 3, 4, and 5 is right because $3^2 + 4^2 = 5^2$. This provided ancient surveyors a practical method for constructing right angles—form a loop of rope with $3 + 4 + 5 = 12$ equally spaced knots, and have assistants hold appropriate knots, stretching the rope taut into a right triangle.

Corollary 13. Distinct points O, P, and Q, such that $(OP)^2 + (OQ)^2 = (PQ)^2$, are noncollinear and form a right angle $\angle POQ$.

Proof. If O, P, and Q were collinear, then one of the equations

$$OP + PQ = OQ \qquad PQ + QO = PO \quad QO + OP = QP$$

would hold. But each of these is inconsistent with $(PQ)^2 = (OP)^2 + (OQ)^2$, so the points must be noncollinear. Find a point Q' such that $\angle POQ'$ is a right angle and $OQ' = OQ$. Pythagoras' theorem and the SSS congruence theorem imply $\Delta POQ \cong \Delta POQ'$. ♦

The following consequence of Pythagoras' theorem is a converse of the triangle inequality: You can construct a triangle with any given edge lengths as long as you don't violate the inequality.

Theorem 14. Let x, y, and z be positive real numbers, each of which is less than the sum of the other two. Then there is a triangle whose edges have lengths x, y, and z.

Proof. Assume $x \ge y \ge z$; the other cases are handled similarly. In case $x = z$, construct an equilateral triangle with edge x. Now suppose $x > z$. Select any points Y and Z such that $YZ = x$. Define

$$t = \frac{z^2 + x^2 - y^2}{2x}.$$

Then $0 < t$ since $x^2 \ge y^2$. From $0 < x - z < y$ follows $0 < y^2 - (x - z)^2 = y^2 - x^2 + 2xz - z^2$, so $z^2 + x^2 - y^2 < 2xz$, and $t < z$. Find the point W in $\overline{YZ}$ with $YW = t$, and let g be a perpendicular to $\overleftrightarrow{YZ}$ through W. Find a point X on g such that $XW = \sqrt{z^2 - t^2}$. Then X, Y, and Z are noncollinear, and by Pythagoras' theorem, $XY = z$ and $XZ = y$. ♦

Alternative area theories

Couldn't we have constructed this area theory *without* new undefined concepts and new axioms? Couldn't we just *define* the area of a triangle by its formula and *define* the area of a polygonal region as the sum of the areas of its constituent triangles? There are two major problems with that process.

First, the triangle area formula—half the product of a base and the corresponding altitude—depends on which edge you select as base. Without this section's axiomatic development, how would you determine that the formula gives the same area no matter which base you select? In figure 3.8.6, this amounts to equations

$$\tfrac{1}{2} AB \cdot CF = \tfrac{1}{2} BC \cdot AD$$
$$(AB)/(BC) = (AD)/(CF).$$

That is, you'd need to show that ΔABD and ΔBCF are similar. But similarity theory hasn't been developed yet. It will be, in the next section, but there it will be *based* on area! Thus, introducing triangle area by definition would require an alternative similarity theory not based on area. That's the way Euclid discussed similarity; but it's much harder than the theory presented in this book.[18]

Figure 3.8.6 Different bases and altitudes

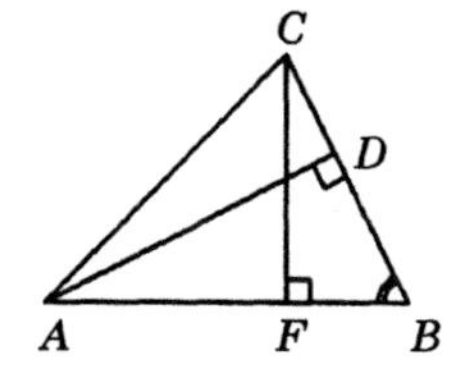

[18] Moise [1963] 1990, chapter 20.

PYTHAGORAS was born around 580 B.C., the son of a wealthy merchant. He grew up on the Greek island Samos, just off the coast of Asia Minor. Little definite is known about his early life, but there's evidence that he studied with the philosophers Thales of Miletus and Pherecydes of Syros. The biographer Iamblichus (who died around A.D. 330) reported that Pythagoras traveled to Egypt, where he studied Egyptian culture and religion for many years, then was captured by invading Persians and taken to Babylon. Evidently after studying Babylonian society and traditions for several years, Pythagoras returned to Samos. He found life there intolerable under a tyrannical government, so he emigrated around 530 B.C. to Croton in southern Italy, where he founded a religious and academic brotherhood. (Parts of Iamblichus' account are fantastic, his chronology of these wanderings is inconsistent, and there is apparently little confirmation from other sources. However, Egyptian and Babylonian ideas strongly influenced the teachings of the brotherhood. The unlikelihood that other Pythagoreans in Italy had directly encountered the Egyptian and Babylonian doctrines lends credence to Iamblichus' story.)

Among the major tenets of the Pythagorean brotherhood were a strict regimen of personal behavior, belief in the immortality and transmigration of souls, a nongeocentric cosmology, and a belief that all knowledge can be reduced to numerical relationships. The Pythagoreans made many deep investigations in music and properties of numbers, and they gained knowledge of much of the geometry later codified by Euclid. Many roots of these sciences had developed hundreds of years earlier in Egypt and Babylon; the exact role the Pythagoreans played in introducing them into Greek culture is still under study.

The Pythagoreans became politically influential in southern Italy, gaining control of several cities. But that brought reaction. After an uprising in 509 B.C. they were banished from Croton, and Pythagoras himself fled to another southern Italian city, Metapontum, where he died around 500 B.C.

During the next decades, the brotherhood did regain some political power, but finally, after widespread unrest, they were driven almost entirely out of Italy around 450 B.C., and subsequent persecution in Greece led to the disappearance of their society. The last vestige of their political hegemony was in the city of Tarentum in southern Italy, where during 400–360 B.C. the Pythagorean Archytas, a friend and colleague of Plato, was prominent in science and government.

Second, to define the area of a polygonal region as the sum of the areas of its constituent triangles, you'd need to prove first—without using new axioms—that the sum doesn't depend on the manner in which the region is divided into triangles. That's possible, but the proof requires as much mathematics as the entire area theory presented here.[19]

---

[19] Ibid., chapter 14.

You can also approach area theory via integrals. Although calculus texts commonly use areas in motivating the definition of the integral, and perhaps even in proving some of its properties, that's not essential. Real analysis texts develop integration theory with no reference to geometry. The advantage of this approach is that it yields immediately the areas of some regions bounded by curves. There are three disadvantages. First, the mathematics is too advanced for elementary geometry. Second, it's difficult to express concisely just what regions have areas. Third, it's hard to formulate and prove the theorem that the area of the union of two regions is the sum of their individual areas, provided they intersect only along their boundaries. Nevertheless, when you need areas for nonpolygonal regions, you have to use integral methods in some form.

3.9 Similarity

Concepts

- Similar triangles
- Similarity ratio
- Invariance of distance ratios under parallel projection
- AA, SAS, and SSS similarity theorems
- Ratio of areas of similar triangles
- Alternate proof of Pythagoras' theorem

This section presents the theory of similar triangles, based on the area theory developed in section 3.8. The approach, much smoother than Euclid's original theory, was introduced by Moise in 1964.[20]

Triangles ΔXYZ and $\Delta X'Y'Z'$ are said to be *similar*, written $\Delta XYZ \sim \Delta X'Y'Z'$, if their corresponding angles are congruent and their corresponding edges proportional—that is,

$$\angle X \cong \angle X' \qquad \angle Y \cong \angle Y' \qquad \angle Z \cong \angle Z'$$

$$\frac{YZ}{Y'Z'} = \frac{ZX}{Z'X'} = \frac{XY}{X'Y'}.$$

This ratio of corresponding edges is called the *similarity ratio*. It's sensitive to the order in which you mention the triangles: if $\Delta XYZ \sim \Delta X'Y'Z'$ with ratio r, then $\Delta X'Y'Z' \sim \Delta XYZ$ with ratio $1/r$. Similarity is a symmetric relation. It's also reflexive: in fact, any two congruent triangles are similar,

[20] Moise and Downs 1964, chapter 12. For a rigorous treatment of Euclid's approach, see Moise [1963] 1990, chapter 20.

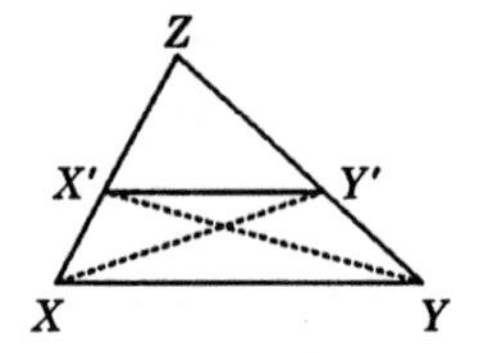

Figure 3.9.1
Proof of theorem 2

with ratio 1. Theorem 1 shows that similarity is transitive, so it's an equivalence relation.

Theorem 1 (Transitivity of similarity). If $\Delta XYZ \sim \Delta X'Y'Z'$ and $\Delta X'Y'Z' \sim \Delta X''Y''Z''$, then $\Delta XYZ \sim \Delta X''Y''Z''$, and the ratio of the third similarity is the product of the ratios of the first two.

Theorem 2. Let X' and Y' be distinct points on edges $\overline{XZ}$ and $\overline{YZ}$ of ΔXYZ. Then $\overleftrightarrow{XY} \parallel \overleftrightarrow{X'Y'}$ if and only if

$$\frac{XX'}{X'Z} = \frac{YY'}{Y'Z}.$$

Proof. (See figure 3.9.1.) Assume $\overleftrightarrow{XY} \parallel \overleftrightarrow{X'Y'}$. If you regard $\overline{XX'}$ and $\overline{X'Z}$ as bases of $\Delta XX'Y'$ and $\Delta X'ZY'$, then these triangles have the same altitude h, hence

$$\frac{\mathrm{a}\Delta XX'Y'}{\mathrm{a}\Delta X'ZY'} = \frac{\frac{1}{2}(XX')h}{\frac{1}{2}(X'Z)h} = \frac{\frac{1}{2}(XX')h}{\frac{1}{2}(X'Z)h} = \frac{XX'}{X'Z}. \tag{1}$$

In the same way,

$$\frac{\mathrm{a}\Delta YY'X'}{\mathrm{a}\Delta Y'ZX'} = \frac{YY'}{Y'Z}. \tag{2}$$

Now regard $\overline{X'Y'}$ as the base of $\Delta XX'Y'$ and $\Delta YY'X'$. These triangles then have the same altitude, namely the distance between $\overleftrightarrow{XY}$ and $\overleftrightarrow{X'Y'}$, hence

$$\mathrm{a}\,\Delta XX'Y' = \mathrm{a}\,\Delta YY'X'. \tag{3}$$

Equations (1) to (3) yield

$$\frac{XX'}{X'Z} = \frac{YY'}{Y'Z}. \tag{4}$$

Conversely, assume equation (4). Find the point Y'' on $\overrightarrow{ZY}$ such that $\overleftrightarrow{XY''} \parallel \overleftrightarrow{X'Y'}$. By the previous paragraph,

$$\frac{XX'}{X'Z} = \frac{Y''Y'}{Y'Z},$$

hence $YY' = Y''Y'$. Thus $Y = Y''$ because $X\text{-}X'\text{-}Z$ implies $Y''\text{-}Y'\text{-}Z$. ♦

Corollary 3 ***(Invariance of distance ratios under parallel projection).*** Let parallel lines g, h, j intersect lines k and l at points X, Y, Z and X', Y', Z'. Then

$$\frac{XY}{YZ} = \frac{X'Y'}{Y'Z'} .$$

Proof. In case $X \neq X'$, draw $\overline{XZ'}$. ♦

Theorem 4 ***(AA similarity theorem).*** Consider ΔXYZ and $\Delta X'Y'Z'$. If $\angle X \cong \angle X'$ and $\angle Y \cong \angle Y'$, then $\Delta XYZ \sim \Delta X'Y'Z'$.

Proof. By the triangle sum theorem, $\angle Z \cong \angle Z'$. Find the points X'' on $\overrightarrow{ZX}$ and Y'' on $\overrightarrow{ZY}$ such that $ZX'' = Z'X'$ and $\overleftrightarrow{X''Y''} \parallel \overleftrightarrow{XY}$. By the alternate interior angles and ASA congruence theorems, $\Delta X''Y''Z \cong \Delta X'Y'Z'$. By theorem 2,

$$\frac{XX''}{X''Z} = \frac{YY''}{Y''Z}$$

$$\frac{XZ}{X'Z'} = \frac{X''Z \pm XX''}{X''Z} = 1 \pm \frac{XX''}{X''Z} = 1 \pm \frac{YY''}{Y''Z} = \frac{Y''Z \pm YY''}{Y''Z} = \frac{YZ}{Y'Z'} .$$

(There are two cases.) An analogous argument with X, Y, and Z permuted yields

$$\frac{YZ}{YZ'} = \frac{XY}{X'Y} . \ ♦$$

Theorem 5 ***(SAS similarity theorem).*** Consider ΔXYZ and $\Delta X'Y'Z'$. If $\angle Z \cong \angle Z'$ and

$$\frac{XZ}{X'Z'} = \frac{YZ}{Y'Z'} ,$$

then $\Delta XYZ \sim \Delta X'Y'Z'$.

Theorem 6 ***(SSS similarity theorem).*** Consider ΔXYZ and $\Delta X'Y'Z'$. If

$$\frac{XY}{X'Y'} = \frac{YZ}{Y'Z'} = \frac{ZX}{Z'X'} ,$$

then $\Delta XYZ \sim \Delta X'Y'Z'$.

Theorem 7. Let $\Delta XYZ \sim \Delta X'Y'Z'$ with ratio r. Then $\dfrac{\mathrm{a}\Delta XYZ}{\mathrm{a}\Delta X'Y'Z'} = r^2$.

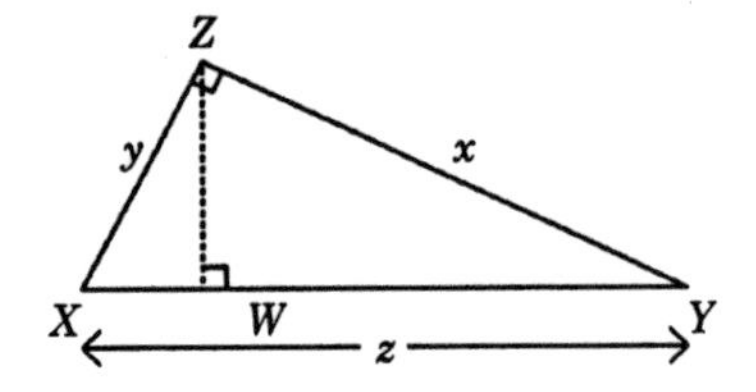

Figure 3.9.2
Theorem 8

Theorem 8. In ΔXYZ, let $\angle Z$ be a right angle and W be the foot of the perpendicular to $\overleftrightarrow{XY}$ through Z. Then X-W-Y, $W \neq X, Y$, and $\Delta XYZ \sim \Delta XZW \sim \Delta ZYW$. (See figure 3.9.2.)

Proof. X-W-Y and $W \neq X, Y$ follow from the exterior angle theorem, and the similarities from the AA similarity theorem. ♦

Theorem 8 affords a proof of Pythagoras' theorem quite different from the area proof given in section 3.8. Let $x = YZ$, $y = ZX$, and $z = XY$ as shown in figure 3.9.2. Similarities $\Delta XYZ \sim \Delta XZW$ and $\Delta XYZ \sim \Delta ZYW$ yield $y/z = (XW)/y$ and $x/z = (WY)/x$, hence $y^2 = (XW)z$ and $x^2 = (WY)z$. Therefore $x^2 + y^2 = (XW + WY)z = z^2$! You might use this proof if you wanted to develop similarity theory and reach Pythagoras' theorem without considering areas.

3.10 Polyhedral volume

Concepts

- Tetrahedra, their faces and interiors; regular tetrahedra
- Tetrahedral regions
- Congruent and similar tetrahedra
- Prisms, pyramids, their bases and altitudes; right and oblique prisms
- Lateral edges and faces
- Parallelepipeds, boxes, and cubes
- Polyhedral regions
- Polyhedral regions associated with prisms and pyramids
- Volume axioms
- Cavalieri's axiom is required
- Volumes of prisms, parallelepipeds, and pyramids

This section presents a theory of volume analogous to that for area developed in section 3.8. Its major results are the volume formulas for prisms and pyramids. Detailed description and analysis of three-dimensional analogs of triangles, quadrilaterals, and polygonal regions is tedious work, only outlined here. You'll probably find this material unfamiliar, and you're unlikely to locate any more complete elementary treatment in the literature. It will challenge your spatial intuition. Most mathematical scientists use only the end results, which they justify by more advanced integral calculus methods. But this version of the theory is useful to specialists in foundations, to teachers who need to cover some of this material in elementary classes, and to software engineers who design programs to manipulate and depict solid figures and compute their volumes.

A *tetrahedron* is an ordered quadruple of noncoplanar points. From them you can construct four triangles. The associated triangular regions are called the *faces* of the tetrahedron. Each face is *opposite* the remaining vertex. Each edge lies on two faces, which determine a corresponding dihedral angle. A tetrahedron has three pairs of *opposite* edges, that don't intersect. (See figure 3.10.1.) A tetrahedron is called *regular* if all its edges are congruent.

Theorem 1. The intersection of the interiors of either pair of opposite dihedral angles of a tetrahedron is the same.

The intersection described in theorem 1 is called the *interior* of the tetrahedron.

Theorem 2. These point sets associated with a tetrahedron T all coincide

1. the set of all points between any two points on faces of T,
2. the union of the interior and faces of T,
3. the intersection of all convex sets that contain the vertices of T.

The set described three ways by theorem 2 is clearly convex; it's called the *tetrahedral region* associated with T.

Tetrahedra $OPQR$ and $O'P'Q'R'$ are called *congruent* if

$$OP = O'P' \qquad OQ = O'Q' \qquad OR = O'R'$$
$$PQ = P'Q' \qquad QR = Q'R' \qquad RP = R'P' .$$

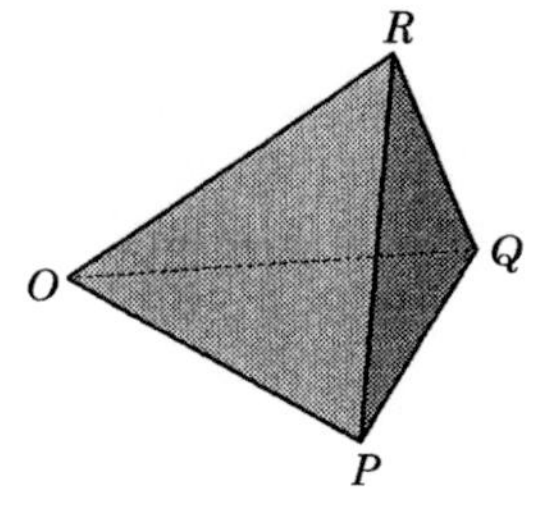

Figure 3.10.1
Tetrahedron

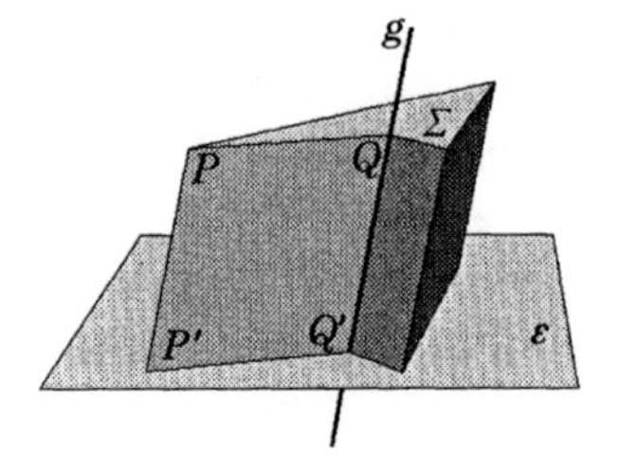

Figure 3.10.2
Oblique nonconvex quadrangular prism

Congruence of tetrahedra is an equivalence relation. Clearly, it implies congruence of all pairs of corresponding faces and edge angles. With some effort, you can also prove that corresponding dihedral angles of congruent tetrahedra have equal measure. It's interesting but tedious to develop a theory analogous to the triangle congruence theorems in section 3.5, to specify exactly which subsets of these equations and congruences imply the rest. Exercise 4.3.8 invites you to try your hand on that. A theory of *similar* tetrahedra analogous to section 3.9 would also be interesting. The author has never seen that worked out. Later in this section, you may need a few rudimentary results like these to analyze prisms and pyramids.

A *prism* is a point set Π determined as in figure 3.10.2 by a polygonal region Σ, a plane ε parallel to the plane of Σ, and a line g that intersects these two planes.[21] For each point P in Σ, let P' denote the intersection of ε with the line through P equal or parallel to g. Let Σ' be the set of all these P'. (Σ' is not visible in the figure.) The prism Π is the union of all such segments $\overline{PP'}$. Σ and Σ' are called its *bases*, the line g is a *directrix* of Π, and the distance between the two planes is its *altitude.* Qualities such as triangularity and squareness of the bases are also attributed to the prism. If Π is triangular or quadrilateral, and P is a vertex of Σ, then segment $\overline{PP'}$ is called a *lateral edge* of Π. If P and Q are adjacent vertices of Σ, as in the figure, then $PQQ'P'$ is a parallelogram, and its associated region is called a *lateral face* of Π.

Theorem 3. If Π is triangular (see figure 3.10.3), then Σ' and the intersection of Π with any plane δ parallel to and between the base planes are triangular regions congruent to Σ.

Proof. It's easy to show that the intersections of the lateral edges with δ or ε form triangles T congruent to Σ. You must also show that the points on the edges or interior to Σ correspond one to one with the points on the edges or interior to T. ♦

---

[21] *Prism* stems from a Greek word meaning *something sawed.*

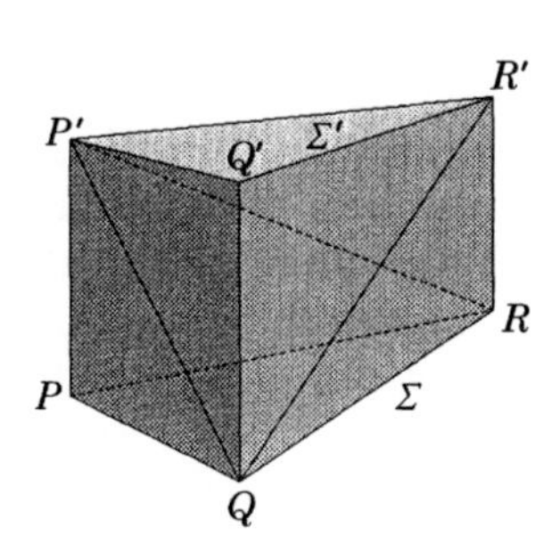

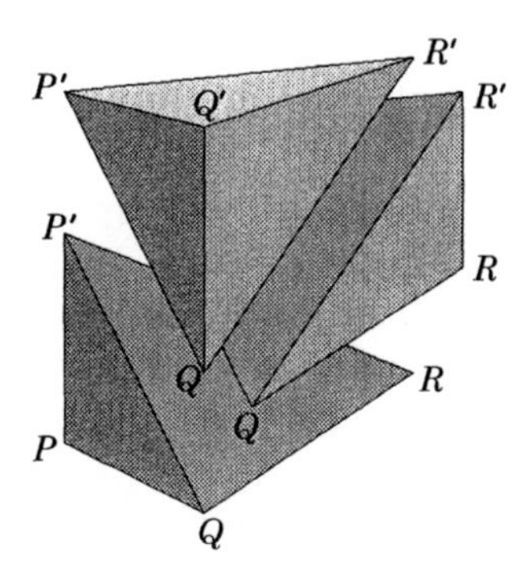

Figure 3.10.3 Building a triangular prism from three tetrahedra

A prism Π based on a parallelogram Σ is called a *parallelepiped.*[22] It has two pairs of opposite lateral faces that lie in parallel planes. You could just as well take either pair of these as bases and one of the edge lines of Σ as directrix; you'd get the same point set Π. Thus, a parallelepiped has three altitudes: the distances between the three pairs of base planes. A parallelepiped has many pairs of equal angles and equal dihedral angles. It's interesting but tedious to develop a complete set of theorems about parallelepipeds analogous to those about parallelograms in section 3.7. Exercise 4.4.5 invites you to try your hand at it.

A prism is called *right* if its directrix is perpendicular to its base planes; otherwise, it's *oblique.* A right rectangular parallelpiped is called a *box.* If all its base and lateral edges are equal, so that its bases and lateral faces are square, it's called a *cube.*

A *polyhedral region* is the union of a finite set of tetrahedral regions, the intersection of any two of which lies in a face plane. Clearly, all tetrahedral regions are polyhedral regions.[23]

Theorem 4. If the intersection of polyhedral regions Π and Ψ lies in the union of finitely many planes, then $\Pi \cup \Psi$ is a polyhedral region.

Theorem 5. All prisms are polyhedral regions. The intersection of a prism with a plane parallel to and between its base planes is a polygonal region. All these have the same area as the base.

---

[22] The latter part of this word is a contraction of *epipedon*, the Greek word for *plane.* That's derived in turn from *epi-* and *pedon*, which mean *upon* and *foot.*

[23] These words stem from the Greek prefixes *tetra-* and *poly-* and for *four* and *many*, and the word *hedra* for *base.*

Proof. As shown in figure 3.10.3, you can build a triangular prism from three tetrahedra. You can build any prism from triangular prisms. ♦

For the axiomatic development of this chapter, volume is another partly undefined concept, like length, angle measure, and area. It's a function that assigns to each polyhedral region Π a positive real number $\mathrm{v}\Pi$ called its *volume*. Further details of the function are undefined. Some properties of volume are postulated by the following *volume axioms* V1 to V4.

Axiom V1. Regions determined by congruent tetrahedra have the same volume.

Axiom V2. If the intersection of two polyhedral regions Π and Ψ lies in the union of finitely many planes, then $\mathrm{v}(\Pi \cup \Psi) = \mathrm{v}\Pi + \mathrm{v}\Psi$.

Axiom V3. The volume of a cube is the cube of one of its edges.

Axiom V4 (Cavalieri's axiom). Polyhedral regions Π and Ψ that lie between planes α and β as in figure 3.10.4 have the same volume if for each plane ε equal to α or β or parallel to and between them, $\Pi \cap \varepsilon$ and $\Psi \cap \varepsilon$ are polygonal regions with the same area.

In figure 3.10.4, $\Pi \cap \varepsilon$ and $\Psi \cap \varepsilon$ are intended to be quadrilateral and triangular regions with the same area for every plane ε considered in axiom V4, so that the polyhedral regions Π and Ψ have the same volume. Their nonhorizontal faces are transparent to ensure visibility of other parts of the figure.

Axiom V4, different in flavor from the area axioms that V1 to V3 imitate, codifies a method invented by Bonaventura Cavalieri in the early 1600s. Cavalieri's axiom provides the justification for theorem 6 and corollary 12 in the volume calculations presented next. Intuitively, it says that if you cut a polyhedral region into infinitesimally thin parallel slices, then you can reshape and reassemble them without affecting the volume, as long

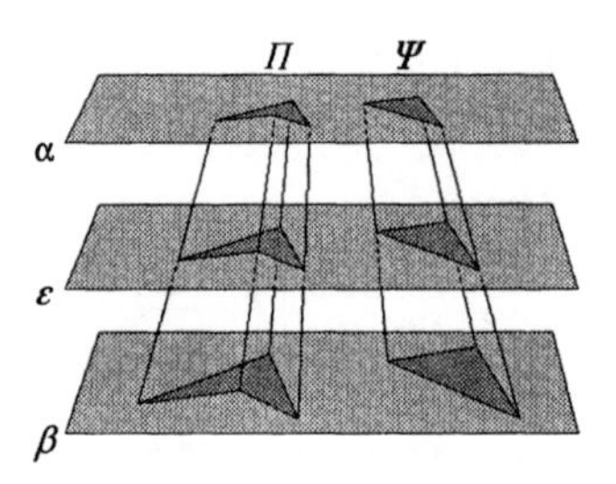

Figure 3.10.4
Cavalieri's axiom

Bonaventura CAVALIERI was born about 1598, to a noble family in Milan whose members had been active in public affairs for two centuries. He entered monastic life in 1615, then studied at the University in Pisa, showing great aptitude. He corresponded voluminously with Galileo Galilei. During 1620–1629 he was stationed in Rome, Milan, Lodi, and Parma, where he taught theology and wrote a book called *Geometria*. In 1629, with Galileo's help, he was appointed Professor of Mathematics at Bologna. He published eleven books altogether. His major achievement was an elaboration of Archimedes' methods, which later led to integration. He also contributed to the development of the theory of conics. Cavalieri died in 1647.

as you keep their areas the same and move them only within their individual planes. (See figure 3.10.4.) Later developments showed that the axiom is really a way of clothing an application of integral calculus in an elementary wrapper. While working out the proofs of the theorems in his 1899 book *Grundlagen der Geometrie*,[24] David Hilbert noted that he seemed unable to base a theory of volume solely on the principles stated in axioms V1 to V3. In an address to the International Congress of Mathematicians in Paris in August 1900 he suggested a list of open problems that might guide mathematics into the twentieth century.[25] History has demonstrated the accuracy of his judgment. His third problem was to determine whether axioms V1 to V3 are sufficient for a theory of volume. It was one of the first to be solved. That same year, Hilbert's student Max Dehn showed that the answer was *no*. Dehn reasoned that computing the volume of a polyhedral region Π on the basis of V1 to V3 alone would require decomposing Π and a cube into equal numbers of tetrahedra that would be congruent in pairs. He found an example Π —a regular tetrahedron—for which that's impossible.[26]

Theorem 6. Prisms with the same base area and altitude have the same volume.

Lemma 7. The volume of a right square prism is its altitude times its base area.

Proof. Let a and b denote its altitude and a base edge. If $a = b$, the result is just axiom V3. Otherwise let a denote the larger of these numbers. Imagine you have some building blocks:

- cubes labeled A, B and Δ with edges a, b, and $a - b$;

[24] Hilbert 1899.
[25] Hilbert 1900.
[26] Dehn 1900.

- three square prisms labeled ABB with altitude a and base edge b;
- three square prisms labeled AAB with altitude b and base edge a.

By dividing the prisms into congruent tetrahedra and applying axioms V1 and V2, you can show that the three ABB prisms have the same volume; call it vABB. Similarly, the three AAB prisms have volume vAAB. Now, assemble all these blocks except Δ into a cube with edge $a + b$. By axioms V2 and V3, $(a+b)^3 = a^3 + 3\text{v}AAB + 3\text{v}ABB + b^3$, hence

$$\text{v}AAB + \text{v}ABB = a^2 b + ab^2. \qquad (1)$$

Second, assemble Δ, B, and all the AAB blocks into a polyhedral region Π. Third, assemble the remaining blocks into a region Π'. Do this deliberately; make $\Pi = \Pi'$! (You're challenged to figure out what this region looks like.) It follows that $(a-b)^3 + b^3 + 3\text{v}AAB = a^3 + 3\text{v}ABB$, hence

$$\text{v}AAB - \text{v}ABB = a^2 b - ab^2. \qquad (2)$$

Equations (1) and (2) imply $\text{v}AAB = a^2 b$ and $\text{v}ABB = ab^2$. One of these or the other yields the desired result, depending on whether b or a is the altitude. ♦

Theorem 8. The volume of any prism is its altitude times its base area.

Proof. Make a square prism with the same altitude and base area. ♦

Corollary 9. The volume of a box is the product of its altitudes.

A *pyramid* is a point set Π determined by a polygonal region Σ and a point O not in the plane of Σ; it consists of all points between O and points in Σ. The region Σ is called its *base*, the point O its *apex*, and the distance between O and the base plane is its *altitude*. Qualities such as triangularity and squareness of the base are also attributed to the pyramid. *Triangular pyramid* and *tetrahedral region* are the same notion; you base the pyramid on any face of the tetrahedron. If Π is triangular or quadrilateral, and P is a base vertex, then segment $\overline{OP}$ is called a *lateral edge* of Π. If P and Q are adjacent base vertices, then triangular region ΔOPQ is called a *lateral face* of Π.

Theorem 10. Let Π be a triangular pyramid with altitude a, ε be its base plane, δ be a plane parallel to ε between ε and the apex O,[27] and

[27] That is, δ should lie between ε and the plane through O parallel to ε. (This applies as well to the statement of theorem 11.)

Figure 3.10.5
Theorem 10:
$OQ = a \qquad OQ' = b$

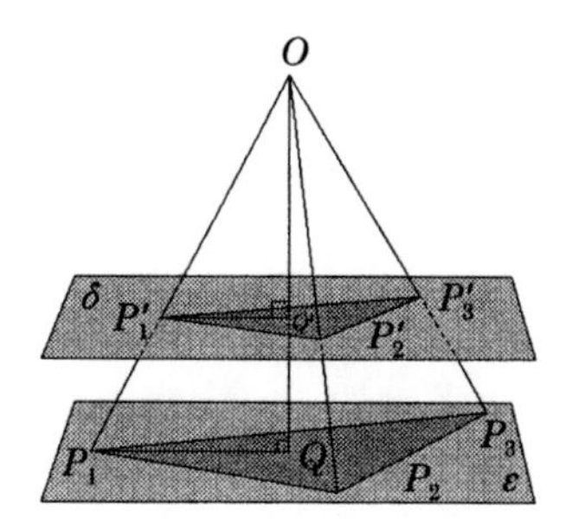

b be the distance from O to δ. Then $\Pi \cap \delta$ is a triangular region similar to the base with ratio b/a. (See figure 3.10.5.)

Proof. Let Q be the foot of the perpendicular from O to the base plane and $Q' = \overline{OQ} \cap \delta$. For $i = 1$ to 3, let P_i be a base vertex and $P_i' = \overline{OP} \cap \delta$. Then $\Delta OP_i'Q' \sim \Delta OP_iQ$ with ratio b/a, so $OP_i' = (b/a)(OP_i)$. Thus $\Delta OP_1'P_2' \sim \Delta OP_1P_2$ with the same ratio, so $P_1'P_2' = (b/a)(P_1P_2)$. The same holds for the corresponding triangles in the other lateral faces, which yields the result. ♦

Theorem 11. Every pyramid Π is a polyhedral region. Let a be its altitude, ε be its base plane, δ be a plane parallel to ε between ε and the apex O,[23] and b be the distance from O to ε. Then $\Pi \cap \delta$ is a polygonal region with area $(b/a)^2$ times the base area.

Proof. Build up the base from triangular regions, and Π from the corresponding triangular pyramids. $\Pi \cap \delta$ is the union of triangular regions similar to the base triangles, with ratio b/a. ♦

Corollary 12. Pyramids with the same base area and altitude have the same volume.

Theorem 13. The volume of a pyramid is one third of its altitude times its base area.

Proof. Make a triangular prism with the same altitude and base area as a given pyramid Π. Label its base triangles ΔPQR and $\Delta P'Q'R'$ as in figure 3.10.3. It's the union of three tetrahedra:

1. base ΔPQR and apex P';
2. base $\Delta P'Q'R'$ and apex Q, or base $\Delta QQ'R'$ and apex P';
3. base $\Delta QRR'$ and apex P'.

Π and (1) have the same volume because they have the same altitude and base area. (1) and (2) have the same volume because they have congruent bases ΔPQR and $\Delta P'Q'R'$ and the same altitude as the prism. (2) and

(3) have the same volume because they have the same apex P' and congruent bases $\Delta QQ'R'$ and $\Delta QRR'$ in the same plane. Thus, Π has one third of the volume of the prism. ♦

Suppose you had worked out a theory of similar tetrahedra as suggested earlier in this section. Then you'd have this result: If two tetrahedra are similar with ratio r, then their volumes have ratio r^3.

3.11 Coordinate geometry

Concepts

- Cartesian coordinate system: origin, axes, and planes
- Distance formula
- Midpoint formula
- Parallelogram law
- Linear parametric equations for lines and planes
- Linear equation for a plane
- Plane coordinate geometry

This section shows how to set up a Cartesian coordinate system, and describes the coordinate geometry concepts corresponding to distance, midpoints, parallelograms, lines, and planes. Proving lemma 5 requires the notion of similarity, so this theory must come after section 3.9. The discussion uses some vector algebra. You may consult appendix C to review that as necessary.

Select a point O to be called the *origin*, and three perpendicular lines through O called the *first*, *second*, and *third coordinate axes*. The axes determine three perpendicular *coordinate planes*; the plane through the ith and jth coordinate axes is called the i,j plane. Select scales c_1 to c_3 for the axes, so that $c_1(O) = c_2(O) = c_3(O) = 0$. The origin, axes, and scales constitute a *Cartesian coordinate system*. To each point P assign three *coordinates* p_1 to p_3 —often called *scalars*—by setting $p_i = c_i(P_i)$, where P_i is the foot of the perpendicular from P to the ith coordinate axis.[28] Until the last two paragraphs of this section, a fixed coordinate system is assumed to be in use.

[28] In informal discussion it's often more convenient to refer to x, y, and z axes and to use notation like x_P, y_P, and z_P for the corresponding coordinates of a point P. That system has two disadvantages—you tend to run out of letters, and it collides with vector component notation. Therefore it's not used in this book. But the author *would* use it for many simple coordinate geometry problems.

Theorem 1. Assignment of coordinates is a one-to-one correspondence between the set of all points and the set of all ordered triples of scalars.

A letter such as P often denotes both a point and its coordinate triple $<p_1, p_2, p_3>$. The latter also serves as a space-saving abbreviation for a column vector when you're using vector algebra:

$$P = <p_1, p_2, p_3> = \begin{bmatrix} p_1 \\ p_2 \\ p_3 \end{bmatrix}.$$

Corresponding upper- and lowercase letters are used whenever possible, and some equations like these are left unstated. In figures in this text, a label such as $P<1,2>$ simultaneously identifies a point and specifies its coordinates. Vector notation provides concise abbreviations for sums, differences, and scalar multiples of columns of coordinates:

$$P \pm Q = \begin{bmatrix} p_1 \pm q_1 \\ p_2 \pm q_2 \\ p_3 \pm q_3 \end{bmatrix} \qquad tP = Pt = \begin{bmatrix} p_1 t \\ p_2 t \\ p_3 t \end{bmatrix}.$$

From Pythagoras' theorem, you can derive easily the following formula for the distance between two points in terms of their coordinates. Then you can apply this formula or theorem 3.9.2 to verify the formula for the coordinates of the midpoint of a segment.

Theorem 2 (Distance formula). The distance between points P and Q is

$$PQ = \sqrt{(p_1 - q_1)^2 + (p_2 - q_2)^2 + (p_3 - q_3)^2}.$$

Thus $(PQ)^2 = (P - Q)\cdot(P - Q)$, the dot product.

The distance $OP = \sqrt{p_1^2 + p_2^2 + p_3^2}$ is often called the *length* or *norm* $\|P\|$ of column vector P. Notice that $PQ = \|P - Q\|$.

Theorem 3 (Midpoint formula). The midpoint of segment $\overline{PQ}$ is $\frac{1}{2}(P + Q) = <\frac{1}{2}(p_1 + q_1), \frac{1}{2}(p_2 + q_2), \frac{1}{2}(p_3 + q_3)>$.

Theorem 4 (Parallelogram law). If the origin O and points P and Q are not collinear, then $R = P + Q$ forms with them a parallelogram $OPRQ$.

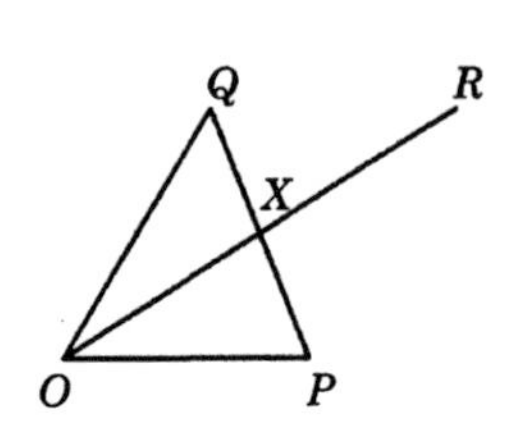

Figure 3.11.1
Proof of theorem 4

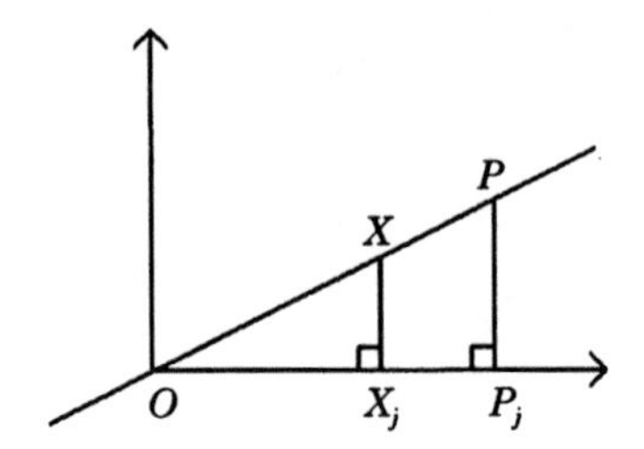

Figure 3.11.2
Proof of lemma 5

Lemma 5. Let P be a point distinct from the origin O. Then a point X lies on $\overleftrightarrow{OP}$ if and only if $X = Pt$ for some scalar t.

Proof. For $j = 1$ to 3 let P_j and X_j be the feet of the perpendiculars from X and P to the jth coordinate axis, as shown in figure 3.11.2. Let $t = OX/OP$. Then X lies on $\overleftrightarrow{OP}$ if and only if $\Delta OPP_j \sim \Delta OXX_j$. That happens just when $OX_j/OP_j = t$, i.e., $X = Pt$. ♦

A system of equations

$$\begin{aligned} x_1 &= p_1 t + q_1 \\ x_2 &= p_2 t + q_2 \\ x_3 &= p_3 t + q_3 \end{aligned} \qquad \text{i.e., } X = Pt + Q,$$

where p_1 to p_3 and q_1 to q_3 are scalars, is called a system of *linear equations in the parameter* t. With vector notation, you can abbreviate it as shown. The equations assign to each scalar t a point $X = \langle x_1, x_2, x_3 \rangle$. The set of all those points is called the *graph* of the equations.

Theorem 6. A point set g is a line if and only if it's the graph of a system $X = Pt + Q$ of linear parametric equations with $P \neq O$. Then g is the line through Q parallel or equal to $\overleftrightarrow{OP}$, and the function $X \to (OP)t$ is a scale for g.

Proof. In case g is a line through the origin O, select a point $P \neq O$ on g. By lemma 5, g is the graph of the equations $X = Pt$.

Suppose g is a line that doesn't pass through O. Let g' be the line through O parallel to g. By the previous paragraph, g' is the graph of some equations $X = Pt$. Select any point Q on g. Consider the equations

$$\begin{aligned} x_1 &= p_1 t + q_1 \\ x_2 &= p_2 t + q_2 \\ x_3 &= p_3 t + q_3 \end{aligned} \qquad \text{i.e., } X = Pt + Q.$$

Let X be a point on g. If $X = Q$, then $X = Pt + Q$ with $t = 0$. Otherwise, there exists t such that the point $X' = Pt$ on g' forms with X, Q, and O a parallelogram $XQOX'$. By the parallelogram law, $X = Pt + Q$. Thus g is a subset of the graph of this system of equations. It's easy to show further that every point in the graph lies on g. Thus each line g coincides with the graph of a system $X = Pt + Q$.

It's also easy to show that, conversely, the graph of an arbitrary system $X = Pt + Q$ is the line through Q equal or parallel to $\overleftrightarrow{OP}$.

To verify that the function $X \to (OP)t$ is a scale for g, note that for any points $X = Pt + Q$ and $Y = Pu + Q$, $(XY)^2$ is the sum of the squares of the differences $(p_j t + q_j) - (p_j u + q_j) = p_j(t - u)$, so

$$
\begin{aligned}
(XY)^2 &= (p_1^2 + p_2^2 + p_3^2)(t - u)^2 = (OP)^2(t - u)^2 \\
XY &= (OP)\,|t - u| = |(OP)t - (OP)u| \;. \blacklozenge
\end{aligned}
$$

By theorem 6, the order of points on g corresponds to the order of corresponding values of $(OP)t$, hence to the order of the t values. Corollary 7 arranges the parametric equations for a line $\overleftrightarrow{PQ}$ to make this correspondence particularly clear.

Corollary 7. If $P \neq Q$, then $X = P + (Q - P)t$ is a system of equations for the line $\overleftrightarrow{PQ}$. Moreover,

$$
\begin{aligned}
t = 0 &\leftrightarrow X = P \qquad & 0 < t &\leftrightarrow X \text{ lies in } \overrightarrow{PQ} \\
t = 1 &\leftrightarrow X = Q \qquad & 0 \le t \le 1 &\leftrightarrow X \text{ lies in } \overline{PQ}\,.
\end{aligned}
$$

Nonzero column vectors P and P' are called *dependent* when one—hence each—is a scalar multiple of the other. You can check that this happens just in case $\overleftrightarrow{OP} = \overleftrightarrow{OP'}$. By theorem 6, those lines coincide just when any two—hence all—lines with equations $X = Pt + Q$ and $X = P't + Q'$ are equal or parallel. For that reason, dependent column vectors are often called *parallel*.

A system of equations

$$
\begin{aligned}
x_1 &= p_1 t + q_1 u + r_1 \\
x_2 &= p_2 t + q_2 u + r_2 \qquad \text{i.e., } X = Pt + Qu + R, \\
x_3 &= p_3 t + q_3 u + r_3
\end{aligned}
$$

is called a system of *linear equations in parameters t and u*. They assign a point X to each pair t, u of scalars. The set of all those points is called the *graph* of the equations.

Theorem 8. A point set ε is a plane if and only if it's the graph of some system $X = Pt + Qu + R$ of linear parametric equations with O, P, and Q noncollinear. In that case, ε is the plane through R parallel or equal to OPQ.

Proof. Suppose ε is a plane and select lines $g \neq h$ in ε that intersect at a point R. By corollary 7, g and h have parametric equations $X = Pt + R$ and $X = Qt + R$ for some points $P, Q \neq O$. By theorem 6, $g \parallel \overleftrightarrow{OP}$ and $h \parallel \overleftrightarrow{OQ}$, so $\varepsilon \parallel OPQ$. A point X lies in ε if and only if it's on a line $\overleftrightarrow{YZ}$ for some points $Y = Pv + R$ and $Z = Qw + R$ on g and h. That is, X lies in ε if and only if

$$X = Y + (Z - Y)s = Pv + R + (Qw - Pv)s = Pv(1 - s) + Qws + R$$

for some scalars s, v, and w. Thus every point in ε lies in the graph of equations $X = Pt + Qu + R$. On the other hand, you can write any such X as $Pv(1 - s) + Qws + R$ with $s = 1/2$, $v = 2t$, and $w = 2u$, so ε coincides with that graph.

Therefore, every plane ε is the graph of some equations $X = Pt + Qu + R$. It's also easy to show that, conversely, the graph of an arbitrary system $X = Pt + Qu + R$ is the plane through R equal or parallel to OPQ. ♦

Corollary 9. If points P, Q, and R are noncollinear, then plane PQR has equations $X = P + (Q - P)t + (R - P)u$.

Another approach, theorems 10 and 11, provides an entirely different algebraic representation for a plane.

Theorem 10. Consider points P and Q distinct from the origin O. Then $\overleftrightarrow{OP} \perp \overleftrightarrow{OQ}$ if and only if the dot product $P \cdot Q = p_1q_1 + p_2q_2 + p_3q_3 = 0$.

Proof. By Pythagoras' theorem and its converse, $\overleftrightarrow{OP} \perp \overleftrightarrow{OQ}$ if and only if $(PQ)^2 = (OP)^2 + (OQ)^2$. That is,

$$(p_1 - q_1)^2 + (p_2 - q_2)^2 + (p_3 - q_3)^2 = (p_1^2 + p_2^2 + p_3^2) + (q_1^2 + q_2^2 + q_3^2)\,.$$

You can simplify this equation to get $p_1q_1 + p_2q_2 + p_3q_3 = 0$. ♦

Nonzero column vectors P and Q are called *perpendicular* when $\overleftrightarrow{OP} \perp \overleftrightarrow{OQ}$, i.e., $P \cdot Q = 0$. By theorem 6, this happens just when intersecting lines with equations $X = Pt + R$ and $X = Qt + S$ are perpendicular. More generally, you can regard $\angle POQ$ as the *angle between* independent nonzero column vectors P and Q. Thus P and Q are perpendicular when the angle between them is right.

An equation

$$p_1x_1 + p_2x_2 + p_3x_3 = b \qquad \text{i.e.,}\quad P \cdot X = b,$$

where p_1 to p_3 and b are scalars, is called a *linear equation* in the coordinates of a point X. Its *graph* is the set of all points whose coordinates satisfy it.

Theorem 11. A point set ε is a plane if and only if it's the graph of some linear equation $P{\cdot}X = b$ with $P \neq O$.

Proof. If ε is a plane through the origin O, select a point $P \neq O$ such that $\overleftrightarrow{OP} \perp \varepsilon$. By theorem 10, a point X lies on ε if and only if $P{\cdot}X = 0$. Thus ε is the graph of a linear equation.

Suppose ε is a plane that doesn't pass through O. Let ε' be the plane through O parallel to ε. By the previous paragraph, ε' is the graph of an equation $P{\cdot}X = 0$. Select any point Q on ε. Consider the equation $P{\cdot}X = P{\cdot}Q$, which clearly holds when $X = Q$. If X is a point in ε distinct from Q, then there's a point X' in ε' such that $XQOX'$ is a parallelogram. By the parallelogram law, $X = Q + X'$. Since X' lies in ε',

$$P{\cdot}X = P{\cdot}(Q + X') = P{\cdot}Q + P{\cdot}X' = P{\cdot}Q + 0 = P{\cdot}Q\,.$$

Thus ε is a subset of the graph of equation $P{\cdot}X = P{\cdot}Q$. It's easy to show that every point in the graph lies in ε. Therefore each plane ε coincides with the graph of some linear equation.

Conversely, consider a linear equation $P{\cdot}X = b$ with $P = \langle p_1, p_2, p_3 \rangle \neq O$. Suppose $p_1 \neq 0$. It's then easy to show that the graph of $P{\cdot}X = b$ is the plane ε through point $\langle b/p_1, 0, 0 \rangle$ parallel or equal to the plane ε' through O perpendicular to $\overleftrightarrow{OP}$. Similar results hold when $p_2 \neq 0$ or $p_3 \neq 0$. Therefore the graph of any linear equation is a plane. ♦

When you set up a Cartesian coordinate system for a problem confined to a plane ε, select the origin in ε and the third axis perpendicular to ε, so that ε is the 1,2 plane. Then $x_3 = 0$ for every point $X = \langle x_1, x_2, x_3 \rangle$ in ε, and you can use the abbreviation $\langle x_1, x_2 \rangle$ for X. Every line g in ε is the graph of a system of linear parametric equations

$$\begin{aligned} x_1 &= p_1 t + q_1 \\ x_2 &= p_2 t + q_2 \qquad \text{i.e., } X = Pt + Q, \\ x_3 &= 0 \end{aligned}$$

where p_1 and p_2 are not both zero. You can ignore the third equation: A point set g in ε is a line if and only if it's the graph of a system of *two* linear parametric equations. Every line g in ε is also the graph of a linear equation $p_1 x_1 + p_2 x_2 + p_3 x_3 = b$. Given p_1 and p_2 you can vary p_3 at will because x_3 is always zero for points on g. Thus you can ignore p_3 altogether and just write the equation in two-dimensional vector notation

$$p_1 x_1 + p_2 x_2 = b \qquad \text{i.e., } P{\cdot}X = b.$$

Section 5.6 continues this development, deriving coordinate geometry techniques for distinguishing the sides of a plane and of a line in a plane, and formulas for triangle area and the volume of a tetrahedron.

3.12 Circles and spheres

Concepts

- Circles, and their interiors and exteriors
- Equation of a circle
- Intersections of lines and circles
- Tangent lines and circles
- Chords and diameters of circles
- Spheres, and their interiors and exteriors
- Equation of a sphere
- Intersections of lines, planes, and spheres
- Tangent lines, planes, and spheres
- Chords and diameters of spheres
- Secant lines

A comprehensive list of definitions and theorems concerning circles and spheres and their elementary properties constitutes this section. No figures and few proofs are included. You should supply all those. The proofs are straightforward; you'll need coordinates only to prove theorems 1 and 11.

Let O be a point in a plane ε and r be a positive real number. Consider the following point sets:

$\{ P \in \varepsilon : OP = r \} = \Gamma$—the *circle* in ε with *center* O and *radius* r,
$\{ P \in \varepsilon : OP < r \}$ —its *interior*,
$\{ P \in \varepsilon : OP > r \}$ —its *exterior*.

A segment between O and a point on the circle Γ is also called a *radius*.

Theorem 1. Set up Cartesian coordinates. Let Γ be the circle in the 1,2 plane with center $Z = \langle z_1, z_2 \rangle$ and radius r. Then a point $X = \langle x_1, x_2 \rangle$ lies

on Γ if $(x_1 - z_1)^2 + (x_2 - z_2)^2 = r^2$,
interior to Γ if $(x_1 - z_1)^2 + (x_2 - z_2)^2 < r^2$,
exterior to Γ if $(x_1 - z_1)^2 + (x_2 - z_2)^2 > r^2$.

Theorem 2. Let P and R be points on or interior to a circle Γ. Then all points $Q \neq P, R$ between P and R lie in the interior of Γ.

Proof. The result is trivial if P, Q, and R are collinear with the center O of Γ. Otherwise $OQ < OP$ or $OQ < OR$ by the hinge and exterior angle theorems. ♦

Corollary 3. The interior of a circle is convex.

Theorem 4. The intersection of a coplanar line g and circle Γ is empty, a point, or exactly two points. The third case holds if and only if g contains a point interior to Γ.

Proof. Theorem 2 implies that $g \cap \Gamma$ cannot contain three distinct points, and that g intersects the interior of Γ whenever $g \cap \Gamma$ contains two distinct points. Conversely, suppose g contains a point Q interior to Γ. If g passes through the center O of Γ, then the two points on g whose distance from O is the radius r of Γ lie on Γ. Suppose O is not on g. Let R be the foot of the perpendicular to g through O; then $OR \le OQ < r$. There exist exactly two points on g whose distance from R is $\sqrt{r^2 - (OR)^2}$. By Pythagoras' theorem, those points lie in $g \cap \Gamma$. ♦

Theorem 5. Consider two distinct coplanar circles Γ and Γ' with centers O and O' and radii r and r'. Exactly one of the following cases holds:

0. $r > OO' + r'$; Γ lies in the exterior of Γ', and Γ' in the interior of Γ.

0′. $r' > OO' + r$; like case (0) with Γ and Γ' interchanged.

1. $r = OO' + r'$; $\Gamma \cap \Gamma'$ is a point X such that O-O'-X; Γ doesn't intersect the interior of Γ', but all points of Γ' except X lie interior to Γ.

1′. $r' = OO' + r$; like case (1) with O, O' and Γ, Γ' interchanged.

2. Each of OO', r, and r' is less than the sum of the other two and $\Gamma \cap \Gamma'$ consists of two distinct points X and Y; O and O' lie on the perpendicular bisector of $\overline{XY}$; and each circle intersects both the interior and the exterior of the other.

3. $OO' = r + r'$; $\Gamma \cap \Gamma'$ is a point X between O and O'; neither circle intersects the interior of the other.

4. $OO' > r + r'$; each circle lies in the exterior of the other.

Proof. The only tricky part is determining $\Gamma \cap \Gamma'$ in case (2). Use theorem 3.8.14 to construct two distinct points X, one on each side of $\overleftrightarrow{OO'}$, such that $XO = r$ and $XO' = r'$. ♦

In case (4) of theorem 5, the circles are said to be *externally tangent*; in cases (1) and (1′), *internally tangent.*

If P and Q are distinct points on a circle Γ, then segment $\overline{PQ}$ is called a *chord* of Γ. A chord through the center of Γ is called a *diameter* of Γ.

Theorem 6. Let $\overline{PQ}$ be a chord of a circle Γ with radius r. Then $PQ \le 2r$, and $PQ = 2r$ if and only if $\overline{PQ}$ is a diameter of Γ.

Theorem 7. Let $\overline{PQ}$ and $\overline{RS}$ be chords of a circle intersecting at a point X. Then any two of the following conditions imply the third:

1. $\overline{RS}$ is a diameter of Γ;
2. $\overleftrightarrow{PQ} \perp \overleftrightarrow{RS}$;
3. X is the midpoint of $\overline{PQ}$.

Theorem 8. Let $\overline{PR}$ and $\overline{P'R'}$ be chords of circles with centers O and O' and the same radius. Let d and d' denote the distances from O and O' to $\overleftrightarrow{PR}$ and $\overleftrightarrow{P'R'}$. Then $PR \le P'R'$ if and only if $d \ge d'$.

A coplanar line and circle are said to be *tangent* if their intersection is a point. A line that intersects a circle in two distinct points is called a *secant* line.[29]

Theorem 9. Let P be a point on a circle Γ with center O in a plane ε. The perpendicular to $\overleftrightarrow{OP}$ through P in ε is the unique tangent to Γ through P.

Theorem 10. Let Γ be a circle with center O in a plane ε, and P be a point in ε exterior to Γ. There are exactly two tangents to Γ through P. If these intersect Γ at points Y and Z, then ΔOYP and ΔOZP are congruent right triangles.

Proof. Let X be the midpoint of $\overline{OP}$. By theorem 5, the intersection of Γ and the circle Γ' in ε with center X and radius OX consists of two points Y and Z noncollinear with O and P. Two pairs $\Delta XOY, \Delta XOZ$ and $\Delta XPY, \Delta XPZ$ of isosceles triangles are congruent by the SSS and SAS theorems. Applying the angle sum theorem to ΔOYP, you get $2\text{m}\angle OYX + 2\text{m}\angle XYP = 180°$, so $\angle OYP$ is right, and $\overleftrightarrow{PY}$ is tangent to Γ by theorem 8. So is $\overleftrightarrow{PZ}$, by a similar argument. If W is a point on Γ such that $\overleftrightarrow{PW}$ is tangent to Γ, then $\angle OWP$ is right by theorem 13, so $XW = OX$ by theorem 3.7.22, hence W is on Γ', so $W = Y$ or $W = Z$. By the SSS theorem, $\Delta OYP \cong \Delta OZP$. ♦

Let O be a point and r be a positive real number. Consider the following point sets:

$\{P : OP = r\} = \Sigma$ —the *sphere* with *center* O and *radius* r,
$\{P : OP < r\}$ —the *interior* of Σ, and
$\{P : OP > r\}$ —the *exterior* of Σ.

A segment between O and a point on the sphere Σ is also called a *radius.*

---

[29] These terms stem from the Latin verbs *tango* and *seco*, meaning *touch* and *cut*.

Theorem 11. Set up Cartesian coordinates. Let Σ be the sphere with center $Z = \langle z_1, z_2, z_3 \rangle$. Then a point $X = \langle x_1, x_2, x_3 \rangle$ lies

on Σ if	$(x_1 - z_1)^2 + (x_2 - z_2)^2 + (x_3 - z_3)^2 = r^2$,
interior to Σ if	$(x_1 - z_1)^2 + (x_2 - z_2)^2 + (x_3 - z_3)^2 < r^2$,
exterior to Σ if	$(x_1 - z_1)^2 + (x_2 - z_2)^2 + (x_3 - z_3)^2 > r^2$.

You should state and prove theorems about spheres analogous to theorem 2, corollary 3, and theorem 4. Theorem 4 has another analog, as follows.

Theorem 12. The intersection of a plane ε and a sphere Σ is empty, a point, or a circle. The last case holds if and only if ε contains a point interior to Σ.

The analog of theorem 5 for spheres Σ and Σ' in place of circles Γ and Γ' is also easy to state and prove. The revised version of case (2) should read as follows:

2. Each of OO', r, and r' is less than the sum of the other two and $\Sigma \cap \Sigma'$ is a circle; O and O' lie on the perpendicular to the plane of this circle through its center; and each sphere intersects both the interior and exterior of the other.

You can define the notions *internally tangent, externally tangent, chord,* and *diameter* for spheres exactly as for circles. Theorem 6 has an obvious analog for spheres. Theorem 7 has two, as follows.

Theorem 13. Let $\overline{PQ}$ be a chord of a sphere Σ, let ε be a plane, and let $\overline{PQ} \cap \varepsilon$ be a point X. Then any two of the following conditions imply the third:

1. ε passes through the center of Σ;
2. $\overleftrightarrow{PQ} \perp \varepsilon$;
3. X is the midpoint of $\overline{PQ}$.

Theorem 14. Let δ be a plane, let Σ be a sphere, and let $\delta \cap \Sigma$ be a circle Γ. Let $\overline{RS}$ be a chord of Σ, and $\delta \cap \overline{RS}$ be a point X. Then any two of the following conditions imply the third:

1. $\overline{RS}$ is a diameter of Σ;
2. $\delta \perp \overleftrightarrow{RS}$;
3. X is the center of Γ.

Theorem 8 has an obvious analog for spheres. Here's another.

Theorem 15. Let Σ and Σ' be spheres with centers O and O' and the same radius. Let δ and δ' be planes whose intersections with Σ and Σ' are circles with radii r and r'. Let d and d' be the distances from O and O' to δ and δ'. Then $r \le r'$ if and only if $d \ge d'$.

A line or a plane is said to be *tangent* to a sphere if their intersection is a point. A line that intersects a sphere but is not tangent is called a *secant*.

Theorem 16. Let P be a point on a sphere Σ with center O. The plane ε perpendicular to $\overleftrightarrow{OP}$ through P is the unique plane tangent to Σ through P. A line through P is tangent to Σ if and only if it lies in ε.

Theorem 17. Let X and Y be points on a sphere Σ and P be a point not on Σ. If $\overleftrightarrow{PX}$ and $\overleftrightarrow{PY}$ are tangent to Σ, then $PX = PY$.

3.13 Arcs and trigonometric functions

Concepts

- Angular scales and angle parameters
- Major and minor arcs, semicircles, and their measures
- Inscribed angles and subtended arcs
- Cosines and sines
- Periodic functions
- Parametric equations for circles and spheres
- Basic identities
- Even and odd functions
- Tangents
- Right triangle trigonometry

This section presents the plane trigonometry you need to analyze right triangles: the simplest properties of the sine, cosine, and tangent. You'll find several conflicting definitions of these functions in various texts. Do you compute with the sine of an *angle*? Or of an *arc*? Or with the sine of the *degree measure of* an angle or arc? Or its *radian* measure? Or with the sine of an *arbitrary real number*? For the program of this book the most convenient choice is to define the trigonometric functions for arbitrary real arguments related to degree measure. That conflicts with the radian measure definitions for analytic trigonometry, which facilitate using calculus to study circular motion. Radian measure is described in section 3.14 but never used in this book. Elementary trigonometry texts often define the functions for angle arguments first, then extend the definitions in stages. That tedious process is shortcut here. Angle measure is extended to incorporate *arbitrary* real numbers in place of measures just between 0° and 180°. This mechanism is used to measure circular arcs, and leads to some useful theorems about arcs. The trigonometric functions are then defined and their most basic properties outlined. The section concludes with a theorem that summarizes their use in analyzing right triangles, and then a simple application. Applications to oblique triangles are covered in detail in section 5.5 and its exercises.

Angle measurement

The degree measure θ of an angle, introduced in section 3.4, lies in the interval $0° < \theta < 180°$. This concept is convenient for elementary geometry because it assigns to each angle a unique measure. But it's inconvenient for angles used to study rotation. The measurement system is elaborated now to allow use of any real number θ —inside that interval or not—to designate an angle. In this context, θ is called an *angle parameter*.

The new system is modeled on the linear scale concept introduced in section 3.2. Instead of relating points on a line g to real numbers, it applies to the rays in a plane ε originating from a point O. To simplify the language, consider a single ray $\overrightarrow{OA}$ as a 0° angle, and when $\overrightarrow{OA'}$ is the opposite ray, call line $\overleftrightarrow{AA'}$ a 180° angle. Thus, in figure 3.13.1, $m\angle AOA = 0°$ and $m\angle AOA' = 180°$.[30]

The arithmetic operations θ div 360 and θ mod 360 are used in these calculations. You can define them for any real number θ through the conditions

$$\theta = 360q + r \quad \begin{cases} q \text{ is an integer} \\ -180 < r \le 180 \end{cases} \quad \begin{aligned} \theta \text{ div } 360 &= q \\ \theta \text{ mod } 360 &= r. \end{aligned}$$ [31]

For example (see figure 3.13.1),

$$\begin{aligned} 390 \text{ mod } 360 &= 30 & \quad -30 \text{ mod } 360 &= -30 \\ 210 \text{ mod } 360 &= -150 & \quad -570 \text{ mod } 360 &= 150\,. \end{aligned}$$

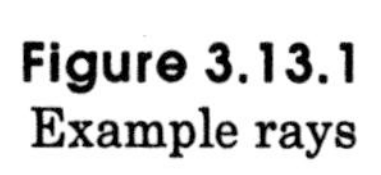

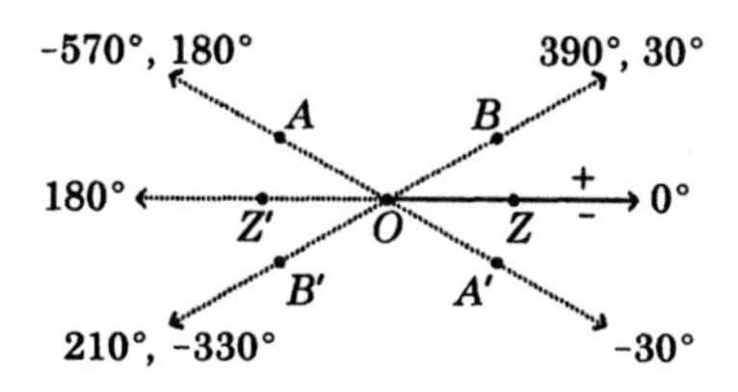

Figure 3.13.1
Example rays

[30] Older texts use the terms *zero* and *straight* angle for angles measuring 0° and 180°.

[31] These operations are examples of *modular* arithmetic: the *modulus* is 360°.

A function p from the set of all real numbers onto the set of rays in a plane ε originating at a point O is called an *angular scale at* O *in* ε if $p(\alpha) = \vec{OA}$ and $p(\beta) = \vec{OB}$ imply

$$|(\alpha - \beta) \bmod 360°| = m\angle AOB.$$

If p is an angular scale, then $p(\alpha) = p(\beta)$ if and only if α and β differ by a multiple of $360°$ —that is, by an *even* multiple of $180°$. In contrast, $p(\alpha)$ and $p(\beta)$ are *opposite* rays just in case α and β differ by an *odd* multiple of $180°$. Theorem 1 summarizes further properties of angular scales:

Theorem 1. Let p be an angular scale at a point O, in some plane ε. Its *initial* ray $\vec{OZ} = p(0°)$ and the ray $\vec{OZ'} = p(180°)$ are opposite. Let $\vec{OU} = p(1°)$. For any number α, with $\alpha^* = \alpha \bmod 360°$ and $\vec{OA} = p(\alpha)$,

$\alpha^* = 180°$ $\Rightarrow \vec{OA} = \vec{OZ'}$;

$0° < \alpha^* < 180°$ $\Rightarrow A$ lies on the same side of $\vec{OZ}$ in ε as U, and $m\angle ZOA = \alpha^*$;

$\alpha^* = 0°$ $\Rightarrow \vec{OA} = \vec{OZ}$;

$-180° < \alpha^* < 0°$ $\Rightarrow A$ lies on the side of $\vec{OZ}$ in ε opposite U and $m\angle ZOA = -\alpha^*$.

Proof. Only the second and fourth statements need verification. Let $\beta = -179°$ and $\vec{OB} = p(\beta)$, so that $\vec{OB}$ is the ray opposite U. Suppose $0° < \alpha^* < 180°$. Then $m\angle ZOA = |(0° - \alpha) \bmod 360°| = \alpha^*$. If A *weren't* on the same side of $\vec{OZ}$ as U, then it would fall on the same side as B, and either A would lie in the interior of $\angle ZOB$ or B would lie in the interior of $\angle ZOA$. That is, either

$$m\angle ZOA + m\angle AOB = m\angle ZOB$$
$$\alpha^* + |(\alpha + 179°) \bmod 360°| = 179°,$$

or

$$m\angle ZOB + m\angle BOA = m\angle ZOA$$
$$179° + |(\alpha + 179°) \bmod 360°| = \alpha^*.$$

Either way,

$$(\alpha + 179°) \bmod 360° = \pm(179° - \alpha),$$

so $(\alpha + 179°) \pm (179° - \alpha)$ would be a multiple of $360°$, which is impossible. Thus in this case A *does* lie on the same side of $\vec{OZ}$ as U.

If $-180° < \alpha^* < 0°$, then $m\angle ZOA = |(0° - \alpha) \bmod 360°| = -\alpha^*$. By the previous paragraph, the ray *opposite* $\vec{OA}$ lies on the *same* side of $\vec{OZ}$ as U, so A itself lies on the *opposite* side. ♦

Theorem 1 provides a method for *constructing* an angular scale at a point O in a plane ε. Choose an *initial* ray $\overrightarrow{OZ}$ in ε, and call one of the sides of $\overrightarrow{OZ}$ in ε *positive* and the other *negative*. For any θ with $0° < \theta \bmod 360° < 180°$, let $p(\theta)$ be the ray $\overrightarrow{OT}$ on the positive side with $m\angle POT = \theta \bmod 360°$. If $\theta \bmod 360° < 0°$, choose the ray $\overrightarrow{OT}$ on the negative side with $m\angle POT = -\theta \bmod 360°$. If $\theta \bmod 360° = 0°$ or $180°$, $p(\theta)$ should be the initial ray or its opposite.

Theorem 2. This function p is in fact an angular scale at O in ε.

Proof. It's tedious to verify equation $|(\alpha - \beta) \bmod 360°| = m\angle AOB$ in all cases. Here are some sample arguments. If A and B are situated as in figure 3.13.1, then

$$\begin{aligned} m\angle AOB &= m\angle ZOA - m\angle ZOB \\ &= \alpha \bmod 360° - \beta \bmod 360° \\ &= (\alpha - \beta) \bmod 360° \end{aligned}$$

$$m\angle BOA = m\angle AOB = -(\beta - \alpha) \bmod 360° = |(\beta - \alpha) \bmod 360°|.$$

If $\overrightarrow{OA'}$ lies opposite $\overrightarrow{OA}$ as in figure 3.13.1 and $\overrightarrow{OA'} = p(\alpha')$, then $\alpha' = \alpha - 180°$ plus a multiple of $360°$, so

$$\begin{aligned} m\angle A'OB &= 180° - m\angle ZOA + m\angle ZOB \\ &= 180° - \alpha \bmod 360° + \beta \bmod 360° \\ &= (180° - \alpha + \beta) \bmod 360° \\ &= |(\alpha - 180° - \beta) \bmod 360°| \\ &= |(\alpha' - \beta) \bmod 360°|. \end{aligned}$$

You can construct arguments for the remaining angles composed of rays in figure 3.13.1. ♦

All uses of angle parameters in this book are applications of theorem 2. The next, final, result about angle measurement, however, is required only for the discussion of rotations in chapter 6.

Theorem 3. Let ζ be any constant. If p is an angular scale at point O in plane ε, then so are the functions $\theta \to p(\zeta + \theta)$ and $\theta \to p(\zeta - \theta)$. Moreover, if p' is also an angular scale at O and $p(\zeta) = p'(0°)$, then $p'(\theta) = p(\zeta + \theta)$ for all θ, or else $p'(\theta) = p(\zeta - \theta)$ for all θ.

Proof. The first conclusion stems from the equations

$$|(\alpha - \beta) \bmod 360°| = \left|\left((\zeta \pm \alpha) - (\zeta \pm \beta)\right) \bmod 360°\right|.$$

To prove the second, consider the angular scale $p'' : \theta \to p(\zeta + \theta)$. It has the same initial ray as p', so by theorem 1 either $p' = p''$ or $p'(\theta) = p''(-\theta)$ for all θ. ♦

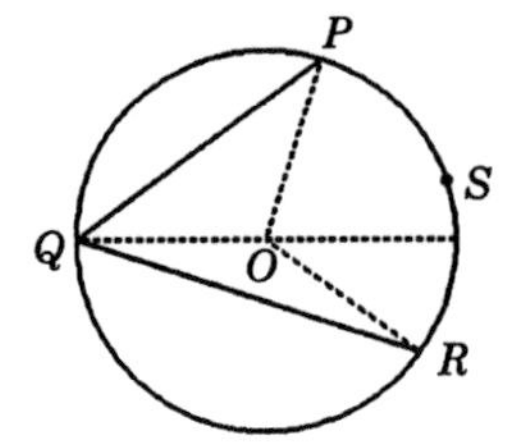

Figure 3.13.2
An inscribed angle
and its subtended arc

Arcs

Two distinct points P and R on a circle Γ determine two subsets of Γ, called *arcs*. Each consists of P, R, and all points of Γ on one side of $\overleftrightarrow{PR}$. They're designated by symbols $\widehat{PQR}$ and $\widehat{PSR}$, where Q and S are any other points in the arcs, as in figure 3.13.2. If $\overline{PR}$ is a diameter, the arcs are called *semicircles*. Otherwise, the one on the same side of $\overleftrightarrow{PR}$ as the center O is called the *major* arc; the other is *minor*. In figure 3.13.2, $\widehat{PQR}$ is the major arc.

You can *measure* arcs of a circle Γ with an angular scale at its center O in its plane. Given angle parameters ψ and ρ corresponding to rays $\overrightarrow{OP}$ and $\overrightarrow{OR}$, let

$$\mu = |(\psi - \rho) \bmod 360°| \qquad \nu = 360° - \mu.$$

One of these is $> 180°$. It's called the *measure* $\mathrm{m}\widehat{PQR}$ of the major arc. The other is the measure of the minor arc. Semicircles are assigned measure $180°$. The single point P and the entire circle are sometimes regarded as arcs $\widehat{PPP}$ and $\widehat{PQP}$ with measures $0°$ and $360°$.

Theorem 4. Let $\widehat{PQR}$ and $\widehat{P'Q'R'}$ be minor arcs of circles with centers O and O' and the same radius. Let d and d' denote the distances from O and O' to $\overleftrightarrow{PR}$ and $\overleftrightarrow{P'R'}$. Then these conditions are equivalent: $\mathrm{m}\angle POR \le \mathrm{m}\angle P'O'R'$, $\mathrm{m}\widehat{PQR} \le \mathrm{m}\widehat{P'Q'R'}$ and $d \ge d'$.

Theorem 5 (Arc addition theorem). Let P, Q, R, S, and T be points on a circle, such that $\widehat{PQR} \cap \widehat{RST}$ is the point R. Then $\widehat{PQR} \cup \widehat{RST} = \widehat{PRT}$ and $\mathrm{m}\widehat{PQR} + \mathrm{m}\widehat{RST} = \mathrm{m}\widehat{PRT}$.

Proof. You need to consider many cases separately and in combination, depending on the relationships of $\mathrm{m}\widehat{PQR}$, $\mathrm{m}\widehat{RST}$, and the measures $0°$, $180°$, and $360°$. ♦

If P, Q, and R are distinct points on a circle Γ, then $\angle PQR$ is said to be *inscribed* in arc $\widehat{PQR}$. If S is any point on Γ in the interior of $\angle PQR$, then arc $\widehat{PSR}$ is said to be *subtended* by $\angle PQR$. (See figure 3.13.2.)

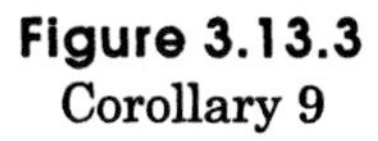

Figure 3.13.3
Corollary 9

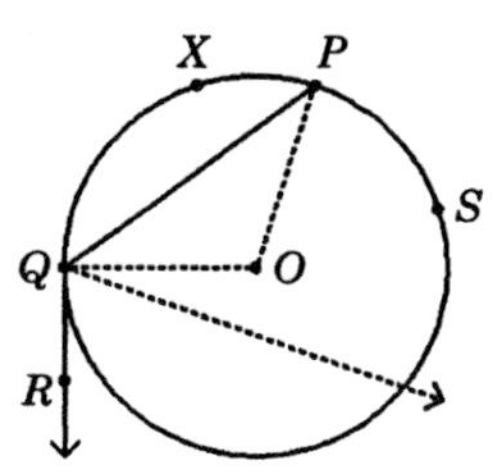

Theorem 6. The measure of an inscribed angle is half that of its subtended arc.

Proof. Consider $\angle PQR$ inscribed in a circle Γ with center O. There are three cases:

1. $\overline{QP}$ or $\overline{QR}$ is a diameter,
2. P and R lie on opposite sides of $\overleftrightarrow{QO}$,
3. P and R lie on the same side of $\overleftrightarrow{QO}$.

Case (1) is almost trivial. The others follow easily from case (1). The dotted lines in figure 3.13.2 suggest the proof for case (2). ♦

Corollary 7. All angles inscribed in the same arc are congruent.

Corollary 8. An angle inscribed in a semicircle is a right angle.

If you move leg $\overrightarrow{QR}$ of $\angle PQR$ in figure 3.13.2, $\text{m}\angle PQR$ remains half of the subtended arc until the line $\overrightarrow{QR}$ becomes tangent to Γ, as in figure 3.13.3. At that stage, you should expect $\text{m}\angle PQR = \frac{1}{2}\,\text{m}\,\widehat{PSQ}$. That's no proof, but the equation *is* valid:

Corollary 9. In figure 3.13.3, $\text{m}\angle PQR = \frac{1}{2}\,\text{m}\,\widehat{PSQ}$.

Proof. If O lies in the interior of $\angle PQR$ as in figure 3.13.3, then

$$\begin{aligned}\text{m}\angle PQR &= 90° + \text{m}\angle PQO \\ &= 90° + \tfrac{1}{2}(180° - \text{m}\angle POQ) \\ &= 180° - \tfrac{1}{2}\,\text{m}\angle POQ \\ &= \tfrac{1}{2}(360° - \text{m}\,\widehat{PXQ}) = \tfrac{1}{2}\,\text{m}\,\widehat{PSQ}.\end{aligned}$$

If O doesn't lie in the interior, you need a slightly different argument. ♦

Trigonometric functions

The trigonometric functions relate angle parameters to coordinates in a plane ε. Consider the origin O and the points $U = \langle 1,0 \rangle$ and $V = \langle 0,1 \rangle$. Choose $\overrightarrow{OU}$ as initial ray and regard as positive the side of the first axis that contains

V. The circle Γ with center O and radius 1 is called the *unit* circle. For any real number θ, consider the intersection $T = <t_1, t_2>$ of Γ and the ray with angle parameter θ. The coordinates $\cos\theta = t_1$ and $\sin\theta = t_2$ are the values of the *cosine* and *sine* functions for the argument θ.[32] (See figure 3.13.4.) This definition immediately yields the *periodic* and *Pythagorean identities*: For $n = 0, \pm 1, \pm 2, \ldots$ and any θ,

$$\cos(\theta + 360°\,n) = \cos\theta \qquad \cos^2\theta + \sin^2\theta = 1.$$
$$\sin(\theta + 360°\,n) = \sin\theta$$

The cosine and sine functions provide in the next three results some basic coordinate methods for representing circles and spheres: their *parametric* equations.

Corollary 11. Theorem 10 also holds for the circle with center $Z = <z_1, z_2>$ and radius r, and the mapping $\theta \to T = <t_1, t_2>$, where

$$t_1 = z_1 + r\cos\theta$$
$$t_2 = z_2 + r\sin\theta.$$

Corollary 12. The function $<\theta, \varphi> \to T = <t_1, t_2, t_3>$, where

$$t_1 = z_1 + r\cos\theta$$
$$t_2 = z_2 + r\sin\theta$$
$$t_3 = z_3 + r\cos\varphi,$$

maps the set $\{<\theta, \varphi>:\ 0° \le \theta < 360°\ \&\ 0° \le \varphi \le 180°\}$ onto the sphere with center $Z = <z_1, z_2, z_3>$ and radius r.

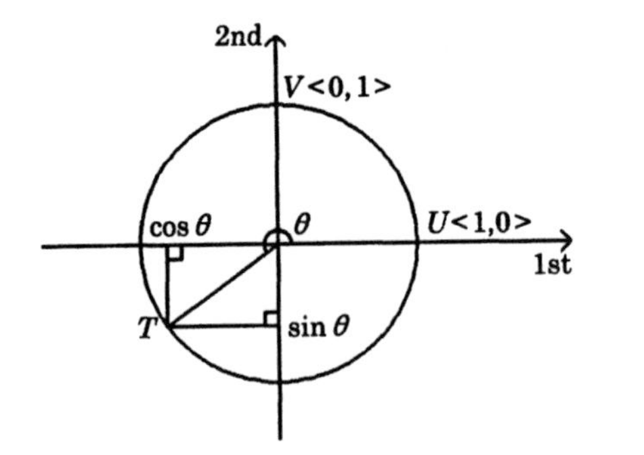

Figure 3.13.4
$\cos\theta$ and $\sin\theta$

[32] *Trigonometric* stems from the Greek words for *triangle measurement*. *Sine* is the English spelling of the Latin word *sinus* that names certain cavities in our cranial bones. Its use in trigonometry is an etymological mystery; see Schwartzman 1994, 200. According to theorem 13, $\cos\theta = \sin(90° - \theta)$. An old term for $90° - \theta$ is the *complement* of θ. And *cosine* means *sine of the complement*.

The next result is easy to see in figure 3.13.4, but messy to prove for all possible cases, because of the periodic definitions.

Theorem 13. For all θ, the sine and cosine satisfy the *cofunction, half-turn, and supplementary-angle identities*

$$\begin{aligned} \cos\theta &= \sin(90° - \theta) & \sin\theta &= \cos(90° - \theta) \\ \cos(\theta + 180°) &= -\cos\theta & \sin(\theta + 180°) &= -\sin\theta \\ \cos(180° - \theta) &= -\cos\theta & \sin(180° - \theta) &= \sin\theta. \end{aligned}$$

Moreover, the cosine is an even function, and the sine is odd:

$$\cos(-\theta) = \cos\theta \qquad \sin(-\theta) = -\sin\theta.$$

One more trigonometric function is useful: the *tangent*. Its value $\tan\theta$ is defined for all $\theta \neq 90° + 180°n$ with $n = 0, \pm 1, \pm 2, \ldots$ by the equation

$$\tan\theta = \frac{\sin\theta}{\cos\theta}.$$

Theorem 14. The tangent is an even function. For $n = 0, \pm 1, \pm 2, \ldots$ and any θ, the following *periodic identity* holds, or else both sides are undefined: $\tan(\theta + 180°n) = \tan\theta$.

Considering an isosceles right triangle and a right triangle with angles measuring $30°$ and $60°$, you can easily determine

$$\begin{aligned} \cos 30° &= \tfrac{1}{2}\sqrt{3} & \sin 30° &= \tfrac{1}{2} & \tan 30° &= \tfrac{1}{3}\sqrt{3} \\ \cos 45° &= \tfrac{1}{2}\sqrt{2} & \sin 45° &= \tfrac{1}{2}\sqrt{2} & \tan 45° &= 1 \\ \cos 60° &= \tfrac{1}{2} & \sin 30° &= \tfrac{1}{2}\sqrt{3} & \tan 60° &= \sqrt{3}\,. \end{aligned}$$

The periodic identities and theorem 11 then yield the values of the cosines, sines, and tangents of some arguments closely related to these. For example,

$$\begin{aligned} \cos 480° &= \cos(120° + 360°) = \cos 120° = \cos(180° - 60°) \\ &= -\cos 60° = -\tfrac{1}{2}. \end{aligned}$$

Using advanced trigonometry and calculus, mathematicians have derived methods for approximating *all* values of the trigonometric functions as closely as required for any application.[33] Moreover, given a value of one of the functions and some coarse information about the corresponding argument θ —for example, given $\sin\theta$ and $90°n \le \theta < 90°(n+1)$ —you can approximate θ as closely as required. Detailed, cumbersome tables of these values and arguments were used for centuries. Now, algorithms are encoded in your calculator to provide them on demand. With the following theorem,

[33] Exercise 5.11.27 develops one method used by ancient mathematicians to approximate sines as accurately as necessary.

they yield easy solutions for many problems concerning right triangles. Theorem 16, a good example, is needed in section 5.5.

Theorem 15. In any right triangle ΔABC with hypotenuse c and legs a and b opposite acute angles at vertices A and B,

$$\cos m\angle A = b/c \qquad a^2 + b^2 = c^2$$
$$\sin m\angle B = a/c \qquad m\angle A + m\angle B = 90°$$
$$\tan m\angle B = b/a.$$

Given two of the quantities a, b, c, $m\angle A$, and $m\angle B$ —but not just the two angle measures—you can find the remaining quantities in the list by using the preceding equations and approximations of the cosine, sine, and tangent values.

Proof. Introduce coordinates with the origin at A and the point $U = \langle 1,0\rangle$ on ray $\vec{AC}$, so that the second coordinate of B is positive. Then use similar triangles. ♦

Theorem 16. Consider figure 3.13.5,[34] where $\widehat{PXQ}$ is an arc of a unit circle with center O, $\theta = m\angle POQ$, and $0° < \theta < 90°$. Then $\sin\theta = QR < PQ < PS = \tan\theta$ and $OS = 1/\cos\theta$.

Proof. $\Delta OQR \sim \Delta OSP$, so

$$\frac{PS}{1} = \frac{PS}{OP} = \frac{QR}{OR} = \frac{\sin\theta}{\cos\theta} \qquad \frac{OS}{1} = \frac{OS}{OP} = \frac{OQ}{OR} = \frac{1}{\cos\theta}.$$

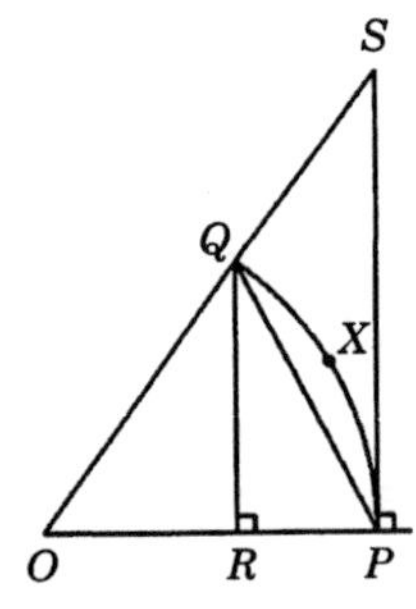

Figure 3.13.5
Theorem 16

[34] Standard texts usually feature three more trigonometric functions: the *secant, cosecant,* and *cotangent,* abbreviated $\sec\theta$, $\csc\theta$, and $\cot\theta$, and defined by the equations $\sec\theta = 1/\cos\theta$, $\csc\theta = \sec(90° - \theta)$, and $\cot\theta = \tan(90° - \theta)$. Once useful for hand calculation with tables of function values, they now merely provide footnote material. The latter two equations explain two of the function names: *Cosecant* and *cotangent* mean *secant* and *tangent of the complement.* (See footnote 29.) The first equation, theorem 16, and figure 3.13.5 illustrate the origin of the function names *tangent* and *secant*: $\tan\theta$ and $\sec\theta$ are the lengths PS and OS of segments that lie in lines tangent and secant to the circle.

By the exterior angle and isosceles triangle theorems, $m\angle S < m\angle OQP = m\angle OPQ < m\angle PQS$. By the hinge theorem, $PQ < PS$. ♦

3.14 π

Concepts

- Polygons inscribed in arcs; regular polygons
- Arc length and circumference
- Proportionality of arc length to radius and to arc measure in degrees
- π
- Archimedes' approximation to π
- Transcendence of π
- Radian measure
- Definition and area of a disk and a sector
- Definition and volume of a cylinder, a cone, and a ball

From the earliest times, geometers considered measuring circumferences of circles. Circles are central to geometry. In practice, you measure one by wrapping it with a string, which you unwind along a ruler. But no concepts corresponding to wrapping and unwinding were incorporated into axiomatic geometry. The only tool for measuring length was a ruler. How can elementary geometry handle this problem? This section shows how to define and measure the length of any arc of a circle. It follows closely the methods of the ancient Greek geometers, introduces the number π as the length of the semicircle with radius 1, and presents Archimedes' approximate value for π. Where the ancients' methods left gaps in the reasoning, more modern techniques involving least upper bounds are employed. The section continues with radian measure and circular area, and concludes with very informal discussions of the volumes of cylinders and spheres.

You can estimate the "length" of an arc $\widehat{PQR}$ of a circle Γ with center O as shown in figure 3.14.1. Choose an integer $n > 0$, set $P_0 = P$, $P_n = R$, and choose points $P_1, \ldots, P_{n-1}$ on $\widehat{PQR}$ distinct from these and each other, so that $m\widehat{POP_i} < m\widehat{POP_{i+1}}$ for $i = 1$ to $n-1$. The union $P_0 \cdots P_n$ of the segments $\overline{P_iP_{i+1}}$ is called a *polygon inscribed* in the arc. The points P_i are called its *vertices*. For $i = 0$ to $n-1$ the segments $\overline{P_iP_{i+1}}$, angles $\angle P_iOP_{i+1}$ and triangles ΔP_iOP_{i+1} are its *edges*, *central* angles and *central* triangles. The sum

$$l(P_0 \cdots P_n) = P_0P_1 + \cdots + P_{n-1}P_n$$

is its *length*. If $P = R$ then $P_0 = P_n$, the arc is the whole circle, and the polygon is said to be *closed*. The length of every polygon inscribed in a given

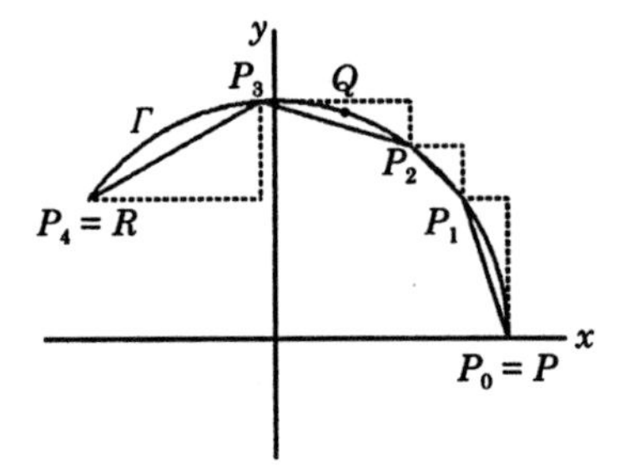

Figure 3.14.1
A polygon
inscribed in an arc

arc should be shorter than the "length" of the arc, and you should be able to *approximate* this arc "length" as closely as you wish by taking n large enough and spacing the vertices evenly. But what number *is* the arc "length"? If you designate two perpendicular lines x and y as in figure 3.14.1, then you can regard each edge of an inscribed polygon as the hypotenuse of a right triangle with legs parallel to x and y. The length of the polygon is less than the sum of the lengths of all the legs, which cannot exceed eight times the radius of Γ. Thus the set of lengths of all polygons inscribed in the arc is *bounded*. Every bounded set of real numbers has a least upper bound. (See appendix B for details of this concept.) That's the number we're looking for. The arc *length* $l(\widehat{PQR})$ is the least upper bound of the lengths of polygons inscribed in $\widehat{PQR}$. The arc length of the whole circle is called its *circumference*.[35]

The first results in this section show that arc length is independent of the position of the arc in the circle, and proportional to the radius.

Theorem 1. The lengths of arcs of equal measure in circles with radii r and r' have the ratio r/r'.

Proof. For each polygon inscribed in one arc you can construct a polygon in the other with congruent central angles. By the SAS similarity theorem, all corresponding central triangles are similar with ratio r/r', so the polygons' lengths stand in that ratio, too. Since multiplication by a fixed positive ratio doesn't change the ordering of numbers, the least upper bounds have the same ratio. ♦

Corollary 2. Arcs of equal measure in circles of equal radius have the same length.

The next results show that arc length behaves the way you should expect when you append one arc to another, and finally that arc length is proportional to arc measure in degrees. Because this concept is really a special case of

---

[35] This word stems from the Latin prefix *circum-* and verb *fero* meaning *around* and *carry*.

the definite integral, the proofs seem overly complex and nongeometric. They use techniques employed by real analysis texts in establishing the foundation of integral calculus.

Lemma 3. Suppose circular arc $\widehat{PRT}$ is the union of arcs $\widehat{PQR}$ and $\widehat{RST}$. Then $l(\widehat{PRT}) = l(\widehat{PQR}) + l(\widehat{RST})$.

Proof. Given a polygon Π inscribed in $\widehat{PRT}$, insert R as a new vertex, if it's not already there, to get a new inscribed polygon Π', which may be longer. Π' consists of two polygons, Φ inscribed in $\widehat{PQR}$ followed by Ψ inscribed in $\widehat{RST}$, so that

$$l(\Pi) \le l(\Pi') = l(\Phi) + l(\Psi) \le l(\widehat{PQR}) + l(\widehat{RST}).$$

Thus $l(\widehat{PQR}) + l(\widehat{RST})$ is an upper bound for the set L of lengths of all polygons inscribed in $\widehat{PRT}$, so $l(\widehat{PRT}) \le l(\widehat{PQR}) + l(\widehat{RST})$.

Suppose $l(\widehat{PRT}) < l(\widehat{PQR}) + l(\widehat{RST})$. Then $l(\widehat{PRT}) = u + v$ for some $u < l(\widehat{PQR})$ and $v < l(\widehat{RST})$. Since u couldn't be an upper bound for the lengths of polygons inscribed in $\widehat{PQR}$, there'd be some polygon Φ inscribed in $\widehat{PQR}$ for which $u < l(\Phi)$. Similarly, there'd be a polygon Ψ inscribed in $\widehat{RST}$ for which $v < l(\Psi)$. You could assemble Φ and Ψ into a single polygon Π inscribed in $\widehat{PRT}$. It would follow that $l(\Pi) = l(\Phi) + l(\Psi) > u + v > l(\widehat{PRT})$, so $l(\widehat{PRT})$ wouldn't be an upper bound for the set L defined in the previous paragraph—contradiction! ♦

Lemma 4. If $\widehat{PQR}$ and $\widehat{P'Q'R'}$ are arcs in the same circle and $\mathrm{m}\widehat{PQR} < \mathrm{m}\widehat{P'Q'R'}$, then $l(\widehat{PQR}) < l(\widehat{P'Q'R'})$.

Proof. Find points S and T on the circle so that $\widehat{PRT}$ and $\widehat{PST}$ are arcs and $\mathrm{m}\widehat{P'Q'R'} = \mathrm{m}\widehat{PRT}$. Then $l(\widehat{PQR}) < l(\widehat{PQR}) + l(\widehat{RST}) = l(\widehat{PRT}) = l(\widehat{P'Q'R'})$. ♦

Theorem 5. The lengths of arcs in a circle Γ have the same ratio as their degree measures.

Proof. Let $\widehat{P_1Q_1R_1}$ and $\widehat{P_2Q_2R_2}$ be arcs in Γ and

$$\frac{\mathrm{m}\widehat{P_1Q_1R_1}}{\mathrm{m}\widehat{P_2Q_2R_2}} = r.$$

It must be shown that

$$\frac{l(\widehat{P_1Q_1R_1})}{l(\widehat{P_2Q_2R_2})} = r.$$

There are several cases, depending on r.

Case 1. If $r = 1$, just apply corollary 2.

Case 2. If r is an integer > 1, then $\widehat{P_1Q_1R_1}$ has the same measure as r arcs of measure $\mathrm{m}\widehat{P_2Q_2R_2}$ placed end to end, so lemma 3 yields the desired result.

Case 3. Suppose r is a rational number > 1 but not an integer. Then there's an arc $\widehat{PQR}$ and integers $n_1, n_2 > 1$ such that $r = n_1/n_2$ and for $i = 1$ and 2, $\widehat{P_iQ_iR_i}$ has the same measure as n_i arcs of measure $\mathrm{m}\widehat{PQR}$ placed end to end. By case 2, $l(\widehat{P_iQ_iR_i}) = n_i\, l(\widehat{PQR})$ for each i, and division yields the desired result.

Case 4. Suppose r is an irrational number > 1. If

$$\frac{l(\widehat{P_1Q_1R_1})}{l(\widehat{P_2Q_2R_2})} < r,$$

then there'd exist some rational number $q > 1$, not an integer, such that

$$\frac{l(\widehat{P_1Q_1R_1})}{l(\widehat{P_2Q_2R_2})} < q < r.$$

You could find an arc $\widehat{PQR}$ such that

$$\frac{\mathrm{m}\widehat{PQR}}{\mathrm{m}\widehat{P_2Q_2R_2}} = q,$$

hence $\mathrm{m}\widehat{PQR} < \mathrm{m}\widehat{P_1Q_1R_1}$, and $l(\widehat{PQR}) < l(\widehat{P_1Q_1R_1})$ by lemma 4. Moreover, by case (3),

$$\frac{l(\widehat{PQR})}{l(\widehat{P_2Q_2R_2})} = q.$$

The previous inequality and equation would imply

$$q < \frac{l(\widehat{P_1Q_1R_1})}{l(\widehat{P_2Q_2R_2})}$$

—contradiction! You can draw a contradiction similarly from the hypothesis

$$\frac{l(\widehat{P_1Q_1R_1})}{l(\widehat{P_2Q_2R_2})} > r,$$

so the desired equation must hold.

Case 5. If $r < 1$, then by cases (2) to (4),

$$\frac{l(\widehat{P_2Q_2R_2})}{l(\widehat{P_1Q_1R_1})} > \frac{1}{r}$$

and the result follows. ♦

Now define π to be the length of a semicircle with radius 1.[36] By proportionality, you get

Theorem 6. The circumference of a circle Γ with radius r is $2\pi r$. The length of a $d°$ arc in Γ is $(\pi/180°)dr$.

About 250 B.C., the Greek mathematician Archimedes approximated π using a circle with radius 1 and inscribed and circumscribed regular polygons.[37] An inscribed *regular* polygon is closed, with all central angles equal. The concept of a *circumscribed* regular polygon is analogous, except that the edges are tangent and the vertices are external to the circle. (See figure 3.14.2.) Archimedes noted that every inscribed regular polygon is shorter than every circumscribed regular polygon, so the circumference 2π of the circle lies between the lengths of any inscribed and circumscribed regular polygons. Computing the lengths of the two polygons with ninety-six edges laboriously, without the benefit of algebraic and decimal notation developed only centuries later, he found $3.140 < \pi < 3.143$. You can pursue these details in exercise 4.9.2.

We now know that π is irrational, so its digits never terminate or repeat. In fact, it's not even a root of any polynomial with rational coefficients. It's that property—called *transcendence*—that prevents any classical Greek ruler and compass construction of a segment of length π.[38]

For a given radius, arc length is proportional to arc measure. That is, there's a constant k such that the length of a $d°$ arc is given by the formula kd. If you selected a fixed radius, you could use arc length to measure angles, instead of degree measure. That would free geometry from the influence of the arbitrarily chosen measure 180° for a semicircle. With radius 1, the arc length technique is called *radian measure.*[39] Since the length of the

Figure 3.14.2
Inscribed and circumscribed regular polygons

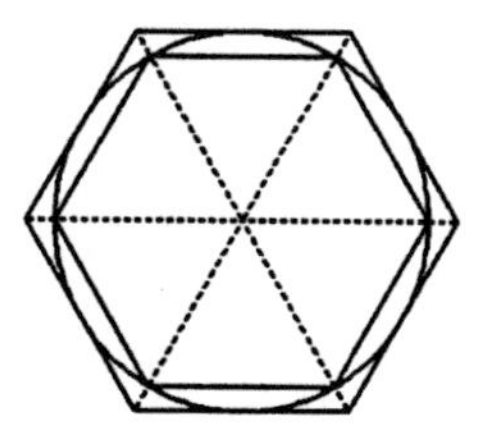

[36] According to Cajori (1919, 158) the first author to use the letter π for this quantity was William Jones (1706, 263). Jones did use π, but without claiming originality.

[37] Archimedes [1912] n.d., 91–98.

[38] Niven (1956, chapter 9) uses advanced calculus to prove that π is transcendental, and Moise (1990, chapter 19) shows that you can't construct with the classical instruments any segment whose length is transcendental.

[39] The word *radian,* which stands for *radi*al *an*gle, originated around 1870 in the lectures and writings of Thomas Muir and James Thomson Sr. (see Muir and Thomson 1910).

ARCHIMEDES was born about 287 B.C. in Syracuse, a Greek kingdom in southern Italy. His father was an astronomer, and his family was personally acquainted with the rulers. Archimedes studied at Alexandria, then returned to Syracuse, devoting his life to mathematics. He became famous for various mechanical inventions. Although he valued his theoretical work much more, practical problems clearly led him into new areas. Among his noted discoveries were the principles of hydrostatics and levers, and many results on centers of gravity. His most important work was the elaboration of the method of exhaustion, which Eudoxus and Euclid had discovered and utilized to some degree. Using the extended method, Archimedes solved many problems involving circles, conics, cones, and spheres that are now handled by integration. Actually, his methods form a foundation for integral calculus. That, however, wouldn't take form for another 1900 years.

Archimedes was killed, against military orders, by a Roman soldier during the massacre following the fall of Syracuse after a siege in 212 B.C. The Romans embellished his tomb with a diagram illustrating his favorite discovery: the curved area of a sphere is equal to that of a tangent cylinder.

$180°$ arc is π, the constant is $k = \pi/180$ and the radian measure of a $d°$ angle is $(\pi/180)\,d$. With degree measure, the constants π and 180 would creep into almost every calculus formula that's related to geometry, so it's usually more convenient to use radian measure. This book, however, doesn't use calculus, so the method of angle measurement is irrelevant. To maintain tradition, it uses degrees.

The first part of this section developed a theory for considering the length of arcs of a circle. Analogously, you can develop a theory for the *area* of a circular region, or *disk*—the union of a circle and its interior. This paragraph sketches that theory. It's convenient to extend the area concept to a *sector*—a part of a disk bounded by an arc and the radii at its endpoints—and to regions related to sectors and polygonal regions by the methods of section 3.8. You can add new axioms postulating that sectors with equal radii and congruent angles have the same area, and that the area of a disk with radius r is the least upper bound of the areas of its inscribed regular polygons. For n edges, the polygon's area is n times an edge times half the altitude a of the triangle. That's half its length times a. As n increases, the polygon's length approaches the circumference $2\pi r$ and a approaches r, so the area approaches $\frac{1}{2}(2\pi r)r = \pi r^2$. You can show that πr^2 is in fact the least upper bound, the area of the disk. A tedious argument like that leading to theorem 5 shows that the area of a sector is proportional to the measure of its arc.

Define a *cylinder* by imitating the definition of a prism in section 3.10. It has two *bases*—disks with the same radius in parallel planes—and consists of all points that lie between points in the bases. Its *altitude* is the distance between the base planes. You can define a *cone* by imitating the definition of a pyramid. Its *base* is a disk, its *apex* a point not in its base plane, and it consists of all points that lie between its apex and a base point. Its *altitude* is the distance between its apex and base plane. Finally, a *ball* is the union of a sphere and its interior.

You can extend the theory of volume to handle regions related to polyhedral regions, cylinders, cones, and balls by the methods of section 3.10. You must modify Cavalieri's axiom to apply to all regions whose intersections with planes have been covered by the previous discussion of areas. That gives the cylinder a volume equal to that of a prism with the same altitude and base area. That is, the *volume of a cylinder* is its altitude times its base area.

Consider a cone and a triangular pyramid with the same apex and base plane. A plane parallel to the base plane, between it and the apex, intersects the cone and the pyramid to form a disk Σ and a triangular region T. Reasoning with similar triangles, you can easily show that the ratios of the areas of Σ and T to the base areas are the same. It follows that the cone and the pyramid have the same volume if they have the same apex and base area. That is, the *volume of a cone* is one third its altitude times its base area.

Finally, the area and volume formulas already presented in this section, used all together, yield the formula for the *volume of a ball* with radius r: it's $\frac{4}{3}\pi r^3$. You can pursue the details of the calculation in exercise 4.8.3.

Chapter 4

Exercises on elementary geometry

Chapter 3 was a fast, perhaps breathtaking, recapitulation of elementary geometry. Most of that material you'd already studied. It had become central to your later pursuit of the field, nevertheless somewhat foggy. A few areas were perhaps new to you. The *learn by doing* precept applies to review as well as initial study. To master elementary geometry, you need to draw the figures and work out the proofs in chapter 3. And you need to do *exercises*, to see how these methods apply in practice. The standard type of routine elementary geometry exercise, fit for beginning students new to the material, isn't appropriate for this book. For those, you can consult a good school text,[1] or a workbook devoted to that kind of exercise.[2] Even better, devise routine exercises yourself. This chapter is a collection of meatier, *nonroutine* problems that complement chapter 3. Most of them require a combination of methods for solution. They're really problems in professional mathematics, not just elementary geometry.

The exercises are collected in this chapter for two reasons. First, some require considerable discussion just to state, and others are clearly peripheral to the main course of the subject. Including them in chapter 3 would distract you from its highly organized flow. Second, the most beneficial problems exercise various combinations of solution methods, and it's sometimes unclear just where in chapter 3 they should be included. Although this chapter is organized roughly the same way as chapter 3, its latter sections have been shuffled to accommodate this particular problem selection.

Some exercises in this chapter complete discussions begun earlier. For example,

---

[1] For example, Moise and Downs 1964.

[2] For example, Rich 1989.

- Exercise 4.5.6 corrects the sophism discussed in section 2.2;
- Exercise 4.6.2 completes the section 3.5 coverage of SSA triangle congruence; and
- Exercise 4.8.3 derives the formula for the volume of a sphere mentioned in section 3.14.

Others introduce entirely new subjects: for example,

- Independence proofs for various axioms in section 4.1;
- Arithmetic, geometric, and harmonic means in 4.5 and 4.6; and
- Spherical trigonometry in section 4.9.

There are a few original exercises here. But for most, the author acknowledges debt to his predecessors. Only a few exercises, with clear heritage, are given footnote citations. Others are old gems, whose origin would be hard to establish. The author has used much of this exercise set for decades. Each year new, different, solutions appear. You're encouraged to dive in, try out ideas old and new, and make these problems *your* old friends.

Hints are provided for a few exercises. Sometimes they suggest an appropriate method. Don't feel constrained to follow that, however, since many problems yield to attack from several directions. Other hints are approximate solutions: for example, "the answer is about 14.14." These help you check the feasibility of your solution. But you need to find the *exact* answer, which isn't obvious, even given the approximation. Indeed, 14.14 is *about* the same as both $11 + \pi$ and $10\sqrt{2}$, but those numbers are different.

4.1 Exercises on the incidence and ruler axioms

Concepts
Models of axioms
Finite geometric systems
Independence of the incidence axioms
Four-dimensional simplices
Ruler placement along a line

The incidence axioms of section 3.1 entail only a small part of Euclidean geometry. It's impossible to single out many applications of that material alone, so the exercises in this section are theoretical.

Axiom I5 states roughly that there are enough points to do *some* three-dimensional geometry. It's worded awkwardly. One reason for that was to facilitate the proof of the more informative lemma 3.1.3 and theorem 3.1.4. Exercise 1 explores alternate wordings. Exercise 2 shows that you can't do without I5, because the remaining axioms can be interpreted to describe a structure—a *model* of those four axioms—that's all in one plane. But according to exercise 3, even *with* I5 you don't get very *much* geometry: I1 to I5 have a model consisting of just four points, so you can't prove from the incidence axioms alone that there are more than four.

Exercise 1. Can you replace axiom I5 by the statement, *There exist four noncoplanar points*? How about, *There exists a plane* ε *and a point not on* ε?

Exercise 2. Consider the mathematical system depicted by figure 4.1.1. It consists of

- three points O, P, and Q;
- three lines $\{O,P\}$, $\{P,Q\}$, and $\{Q,O\}$;
- one plane $\{O,P,Q\}$.

Show that this system satisfies all the incidence axioms except I5.

Exercise 3. Construct a system with four points, six lines, and four planes, that satisfies *all* the incidence axioms.

Exercises 2 and 4 to 6 together demonstrate that the incidence axioms are *independent*: None is derivable from the others. For each axiom $\mathscr{A}$ there's a model in which $\mathscr{A}$ is false but the rest are true. If you could derive $\mathscr{A}$ from the other axioms, it would also be true in that model.

Exercise 4, Part 1. Consider the system depicted in figure 4.1.2 with four points, five lines, and four planes. Which incidence axioms does it satisfy, and which not? *Careful*: Your answer should depend on whether you interpret the word *three* in I2 as meaning *three distinct*.

Part 2. Modify the system of exercise 3 to produce one with four points, six lines, and three planes, that satisfies all the incidence axioms except I2.

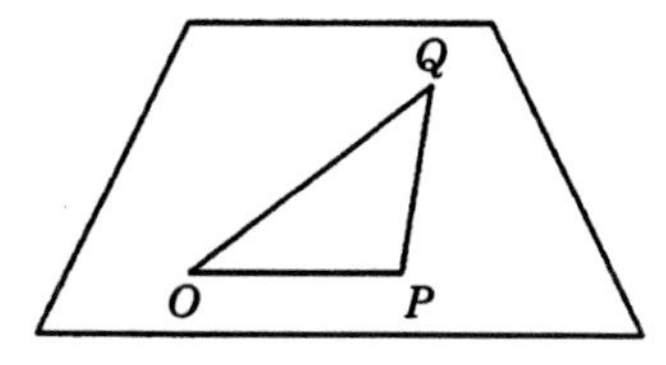

Figure 4.1.1 Finite plane for exercise 2

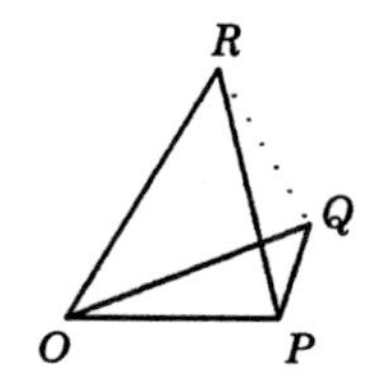

Figure 4.1.2 Finite model for exercise 4, part 1

Exercise 5. Consider a mathematical system with five points, ten lines each containing just two points, and ten planes each containing just three points. Show that it satisfies all the incidence axioms except I4. Show how to interpret this system as a *simplex*—the simplest polyhedron—in four-dimensional analytic geometry.

Exercise 6. Construct a mathematical system that satisfies all the incidence axioms except I3. *Suggestion*: You can do it by adding two points and some lines and planes to the system of exercise 3, and adjusting the content of some of the original lines and planes.

Another interesting project along these lines would be to sp lit I1 and I2 each into two axioms (they consist of two sentences each), and explore the independence of the resulting system.

The ruler axiom of section 3.2 excludes finite models like those considered in the previous exercises, since it states that the set of points on any line corresponds to the infinite set of real numbers and shares many properties with it. It says that you can establish such a correspondence—a scale—in at least one way. Exercise 7 shows what other possibilities there are. These relate to the ways you can place a ruler along a line in familiar experience.

Exercise 7. Given a scale c for a line g, and real constants a and b with $b = \pm 1$, show that the function d defined by setting $d(P) = a + bc(P)$, for all points P on g, is another scale. Conversely, show that if d is any scale for g, then there are constants a and b such that $b = \pm 1$ and $d(P) = a + bc(P)$ for all P.

4.2 Exercises related to Pasch's axiom

Concepts

Trihedral angles, their vertices and interiors
Lines and planes in general position
Partitioning a plane by lines
Partitioning space by planes
Recurrence formulas
Bounded sets
Difference equations

An ordered triple $\mathcal{T} = < \vec{OP}, \vec{OQ}, \vec{OR} >$ of three noncoplanar rays with a common origin O is called a *trihedral angle*. O is its *vertex*. Clearly, $\mathcal{T}$ determines three angles $\angle POQ$, $\angle QOR$, and $\angle ROP$. (It might seem simpler to define $\mathscr{T}$ as those angles' union, but that wouldn't let you distinguish one ray of $\mathscr{T}$ from another, which you must do to discuss congruence in section 4.3.)

Exercise 1. Show that a trihedral angle $\mathscr{A}$ determines three dihedral angles, the interiors of any two of which have the same intersection. That intersection is called the *interior* of $\mathscr{A}$.

With n distinct points you can divide a line into two rays and $n - 1$ segments that overlap only at common ends. The remaining exercises in this section generalize that familiar result to two and three dimensions. Exercise 6 introduces an algebraic method for solving polynomial difference equations that arise in this study. Give as much detail as possible in your geometric proofs, but don't sacrifice readability and grace. You should be able to confine your methods to those discussed in sections 3.1 to 3.3. These results depend only on the incidence, ruler, and Pasch axioms.

The two-dimensional generalization of the notion of distinct collinear points is that of coplanar lines *in general position*: Each two intersect, but no three pass through the same point. For an example, see figure 4.2.1.

Exercise 2. Show that a finite set $\mathscr{S}$ of n coplanar lines in general position divides their plane ε into a finite number of disjoint partitions. One of these is the union of $\mathscr{S}$, darkened in figure 4.2.1. Each of the others is the intersection of a finite number of sides in ε of members of $\mathscr{S}$. Let p_n be the number of partitions. Show that $p_{n+1} = p_n + n + 1$ for every $n > 0$. Compute p_{10}. *Suggestion*: Starting with no lines and one partition, add lines one by one. Consider the new partitions created when you add the n+1st line. Figure 4.2.1 shows that $p_4 = 12$. You should find $p_{10} = 57$.

A formula such as $p_{n+1} = p_n + n + 1$ that expresses each entry of a sequence $p_1, p_2, \ldots$ in terms of one or more of its predecessors is called a *recurrence* formula.

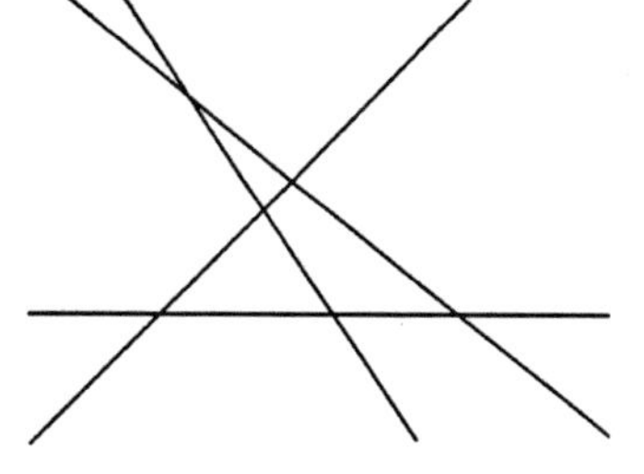

Figure 4.2.1 Coplanar lines in general position

Exercise 3. Express p_n, for $n > 0$, as a polynomial in terms of n. Use this formula to verify your value of p_{10}. Then compute p_{100}. How does the ratio p_{2n}/p_n behave as n increases? *Suggestion*: You should find $p_{100} = 5052$.

A point set Σ is called *bounded* if there's a number b such that $PQ < b$ for any points P and Q in Σ.

Exercise 4. In general, how many of the partitions in exercise 2 are *un*bounded? How many for $n = 100$? *Suggestion*: For $n = 100$ you should find 201 unbounded partitions.

A family of planes is said to be *in general position* if the intersection of each three is a point, but no four pass through the same point.

Exercise 5. Show that a finite set $\mathscr{S}$ of planes in general position divides space into a finite number of disjoint partitions. One of these is the union of $\mathscr{S}$. Each of the others is the intersection of a finite number of sides of members of $\mathscr{S}$. Let q_n be the number of partitions. Derive the recurrence formula $q_{n+1} = q_n + p_n - 1$. Compute q_{10}. *Suggestion*: You should find $q_{10} = 177$.

An integer constant c and a formula that defines a sequence $d_1, d_2, \ldots$ determine a *difference equation* problem: to find a formula for a sequence $f_1, f_2, \ldots$ such that $f_1 = c$ and $f_{k+1} - f_k = d_k$ for all k.

Exercise 6. Suppose $d_1, d_2, \ldots$ are given by a polynomial of degree n: There are coefficients a_0 to a_n such that $d_k = a_0 + a_1 k + a_2 k^2 + \cdots + a_n k^n$ for all k. Show how to find a polynomial solution f of degree $n + 1$ for the difference equation problem $f_1 = c$ and $f_{k+1} - f_k = d_k$ by solving a system of linear equations for the coefficients of f.

Exercise 7. Use the technique you developed in exercise 6 to find a formula that expresses q_n for $n > 0$, as defined in exercise 5, in terms of n. Use it to verify your value of q_{10}. Then compute q_{100}. How does the ratio q_{2n}/q_n behave as n increases? *Suggestion*: You should find $q_{100} = 166752$.

Exercise 8. In general, how many of the partitions in exercise 5 are unbounded? How many for $n = 100$? *Suggestion*: For $n = 100$ you should find 9903 unbounded partitions.

This section's numerical results suggest a guideline for developing algorithms for geometric software. Suppose you've used lines or planes to divide a plane or space into partitions $\Sigma_1, \Sigma_2, \ldots$ or $T_1, T_2, \ldots,$ and you know that only a few of the partitions have some specific property. If you must perform some operation on just those partitions, then you should try to avoid considering *all* the partitions and testing each in turn for that property. There are so many that such an algorithm could be too inefficient. You should look for some quicker way to identify the appropriate partitions.

Mathematicians have extended and modified the techniques involved in this section to give more detail and to apply to related problems.[3]

4.3 Exercises on congruence and perpendicularity

Concepts

Angle between a line and a plane
Congruent trihedral angles
Congruent tetrahedra

The first two exercises in this section are simple applications of the theory in sections 3.5 and 3.6. Exercises 3 and 4 are a little more involved; those results are used often in more complex arguments about distances in triangles and about perpendicularity in three dimensions. You may find exercise 5 the most challenging in this section. Although stated abstractly, it has a very concrete application in operations research:

> Your express company needs to drive four vans each morning from distribution center X to retailers A to D. Assuming that only distances are relevant, where should you locate X to minimize fuel costs?

The remaining exercises lead to a three-dimensional analog of the triangle congruence theory developed in section 3.5. The mathematics isn't difficult, but this study requires attention to organization, and facility with three-dimensional examples. Exercises 7 and 8 are open-ended. The author has never seen a three-dimensional congruence theory published in detail.

Virtually all exercises in the rest of this chapter routinely use the congruence and perpendicularity notions from chapter 3 that are stressed here.

Exercise 1. In figure 4.3.1, X and Y are distinct points in the intersection of distinct planes α and β, A is a point in α but not β, and B is a point in β but not α. Also, $AX = BY$ and $AY = BX$. Prove that $\angle AXB \cong \angle AYB$.

---

[3] See Wetzel 1978 for a summary, and follow its citations for more detail.

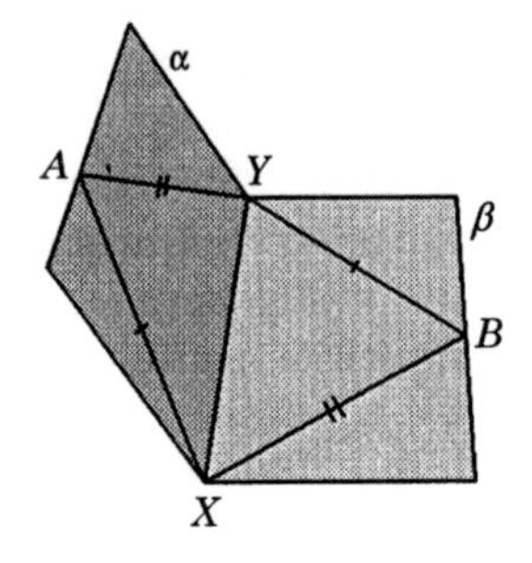

Figure 4.3.1
Exercise 1

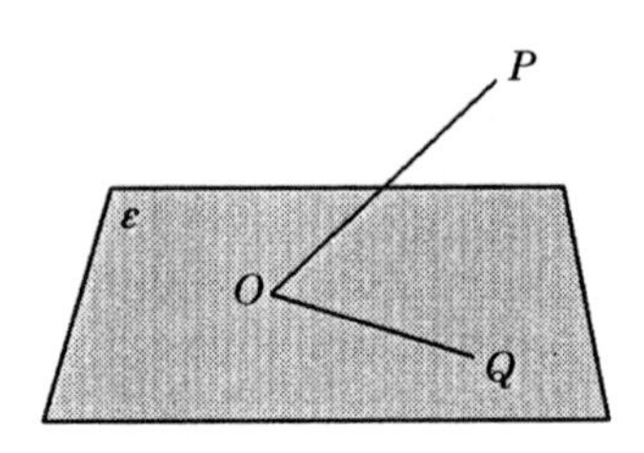

Figure 4.3.2
Exercise 2

Exercise 2. In figure 4.3.2, O is a point in a plane ε, and P is a point not in ε. Line $\overleftrightarrow{OP}$ is not perpendicular to ε. Find a point $Q \neq O$ in ε so that $m\angle POQ$ is as small as possible.

The smallest angle in exercise 2 is called the angle *between* $\overleftrightarrow{OP}$ *and* ε.

Exercise 3, Part 1. Prove that if point X is between but different from vertices A and B of ΔABC, then $CX < AC$ or $CX < BC$.

Part 2. Prove that if point X lies inside ΔABC, then $AX + BX < AC + BC$.

Exercise 4. Without using the parallel axiom prove that any three of the following conditions on noncoplanar points A to D implies the fourth:

$\angle B$ of ΔABC is right, $\angle C$ of ΔACD is right,
$\angle B$ of ΔABD is right, $\angle C$ of ΔBCD is right.[4]

Exercise 5. Given four coplanar points A to D, find a point X such that $AX + BX + CX + DX$ is as small as possible. *Suggestion*: Treat various configurations separately. For example, consider

1. all points on a line,
2. D on the edge of ΔABC opposite B,
3. $ABCD$ a convex quadrilateral, and
4. D in the interior of ΔABC.

Case 4 may prove troublesome. Try subdividing the plane as in figure 4.2.1 and considering X in various partitions.

---

[4] This form of exercise 4 is due to H. N. Gupta (1967).

Two trihedral angles $\mathcal{T} = <\vec{OP}, \vec{OQ}, \vec{OR}>$ and $\mathcal{T}' = <\vec{O'P'}, \vec{O'Q'}, \vec{O'R'}>$ are called *congruent* if the angles $\angle POQ, \ldots$ of $\mathcal{T}$ are congruent to the corresponding angles $\angle P'O'Q', \ldots$ of $\mathcal{T}'$ and corresponding dihedral angles of $\mathcal{T}$ and $\mathcal{T}'$ have the same measure. Trihedral angle congruence is thus conveyed by six equations stating that corresponding angles and dihedral angles have the same measure.

Exercise 6. Prove that if $<\vec{OP}, \vec{OQ}, \vec{OR}>$ is a trihedral angle, then $\text{m}\angle POQ + \text{m}\angle QOR > \text{m}\angle ROP$. *Suggestion*: You may want to consider separately cases $\text{m}\angle POQ \geq \text{m}\angle ROP$ and $\text{m}\angle POQ < \text{m}\angle ROP$.

Exercise 7. Find minimal sets of the six trihedral congruence equations that imply the rest. *Suggestion*: Perhaps you can prove that two trihedral angles are congruent if their three pairs of corresponding dihedral angles are. If so, can you get the same result from an even smaller set of equations? Try other possibilities, too.

Two tetrahedra are called *congruent* if their four pairs of corresponding trihedral angles and six pairs of corresponding edges are congruent. Tetrahedral congruence is thus conveyed by six equations stating that corresponding edges have the same length, and sixteen more stating that pairs of corresponding angles have the same measure (twelve pairs of angles determined by vertices, and four pairs of dihedral angles).

Exercise 8. Find minimal sets of the twenty-two tetrahedral congruence equations that imply the rest. Try to organize your results like the triangle congruence theory.

4.4 Exercises involving the parallel axiom

Concepts
Angle chasing
Properties of parallelepipeds

Except for the last, the exercises in this section aren't *about* parallel lines. But to solve them, you'll need to use results that depend on the parallel axiom. The solution of exercise 1 is an elementary example of the *angle chasing* technique. Exercise 2 looks at first like it might yield to the same method. But it's deceptive and troubling—one of the simpler instances of a genre of problems in which angle chasing seems like tail chasing. The remaining exercises are three-dimensional. The last, which calls for a three-dimensional analog of the theory of parallelograms developed in section 3.7, is open-ended, although its mathematics is not hard. The author has never seen such a theory worked out in detail.

Exercise 1. Which segment in figure 4.4.1 is shortest? Prove that your choice is correct.

Exercise 2. Suppose E is a point inside square $ABCD$, and $m\angle EAB = 15° = m\angle EBA$, as shown in figure 4.4.2. Without using trigonometry, show that ΔCDE is equilateral.

Exercise 3, Part 1. Suppose α, β, and γ are planes intersecting in distinct lines g, h, and k as shown in figure 4.4.3. Using only the incidence axioms, prove that either the three lines are concurrent, or no two of them intersect.

Part 2. Suppose α, β, γ, δ, and ε are planes intersecting in distinct lines g, h, k, and l as shown in figure 4.4.4. Prove that g and l are coplanar. How are you stymied when you try to prove this statement using only the incidence axioms?

Arthur Winternitz (1940, section II.2) showed by a circuitous argument that the result in exercise 3, part 2, holds even in hyperbolic non-Euclidean geometry—the theory you get by replacing the parallel axiom by its negation. Thus, there ought to be a solution of part 2 that uses only the incidence, ruler, Pasch, and congruence axioms. You're challenged to find one. Winternitz showed that you can't derive it from just the incidence axioms.

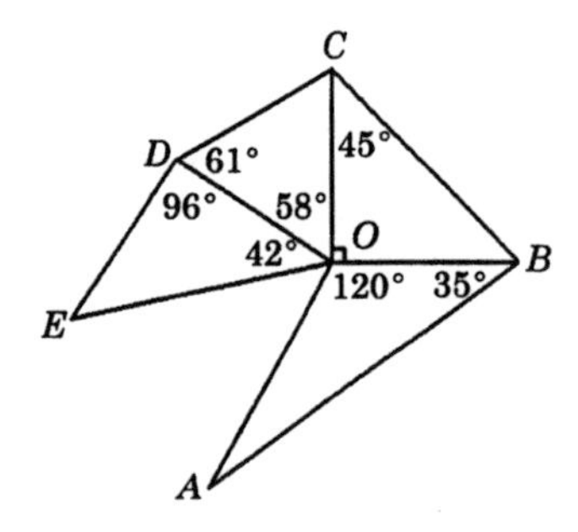

Figure 4.4.1
Exercise 1

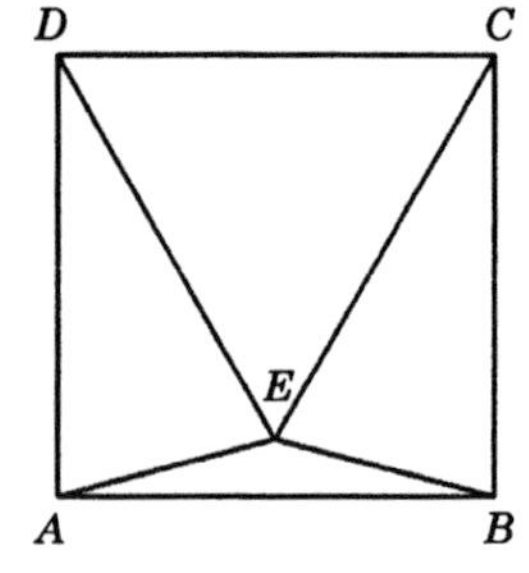

Figure 4.4.2
Exercise 2

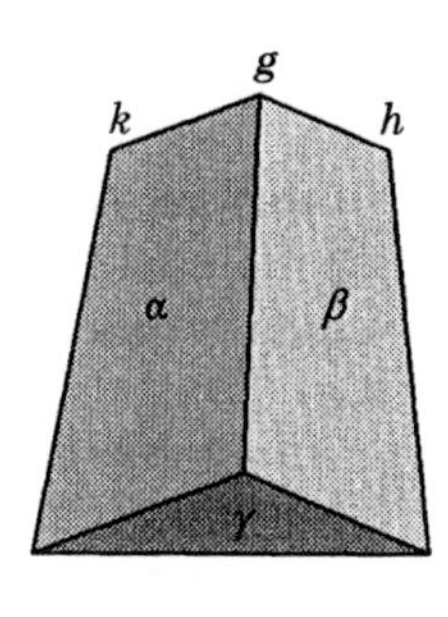

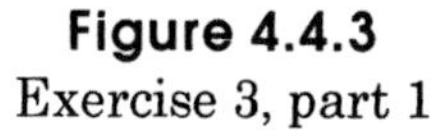

Figure 4.4.3
Exercise 3, part 1

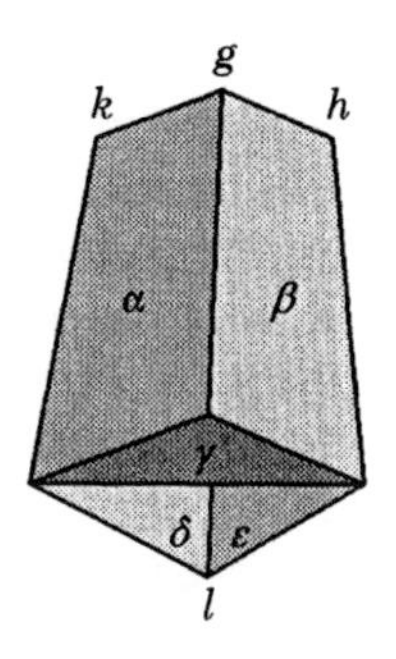

Figure 4.4.4
Exercise 3, part 2

Exercise 4. Prove that two noncoplanar lines have exactly one common perpendicular. *Suggestion*: Try building a right prism whose bases are quadrilaterals $\mathcal{Q}$ and $\mathcal{R}$ with two right angles each, such that $\mathcal{Q}$ has an edge on one given line and $\mathcal{R}$ has one on the other.

Exercise 5. Formulate and prove a set of theorems about parallelepipeds analogous to theorems 3.7.17–3.7.20 about parallelograms. More detail may be appropriate, because there are more concepts in three dimensions:

Two dimensions	*Three dimensions*
parallelogram	parallelepiped
	right prism with parallelogram bases
	rectangular prism
rectangle	box
	prism with two rhombic bases
	right rhombic prism
rhombus	cuboid (all faces rhombic)
square	cube

4.5 Exercises on similarity and Pythagoras' theorem

Concepts
Arithmetic, geometric, and harmonic means
Right-triangle trigonometry applications
Correcting the section 2.2 sophism

This section contains exercises that use similarity and Pythagoras' theorem in various ways. In section 4.6 you'll find more exercises that stress

Pythagoras' theorem. The techniques emphasized here will be used routinely in the rest of this chapter.

Exercise 1 demonstrates geometrically some properties of three *mean* operations used in many applications. The average $m = \frac{1}{2}(x+y)$ of two numbers x and y is called their *arithmetic* mean because the triple $<x,m,y>$ forms an *arithmetic* series: $y - m = m - x$. The *geometric* mean of x and y is the number g such that $<x,g,y>$ forms a *geometric* series: $y/g = g/x$, so $g^2 = xy$, hence $g = \sqrt{xy}$. The *harmonic* mean of x and y is the number h whose reciprocal is the average of those of x and y:

$$\frac{1}{h} = \frac{1}{2}\left(\frac{1}{x} + \frac{1}{y}\right).$$

Exercise 1, Part 1. Given x and y with $x < y$ as in figure 4.5.1, show that $h = 2z$ is their harmonic mean. Show geometrically that $x < h < m$. Show algebraically that $<h,g,m>$ forms a geometric series, so that $x < h < g < m < y$.

Part 2. Let $ABCD$ be a trapezoid with bases of length $AB = x$ and $CD = y$. Let the line parallel to the bases through the intersection of the diagonals meet the other edges at points E and F. Show that EF is the harmonic mean h.

Exercise 2, Part 1. Prove that if A to D are distinct noncollinear (not necessarily coplanar) points, and

A' is the midpoint of $\overline{AB}$, B' is the midpoint of $\overline{BC}$,
C' is the midpoint of $\overline{CD}$, D' is the midpoint of $\overline{DA}$,

then $A'B'C'D'$ is a parallelogram or A' to D' are collinear and $A'B' = C'D'$.

Part 2. Prove that if A to F are distinct points, no four of which are collinear,

A' is the midpoint of $\overline{AB}$, D' is the midpoint of $\overline{DE}$,
B' is the midpoint of $\overline{BC}$, E' is the midpoint of $\overline{EF}$,
C' is the midpoint of $\overline{CD}$, F' is the midpoint of $\overline{FA}$,

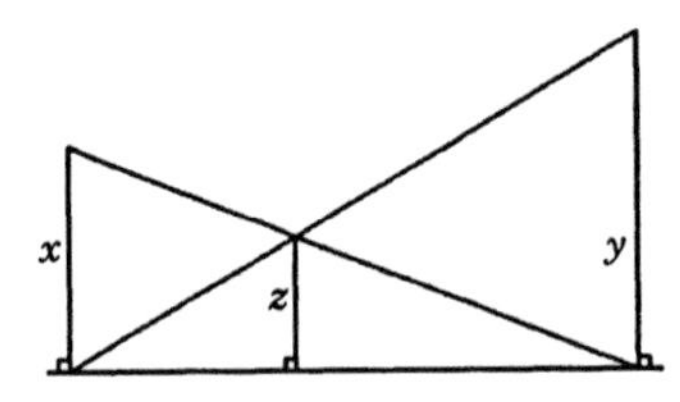

Figure 4.5.1 Constructing the harmonic mean

and O is a point such that $A'B'C'O$ is a parallelogram, then $D'E'F'O$ is also a parallelogram.

Exercise 3. The measures of the angles and edges of a triangle are sometimes called its *parts*. Construct two incongruent triangles such that five parts of one equal five of the other. Describe all possible ways to do this.

Theorem 3.13.15 provided a summary of right-triangle trigonometry. Exercise 4 demonstrates all cases of its use in practical applications. Exercise 5 shows a typical application: measuring some particularly symmetric figures that you'll study in more detail in chapter 8.

Exercise 4. Consider right triangle ΔABC with legs $a = BC$, $b = CA$, hypotenuse $c = AB$, and acute angles $\angle A$ and $\angle B$.

Part 1. Given $a = 10$ and $b = 20$, find c, $m\angle A$, and $m\angle B$.
Part 2. Given $b = 20$ and $c = 30$, find a, $m\angle A$, and $m\angle B$.
Part 3. Given $a = 10$ and $m\angle A = 40°$, find b, c, and $m\angle B$.
Part 4. Given $b = 20$ and $m\angle A = 40°$, find a, c, and $m\angle B$.
Part 5. Given $c = 30$ and $m\angle A = 40°$, find a, b, and $m\angle B$.

Exercise 5. Find the measures of the dihedral angles formed by the faces of a regular tetrahedron and of a regular octahedron. The latter polyhedron consists of two square pyramids with a common base and eight equilateral faces. *Suggestion*: To gain insight and check the reasonableness of your solutions, you may want to make stiff cardboard models. More precise and robust models of these and other commonly studied polyhedra are available as kits. You glue together plastic face tiles. To fit precisely, they should be manufactured with the edges beveled to produce the correct dihedral angles. (See the discussion in section 8.5 under the heading

Exercise 6 is the result you need to correct the sophism in section 2.2. It's included here only because part of the solution—already given in 2.2—uses Pythagoras' theorem. The argument you must supply here uses only methods covered through section 3.7.

Exercise 6. Consider ΔABC with $CA < AB$. Prove that the bisector of $\angle A$ meets the perpendicular bisector of $\overline{BC}$ at a point X on the opposite side of $\overleftrightarrow{BC}$ from A. Moreover, the feet Y and Z of the perpendiculars from X to $\overleftrightarrow{CA}$ and $\overleftrightarrow{AB}$ lie outside and inside the triangle legs, respectively.

4.6 Exercises on circles and spheres, part 1

Concepts

Geometric interpretation of arithmetic, geometric, and harmonic means
SSA congruence data
Angles between tangents and secants to circles
External and internal tangents to disjoint circles

This section and the following three all contain exercises involving circles and spheres. Those here are general in scope. Sections 4.7 and 4.8 emphasize computation of areas and volumes of regions constructed from circles and spheres. Section 4.9 contains exercises on plane and spherical trigonometry, the value of π, and the curved areas of cylinders, cones, and spheres.

Exercise 1 continues the study of arithmetic, geometric, and harmonic means begun in exercise 4.5.1. It was included in the work of the Alexandrian mathematician Pappus about A.D. 320.[5]

Exercise 1. Given $x < y$, locate points A to C so that A-B-C, $AB = x$, and $BC = y$. Let O be the midpoint of $\overline{AC}$, and D be the intersection of the perpendicular to $\overleftrightarrow{AC}$ through B with a semicircle whose diameter is $\overline{AC}$. Let E be the foot of the perpendicular to $\overleftrightarrow{OD}$ through B. Among the segments determined by points A to E and O, find segments of length

- h the harmonic mean of x and y,
- g their geometric mean, and
- m their arithmetic mean.

Point out geometric meanings of the inequalities $x < h < g < m < y$.

Although exercise 2 isn't *about* circles, you may want to use some of their properties to solve it. This result completes the triangle congruence theory begun in section 3.5. There you reviewed the SAS, ASA, SSS, and SAA theorems. An example indicated that there's no analogous SSA theorem. Nevertheless, as exercise 2 shows, SSA data can be useful. Sometimes they're enough to determine all the other parts of the triangle. In the other cases, you can *almost* do that; you've only two choices.

Exercise 2. Suppose you have SSA data about ΔABC: $a = BC$, $b = CA$, and $m\angle A$.

---

[5] See Eves 1963, volume 1, problem 1.5-17.

Part 1. Suppose $\angle A$ is obtuse or right. Prove that $a > b$ and that all triangles with this same SSA data are congruent to ΔABC.

Part 2. Suppose $\angle A$ is acute. Let d be the distance from C to $\overleftrightarrow{AB}$. Show that $d < b$ and $d \le a$. Prove that if $a = d$ or $a \ge b$, then all triangles with this same SSA data are congruent to ΔABC. Show that if $d < a < b$, then there's a triangle T incongruent with ΔABC such that all triangles with this same SSA data are congruent to ΔABC or to T.

Exercises 3 and 4 seem to relate to church window design and to packing balls in a box. They're really exercises on Pythagoras' theorem.

Exercise 3. In figure 4.6.1, arcs $\widehat{AGC}$ and $\widehat{BHC}$ have centers B and A. The semicircles with centers D and F on $\overline{AB}$ have radius $\frac{1}{4}\,s$, where $s = AB$. The circle with center E is tangent to the first two arcs and the semicircles. Find its radius r in terms of s.

Exercise 4, Part 1. In figure 4.6.2, called a *quincunx*,[6] the square has edge s and the circles, radius r. Find r in terms of s. *Suggestion*: $r \approx 0.207\,s$.

Part 2. Same as part 1, with nine spheres placed inside a cube. *Suggestion*: $r \approx 0.232\,s$.

This section concludes with two exercises on tangents and secants to circles.

Exercise 5. Investigate the angles between two tangents to a circle, between two secants, and between a tangent and a secant. Include all five cases shown in figure 4.6.3. In each case derive a formula for the measure of $\angle A$ in terms

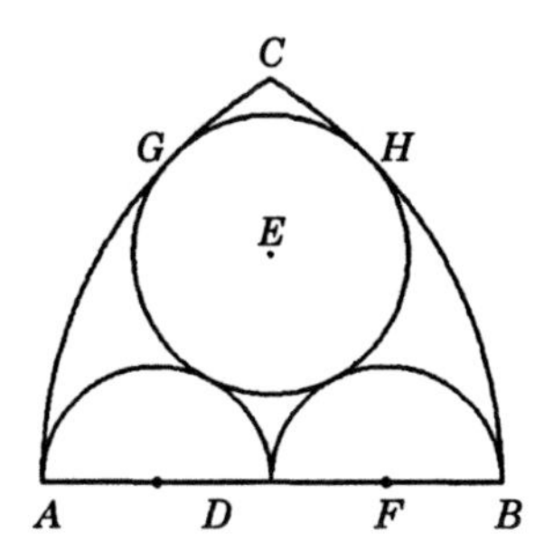

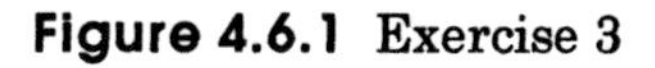
Figure 4.6.1 Exercise 3

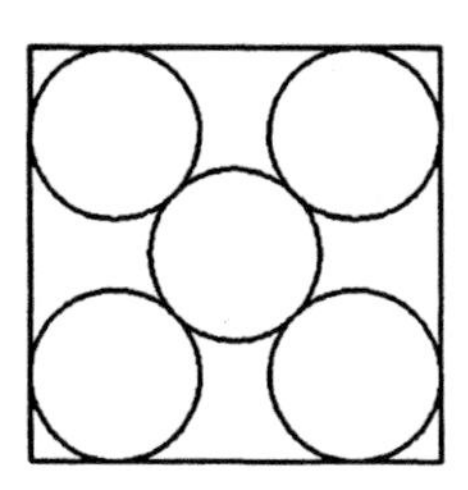
Figure 4.6.2 Quincunx

---

[6] This word is a Latin compound: *quinque* + *uncia* = *five* + *ounce* or *twelfth*.

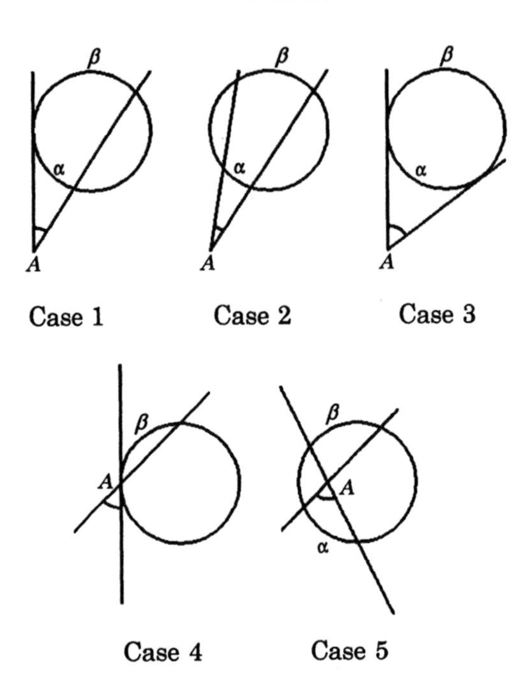

Figure 4.6.3 Exercise 5

of the measures α and β of the indicated arcs. *Suggestion*: More general cases are best derived from more special ones. You may have to consider separate cases depending on the location of the center with respect to $\angle A$. Methods from the proof of theorem 3.13.6 may help.

Exercise 6. Lines $\overleftrightarrow{AB}$ and $\overleftrightarrow{CD}$ are *externally* tangent to disjoint circles Γ and Δ at A, B and C, D in figure 4.6.4. Line $\overleftrightarrow{B'C'}$ is *internally* tangent at C' and B'. The internal tangent intersects the others at P and Q. Prove that $AB = CD = PQ$. *Suggestion*: Methods appropriate for solving this exercise are used later to prove theorem 5.9.2.

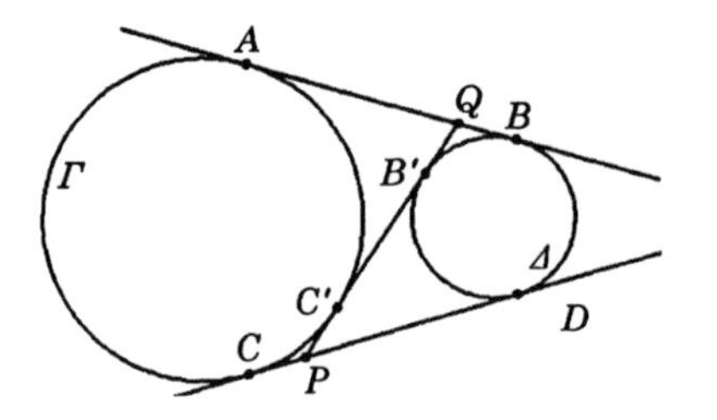

Figure 4.6.4
Exercise 6

4.7 Exercises on area

Concepts
- Dissecting polygonal regions
- Kites and lunes
- Interior and boundary of a polygonal region
- Open and closed subsets of a plane

This section contains exercises on polygonal regions and regions constructe d from circles, emphasizing area. In exercises 3 to 8, the main technique required is *dissection*: splitting a given region into pieces that you can handle with the area theory developed in sections 3.8 and 3.14. All exercises except the last two involve area calculations. The last ones let you develop the concepts of boundary and interior of a polygonal region. Although polygonal regions are built from triangular ones, and the interior of a triangular region is a simple idea, you'll find it hard to pin down the corresponding notion for polygonal regions in general. The exercises provide a workable definition, which you'll need in chapter 8.

Exercise 1. Figure 4.7.1 is an isosceles trapezoid $ABCD$ with $\overleftrightarrow{AB} \parallel \overleftrightarrow{CD}$, $AB = b_1$, $CD = b_2$, $BC = c = DA$, and a point E on $\overline{BC}$. Find $d = BE$ so that $\overline{AE}$ splits $ABCD$ into equal areas.

Exercise 2. Consider a tetrahedron with three mutually perpendicular faces. Prove that the sum of the squares of their areas equals the square of the area of the other face.

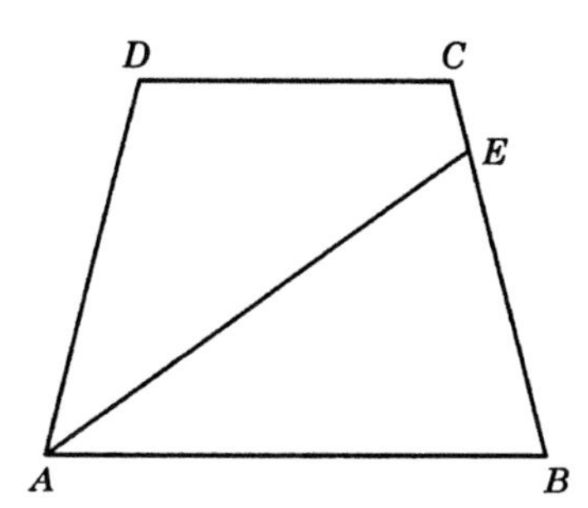

Figure 4.7.1
Exercise 1

Figure 4.7.2
Kürschak's tile

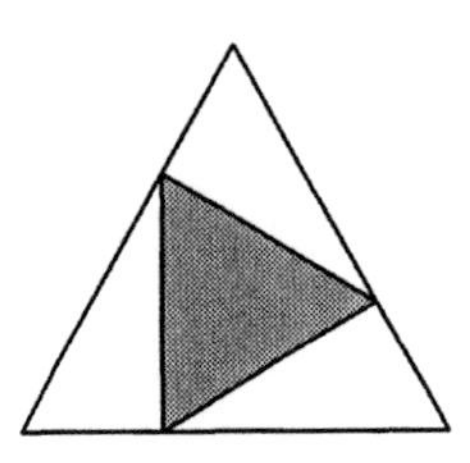

Figure 4.7.3
Exercise 4

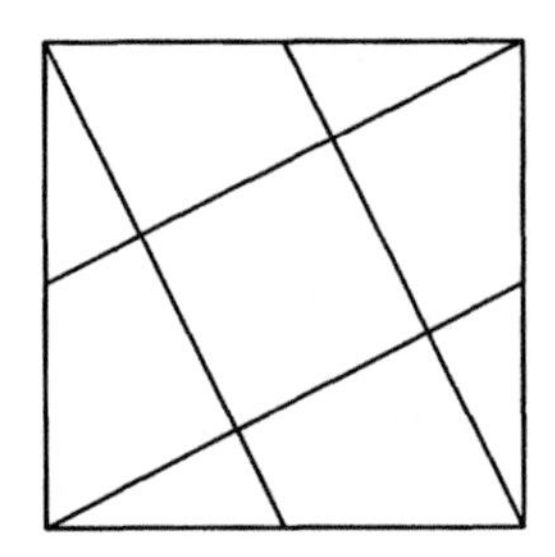

Figure 4.7.4
Exercise 5

Exercise 3. Using figure 4.7.2, compute the area of the regular dodecagon with radius 1.[7]

Exercise 4. In figure 4.7.3, each edge of an outer equilateral triangle is divided in the ratio 2 to 1. Find the area of the shaded inner triangle, in terms of the edge s of the outer one. *Suggestion*: It's approximately $0.14s^2$.

Exercise 5. In figure 4.7.4, segments join the vertices of a square to midpoints of its sides. Prove that the quadrilateral in the middle is a square, and determine the ratio of its area to that of the outer square. Find an analogous result for a similar situation in which the outer figure is a rectangle. *Suggestion*: The only computation in your solution should be counting.

Exercise 6. In figure 4.7.5, some vertices of a square have been joined to edge midpoints to form a *kite*: a quadrilateral (shaded) with two adjacent pairs of congruent edges. Find its area, in terms of the edge s of the square. *Suggestion*: It's about $0.27s^2$.

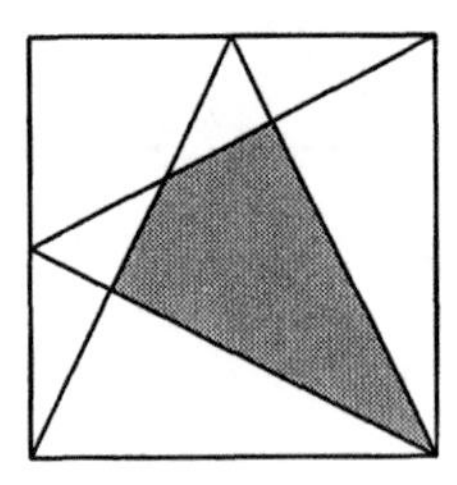

Figure 4.7.5
Exercise 6: kite

[7] This exercise is due to J. Kürschak. See Alexanderson and Seydel 1978.

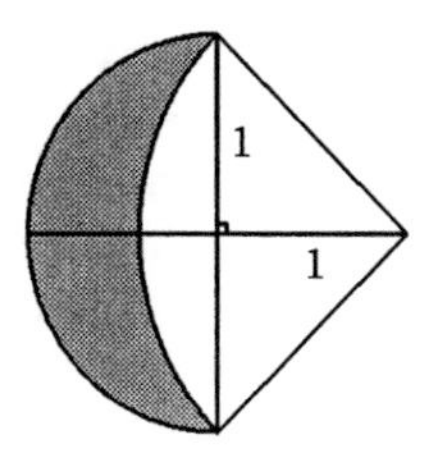

Figure 4.7.6
Exercise 7

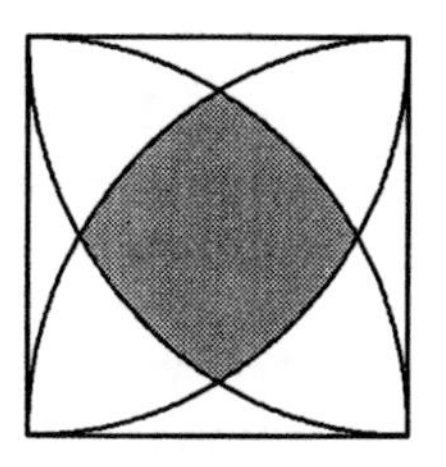

Figure 4.7.7
Exercise 8

Exercise 7. Compute the area of the shaded region in figure 4.7.6. Called a *lune*,[8] it lies "between" arcs of two circles with radius 1. The solution may surprise you.

Exercise 8. Compute the area of the shaded region in figure 4.7.7. The length of an edge of the square is 1. *Suggestion*: This exercise is related to exercise 3; the area is approximately 0.32.

Even though a polygonal region Π is the union of regions corresponding to triangles T_1 to T_n, and the notion of the interior of a triangular region is simple, that of the interior of Π is surprisingly hard to make precise. The section's concluding exercises will force you to confront that difficulty, and will provide a definition you can use later, in chapter 8. For the moment, trust your intuition to answer correctly whether a given point P is interior to Π. The problem is to describe in words how your intuition works.

Exercise 9, Part 1. Is the interior of Π always the union of the interiors of T_1 to T_n? Using $n = 2$, present an example where this is true and a counterexample where it's not. *Suggestion*: Your counterexample will probably involve triangles that intersect along an edge.

Part 2. Does the interior of Π ever consist of the union of the interiors of T_1 to T_n and the points on their common edges? Using $n = 2$, present an example where this is true and a counterexample where it's not.[9] *Suggestion*: Your counterexample will probably involve a shared vertex.

Part 3. Does the interior of Π always consist of the union of the interiors of T_1 to T_n and the points interior to common edges? Using $n = 3$ and triangles T_i that intersect, present an example where this is true and a counterexample where it's not.

---

[8] From the Latin word *luna* for *moon*.

[9] Does it make any difference whether you define *polygonal region* as in section 3.8, note 13, so that the T_i intersect, if at all, only along *entire* edges?

You should be pleased that your intuition works in complicated situations, but troubled that it's hard to describe. You *will* need a precise definition of the interior of a polygonal region Π in a plane ε. You may have noticed that part of the difficulty lies in deciding whether some points are on its boundary. In fact, the problems of defining interior and boundary are equivalent, because the interior of Π should be the set of its nonboundary points, and its boundary should be the set of its noninterior points. Here's a precise definition: a point P lies on the *boundary* of Π just in case

> for every triangle T in ε, if P is interior to T, then there exist points U and V interior to T such that U lies in Π but V does not.

You can use this to define the *interior* of Π precisely, as the set of all nonboundary points in Π.

Exercise 10, Part 1. Show that you could replace the word *triangle* in this definition by *circle* or *square.*

Part 2. Show that every boundary point of Π lies on an edge of one of the triangles T_i.

Part 3. Show that a point P lies in the interior of Π if and only if P is interior to some triangle, all of whose interior points belong to Π.

Part 4. First, a definition: A set Γ of points in a plane ε is called *open* in ε if every point P of Γ is interior to some triangle in ε, all of whose interior points belong to Γ. Part 3 showed that the interior of a polygonal region Π in ε is open in ε. Now show that the boundary of Π contains no set that's open in ε.

Part 5. A set Σ of points in ε is called *closed* in ε if its relative complement—the set of points in ε but *not* in Σ' —is open. Show that Π is closed.

4.8 Exercises on volume

Concepts

- Dissecting volumes
- Prismoids and antiprisms
- Volumes of spheres and spherical caps
- Interior and boundary of a polyhedral region
- Open and closed point sets

Like most exercises in the previous section, exercises 1 and 2 demonstrate *dissection*: splitting a given region into pieces that you can handle with the volume theory developed in sections 3.10 and 3.14. Exercise 3 provides the proof for the last result mentioned in section 3.14: the formula for the volume of a sphere. Exercise 4 considers some volumes related to that of a sphere. Exercise 5 extends to three dimensions the discussion of interior and boundary points and open and closed sets presented for plane geometry in exercises 4.7.9 and 4.7.10.

Exercise 1. Make a polyhedron—called a *prismoid*—with two dissimilar rectangular faces $ABCD$ and $A'B'C'D'$ in parallel planes, so that pairs of corresponding edges lie in parallel lines. The remaining four lateral faces are trapezoids whose bases are corresponding edges of the rectangles. Find its volume in terms of the rectangles' edges a, b and a', b' and the distance h between the base planes. *Suggestion*: If $a, b, a', b', h = 2, 3, 4, 5, 6$ then the volume is 74.

Exercise 2. Make cutouts like the two parts of figure 4.8.1, so that all edges a to h and those of the internal squares have the same length s. Make a polyhedron—called a *square antiprism*—by folding along the dotted lines and taping similarly lettered edges together. Find its volume. *Suggestion*: Dissect the antiprism into several pyramids. If $s = 1$, their total volume is approximately 0.957.

Exercise 3. A plane through a point O divides a sphere with center O and radius r into two hemispheres. On their common circular base, construct a cylinder with altitude r. Using Cavalieri's axiom, show that the volume of the region between the cylinder and one of the hemispheres is the same as that of a certain cone. Apply the volume formulas for the cylinder and cone to derive the formula for the volume of the sphere.

Exercise 4, Part 1. Let $0 \le h < r$. A plane at distance h from O divides the sphere with center O and radius r into two parts. The smaller is called

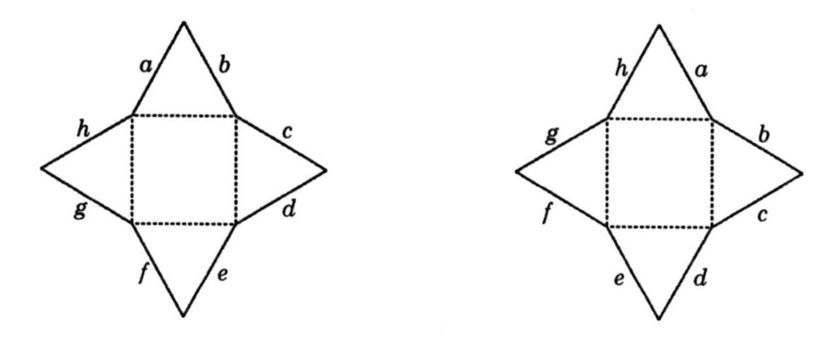

Figure 4.8.1 Antiprism cutouts for exercise 2

a *spherical cap*. Without using calculus, find a formula for its volume. *Suggestion*: If $r = 1$ and $h = 1/2$, the cap's volume is about 0.65.

Part 2. Without using calculus, find the volume of material removed from of a hole with radius $1/2$ drilled through the center of a sphere with radius 1. *Suggestion*: The hole's volume is about 0.47.

Exercise 5. Construct and solve a three-dimensional analog of exercises 4.7.9 and 4.7.10. This must include precise definitions of the interior and boundary of a polyhedral region, and of open and closed sets in three-dimensional space.

4.9 Exercises on circles and spheres, part 2

Concepts

- cos 36°
- Archimedes' approximation to π
- Great circles and antipodal points on a sphere
- Measuring great circular arcs and angles between them
- Spherical triangles
- Trigonometry of right spherical triangles
- Curved area of cylinders, cones, spheres, and spherical triangles

This section contains some specialized problems on circles and spheres. Exercise 1 adds a new angle, 36°, to those for which elementary geometry yields the values of the trigonometric functions. Later exercises work out Archimedes' approximation to π, the beginnings of spherical trigonometry, and some example curved area calculations.

Exercise 1, Part 1. Consider a regular pentagon $ABCDE$ with edge 1. Show that two diagonals $\overline{AC}$ and $\overline{BD}$ meet at the point P for which $APDE$ is a parallelogram. Let $x = PC$. Use similar and congruent triangles to show that

$$x = \frac{1}{1+x}.$$

Compute the value of x. Finally, use x and ΔBCP to compute $\cos 36°$.

Part 2. Consider a regular pentagon, hexagon, and decagon with the same radius. Show that you can make a right triangle from segments congruent to their edges.[10]

[10] Part 2 is proposition 10 of book XIII of Euclid's *Elements* ([1908] 1956).

In exercise 2 you'll carry out Archimedes' approximation[11] to π as described earlier in section 3.14. You'll see that if you can perform the familiar arithmetic operations and calculate square roots, then you can compute π as accurately as you might wish. Exercise 3, a sidelight, shows that our idea of circle geometry and π might be quite different if we considered only the geometry of the surface of the earth.

Exercise 2. Consider regular polygons with n edges of length s_n and t_n, inscribed in and circumscribed about a circle with radius 1. Thus, $\frac{1}{2}ns_n < \pi < \frac{1}{2}nt_n$. Compute s_3 and t_3. Find formulas for t_n and s_{2n} in terms of s_n. Use these formulas to compute $\frac{1}{2}ns_n$ and the error bound $\varepsilon_n = \frac{1}{2}nt_n - \frac{1}{2}ns_n$ for $n = 6, 12, 24, 48$, and 96. What's the resulting approximation of π and how accurate is it guaranteed to be? This is Archimedes' result. How big must n be to guarantee that the error is at most 10^{-6}?

Exercise 3. Consider a sphere Σ with center O and radius R. A plane ε not through O intersects Σ in a circle Γ with center P and radius r. The perpendicular to ε through P meets Σ at a point Q on the side of ε opposite O. Let X be any point on Γ. You could call the length r' of arc $\overset{\frown}{QX}$ the "radius of the circle measured *on* the sphere" and the ratio of the circumference of Γ to $2r'$ as an "approximation to π measured on Σ." This approximation depends on the circle. What distance r' on the earth (a sphere) gives the approximation $\pi \approx 3$?

In exercise 4 you'll work out the beginning of spherical trigonometry. Exercise 8 and some exercises in section 5.11 pursue the subject further. First, you must understand what a spherical triangle is and how you measure it. A *great circle* on a sphere Σ is its intersection with a plane through its center O. Any two points P and Q on Σ that aren't collinear with O determine a great circle Γ. Using calculus, it's shown that the shortest path between these points that lies entirely in Σ is the minor arc they define on Γ. (Two *antipodal* points—distinct points on Σ collinear with O —determine many great circles, and on any of these, either semicircle is a shortest path between them.) Thus, to develop geometry of the sphere analogous to that of the plane, it's useful to make arcs of great circles play the role of line segments. Instead of basing spherical geometry on the *length* of these arcs though, it's simpler to use their *degree measures*. That lets you state many results without mentioning the sphere's radius.

Consider points A, B, and C on Σ that aren't coplanar with O. You can define the *angle* between the intersecting great circle arcs $\overset{\frown}{AB}$ and $\overset{\frown}{BC}$ as the dihedral angle formed by the sides of $\overset{\leftrightarrow}{OB}$ in the planes ABO

[11] See van der Waerden 1963, chapter 7.

and BCO in which the arcs lie. These points also define a *spherical triangle* ΔABC whose edges are the minor arcs $\overset{\frown}{AB}$, $\overset{\frown}{BC}$, and $\overset{\frown}{CA}$ of the great circles. (This is ambiguous if two of these points are antipodal. You must use more detailed language to handle that case.) A spherical triangle ΔABC has six measurements:

$$\begin{array}{lll} \mathrm{m}\angle A & \mathrm{m}\angle B & \mathrm{m}\angle C \\ a = \mathrm{m}\overset{\frown}{BC} & b = \mathrm{m}\overset{\frown}{CA} & c = \mathrm{m}\overset{\frown}{AB}\,. \end{array}$$

Exercise 4 develops equations for these measurements in a *right* spherical triangle with $\mathrm{m}\angle C = 90°$.

Exercise 4. Prove that in this right spherical ΔABC, $\cos a \cos b = \cos c$ and

$$\sin \mathrm{m}\angle A = \frac{\sin a}{\sin c} \qquad \cos \mathrm{m}\angle A = \frac{\cos a \sin b}{\sin c}$$

$$\sin \mathrm{m}\angle B = \frac{\sin b}{\sin c} \qquad \cos \mathrm{m}\angle B = \frac{\cos b \sin a}{\sin c}\,.$$

Construct examples like those in exercise 4.5.4 to show how to find any three of $\mathrm{m}\angle A$, $\mathrm{m}\angle B$, a, b, and c given the remaining two. *Suggestion*: You may find exercise 4.3.4 helpful.

Exercises 5 to 7 develop informally a theory for the curved areas of cylinders, cones, and spheres. No axiomatic system in this book supports this; it's more delicate than the notions discussed in chapter 3. You'll derive various formulas here by means that seem reasonable, some by more than one route. But you'll find no compelling argument that different derivations will *always* yield the same result. A broader, consistent, and more robust theory can be based on calculus.

Exercise 5. You can "unroll" the curved area of a cylinder or cone so it lies flat. Whatever the curved area is, it should be the same as the flat area. Derive formulas for these curved areas in terms of the base radius r and altitude h.

Archimedes reasoned[12] that if he should divide the interior of a sphere with center O and radius r into very many regions like pyramids with apex O but with spherical triangular "bases," then their total volume $\frac{4}{3}\pi r^3$ would be approximately the sum of the volumes of the corresponding triangular pyramids, which is $\frac{1}{3}r$ times their total base area. That area would be approximately the same as the curved area a of the sphere. Thus,

[12] See van der Waerden 1963, Chapter 7.

he argued, $\frac{4}{3}\pi r^3 = \frac{1}{3}ra$, hence $a = 4\pi r^2$, provided the errors in the two approximations both disappear as the number of pyramids increases.

Exercise 6. Present two more arguments for this formula. One, used in elementary texts now, is based on the difference of the volumes of two spheres of nearly equal radii. Archimedes gave the other,[3] which he based on polygonal approximations to a circle and on the curved area formula for a cone.

Exercise 7, Part 1. Derive a formula for the curved area of a cap of a sphere with radius r formed by a plane at distance $h < r$ from its center.

Part 2. Derive a formula for the curved area of the region of a sphere between two parallel planes at distances h and $h + w < r$ from its center.

Part 3. The state of Wyoming, a rectangle on a Mercator map, extends between latitudes 41° N and 45° N and longitudes 104°3′ W and 111°3′ W. What's its area? You may assume the earth is a sphere with radius $r = 3959$ miles.[13]

Part 4. Colorado, also rectangular, borders Wyoming to the south. Its edge arcs have exactly the same degree measures as Wyoming's, but it's shifted 2° eastward. What's its area? Why don't these areas agree exactly with the official figures in a reference book?

Exercise 8. Consider a spherical ΔABC on a sphere Σ with center O and radius r. Planes AOB and AOC intersect the sphere in two great circles that divide it into four regions called *lunes*. The A lune is the one that contains arc $\widehat{BC}$. What's its area, in terms of $m\angle A$ and r? You can handle the B and C lunes similarly. Show that the sum of the areas of these three lunes is half that of the sphere plus twice the area of spherical ΔABC. Derive a formula for the area of the spherical triangle in terms of r and the sum m of the measures of its angles. Finally, prove that $180° < m < 540°$. These inequalities constitute the *spherical triangle sum theorem*.

4.10 Exercises on coordinate geometry

Concepts

Constructing general, robust conceptual tool kits that are easy to use
Analytic geometry tool kit

Exercises 1 to 6 constitute a review of elementary three-dimensional coordinate geometry. Solving them, you'll construct a tool kit of analytic formulas and

[13] $1° = 60'$ = sixty *minutes*; $1' = 60''$ = sixty *seconds*.

methods corresponding to many of the synthetic concepts in chapter 3. Here, phrases such as *point P*, *line g*, and *plane ε* mean *coordinates of P*, *linear parametric equations for g*, and *linear equation* or *linear parametric equations for ε*. Exercise 7 adapts the tools to plane geometry. For this tool kit, you should use the theory and notation summarized in appendix C for vector algebra and systems of linear equations. You may also want to consult a standard analytic geometry text.[14] For each part of these exercises, you should describe the corresponding method in detail and provide enough examples to demonstrate and check it in all relevant cases.

Guidelines for constructing conceptual tool kits have emerged from recent software engineering projects. They apply to mathematical as well as programming practice. Construct a tool kit for a whole field of applications, but independent of any particular one. Make your tools general, so that you can use them for *every* problem you encounter in that field. That way, you'll retain familiarity with them. Employ uniform methods across the entire tool kit, so it's easy to use and document. Write your documentation intending that your tools be used by others with different backgrounds, and by yourself later when you don't remember details of the underlying theory. Finally, develop your tool kit ahead of time, when you're concentrating on its theory and all its details. You won't be distracted by the immediate need for some special case. Having spent this effort in constructing your tools, you'll feel justified to continue with thorough documentation that could prove invaluable later. If you follow these guidelines, you'll create an intellectual tool kit that's easy, effective, and safe to use in a wide variety of applications.

Many tasks of the tool kit you'll construct are best accomplished by turning them into problems about systems of linear equations, usually nonsquare. You should become adept with the elimination techniques described in appendix C for determining whether a system has a solution, for computing a unique solution, and for expressing infinitely many solutions in terms of parameters. Use them in the tool kit.

Standard texts often use vector cross products for some of these tasks. Cross products aren't covered in appendix C. Instead, you should seek methods based on the elimination techniques just mentioned. Cross products are described in detail in section 5.6, and in exercise 5.11.33 you'll be invited to use them to streamline your tools.

Exercise 1. Describe in detail coordinate geometry methods for the following tasks. For each task specify the proper form for the input and for the output, and present an *algorithm*—a computational procedure for hand calculation. It should work for all possible input cases. Provide an example of each case, show how your algorithm handles it, display your output, and check it.

---

[14] For example, Thomas and Finney [1951] 1979, chapter 11.

Task: *given*	*determine or find*
1. points P and Q	the distance PQ;
2. points $P \neq Q$	the line $\vec{PQ}$;
3. points $P \neq Q$ and a scalar $t > 0$	point X on $\vec{PQ}$ with $XQ/PQ = t$;
4. three points	whether they're collinear;
5. three distinct collinear points	which one lies between the others;
6. noncollinear points P, Q, R	point X of parallelogram $PQRX$;
7. four points	whether they're coplanar;
8. noncollinear points P, Q, R	plane PQR.

Suggestion: As a style guide, here's a solution for task 7, the one algorithm among these eight with a traditional formulation that's elegant but not straightforward. It's presented as a problem of determining whether a linear system has a solution; algorithms for other tasks might involve computing such a solution.

Task 7 algorithm. Given points P, Q, R, S as columns of coordinates, append entry -1 to each, to build columns P', Q', R', S' of length four. The given points are coplanar just in case the matrix $[P'\ Q'\ R'\ S']$ with those columns is singular.

Explanation. If the points are coplanar, there's a nonzero vector T and a scalar t_4 such that $T^tX = t_4$ when $X = P, Q, R, S$. Build a row T' of length four by appending entry t_4 to T. Then $T'^tX' = 0$ for these X. Conversely, if these four equations hold for any nonzero row T' of length four, then the first three entries of T' constitute a nonzero row T of length three such that $T^tX = t_4$ for these X. Finally, the four equations $T'^tX' = 0$ are together equivalent to one: $T'^t[P'\ Q'\ R'\ S'] = O$; and that has a nonzero solution just in case this matrix is singular.

Example. The unit points on the three axes are *not* coplanar with the center of the cube they form with the origin, but they *are* coplanar with their centroid $<\frac{1}{3}, \frac{1}{3}, \frac{1}{3}>$, because

$$\det\begin{bmatrix} 1 & 0 & 0 & \frac{1}{2} \\ 0 & 1 & 0 & \frac{1}{2} \\ 0 & 0 & 1 & \frac{1}{2} \\ -1 & -1 & -1 & -1 \end{bmatrix} = -\tfrac{1}{2} \neq 0 \qquad \det\begin{bmatrix} 1 & 0 & 0 & \frac{1}{3} \\ 0 & 1 & 0 & \frac{1}{3} \\ 0 & 0 & 1 & \frac{1}{3} \\ -1 & -1 & -1 & -1 \end{bmatrix} = 0.$$

Exercise 2. Continue exercise 1 for the following tasks. Assume that algorithms are available for all tasks in exercise 1.

Task:	*given*	*determine or find*
9.	lines g and h	whether $g = h$;
10.	lines $g \neq h$	whether they're intersecting, parallel, or noncoplanar;
11.	intersecting lines $g \neq h$	the point $g \cap h$;
12.	point P not on line g	the parallel to g through P;
13.	point P not on line g	the plane through P and g;
14.	intersecting or parallel lines $g \neq h$	their common plane.

Exercise 3. Continue exercises 1 and 2 for the following tasks. Assume that algorithms are available for all previous tasks.

Task:	*given*	*determine or find*
15.	planes δ and ε	whether $\delta = \varepsilon$;
16.	planes $\delta \neq \varepsilon$	whether they intersect;
17.	intersecting planes $\delta \neq \varepsilon$....	the line $\delta \cap \varepsilon$;
18.	a line and a plane	whether they're incident, intersecting, or parallel;
19.	line g intersecting plane ε	point $g \cap \varepsilon$;
20.	point P not on plane ε	the plane $\delta \parallel \varepsilon$ through P;
21.	line g parallel to plane ε ...	the plane $\delta \parallel \varepsilon$ through g.

Exercise 4. Continue exercises 1 to 3 for the following tasks. Assume that algorithms are available for all previous tasks.

Task:	*given*	*determine or find*
22.	intersecting lines $g \neq h$	whether $g \perp h$;
23.	point P not on line g	the line $k \perp g$ through P;
24.	point P on line g in .... plane ε	the line $k \perp g$ through P in ε;
25.	an intersecting line and plane	whether they're $\perp$;
26.	point P and line g	the plane $\varepsilon \perp g$ through P;
27.	point P and plane ε	the line $k \perp \varepsilon$ through P;
28.	planes $\delta \neq \varepsilon$	whether they're $\perp$;
29.	line and plane g, δ with $g \not\perp \delta$	the plane $\varepsilon \perp \delta$ through g;
30.	intersecting lines $g \neq h$	the line $k \perp g, h$;
31.	noncoplanar lines g, h	the line $k \perp g, h$.

Exercise 5. Find coordinate formulas for the distance between

1. a point and a line,
2. two parallel lines,
3. a point and a plane,
4. a parallel line and plane,
5. two parallel planes,
6. two noncoplanar lines.

Exercise 6, Part 1. How do you find a linear equation for a plane ε if you're given a system of linear parametric equations?

Part 2. How do you find such a system if you're given a linear equation for ε?

Exercise 7. Extract from exercises 1 to 6 the material that applies to plane geometry, and present it using two-dimensional analytic formulas.

The last exercise of this chapter investigates an alternate definition for the interior of a tetrahedron. Clearly, a point lies in the interior if it's between, but different from, two points on opposite edges. Is the converse true? The problem is included here because analytic methods may work better than synthetic ones.

Exercise 8. Select two opposite edges of a tetrahedron T. Does every point interior to T lie between points on these edges?

Chapter

5

Some triangle and circle geometry

This chapter's aim is to demonstrate the power of the methods of elementary Euclidean geometry developed in chapter 3. In most contemporary applications of geometry, these techniques have been supplanted by coordinate methods. In past centuries, though, geometers honed them to a high degree of sophistication. The results they attained are beautiful and often amazing. Although the bulk of that material lies outside the mainstream of contemporary mathematics, some results—for example, the trigonometry in sections 5.5 and 5.6—play critical roles in particular fields. Many books are devoted entirely to triangle and circle geometry. The best is *Advanced Euclidean geometry*, by Roger A. Johnson ([1929] 1960). The most readable—by far—is *Geometry revisited*, by H. S. M. Coxeter and Samuel L. Greitzer (1967). The bibliography includes yet other sources.

Chapter 5 presents a sampler of this theory, tailored to include some of the most useful theorems, some of the most beautiful, and one example of a really deep and amazing result.

Advanced Euclidean geometry is organized like a marvelous rug with vivid pictorial designs created from threads interwoven in several directions. You can chart many routes through parts of it by following various strands. Or you can step directly to one picture, examine it in detail, and follow some of these paths to other parts of the subject. Once you start such an excursion, you'll find it hard to stop!

This chapter's tour examines a few neighboring regions of the carpet. You'll see that every one is closely related to some of the others. Some of the theorems were selected because of their importance for other areas of geometry. You've already used Desargues' theorem—section 5.3—to settle a question in perspective drawing in section 1.1, and you've used the trigonometry in section 5.5 in many contexts. The material at the end of the chapter is aimed straight at a wonderful result, Feuerbach's theorem about five circles related to a triangle. This tour de force shows the depth you can achieve with these methods.

En route you'll encounter more strands of advanced Euclidean geometry that you could follow to many fascinating parts of the subject. To limit the excursions, these strands are generally abandoned quickly. The exercises in section 5.11 will pursue some of them. Return to others as you please in later years, and follow them into the vast literature of this fascinating part of geometry!

5.1 Four concurrence theorems

Concepts

- Standard triangle terminology and notation
- Edge bisectors of ΔABC
- Circumcircle, circumcenter O, and circumradius R
- Angle bisectors
- Incircle, incenter I, and inradius r
- Medians and centroid G
- Altitudes and orthocenter H

The theorems in this section often constitute the most intricate part of an elementary geometry text. They're related to almost all advanced work in triangle and circle geometry.

For triangle ΔABC as in figure 5.1.1, let A', B', and C' denote the midpoints of the edges a, b, and c opposite vertices A, B, and C. That notation is used throughout this chapter. The perpendicular bisectors of a and b meet at a point O; if they coincided or were parallel, the transitivity theorem for parallel lines and the alternate interior angles theorem would imply $a \parallel b$. By the perpendicular bisector theorem, O is equidistant from

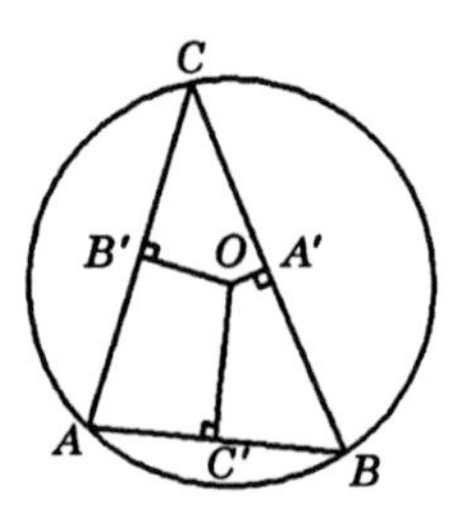

Figure 5.1.1 ΔABC, its circumcenter O, and its circumcircle

B and C and from C and A, hence from A and B, so it lies on the perpendicular bisector of c. This proves

***Theorem 1* (*Edge bisectors theorem*).** The perpendicular bisectors of the edges of ΔABC meet at a point O equidistant from its vertices.

O is called the *circumcenter* of ΔABC. The circle with center O through A, B, and C is the *circumcircle* of ΔABC. Its radius R is the *circumradius*.

***Theorem 2*.** Any three noncollinear points lie on a unique circle.

Proof. Theorem 1 shows that the circle exists. To prove its uniqueness, note that the center of any circle through the vertices of a triangle is equidistant from them, hence must lie on the edge bisectors, so must coincide with the circumcenter. ♦

The *bisector* of angle $\angle POQ$ is the line g through its vertex O such that, if R is a point on g inside $\angle POQ$, then $\angle POR \cong \angle ROQ$. An argument suggested by figure 5.1.2 shows that the bisector consists of O and all points equidistant from lines $\overleftrightarrow{OP}$ and $\overleftrightarrow{OQ}$ that are interior to $\angle POQ$ or its vertical counterpart.

Now consider the bisectors of the angles of ΔABC, as in figure 5.1.3. Since the ends of edge a lie on different rays of $\angle A$, that angle's bisector meets a at some point X. For the same reason, the bisector of $\angle C$ meets segment $\overline{AX}$. The bisectors' intersection I is equidistant from edge lines $\overleftrightarrow{AB}$ and $\overleftrightarrow{CA}$ and from $\overleftrightarrow{CA}$ and $\overleftrightarrow{BC}$, hence from $\overleftrightarrow{BC}$ and $\overleftrightarrow{AB}$, so it lies on the bisector of $\angle B$. This proves

***Theorem 3* (*Angle bisectors theorem*).** The angle bisectors of ΔABC meet at an interior point I equidistant from the edge lines.

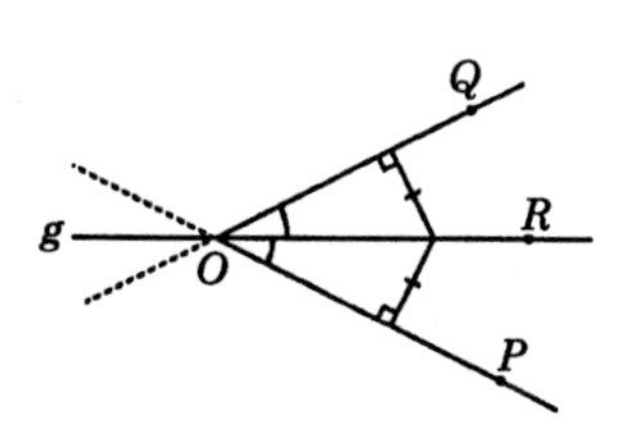

Figure 5.1.2
Bisector of $\angle POQ$

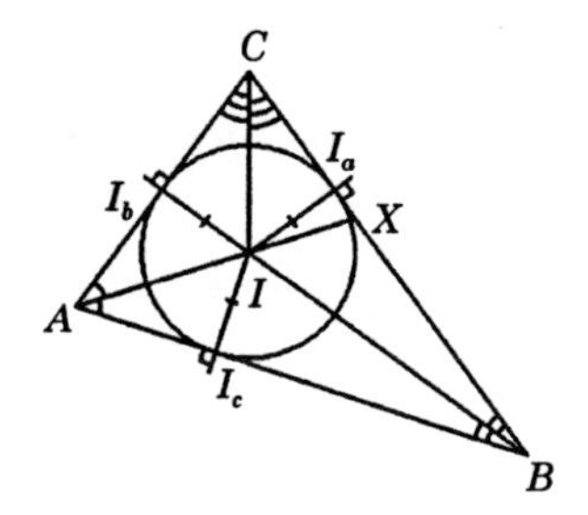

Figure 5.1.3 ΔABC, its incenter I, and its incircle

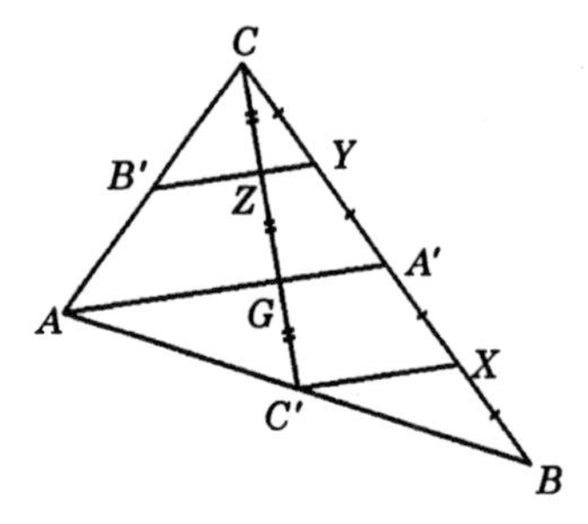

Figure 5.1.4 Proof of lemma 4: ΔABC, two medians, and centroid G

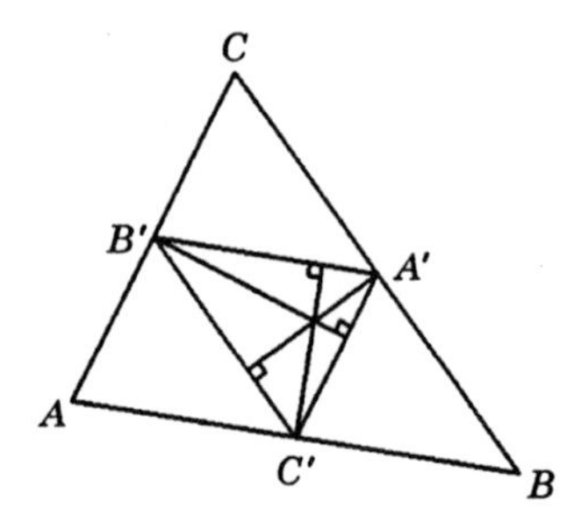

Figure 5.1.5 ΔABC and its medial triangle $\Delta A'B'C'$

I is called the *incenter* of ΔABC. The common distance r from I to the edge lines is the triangle's *inradius*. The circle with center I and radius r is the *incircle*. It's tangent to the edge lines at points I_a, I_b, and I_c within edges a, b, and c. You can show easily that the incircle is the only circle tangent to all three lines at points interior to the edges. Three more circles are tangent to these lines, however, at points outside the edges. They're studied in detail in section 5.9.

Segments $\overline{AA'}$, $\overline{BB'}$, and $\overline{CC'}$ in ΔABC are called its *medians*.

Lemma 4. $\overline{AA'}$ and $\overline{CC'}$ meet at a point G two thirds of the way from C to C'.

Proof. As shown in figure 5.1.4, C and C' lie on different legs of $\angle A$ and $\overrightarrow{AA'}$ lies in the interior of $\angle A$, so $\overline{AA'}$ and $\overline{CC'}$ meet. Let X and Y be the midpoints of $\overline{BA'}$ and $\overline{CA'}$. By the same argument, $\overline{YB'}$ and $\overline{CC'}$ meet at a point Z. Then $\overleftrightarrow{YB'} \parallel \overleftrightarrow{AA'}$ since Y and B' are edge midpoints of $\Delta AA'C$. Similarly, $\overleftrightarrow{AA'} \parallel \overleftrightarrow{XC'}$, hence $CZ = ZG = GC'$. ♦

Theorem 5 (Medians theorem). The medians of ΔABC meet at a point G two thirds of the way from any vertex to the midpoint of the opposite side.

Proof. Apply lemma 4 to ΔBAC: $\overline{BB'}$ and $\overline{CC'}$ meet at the same point G as $\overline{AA'}$ and $\overline{CC'}$. ♦

G is called the *centroid* of ΔABC. $\Delta A'B'C'$ is its *medial* triangle. The following properties of these triangles, shown in figure 5.1.5, are easy to verify.

Theorem 6. The edges of the medial triangle are parallel to those of ΔABC and half their length. Also,

$$\Delta ABC \sim \Delta A'B'C' \cong \Delta AB'C' \cong \Delta A'BC' \cong \Delta A'B'C.$$

The medians of ΔABC contain those of the medial triangle, hence these triangles have the same centroid. The edge bisectors of ΔABC contain the altitudes of the medial triangle.

The feet of the altitudes of ΔABC through vertices A, B, and C are denoted by D, E, and F. From a given ΔABC it's simple to construct a new triangle whose medial triangle is ΔABC. Since the altitudes of ΔABC lie in the edge bisectors of the new triangle, the altitude lines are concurrent. This proves

***Corollary 7* (*Altitudes theorem*).** The altitude lines of triangle ΔABC meet at a point H.

H is called the *orthocenter*[1] of ΔABC. It's shown in figure 5.1.6.

***Corollary 8*.** The circumcenter of a triangle is the orthocenter of its medial triangle.

A set of lines that pass through a single point is called *concurrent*.[2] Euclid described the concurrence of the edge bisectors and angle bisectors of a triangle.[3] He didn't mention concurrence of altitudes or medians. But the altitudes' concurrence is so closely related to that of the angle bisectors that it must have been familiar then. The medians' concurrence is included as a very elementary result in Archimedes' treatise "On plane equilibria."[4] The terminology *circumcircle*, *inradius*, etc., was introduced in the late

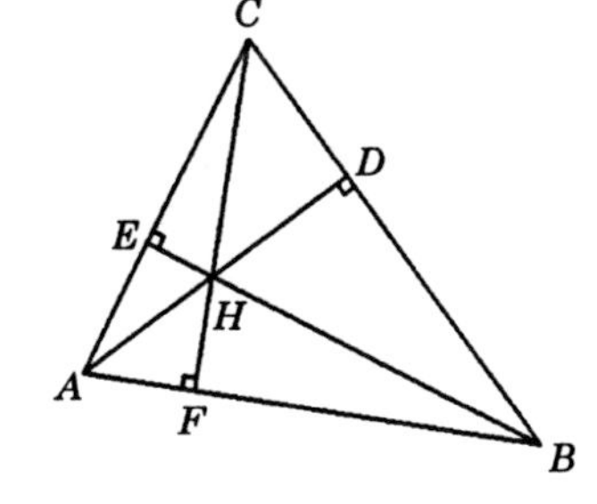

Figure 5.1.6 ΔABC and its orthocenter H

1 The prefix *ortho-* occurs in many areas of mathematics. It stems from a Greek word for *upright*. Here that refers to the altitudes.

2 From the Latin verb *curro* and prefix *con-*, which mean *run* and *together*.

3 Euclid [1908] 1956, book IV, propositions 4 and 5.

4 Archimedes [1912] n.d., book I, proposition 14.

nineteenth century.[5] At that time the following notation for points and distances related to ΔABC was standardized:

$A\ B\ C$ .... vertices,
$a\ b\ c$ ..... opposite edges,
$A'\ B'\ C'$ ... midpoints of opposite edges,
$D\ E\ F$ .... feet of corresponding altitudes,
$I_a\ I_b\ I_c$ .... opposite incircle tangency points,
$G\ H$ ...... centroid, orthocenter,
$O\ I$ .... circumcenter, incenter,
$R\ r$ .... circumradius, inradius.

Further standard notation is introduced later in this chapter.

5.2 Menelaus' theorem

Concepts
Directed distances
Menelaus' product
Menelaus' theorem

Menelaus' theorem is included in this chapter for two reasons. In the next section, it leads to Desargues' theorem, which was used in section 1.1 to settle a question in perspective drawing. In section 5.4, Menelaus yields Ceva's theorem, a generalization of three of the four concurrence theorems that started this chapter.

Menelaus' and several related theorems involve both the order of points on various lines and some ratios of distances between them. To consider these simultaneously, it's convenient to use coordinate geometry on one line g at a time. Give g a scale c. In place of the distance $PQ = |c(P) - c(Q)|$ between points P and Q on g consider their *directed distance*

$$P \text{ to } Q = c(Q) - c(P),$$

so that

$$P \text{ to } Q = -(Q \text{ to } P) \qquad PQ = |P \text{ to } Q|.$$

Although the scale c is suppressed in this notation, you need it to compute the directed distance. Most authors use the same symbol PQ for directed and undirected distances; you must rely on the context to distinguish the concepts. However, directed distances see only limited use in this text, and

[5] See W. H. H. H. 1883.

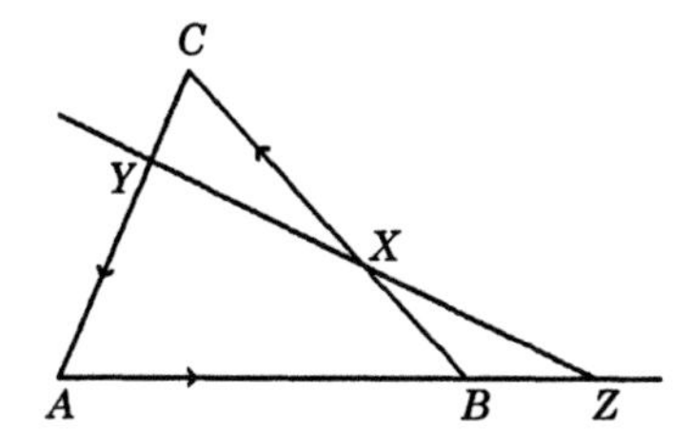

Figure 5.2.1
Menelaus' theorem

it seems worthwhile to make them visibly different. In a diagram, you can indicate the use of an unnamed scale for a line g by drawing an arrow from some point on g to a point with a larger coordinate, like this: $\longrightarrow$.

Directed distances usually appear by twos, in ratios. That explains why the scale is suppressed from the notation: For any points P, Q, R, and S on g with $R \neq S$, the ratio $(P \text{ to } Q)/(R \text{ to } S)$ doesn't depend on the scale. That is, if d is another scale for g, then

$$\frac{P \text{ to } Q}{R \text{ to } S} = \frac{c(Q) - c(P)}{c(S) - c(R)} = \frac{d(Q) - d(P)}{d(S) - d(R)}.$$

It's easy to prove that; see exercise 4.1.7. Signs of ratios like this are often used to convey information about the order of points on g. For example, if P, Q, and R lie on g and $Q \neq R$, then

$$\frac{P \text{ to } Q}{Q \text{ to } R} \geq 0 \leftrightarrow P\text{-}Q\text{-}R.$$

Menelaus' theorem involves a product of ratios of directed distances between points on the edge lines of a triangle ΔABC. Assign scales to lines $\overleftrightarrow{BC}$, $\overleftrightarrow{CA}$, and $\overleftrightarrow{AB}$, as indicated by the arrows in figure 5.2.1. Then select points X to Z on these lines but distinct from the vertices. The product

$$\frac{A \text{ to } Z}{Z \text{ to } B} \, \frac{B \text{ to } X}{X \text{ to } C} \, \frac{C \text{ to } Y}{Y \text{ to } A}$$

of ratios of two directed distances on each line is called *Menelaus' product*. It doesn't depend on the choice of scales. Its sign gives information about how many of the points X, Y, and Z may lie between the vertices. The product is positive if all three lie within the edges or just one does, but negative if just two lie within the edges.

***Theorem 1* (*Menelaus' theorem*).** X, Y, and Z are collinear if and only if Menelaus' product equals -1.

Proof. Suppose X, Y, and Z lie on a line g, as in figure 5.2.2. Find a line $h \neq \overleftrightarrow{AB}, \overleftrightarrow{CA}$ through A not parallel to $\overleftrightarrow{BC}$ nor g. Construct lines j and k parallel to h through B and C. Then h, j, and k intersect

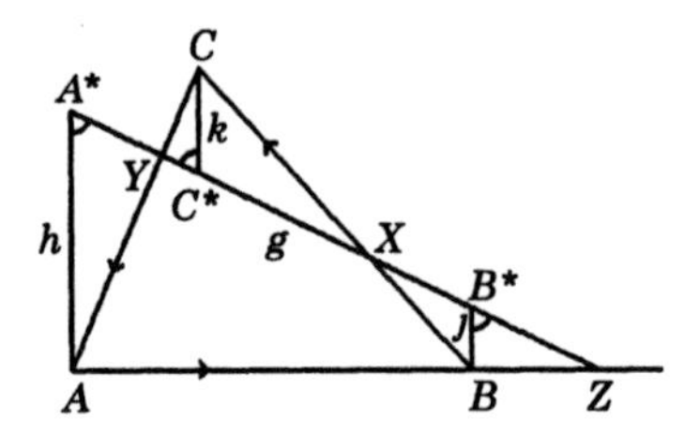

Figure 5.2.2
Proof of Menelaus' theorem

g at points A^*, B^*, and C^* that form three pairs of similar triangles with associated ratios as follows:

$$\Delta AZA^* \sim \Delta BZB^* \qquad \Delta BXB^* \sim \Delta CXC^* \qquad \Delta CYC^* \sim \Delta AYA^*$$

$$\frac{A \text{ to } Z}{Z \text{ to } B} = -\frac{AA^*}{BB^*} \qquad \frac{B \text{ to } X}{X \text{ to } C} = \frac{BB^*}{CC^*} \qquad \frac{C \text{ to } Y}{Y \text{ to } A} = \frac{CC^*}{AA^*}.$$

These equations imply

$$\frac{A \text{ to } Z}{Z \text{ to } B}\,\frac{B \text{ to } X}{X \text{ to } C}\,\frac{C \text{ to } Y}{Y \text{ to } A} = -1.$$

This computation depends on the positions of X, Y, and Z. Since they're collinear, they can't all lie within the edges, nor can just one of them. Thus, just one or all of them must lie outside ΔABC. Figure 5.2.2 shows the case when just one—namely, Z —lies outside. If all three lay outside, then all three ratios of directed distances would be negative, and their product would still equal -1.

Conversely, suppose Menelaus' product equals -1. Then all three or just one of X, Y, and Z must lie outside ΔABC. You can assume that X and Y both lie within the edges, as in figure 5.2.2, or both lie outside. Let $g = \overleftrightarrow{XY}$. If $g \parallel c$, then

$$\frac{B \text{ to } X}{X \text{ to } C} = \frac{Y \text{ to } A}{C \text{ to } Y}$$

Mathematics flourished at Alexandria for many centuries after Euclid, waning gradually with the rise of Christianity and its antipathy toward Greek culture. MENELAUS worked there around A.D. 100. Principally an astronomer, he studied the precession of the equinoxes and other phenomena. To further that work he founded the subject of spherical trigonometry with a book, *Sphaerica*. Menelaus authored several other texts on geometry and astronomy as well. This section's theorem carrying his name is related to his work in spherical trigonometry, but here it's used as a key to some important theorems in plane geometry.

hence

$$\frac{A \text{ to } Z}{Z \text{ to } B} = -1,$$

and it would follow that A to $Z = B$ to Z, hence $A = B$ —contradiction! Therefore, g must intersect c at a point $Z' \neq A, B$. By the previous paragraph,

$$\frac{A \text{ to } Z'}{Z' \text{ to } B} \frac{B \text{ to } X}{X \text{ to } C} \frac{C \text{ to } Y}{Y \text{ to } A} = -1,$$

hence

$$\frac{A \text{ to } Z'}{Z' \text{ to } B} = \frac{A \text{ to } Z}{Z \text{ to } B}.$$

This equation clearly implies $Z = Z'$, so Z also falls on g. ♦

5.3 Desargues' theorem

Concepts
- Pencils of lines
- Euclidean forms of Desargues' theorem
- Does Desargues' theorem require metric notions?

Section 1.1 used a form of Desargues' theorem to solve a fundamental problem in descriptive geometry. Desargues' theorem underlies various fundamental techniques in that field; in fact, it was originally discovered in that context.[6] There are several forms of the theorem in Euclidean geometry. They can all be interpreted as specific instances of a single result in projective geometry, the theory that was developed to serve as a basis for the techniques of perspective drawing. That unified form of the theorem plays a central role in the modern axiomatic development of projective geometry.

In this section, however, Desargues' theorem is considered only in the context of Euclidean geometry, hence some forms must be treated separately. Some of the distinctions have to do with situations where three lines are parallel or have a common intersection. To simplify the terminology, the families of all lines parallel to a given line and of all lines through a given point are called parallel and concurrent *pencils*.[7] A set of lines is called *copencilar* if they all belong to the same pencil. Several styles of proof are

---

[6] The original publication is available in translation: Desargues 1648.

[7] This usage stems from the Latin word *penicillus* for *brush*.

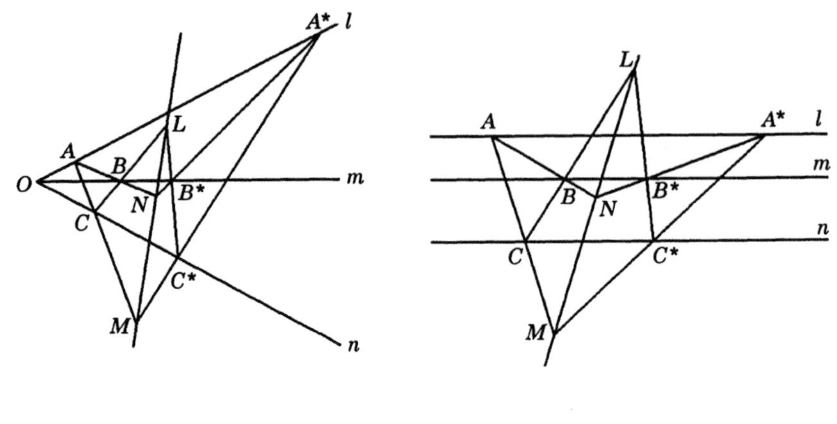

Figure 5.3.1 First form of Desargues' theorem

useful for various forms of Desargues' theorem. The first form considered here can be proved neatly from Menelaus' theorem.

Theorem 1 (First form of Desargues' theorem and its converse). Consider coplanar triangles ΔABC and $\Delta A^*B^*C^*$ as in figure 5.3.1, with three distinct lines l, m, and n and three distinct points L, M, and N such that

$$\begin{array}{l} A \text{ and } A^* \text{ lie on } l \text{ but not on } m \text{ nor } n, \\ B \text{ and } B^* \text{ lie on } m \text{ but not on } n \text{ nor } l, \\ C \text{ and } C^* \text{ lie on } n \text{ but not on } l \text{ nor } m, \end{array}$$

$$L = \overleftrightarrow{BC} \cap \overleftrightarrow{B^*C^*}, \quad M = \overleftrightarrow{CA} \cap \overleftrightarrow{C^*A^*}, \quad N = \overleftrightarrow{AB} \cap \overleftrightarrow{A^*B^*}.$$

Then l, m, and n are copencilar if and only if L, M, and N are collinear.

Proof. The hypotheses imply $A \neq A^*$, $B \neq B^*$, and $C \neq C^*$. In order to use directed distances, assign scales to all lines under consideration. Suppose l, m, and n all pass through a point O. Apply Menelaus' theorem first to the collinear points L, B^*, and C^* on the edge lines of ΔOBC to get

$$\frac{O \text{ to } B^*}{B^* \text{ to } B} \, \frac{B \text{ to } L}{L \text{ to } C} \, \frac{C \text{ to } C^*}{C^* \text{ to } O} = -1,$$

then to the collinear points M, C^*, and A^* on the edge lines of ΔOAC to get

$$\frac{C \text{ to } M}{M \text{ to } A} \, \frac{A \text{ to } A^*}{A^* \text{ to } O} \, \frac{O \text{ to } C^*}{C^* \text{ to } C} = -1,$$

Gérard DESARGUES was born in 1593 in Lyon, the son of a minor government official. Nothing is known of his education. By 1626 he'd moved to Paris and developed a reputation as a mathematician and engineer, particularly interested in problems arising in graphic arts and architecture. There he met Descartes. They became lifelong friends, and influenced each other's mathematical work. Richelieu appointed Desargues an army engineer officer in 1628. He assisted in the design of the fortifications and dikes at La Rochelle. During the late 1620s and 1630s, Desargues researched and gave mathematics lectures in Paris. His principal interest was the analysis and application of perspective techniques, and he pioneered the projective theory of conic sections. His work was published in the late 1630s and 1640s, but in scattered fashion, and sometimes only as appendices to his students' works. His style was novel and overly concise. For these reasons, and because of the great attention given to the 1637 publication of Descartes' elements of analytic geometry, Desargues' work received little note during his lifetime and in fact for the next two hundred years. His student Blaise Pascal, however, achieved fame with a treatise on conic sections based on Desargues' work. Desargues continued to practice engineering and architecture until he retired to his country estate near Lyon in 1650. He died there in 1662.

and to the collinear points N, A^*, and B^* on the edge lines of ΔOAB to get

$$\frac{O \text{ to } A^*}{A^* \text{ to } A} \; \frac{A \text{ to } N}{N \text{ to } B} \; \frac{B \text{ to } B^*}{B^* \text{ to } O} = -1 .$$

Now multiply these three equations and simplify to get

$$\frac{A \text{ to } N}{N \text{ to } B} \; \frac{B \text{ to } L}{L \text{ to } C} \; \frac{C \text{ to } M}{M \text{ to } A} = -1 .$$

By Menelaus' theorem, points N, L, and M on the edge lines of ΔABC are collinear. The proof that concurrence of l, m, and n implies collinearity of L, M, and N is complete.

Next, suppose that L, M, and N are collinear but l, m, and n aren't all parallel. Then two of these lines, say n and l, intersect at a point O. Collinearity of B^*, O, and B follows by the previous paragraph's argument with

A, B, C, A^*, B^*, C^*, L, M, N, and O replaced by
C, L, C^*, A, N, A^*, B^*, O, B, and M.

That is, O lies on m as well. Thus, collinearity of L, M, and N implies copencilarity of l, m, and n.

Finally, suppose that l, m, and n are parallel. Considering similar triangles, you get

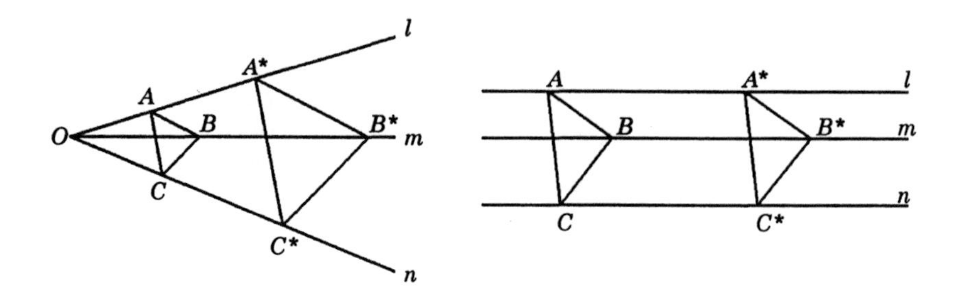

Concurrent case Parallel case

Figure 5.3.2 Second form of Desargues' theorem

$$\frac{A \text{ to } N}{N \text{ to } B}\,\frac{B \text{ to } L}{L \text{ to } C}\,\frac{C \text{ to } M}{M \text{ to } A} = \left[-\frac{A \text{ to } A^*}{B \text{ to } B^*}\right]\left[-\frac{B \text{ to } B^*}{C \text{ to } C^*}\right]\left[-\frac{C \text{ to } C^*}{A \text{ to } A^*}\right] = -1.$$

By Menelaus' theorem, points N, L, and M on the edge lines of ΔABC are collinear. ♦

In the second form of Desargues' theorem, ΔABC and $\Delta A^*B^*C^*$ have parallel edges. You can supply the proof by considering similar triangles.

Theorem 2 **(*Second form of Desargues' theorem and its converse*).** Consider coplanar triangles ΔABC and $\Delta A^*B^*C^*$ as in figure 5.3.2, with three distinct lines l, m, and n such that

A and A^* lie on l but not on m nor n,
B and B^* lie on m but not on n nor l,
C and C^* lie on n but not on l nor m,

$\overleftrightarrow{AB} \parallel \overleftrightarrow{A^*B^*}$, $\overleftrightarrow{BC} \parallel \overleftrightarrow{B^*C^*}$.

Then l, m, and n are copencilar if and only if $\overleftrightarrow{CA} \parallel \overleftrightarrow{C^*A^*}$.

The statements of Desargues' theorem involve only incidence notions (collinearity, concurrence, parallelism, etc.), but the proofs given and suggested above involve metric notions (directed distances, similar triangles, etc.). When the figures are three-dimensional, however, as in the following form of the theorem, you can avoid metric techniques.

Theorem 3 **(*Third form of Desargues' theorem and its converse*).** Consider noncoplanar triangles ΔABC and $\Delta A^*B^*C^*$ with three distinct lines l, m, and n and three distinct points L, M, and N such that

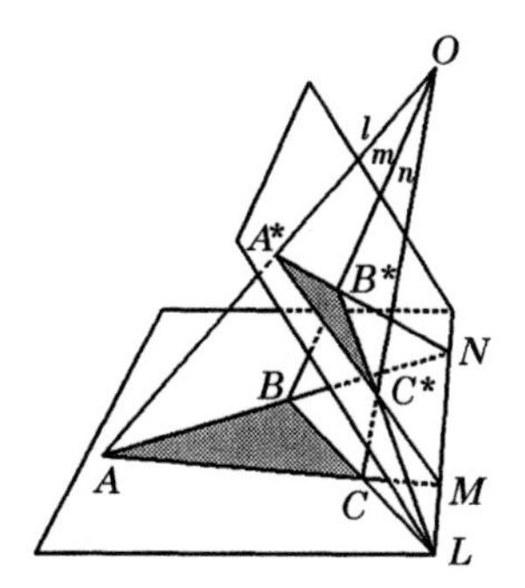

Figure 5.3.3 Third form of Desargues' theorem: noncoplanar case

A and A^* lie on l but not on m nor n,
B and B^* lie on m but not on n nor l,
C and C^* lie on n but not on l nor m,

$$L = \overleftrightarrow{BC} \cap \overleftrightarrow{B^*C^*},\ M = \overleftrightarrow{CA} \cap \overleftrightarrow{C^*A^*},\ N = \overleftrightarrow{AB} \cap \overleftrightarrow{A^*B^*}.$$

Then l, m, and n are copencilar if and only if L, M, and N are collinear.

Proof. If l, m, and n are copencilar, then L, M, and N are collinear because they all lie in the planes of both triangles. See figure 5.3.3 for the concurrent case; you may draw a figure for the parallel case. You may also supply the proof that collinearity of L, M, and N implies copencilarity of l, m, and n. ♦

In 1907, Alfred North Whitehead showed how to use a three-dimensional construction to prove the coplanar case of Desargues' theorem—theorem 1—without using metric concepts.[8] Suppose ΔABC and $\Delta A^*B^*C^*$ lie in the same plane ε, and l, m, and n intersect in a common point O, as in figure 5.3.4. (To enhance the figure's legibility, lines $\overleftrightarrow{CA}$ and $\overleftrightarrow{C^*A^*}$ were omitted.) There's no loss in assuming that A lies between O and A^* as in the figure—otherwise, just relabel the triangles. Select a point P not on ε and a point R between but different from O and P. The plane $\delta = LMR$ intersects $\overleftrightarrow{PA}$, $\overleftrightarrow{PB}$, and $\overleftrightarrow{PC}$ at points A'', B'', and C'' between but different from P and A, B, and C. Lines $\overleftrightarrow{A^*A''}$ and $\overleftrightarrow{OP}$ intersect at a point Q between but different from O and P. Several triples of points are collinear because they lie in two distinct planes:

B'', C'', L	in	δ	and	BCP
C'', A'', M	in	δ	and	CAP
C^*, C'', Q	in	$A^*A''M$	and	C^*OP
B^*, B'', Q	in	$C^*C''L$	and	B^*OP.

[8] Whitehead [1907] 1971, 16–17.

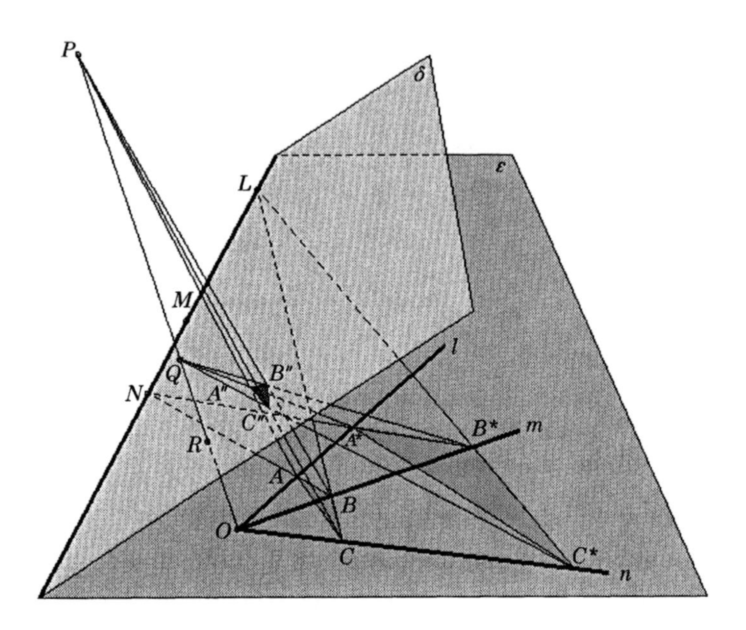

Figure 5.3.4
Third form of Desargues' theorem: coplanar case

By the proof of noncoplanar case of theorem 3, lines $\overleftrightarrow{AB}$ and $\overleftrightarrow{A''B''}$ must be parallel or intersect in a point on $\overleftrightarrow{LM}$, and lines $\overleftrightarrow{A^*B^*}$ and $\overleftrightarrow{A''B''}$ must be parallel or intersect in a point on $\overleftrightarrow{LM}$. One of $\overleftrightarrow{AB} \parallel \overleftrightarrow{A''B''}$ and $\overleftrightarrow{A^*B^*} \parallel \overleftrightarrow{A''B''}$ must fail because $\overleftrightarrow{AB}$ and $\overleftrightarrow{A^*B^*}$ aren't parallel. Assume $\overleftrightarrow{AB}$ and $\overleftrightarrow{A''B''}$ intersect at a point N^* on $\overleftrightarrow{LM}$; you can handle the other possibility similarly. Points A^*, B^*, and N^* are collinear because they all lie in the distinct planes δ and $A''B''Q$. This implies $N = N^*$, hence N lies in $\overleftrightarrow{LM}$. The proof that concurrence of l, m, and n implies collinearity of L, M, and N is complete. You may supply a similar nonmetric proof that L, M, and N are collinear if l, m, and n are parallel. The proof of theorem 1 included a nonmetric argument that in this planar situation, collinearity of L, M, and N implies copencilarity of l, m, and n.

Several years before Whitehead's work, David Hilbert (1899, section 23) had already shown that *every* proof of the plane Desargues theorem requires either metric concepts or a three-dimensional construction.

Alfred North WHITEHEAD was born in 1861 in Ramsgate. His father was a schoolteacher and Anglican clergyman. Alfred's brother became a bishop. In grammar school, Alfred excelled in sports and mathematics. In 1880, he entered Trinity College, Cambridge; all his courses there were in mathematics. Upon graduation he became a fellow of the college, and soon was appointed senior lecturer in mathematics. In 1890 he married Evelyn Wade; they had three children.

Whitehead's early mathematical work had to do with group theory and algebra. He published a book *Universal algebra* that developed in a roundabout way material that would later be included in the theories of linear and multilinear algebra. Around 1900 he began working more in foundations of mathematics, and published two small works on axiom systems for projective and descriptive geometry.

During this period at Cambridge he guided the work of Bertrand Russell as a student, including Russell's dissertation on foundations of geometry. (Whitehead's father had been the Russells' vicar, and in that capacity had been called on to convince the young boy that the earth was round!) Early in the new century they began collaboration on *Principia mathematica*, a monumental work that attempted to demonstrate the *logistic thesis*, that all mathematics can be derived from logical principles alone. A major achievement in mathematical logic, the book provided a framework for much significant research, although Gödel showed later that logicism was inadequate as a philosophy of mathematics. (No system like *Principia* can entail all true arithmetic statements but no false ones.)

By 1910, Whitehead's interests had turned almost entirely to philosophy of science, and he moved to the University of London. During his London period he produced three books on foundations of physics, especially relativity theory. Influenced by the tragedies of World War I, including the death of his son, an aviator, he turned even more to philosophical studies. He crossed the Atlantic in 1924 to become professor of philosophy at Harvard. There he wrote a series of works on philosophy of science, education, and religion, finally retiring in 1937 at age 76. Whitehead died in 1947.

5.4 Ceva's theorem

Concepts

Ceva's theorem
Deriving the medians, altitudes, and angle bisectors theorems
Proving Ceva's theorem by mechanics

A generalization of three of the four concurrence theorems in section 5.1 was found about nineteen centuries after them. It's phrased in terms of Menelaus' product:

***Theorem 1* (*Ceva's theorem*).** In order to use directed distances, assign scales to the edge lines of ΔABC. If points X, Y, and Z lie on these lines and are distinct from the vertices, then

$$\frac{A \text{ to } Z}{Z \text{ to } B}\,\frac{B \text{ to } X}{X \text{ to } C}\,\frac{C \text{ to } Y}{Y \text{ to } A} = 1$$

if and only if lines $l = \overleftrightarrow{AX}$, $m = \overleftrightarrow{BY}$, and $n = \overleftrightarrow{CZ}$ are copencilar.

Proof. Each of these statements will be proved in turn:

i. if X, Y, and Z lie within the edges, then l, m, and n aren't all parallel; if they concur, then Menelaus' product equals 1;

ii. if X, Y, and Z lie within the edges and Menelaus' product equals 1, then l, m, and n concur;

iii. if X and Y lie outside the edges and Z within, and l, m, and n concur, then Menelaus' product equals 1;

iv. if X and Y lie outside the edges and Z within, and l, m, and n are parallel, then Menelaus' product equals 1;

v. if X and Y lie outside the edges and Z within, and Menelaus' product equals 1, then l, m, and n are copencilar;

vi. if just one or all three of X, Y, and Z lie within the edges, then Menelaus' product is negative and l, m, and n aren't copencilar.

Parts (*iii*) to (*v*) cover all cases where exactly one of X, Y, and Z lies within an edge of ΔABC; if X or Y instead of Z lies inside, just relabel the triangle.

Part (*i*). Suppose X, Y, and Z lie within the edges, as in figure 5.4.1. You can easily see that l, m, and n can't be all parallel. Suppose they meet at a point W. Then W lies inside the triangle. Apply Menelaus' theorem to the collinear points Z, C, and W on the edge lines of ΔABX:

$$\frac{A \text{ to } Z}{Z \text{ to } B}\,\frac{B \text{ to } C}{C \text{ to } X}\,\frac{X \text{ to } W}{W \text{ to } A} = -1.$$

Apply it again to the collinear points W, B, and Y on the edge lines of ΔAXC:

$$\frac{A \text{ to } W}{W \text{ to } X}\,\frac{X \text{ to } B}{B \text{ to } C}\,\frac{C \text{ to } Y}{Y \text{ to } A} = -1.$$

Multiply these equations to get

$$\frac{A \text{ to } Z}{Z \text{ to } B}\,\frac{B \text{ to } X}{X \text{ to } C}\,\frac{C \text{ to } Y}{Y \text{ to } A} =$$

Figure 5.4.1 Ceva's theorem, parts (i) and (ii)

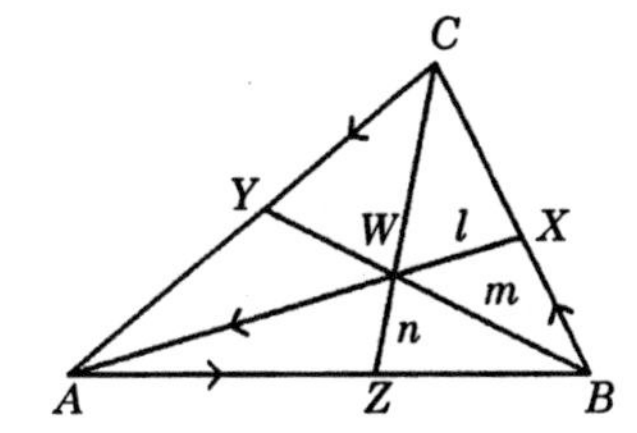

$$\left[\frac{A \text{ to } Z}{Z \text{ to } B}\,\frac{B \text{ to } C}{C \text{ to } X}\,\frac{X \text{ to } W}{W \text{ to } A}\right]\left[\frac{A \text{ to } W}{W \text{ to } X}\,\frac{X \text{ to } B}{B \text{ to } C}\,\frac{C \text{ to } Y}{Y \text{ to } A}\right] = (-1)(-1) = 1.$$

Part (ii). Suppose X, Y, and Z lie within the edges, as in figure 5.4.1. Then lines l and m meet at a point W inside the triangle, and line $\overleftrightarrow{CW}$ meets edge $\overline{AB}$ at a point Z'. By part (i),

$$\frac{A \text{ to } Z'}{Z' \text{ to } B}\,\frac{B \text{ to } X}{X \text{ to } C}\,\frac{C \text{ to } Y}{Y \text{ to } A} = 1.$$

If Menelaus' product equals 1, i.e.

$$\frac{A \text{ to } Z}{Z \text{ to } B}\,\frac{B \text{ to } X}{X \text{ to } C}\,\frac{C \text{ to } Y}{Y \text{ to } A} = 1,$$

then

$$\frac{A \text{ to } Z}{Z \text{ to } B} = \frac{A \text{ to } Z'}{Z' \text{ to } B}.$$

This implies $Z = Z'$, so W lies on n as well as l and m.

Part (iii). Suppose X and Y lie outside the edges and Z within, and l, m, and n meet at a point W, as in figure 5.4.2. Then W lies outside the triangle. The proof for part (i) is valid in this case as well.

Part (iv). Suppose X and Y lie outside the edges and Z within, and l, m, and n are parallel, as in figure 5.4.3. Reasoning with similar triangles, you get

$$\frac{A \text{ to } Z}{Z \text{ to } B}\,\frac{B \text{ to } X}{X \text{ to } C}\,\frac{C \text{ to } Y}{Y \text{ to } A} = \frac{A \text{ to } Z}{Z \text{ to } B}\,\frac{B \text{ to } A}{A \text{ to } Z}\,\frac{Z \text{ to } B}{B \text{ to } A} = 1.$$

Part (v). Suppose X and Y lie outside the edges and Z within, and l, m, and n aren't all parallel. Then at least two of these lines meet; suppose l meets m at a point W. Then W lies outside the triangle, and the lines $\overleftrightarrow{CW}$ and $\overleftrightarrow{AB}$ meet at a point Z'. Proceed as in part (ii) to show that $Z = Z'$, hence W lies on n. The arguments for l or m meeting n are similar.

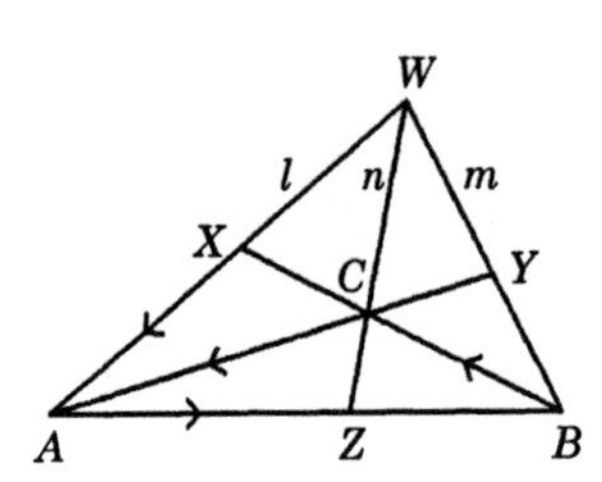

Figure 5.4.2 Ceva's theorem, part (*iii*)

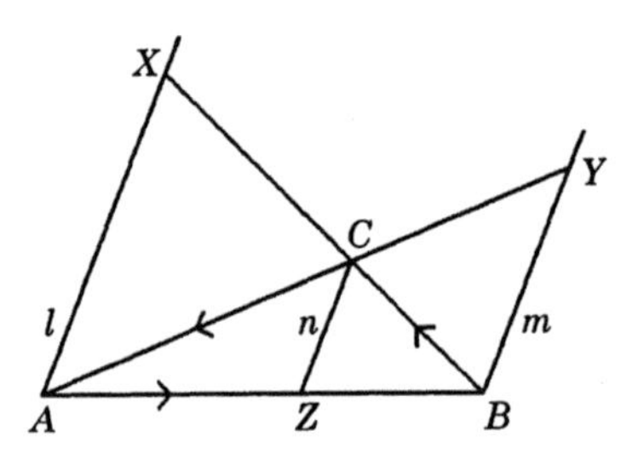

Figure 5.4.3 Ceva's theorem, part (*iv*)

Part (*vi*). If just one or all three of X, Y, and Z lie inside the edges, then Menelaus' product is negative, and you can verify that l, m, and n can't be parallel or concurrent. ♦

Ceva's theorem directly implies concurrence of the medians of ΔABC. Let X, Y, and Z be the edge midpoints A', B', and C', so that

$$\frac{A \text{ to } Z}{Z \text{ to } B}\,\frac{B \text{ to } X}{X \text{ to } C}\,\frac{C \text{ to } Y}{Y \text{ to } A} = \frac{A \text{ to } C'}{C' \text{ to } B}\,\frac{B \text{ to } A'}{A' \text{ to } C}\,\frac{C \text{ to } B'}{B' \text{ to } A} = 1 \cdot 1 \cdot 1 = 1.$$

By Ceva's theorem the medians are copencilar. They can't be parallel because the midpoints lie within the edges, so they must concur.

To derive the altitudes' concurrence from Ceva's theorem, first label the vertices of ΔABC so that $\angle A$ and $\angle B$ are acute. Then construct the circle with diameter $\overline{AB}$, as in figure 5.4.4. The feet D and E of the altitudes through A and B lie on this circle. Inscribed angles $\angle CAD$ and $\angle CBE$ are congruent because they subtend the same arc $\widehat{DJE}$, hence $\Delta ADC \sim \Delta BEC$, so

$$\frac{CE}{DC} = \frac{BC}{CA}\,.$$

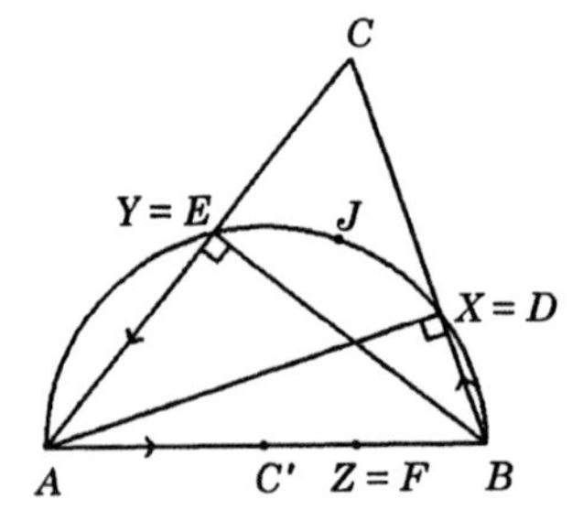

Figure 5.4.4 Deriving the altitudes' concurrence from Ceva's theorem

Giovanni CEVA was born around 1647 in Milan, to a wealthy family. He studied in Pisa, and later in life received a regular stipend from the archduke of Tuscany as a sort of court mathematician. In his 1678 book *De lineis rectis* is found the triangle theorem known by his name, and many elaborations, treated in the framework of mechanics. In several other books, he continued detailed studies in geometrical mechanics, many of which would be supplanted soon by calculus applications. He died in 1734. His Jesuit brother Tomasso was also a mathematician, and a poet.

Similarly,

$$\frac{BD}{FB} = \frac{AB}{BC} \qquad \frac{AF}{EA} = \frac{CA}{AB},$$

where F is the foot of the altitude through C. Now, choose scales for the edge lines as in figure 5.4.4, and let $X, Y, Z = D, E, F$. If $\angle C$ is acute, as in the figure,

$$\frac{A \text{ to } Z}{Z \text{ to } B}\,\frac{B \text{ to } X}{X \text{ to } C}\,\frac{C \text{ to } Y}{Y \text{ to } A} = \frac{AF}{FB}\,\frac{BD}{DC}\,\frac{CE}{EA}$$

$$= \frac{AF}{EA}\,\frac{BD}{FB}\,\frac{CE}{DC} = \frac{CA}{AB}\,\frac{AB}{BC}\,\frac{BC}{CA} = 1.$$

If $\angle C$ is obtuse, then you must place minus signs before two of the fractions in each of these products, but you get the same result. By Ceva's theorem, the altitudes are copencilar. Since they're perpendicular to the edges of a triangle, they can't be parallel, so they must concur.

To derive the concurrence of the angle bisectors of ΔABC from Ceva's theorem, you need a preliminary result. As in figure 5.4.5, the bisector g of $\angle A$ meets the opposite edge at a point X. Let B^* and C^* be the feet of the perpendiculars to g through B and C. If $B^* \neq X \neq C^*$, then $\Delta XBB^* \sim \Delta XCC^*$ and $\Delta ABB^* \sim \Delta ACC^*$. From these similarities it follows that

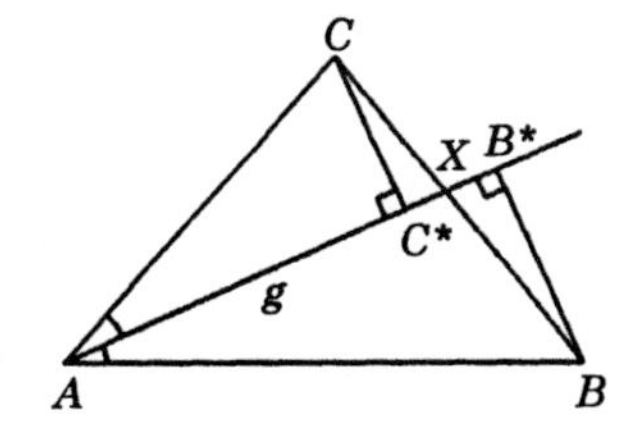

Figure 5.4.5 Deriving the angle bisectors' concurrence from Ceva's theorem

$$\frac{B \text{ to } X}{X \text{ to } C} = \frac{BB^*}{CC^*} = \frac{AB}{CA}\,.$$

On the other hand, if $B^* = X$ or $C^* = X$, then $\overleftrightarrow{BC} \perp g$, ΔABC is isosceles, and you get the same result:

$$\frac{B \text{ to } X}{X \text{ to } C} = 1 = \frac{AB}{CA}\,.$$

By the same argument, the bisectors of $\angle B$ and $\angle C$ meet the opposite edges at points Y and Z such that

$$\frac{C \text{ to } Y}{Y \text{ to } A} = \frac{BC}{AB} \qquad \frac{A \text{ to } Z}{Z \text{ to } B} = \frac{CA}{BC}\,.$$

The previous three equations entail

$$\frac{A \text{ to } Z}{Z \text{ to } B}\,\frac{B \text{ to } X}{X \text{ to } C}\,\frac{C \text{ to } Y}{Y \text{ to } A} = \frac{CA}{BC}\,\frac{AB}{CA}\,\frac{BC}{AB} = 1.$$

By Ceva's theorem, the angle bisectors are copencilar. They can't be parallel because X, Y, and Z lie within the edges, so they must concur.

The original derivation of Ceva's theorem was quite different from that given in this section. Ceva didn't base it directly on Euclid; instead, he used some principles governing the center of gravity of a system of weights. Given points X, Y, and Z within the edges of ΔABC opposite angles A, B, and C, Ceva assigned weights to the vertices as follows:

Vertex	*Weight*
A	$CY \cdot BX$
B	$YA \cdot XC$
C	$YA \cdot BX$.

(See figure 5.4.1.) Assuming that the supporting triangle has no mass, he then computed the center of gravity of the system of three weights. In such a computation, you can replace a subsystem of two weights u and v by a single weight $u + v$ at its center of gravity. This is the point P about which the moments of the weights u and v are opposite; they're on opposite sides of P, and the products of u and v by their distances from P are equal. Since $BX \cdot (YA \cdot XC) = XC \cdot (YA \cdot BX)$, the moments about X of the weights at B and C are opposite, so X is their center of gravity. You can replace those two weights by a single one at X. It follows that the center of gravity of the original system coincides with that of the new one consisting of weights at A and X, hence it lies on $\overline{AX}$. Similarly, since $CY \cdot (YA \cdot BX) = YA \cdot (CY \cdot BX)$, you can replace the weights at C and A by a single one at Y, so the center of gravity of the original system lies on $\overline{BY}$. Thus, it's the intersection of $\overline{AX}$ and $\overline{BY}$. Now, Ceva reasoned, segments $\overline{AX}$, $\overline{BY}$, and $\overline{CZ}$ are concurrent if and only if the center of gravity of the original system also falls on $\overline{CZ}$. That happens just in case you can replace the weights at A and B by a single weight at Z, i.e., just when their moments about Z are opposite: $AZ \cdot (CY \cdot BX) = ZB \cdot (YA \cdot XC)$. This condition simply says Menelaus' product equals 1:

$$\frac{A \text{ to } Z}{Z \text{ to } B} \frac{B \text{ to } X}{X \text{ to } C} \frac{C \text{ to } Y}{Y \text{ to } A} = 1.$$

This argument, of course, is not supported by the axiomatic foundation of chapter 3. But that foundation can be extended to describe the behavior of systems of weights, and justify the argument.

If $X, Y, Z = A', B', C'$ in the preceding argument, so that figure 5.4.1 represents the concurrence of the medians of ΔABC, then the weights assigned $A, B,$ and C are all equal. Thus, in that case, Ceva's argument is identical to one used often in mechanics courses to demonstrate the medians' concurrence.

5.5 Trigonometry

Concepts

- Law of sines
- Law of cosines
- Determining other parts of a triangle from ASA, SAS, or SSS data
- Using SSA data
- Cosine and sine sum and difference formulas
- Cosine and sine double and half angle formulas
- SAS and ASA area formulas; Hero's SSS area formula
- Area formulas involving the circumradius
- Inequalities for approximating sine values

This section completes the study of plane trigonometry begun in chapter 3. In addition to the material you'd find in standard trigonometry texts, it presents several striking area formulas that go beyond what you'll find there. The section concludes with a method for approximating sine values, which also plays a major role in applying calculus to trigonometry problems.

Classical trigonometry shows how to determine some parts of ΔABC —measures of $\angle A$, $\angle B$, and $\angle C$ and lengths a, b, and c of the opposite edges—when you're given others. Theorem 3.13.15 and exercise 4.5.4 summarized this for the case where $\angle C$ is right. In that situation, given any two of a, b, c, $m\angle A$, and $m\angle B$, but not just the two angle measures, you can use Pythagoras' theorem, some simple equations involving the trigonometric functions, and computed values of these functions to find the remaining parts. For oblique triangles, you need more sophisticated methods: the laws of sines and cosines presented in this section. By the triangle congruence theory in section 3.5, all parts of a triangle are fixed once you have SAS, SSS, or ASA data. Using the laws of sines and cosines you can actually compute the remaining parts.

Law of sines

The law of sines is usually phrased, *the edges of a triangle are proportional to the sines of the opposite angles.* Here's a more complete version of the law, which identifies the proportionality factor:

Theorem 1 (Law of sines). If R is the circumradius of triangle ΔABC, then

$$\frac{a}{\sin m\angle A} = \frac{b}{\sin m\angle B} = \frac{c}{\sin m\angle C} = 2R.$$

Proof. You must consider four cases, according to the situation of A and C and diameter $\overline{BX}$ of the circumcircle.

Case 1: If A and C lie on different sides of the diameter, as in figure 5.5.1, then $\angle X \cong \angle A$ because they're both inscribed in arc $\widehat{CAB}$, hence $a/(2R) = a/BX = \sin m\angle X = \sin m\angle A$.

Case 2: If A and C lie on the same side, as in figure 5.5.2, the same equation holds because $m\angle X + m\angle A = 180°$.

Case 3: If A lies on the diameter, as in figure 5.5.3, then $a/(2R) = a/AB = \sin m\angle A$.

Case 4: If C lies on the diameter, proceed as in case (3). ♦

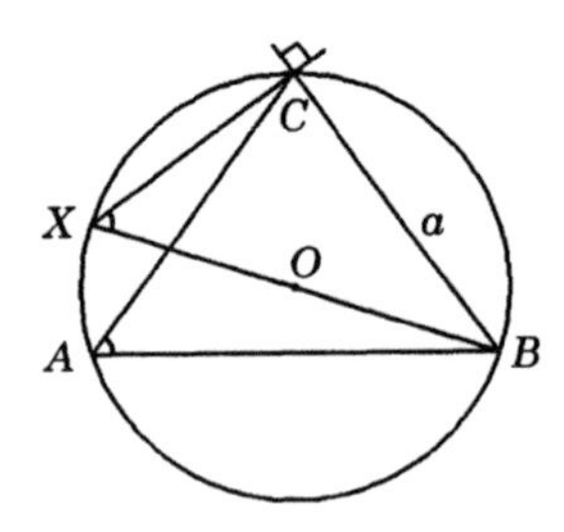

Figure 5.5.1 Proving the law of sines, case (1)

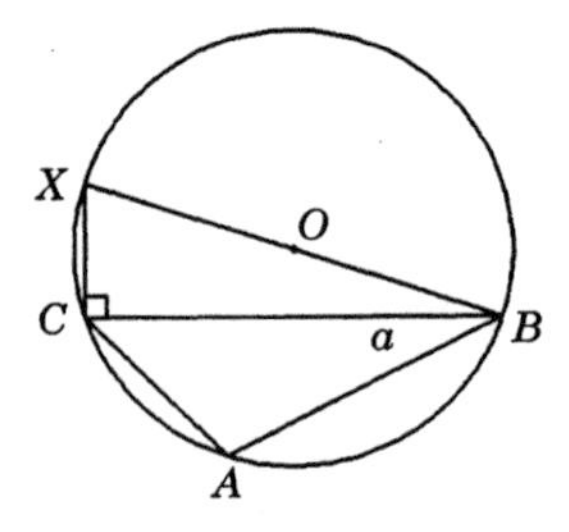

Figure 5.5.2 Proving the law of sines, case (2)

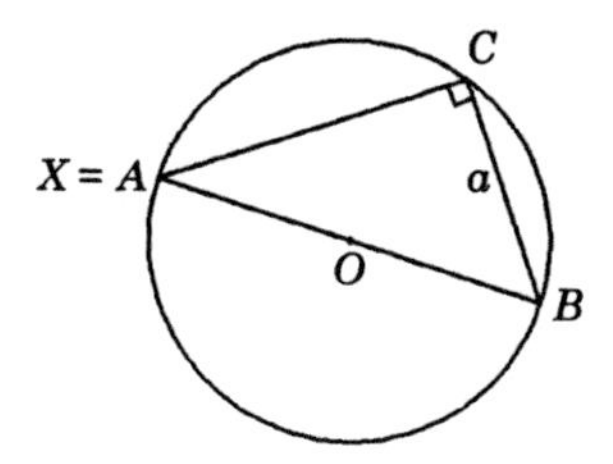

Figure 5.5.3 Proving the law of sines, case (3)

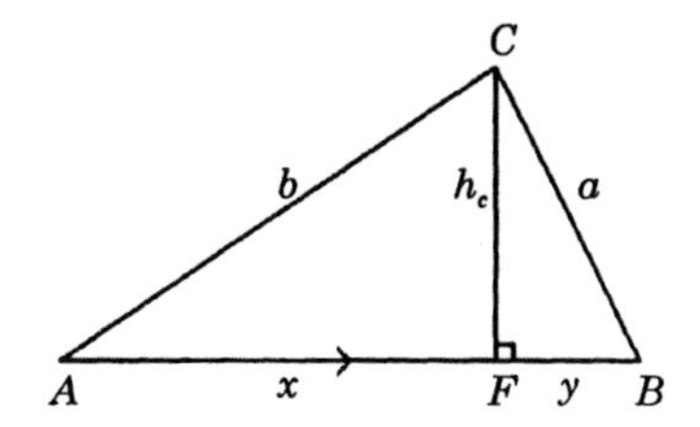

Figure 5.5.4 Proving the law of cosines

Law of cosines

The law of cosines results from applying Pythagoras' theorem twice:

***Theorem 2* (*Law of cosines*).** In ΔABC, $a^2 = b^2 + c^2 - 2bc \cos m\angle A$.

Proof. Assign a scale to $\overleftrightarrow{AB}$ so that $c = (A \text{ to } B)$. As in figure 5.5.4, let F be the foot of the altitude through C and $h_c = CF$. Set $x = (A \text{ to } F)$ and $y = (B \text{ to } F)$, so that $c = x + y$. Then $x/b = \cos m\angle A$ and

$$\begin{aligned} a^2 &= h_c^2 + y^2 \\ &= b^2 - x^2 + y^2 \\ &= b^2 - x^2 + (c - x^2) \\ &= b^2 + c^2 - 2cx \\ &= b^2 + c^2 - 2bc \cos m\angle A. \ \blacklozenge \end{aligned}$$

By relabeling the triangle, you can derive analogous equations

$$\begin{aligned} b^2 &= c^2 + a^2 - 2\,ca \cos m\angle B \\ c^2 &= a^2 + b^2 - 2\,ab \cos m\angle C. \end{aligned}$$

Finding unknown parts of a triangle from SSS, SAS, or ASA data

If you know SSS data for ΔABC, you can use the law of cosines to find its angle measures. For example,

$$\cos m\angle A = \frac{-a^2 + b^2 + c^2}{2bc},$$

and you can determine $m\angle A$ from the cosine value. You can find the other measures from the analogous equations just stated.

If you know SAS data—for example c, $m\angle A$, and b —then you can find the unknown edge from the law of cosines:

$$a = \sqrt{b^2 + c^2 - 2bc \cos \mathrm{m}\angle A}\ .$$

Now you have SSS data, so you can proceed as in the previous paragraph to find another angle measure, for example $\mathrm{m}\angle B$. Then $\mathrm{m}\angle C = 180° - \mathrm{m}\angle A - \mathrm{m}\angle B$.

If you know ASA data—for example $\mathrm{m}\angle A$, b, and $\mathrm{m}\angle C$ —then $\mathrm{m}\angle C = 180° - \mathrm{m}\angle A - \mathrm{m}\angle B$. Use the law of sines to find the unknown edges:

$$a = \frac{b \sin \mathrm{m}\angle A}{\sin \mathrm{m}\angle B} \qquad c = \frac{b \sin \mathrm{m}\angle C}{\sin \mathrm{m}\angle B}\ .$$

Section 3.5 showed that SSA data don't necessarily allow you to compute the remaining parts. But according to exercise 4.6.2 there are at most two possibilities. Exercise 5.11.22 will lead you through the corresponding computation.

Addition, subtraction, double and half angle formulas

The familiar formulas for sines and cosines of sums and differences of angle measures are closely interrelated. It's efficient to derive first the cosine difference formula from the law of cosines, then get the others from that.

Theorem 4. For any a and b, $\cos(a - b) = \cos a \cos b + \sin a \sin b$.

Proof. If $a - b = 0°$, this merely says $1 = 1$. Now assume that $0° < a - b < 180°$. By the law of cosines and the distance formula, the square of the distance between points $A = \langle \cos a, \sin a \rangle$ and $B = \langle \cos b, \sin b \rangle$ in figure 5.5.5 is

$$1^2 + 1^2 - 2 \cdot 1 \cdot 1 \cos(a - b) = (\cos a - \cos b)^2 + (\sin a - \sin b)^2,$$

hence

$$\begin{aligned}
2 - 2\cos(a - b) &= \\
&= \cos^2 a - 2\cos a \cos b + \cos^2 b + \sin^2 a - 2 \sin a \sin b + \sin^2 b \\
&= (\cos^2 a + \sin^2 a) + (\cos^2 b + \sin^2 b) - 2(\cos a \cos b + \sin a \sin b) \\
&= 2 - 2(\cos a \cos b + \sin a \sin b),
\end{aligned}$$

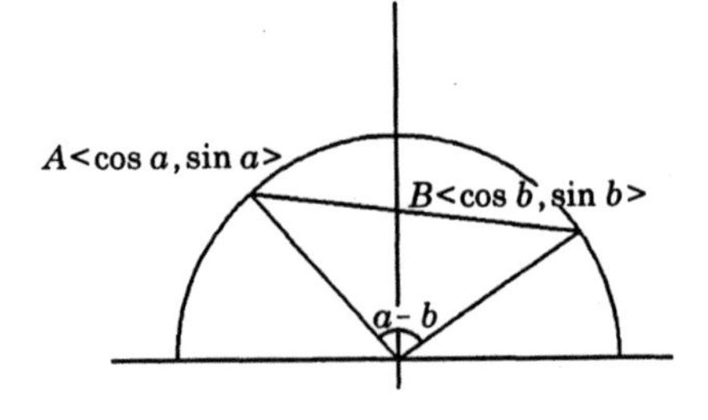

Figure 5.5.5
Proving theorem 4

which yields the desired formula. Finally, consider the general case. Given any a and b, find the integer n such that $0° \le a - b + 180°n < 180°$. By the equations

$$\cos(t + 180°) = -\cos t \qquad \sin(t + 180°) = -\sin t$$

and the case just considered,

$$\begin{aligned}\cos(a - b) &= (-1)^n \cos(a - b + 180°n) = (-1)^n \cos(a + 180°n - b)\\ &= (-1)^n [\cos(a + 180°n) \cos b + \sin(a + 180°n) \sin b]\\ &= (-1)^n \cos(a + 180°n) \cos b + (-1)^n \sin(a + 180°n) \sin b\\ &= \cos a \cos b + \sin a \sin b. \blacklozenge\end{aligned}$$

***Corollary 5* (*Cosine and sine sum and difference formulas*).** For any a and b,

$$\begin{aligned}\cos(a \pm b) &= \cos a \cos b \mp \sin a \sin b\\ \sin(a \pm b) &= \sin a \cos b \pm \cos a \sin b.\end{aligned}$$

Proof. You can derive the cosine sum formula by substituting $-b$ for b in theorem 4. To get the sine formulas, use the cofunction identities:

$$\begin{aligned}\sin(a \pm b) &= \cos[90° - (a \pm b)] = \cos[(90° - a) \mp b]\\ &= \cos(90° - a) \cos b \pm \sin(90° - a) \sin b\\ &= \sin a \cos b \pm \cos a \sin b. \blacklozenge\end{aligned}$$

***Corollary 6* (*Cosine and sine double angle formulas*).** For any a,

$$\begin{aligned}\sin 2a &= 2 \sin a \cos a\\ \cos 2a &= \cos^2 a - \sin^2 a = 1 - 2\sin^2 a = 2\cos^2 a - 1.\end{aligned}$$

Proof. Set $b = a$ in the addition formulas and simplify, using the Pythagorean identity. ♦

***Corollary 7* (*Cosine and sine half angle formulas*).** For any a,

$$\cos^2(\tfrac{1}{2} a) = \frac{1 + \cos a}{2} \qquad \sin^2(\tfrac{1}{2} a) = \frac{1 - \cos a}{2}.$$

Proof. $\cos(2 \cdot \tfrac{1}{2} a) = 1 - 2\sin^2(\tfrac{1}{2} a) = 2\cos^2(\tfrac{1}{2} a) - 1$ by corollary 6. ♦

Area formulas

Since you can determine all parts of a triangle once you know SAS, SSS, or ASA data, you can find its area. The next two theorems give the three formulas.

***Theorem 8* (*SAS and ASA area formulas*).** The area of ΔABC is

$$\text{a}\Delta ABC = \tfrac{1}{2}\, bc \sin \text{m}\angle A = \frac{c^2 \sin \text{m}\angle A \sin \text{m}\angle B}{2 \sin \text{m}\angle C}.$$

Proof. As in figure 5.5.4, let F be the foot of the altitude through C and $h_c = CF$. Then $h_c/b = \sin \text{m}\angle A$ and $\text{a}\Delta ABC = \tfrac{1}{2}\, ch_c = \tfrac{1}{2}\, bc \sin \text{m}\angle A$. Now use the law of sines:

$$\begin{aligned}\tfrac{1}{2}\, bc \sin \text{m}\angle A &= \tfrac{1}{2}\, bc \sin \text{m}\angle \text{A} \frac{\sin \text{m}\angle B}{\sin \text{m}\angle B} \\ &= \frac{\tfrac{1}{2} b}{\sin \text{m}\angle B}\, c \sin \text{m}\angle A \sin \text{m}\angle B \\ &= \frac{\tfrac{1}{2} c}{\sin \text{m}\angle C}\, c \sin \text{m}\angle A \sin \text{m}\angle B. \blacklozenge\end{aligned}$$

The SSS triangle area formula is phrased in terms of the *semiperimeter*[9] $s = \tfrac{1}{2}(a + b + c)$ of ΔABC. (Its *perimeter* is the sum of the lengths of its edges.) First, here's an algebraic result that you'll need to derive the SSS formula and another one later. It's stated specifically for $\angle A$, but analogous formulas hold for the other angles.

***Lemma 9*.** $\cos \tfrac{1}{2}\, \text{m}\angle A = \dfrac{s(s-a)}{bc}$ and $\sin \tfrac{1}{2}\, \text{m}\angle A = \dfrac{(s-b)(s-c)}{bc}$.

Proof. By the sine half angle formula and the law of cosines,

$$\begin{aligned}\sin \tfrac{1}{2}\, \text{m}\angle A &= \frac{1 - \cos \text{m}\angle A}{2} = \frac{1}{2}\left[1 - \frac{-a^2 + b^2 + c^2}{2bc}\right] \\ &= \frac{2bc + a^2 - b^2 - c^2}{4bc} = \frac{a^2 - (b-c)^2}{4bc} \\ &= \frac{1}{bc}\,\frac{a+b-c}{2}\,\frac{a-b+c}{2} = \frac{(s-c)(s-b)}{bc}.\end{aligned}$$

You can derive the other formula similarly. ♦

***Theorem 10* (*Hero's SSS area formula*).** The area of ΔABC is

$$\text{a}\Delta ABC = \sqrt{s(s-a)(s-b)(s-c)}.$$

Proof. By the SAS area formula and the Pythagorean identity,

---

[9] This word stems from the Greek prefixes *semi-* and *peri-*, meaning *half* and *around*, and the verb *metron*, to *measure*.

$$\text{a}\Delta ABC = \tfrac{1}{2}\, bc \sin \text{m}\angle A = \tfrac{1}{2}\, bc\sqrt{1 - \cos^2 \text{m}\angle A}$$

$$= bc\sqrt{\frac{1 + \cos \text{m}\angle A}{2}\,\frac{1 - \cos \text{m}\angle A}{2}}.$$

Now substitute the lemma 9 formulas and simplify. ♦

You can find in most trigonometry texts the three area formulas just presented. However, when you combine them with the version of the law of sines given in theorem 1, you get some striking and unfamiliar area formulas involving the circumradius.

Theorem 11. The area of ΔABC is

$$\text{a}\Delta ABC = \frac{abc}{4R} = 2R^2 \sin \text{m}\angle A \sin \text{m}\angle B \sin \text{m}\angle C$$

$$= 4Rs \sin \tfrac{1}{2}\, \text{m}\angle A \sin \tfrac{1}{2}\, \text{m}\angle B \sin \tfrac{1}{2}\, \text{m}\angle C.$$

Proof. By theorem 8 and the law of sines,

$$\text{a}\Delta ABC = \tfrac{1}{2}\, bc \sin \text{m}\angle A = \tfrac{1}{2}\, abc\frac{\sin \text{m}\angle A}{a} = \tfrac{1}{2}\, abc\frac{1}{2R}$$

$$\text{a}\Delta ABC = \tfrac{1}{2}\, bc \sin \text{m}\angle A$$

$$= \tfrac{1}{2}\,\frac{b}{\sin \text{m}\angle B}\,\frac{c}{\sin \text{m}\angle C} \sin \text{m}\angle A \sin \text{m}\angle B \sin \text{m}\angle C$$

$$= \tfrac{1}{2}\,(2R)^2 \sin \text{m}\angle A \sin \text{m}\angle B \sin \text{m}\angle C.$$

Now by Hero's formula and the lemma 9 formulas for all three angles,

$$(\text{a}\Delta ABC)^4 = [s(s-a)(s-b)(s-c)]^2$$

$$= s^2(abc)^2\,\frac{(s-b)(s-c)}{bc}\,\frac{(s-c)(s-a)}{ca}\,\frac{(s-a)(s-b)}{ab}$$

$$= s^2(4R\,\text{a}\Delta ABC)^2\,\frac{1-\cos \text{m}\angle A}{2}\,\frac{1-\cos \text{m}\angle B}{2}\,\frac{1-\cos \text{m}\angle C}{2}$$

$$(\text{a}\Delta ABC)^2 = 16R^2s^2 \sin^2(\tfrac{1}{2}\,\text{m}\angle A)\sin^2(\tfrac{1}{2}\,\text{m}\angle B)\sin^2(\tfrac{1}{2}\,\text{m}\angle C).$$ ♦

HERO flourished in Alexandria around A.D. 62. No biographical information about him is available. He wrote thirteen books on mathematics, science, and engineering. His triangle area formula occurs in one on mensuration, entitled *Metrica*.

Inequalities for approximating sine values

So far, this book has computed values $\sin\theta$ for only a few arguments θ. Section 3.13 considered $\theta = 0°$, $30°$, $45°$, and closely related angles, and exercise 4.9.1 added $36°$ to this list. How are other values computed? You can apply the half angle formula to compute $\sin\theta$ for $\theta = 18° = \frac{1}{2}\cdot 36°$, the difference formula to handle $15° = 45° - 30°$ and $3° = 18° - 15°$, and the double angle and difference formulas to compute $\sin 3n°$ for each integer n. The half angle formula yields $\sin\theta$ for all integral multiples θ of $1\frac{1}{2}°$, $\frac{3}{4}°$, etc. The Pythagorean identity $\sin^2\theta + \cos^2\theta = 1$ and the definition $\tan\theta = \sin\theta/\cos\theta$ provide the corresponding cosine and tangent values. How are other values computed, to enable us to use effectively the trigonometric formulas derived in this section?

With some reflection, you'll see that the only ways to use previously introduced methods to extend the list of arguments θ for which $\sin\theta$ is known are

- to continue using the half angle and sum formulas to reach sums θ of terms of the form $3°/2^n$, and
- to find geometric constructions analogous to exercise 4.9.1 for new θ.

The latter feat is possible for a few θ, but those aren't easily tabulated, it's hard to carry out the construction and the algebra, and hard to use the results. After centuries of investigation, mathematicians determined in the mid-1800s that these methods will *never* yield algebraic formulas for trigonometric function values for *some* arguments: for example, $\sin 1°$ or $\sin 2°$.[10] Even two thousand years ago, however, Alexandrian mathematicians had discovered how to *approximate* values $\sin\theta$ with any desired degree of accuracy.[11] You can reproduce some of their work in exercise 5.11.27.

Approximation requires manipulation with *inequalities*, not just equations. For given arguments θ you must calculate numbers y then show that $|y - \sin\theta|$ is less than some tolerance, so that you can compute with y in place of $\sin\theta$ without making unacceptable errors. Some trigonometric inequalities are simple; for example, if $0° < \theta < 90°$ then $0 < \sin\theta < 1$ and $0 < \cos\theta < 1$, hence $\sin\theta < \tan\theta$ because $\tan\theta = \sin\theta/\cos\theta$. Theorem 12 is a refinement of this last inequality; it will be used to derive an approximation for $\sin 1°$ and estimate the error. Controlled approximation is an essential calculus technique, and standard calculus texts derive from theorem 12 the formulas for the derivatives of the trigonometric functions.[12]

[10] Moise ([1963] 1990, section 19.11) presents that result. It requires algebraic techniques beyond those assumed for or covered in this book.

[11] See van der Waerden 1963, chapter VII.

[12] For example, see Thomas and Finney 1979, section 2.10.

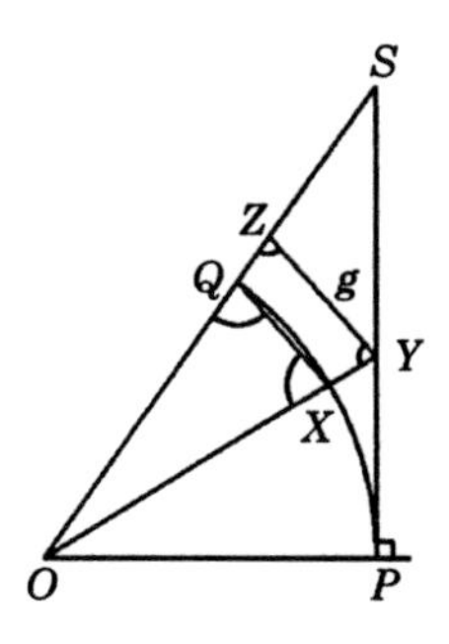

Figure 5.5.6
Proving theorem 12

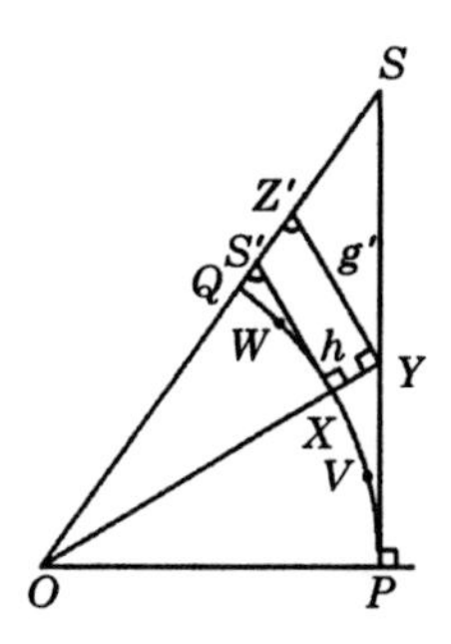

Figure 5.5.7 Proving theorem 12, continued

Theorem 12. If $0° < \theta < 90°$, then $\sin\theta < \pi\theta/180° < \tan\theta$.[13]

Proof. Consider figure 3.13.5 and theorem 3.13.16: $\sin\theta = QR < PQ$, the length of a polygon inscribed in $\widehat{PXQ}$, and $PS = \tan\theta$. Thus $PQ < l(\widehat{PXQ})$, the least upper bound of such lengths. By theorem 3.14.6, $l(\widehat{PXQ}) = \pi\theta/180°$. That establishes the first of the desired inequalities.

Now consider the related figure 5.5.6. Given $X \neq Q$ in the arc, find Y, then line $g \parallel \overleftrightarrow{QX}$ through Y, and the intersection Z, so that $\Delta OXQ \sim \Delta OYZ$ and $XQ < YZ$. By the exterior angle theorem, $m\angle YSZ < m\angle YZO = m\angle OYZ < m\angle YZS$, hence $YZ < YS$. Repeat this argument with $\overline{XQ}$ replaced by each edge, in turn, of a polygon Π inscribed in $\widehat{PXQ}$; it shows that the length $l(\Pi) < PS = \tan\theta$. That is, PS is an upper bound for the lengths of polygons inscribed in $\widehat{PXQ}$. Since the arc length is the *least* upper bound, $l(\widehat{PXQ}) \le \tan\theta$. Could these be equal?

To show that equality is impossible, consider figure 5.5.7; select any point $X \neq P, Q$ in the arc, and leave it fixed. Find Y and the perpendiculars h and g' to $\overrightarrow{OX}$ through X and Y, then their intersections S' and Z' with $\overrightarrow{OQ}$, so that $\Delta OXS' \sim \Delta OYZ'$ and $XS' < YZ'$. By the exterior angle theorem, $m\angle YSZ' < m\angle OZ'Y = m\angle OS'X < 90° < m\angle YZ'S$, hence $YZ' < YS$ and therefore $XS' < YS$. By the previous paragraph applied to $\widehat{PVX}$ and $\widehat{XWQ}$,

$$\begin{aligned} l(\widehat{PXQ}) &= l(\widehat{PVX}) + l(\widehat{XWQ}) \\ &\le PY + XS' < PY + YS = PS = \tan\theta. \ \blacklozenge \end{aligned}$$

When θ is very small, $\cos\theta$ is only slightly smaller than 1, hence $\sin\theta/\cos\theta = \tan\theta$ is only slightly larger than $\sin\theta$, and by theorem 12, $\pi\theta/180°$ lies between them. This number, which is certainly easy to compute,

[13] If you use radian measure, these inequalities become $\sin\theta < \theta < \tan\theta$.

should be a close approximation to either one. How close is it? You can use theorem 12 to estimate the approximation error, as follows. The inequality $\pi\theta/180° < \tan\theta$ and the sine half angle formula imply

$$\begin{aligned}\pi\theta/180° - \sin\theta &< \tan\theta - \sin\theta \\ &= (1 - \cos\theta)\tan\theta = 2\sin^2(\tfrac{1}{2}\theta)\tan\theta. \qquad (1)\end{aligned}$$

Apply theorem 12 again with $\frac{1}{2}\theta$ in place of θ:

$$2\sin^2(\tfrac{1}{2}\theta)\tan\theta < 2(\pi\cdot\tfrac{1}{2}\theta/180°)^2\tan\theta. \qquad (2)$$

Now suppose you know $\tan\theta_1$ roughly for some θ_1 slightly larger than θ. Since the sine is an increasing function and the cosine is decreasing, the tangent is increasing, so

$$\tan\theta < \tan\theta_1. \qquad (3)$$

For example, consider $\theta = 1°$ and $\theta_1 = 1\tfrac{1}{2}°$. As mentioned earlier, you can compute $\tan 1\tfrac{1}{2}°$ by using trigonometric formulas. Rounding up to one significant digit, you'll find $\tan\theta_1 \approx 0.03$. Inequalities (1) to (3) with $\theta = 1°$ and $\theta_1 = 1\tfrac{1}{2}°$ yield an estimate of the approximation error:

$$\pi\theta/180° - \sin\theta < 2(\pi\cdot\tfrac{1}{2}\theta/180°)^2\tan\theta_1 \approx 0.000004.$$

Therefore $\sin 1° \approx \pi/180 \approx 0.0174533$, with the first five decimal places guaranteed correct. (Correct to seven decimals, $\sin 1° \approx 0.0174524$.)

Current programs for approximating sine values use a variant of this method for small arguments, but apply more recently developed calculus techniques for estimating the approximation error. For larger arguments they employ the sine and cosine sum and difference formulas, with elaborate attention to round-off error.

5.6 Vector products

Concepts

- Dot products
- Cross products
- Distinguishing sides of lines and planes
- Determinant criteria for collinearity and coplanarity
- Determinant formulas for triangle area, and volume of a tetrahedron
- When do four planes, or three lines in a plane, have a common point?

This section is a digression. It covers the geometric aspects of vector dot and cross products. These are used occasionally in the rest of the book. Since they're sometimes not covered in more elementary courses, and depend on some trigonometry presented in section 5.5, it seemed inappropriate merely to refer to them in appendix C. They're developed fully here, using the vector algebra notation reviewed in the appendix.

Dot products

You can use the law of cosines with a coordinate system and vector algebra to measure angles. Regard points $P = \langle p_1, p_2, p_3 \rangle$ and $Q = \langle q_1, q_2, q_3 \rangle$ as vectors, with dot product $P \cdot Q = p_1 q_1 + p_2 q_2 + p_3 q_3$. Theorem 1 gives a formula for $P \cdot Q$, which you can turn into one for $m\angle POQ$ if P and Q aren't collinear with the origin O. Corollary 2 presents a general angle formula.

Theorem 1. If points P and Q are noncollinear with the origin O, then $P \cdot Q = (OP)(OQ) \cos m\angle POQ$.

Proof. By the law of cosines:

$$(PQ)^2 = (OP)^2 + (OQ)^2 - 2(OP)(OQ) \cos m\angle POQ$$

$$(p_1 - q_1)^2 + (p_2 - q_2)^2 + (p_3 - q_3)^2$$
$$= p_1^2 + p_2^2 + p_3^2 + q_1^2 + q_2^2 + q_3^2 - 2(OP)(OQ) \cos m\angle POQ$$

$$-2p_1 q_1 - 2p_2 q_2 - 2p_3 q_3 = -2(OP)(OQ) \cos m\angle POQ. \blacklozenge$$

Corollary 2 (Angle formula). If P, X, and V are noncollinear points, then

$$\cos m\angle PVX = \frac{(P - V) \cdot (X - V)}{(VP)(VX)}.$$

Proof. Let $P' = P - V$ and $X' = X - V$. By the parallelogram law, $OP' = VP$, $OX' = VX$, and $PX = P'X'$, so $\Delta P'OX' \cong \Delta PVX$ and $m\angle PVX = m\angle P'OX'$. ♦

For the rest of this heading, work in a fixed plane ε with a two-dimensional coordinate system.

The angle formula leads to some convenient procedures for determining whether two points X and Y lie on different sides of a line $g = \overleftrightarrow{VP}$. First, consider the point $R = V + \langle v_2 - p_2, p_1 - v_1 \rangle$ shown in figure 5.6.1. It has two important properties:

$$VR = \sqrt{(v_2 - p_2)^2 + (p_1 - v_1)^2} = VP$$

$$(PR)^2 = [(p_1 - v_1) - (v_2 - p_2)]^2 + [(p_2 - v_2) - (p_1 - v_1)]^2$$
$$= 2(p_1 - v_1)^2 + 2(p_2 - v_2)^2$$
$$= 2(VP)^2 = (VP)^2 + (VR)^2.$$

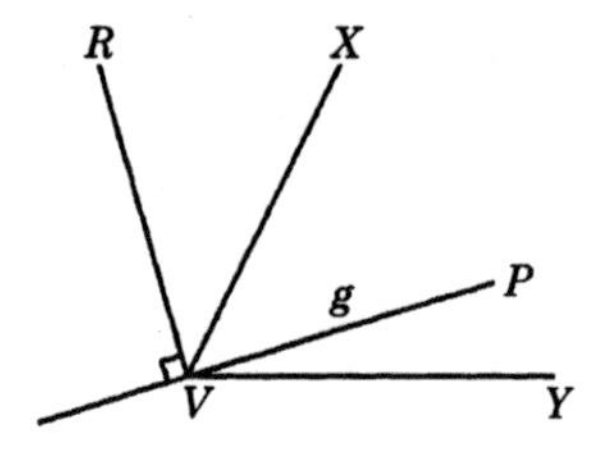

Figure 5.6.1
Proof of theorem 3

By the converse of Pythagoras' theorem, $\vec{VP} \perp \vec{VR}$. Now, X and R lie on the same side of g just in case X is in $\vec{VR}$ or $\angle RVX$ is acute. By the angle formula, that happens just when

$$\begin{aligned} 0 &< (R - V) \cdot (X - V) \\ &= (v_2 - p_2)(x_1 - v_1) + (p_1 - v_1)(x_2 - v_2) \\ &= (p_1 - v_1)(x_2 - v_2) - (p_2 - v_2)(x_1 - v_1) \\ &= \det \begin{bmatrix} p_1 - v_1 & x_1 - v_1 \\ p_2 - v_2 & x_2 - v_2 \end{bmatrix}. \end{aligned}$$

That determinant is often abbreviated $\det[P - V, X - V]$. Notice that X lies on g just in case $X = V$ or $\angle RVX$ is right—that is, just when the determinant is zero. This discussion has demonstrated

Theorem 3. A point X lies on a line $g = \vec{VP}$ just in case

$$\det[P - V, X - V] = 0.$$

Points X and Y lie on different sides of g just when this determinant and $\det[P - V, Y - V]$ have different signs.

In theorem 3, *which* sign corresponds to *which* side of g depends on the choice of points P and V. If you change them, either both determinants change sign, or neither does.

Let $S = <s_1, s_2>$ denote the vector $R - V$ considered earlier, and let $t = S \cdot V$. A point X lies on g if and only if $(R - V) \cdot (X - V) = 0$. That equation is equivalent to $S \cdot X = S \cdot V$, i.e., $s_1 x_1 + s_2 x_2 = t$, a linear equation for g. Thus, you can interpret the previous discussion as

Theorem 4. Points Y and Z lie on different sides of the line with equation $s_1 x_1 + s_2 x_2 = t$ just when one of $s_1 y_1 + s_2 y_2$ and $s_1 z_1 + s_2 z_2$ is greater than t and the other is less.

The determinant in theorem 3 provides a convenient formula for the area of a triangle:

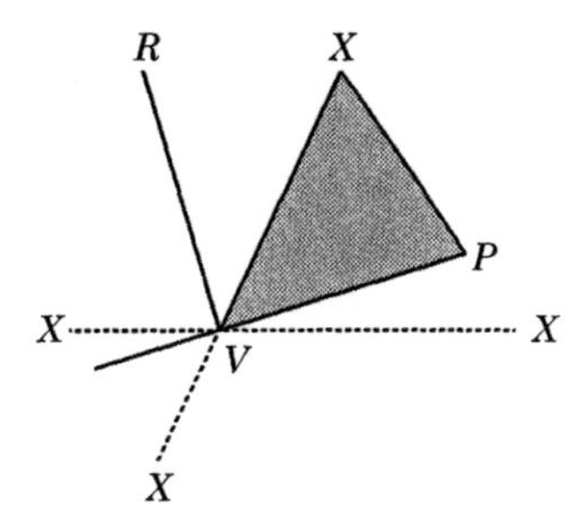

Figure 5.6.2
Theorem 5

Theorem 5. If points P, V, and X are noncollinear, then the area of ΔPVX is

$$a\Delta PVX = \pm \tfrac{1}{2} \det [P - V, X - V].$$

The + sign is correct just when X lies on the same side of $\overleftrightarrow{VP}$ as the point R considered earlier.

Proof. If X lies on the same side of $\overleftrightarrow{VP}$ as R, indicated by the solid line in figure 5.6.2, then $m\angle PVX = 90° \pm m\angle RVX$, so $\sin m\angle PVX = \cos m\angle RVX$. Otherwise, as indicated by the dotted lines, $m\angle PVX = m\angle RVX - 90°$ or $m\angle PVX + m\angle RVX = 270°$; in those cases $\sin m\angle PVX = -\cos m\angle RVX$. By the SAS area formula, equation $VP = VR$, and the angle formula,

$$\begin{aligned} a\Delta PVX &= \tfrac{1}{2}(VP)(VX)\sin m\angle PVX \\ &= \pm \tfrac{1}{2}(VR)(VX)\cos m\angle RVX \\ &= \pm \tfrac{1}{2}(R - V)\cdot(X - V) \\ &= \pm \tfrac{1}{2}\det[P - V, X - V]. \blacklozenge \end{aligned}$$

Cross products

The *cross product* W of vectors X and Y is used only in three dimensions:

$$W = \begin{bmatrix} w_1 \\ w_2 \\ w_3 \end{bmatrix} = X \times Y \qquad \begin{aligned} w_1 &= \det\begin{bmatrix} x_2 & y_2 \\ x_3 & y_3 \end{bmatrix} \\ w_2 &= \det\begin{bmatrix} x_3 & y_3 \\ x_1 & y_1 \end{bmatrix} \\ w_3 &= \det\begin{bmatrix} x_1 & y_1 \\ x_2 & y_2 \end{bmatrix}. \end{aligned}$$

You should verify its most basic properties algebraically:

(a) $X \times Y = O$ if and only if X and Y are collinear with O;
(b) $X \cdot (X \times Y) = 0 = (X \times Y) \cdot Y$;
(c) $X \times Y = -(Y \times X)$;
(d) $(sX) \times Y = s(X \times Y) = X \times (sY)$;
(e) $X \times (Y \pm Z) = (X \times Y) \pm (X \times Z)$;
(f) $$(X \times Y) \cdot Z = \det \begin{bmatrix} x_1 & y_1 & z_1 \\ x_2 & y_2 & z_2 \\ x_3 & y_3 & z_3 \end{bmatrix}.$$

This determinant is often abbreviated $\det[X, Y, Z]$.

Now set up a fixed coordinate system with origin O. If points X and Y are noncollinear with O, then according to property (b) and the converse of Pythagoras' theorem, $X \times Y$ falls on the perpendicular k to plane OXY through O. The next results show that it's one of the two points that lie on k at a certain distance from O. (Which one depends on your coordinate system.) You can verify lemma 6 directly through algebra.

Lemma 6. If points X and Y are noncollinear with the origin O, and $Z = X \times Y$, then $(OZ)^2 = (OX)^2(OY)^2 - (X \cdot Y)^2$. Therefore, if $X \cdot Y = 0$, then $OZ = (OX)(OY)$.

Theorem 7. If points X and Y are noncollinear with the origin and $Z = X \times Y$, then $OZ = (OX)(OY) \sin m\angle XOY$.

Proof. Construct line $h \perp \vec{OX}$ through O in OXY, and let X' and Y' be the feet of perpendiculars from Y to $\vec{OX}$ and h. Then $OX'YY'$ is a rectangle, so $OY' = OY \sin m\angle XOY$ and $Y = X' + Y'$. Therefore

$$\begin{aligned} Z &= X \times Y = X \times (X' + Y') = (X \times X') + (X \times Y') && \text{—by property (c)} \\ &= X \times Y' && \text{—by property (a)} \\ OZ &= (OX)(OY') && \text{—by lemma 8} \\ &= (OX)(OY) \sin m\angle XOY. \blacklozenge \end{aligned}$$

The cross product leads to some convenient procedures for determining whether two points X and Y lie on different sides of a plane $\varepsilon = VPQ$. First, consider point

$$R = V + (P - V) \times (Q - V).$$

You can check by using parallelograms that $P - V$ and $Q - V$ are noncollinear with O, so $R - V \neq O$ and $\vec{VR} \perp \vec{VP}$ and $\vec{VR} \perp \vec{VQ}$, hence $\vec{VR} \perp \varepsilon$. Points X and R lie on the same side of ε just in case X is in $\vec{VR}$ or $\angle XVR$ is acute. By the angle formula and property (f), that happens just when

$$\begin{aligned} 0 &< (R - V) \cdot (X - V) \\ &= [(P - V) \times (Q - V)] \cdot (X - V) \\ &= \det [P - V, Q - V, X - V]. \end{aligned}$$

Notice that X lies on ε just in case $X = V$ or $\angle RVX$ is right—that is, just when the determinant is zero. This discussion has demonstrated

Theorem 8. A point X lies on a plane $\varepsilon = VPQ$ just when

$$\det [P - V, Q - V, X - V] = 0.$$

Points X and Y lie on different sides of ε just when this determinant and $\det [P - V, Q - V, Y - V]$ have different signs.

In theorem 8, *which* sign corresponds to *which* side of ε depends on the choice of points P, Q, and V. If you change them, either both determinants change sign, or neither does.

Let $S = \langle s_1, s_2, s_3 \rangle$ denote the vector $R - V$ considered earlier, and let $t = S \cdot V$. A point X lies on ε if and only if $(R - V) \cdot (X - V) = 0$. That equation is equivalent to $S \cdot X = S \cdot V$, i.e., $s_1 x_1 + s_2 x_2 + s_3 x_3 = t$, a linear equation for ε. Thus, you can interpret the previous discussion as

Theorem 9. Points Y and Z lie on different sides of the plane with equation $s_1 x_1 + s_2 x_2 + s_3 x_3 = t$ just when one of $s_1 y_1 + s_2 y_2 + s_3 y_3$ and $s_1 z_1 + s_2 z_2 + s_3 z_3$ is greater than t and the other is less.

The determinant in theorem 8 provides a convenient formula for the volume of a tetrahedron:

Theorem 10. If points P, Q, V, and X are noncoplanar, then

$$\mathrm{v}\, PQVX = \pm\tfrac{1}{6} \det [P - V, Q - V, X - V].$$

The $+$ sign is correct just when X lies on the same side of VPQ as the point R considered earlier.

Proof. Let W be the foot of the perpendicular from X to VPQ. Then $\mathrm{m}\angle WVX = 90° - \mathrm{m}\angle RVX$ if X and R lie on the same side of VPQ, else $\mathrm{m}\angle WVX = 90° + \mathrm{m}\angle RVX$. Therefore,

$$\begin{aligned} \mathrm{v}\, PQVX &= \tfrac{1}{3} (\mathrm{a}\, \Delta VPQ)(WX) && \text{—by theorem 3.10.13} \\ &= \tfrac{1}{3} [\tfrac{1}{2} (VP)(VQ) \sin \mathrm{m}\angle PVQ](WX) && \text{—by the SAS area formula} \\ &= \tfrac{1}{6} (VR)(WX) && \text{—by theorem 7} \\ &= \tfrac{1}{6} (VR)(VX) \sin \mathrm{m}\angle WVX \\ &= \pm \tfrac{1}{6} (VR)(VX) \cos \mathrm{m}\angle RVX \\ &= \pm \tfrac{1}{6} (R - V) \cdot (X - V) && \text{—by the angle formula} \\ &= \pm \tfrac{1}{6} [(P - V) \times (Q - V)] \cdot (X - V) \\ &= \pm \tfrac{1}{6} \det [P - V, Q - V, X - V] && \text{—by property (f).} \ \blacklozenge \end{aligned}$$

When do four planes, or three lines in a plane, have a common point?

You've seen how to use determinants to tell whether three points lie on a line in plane analytic geometry or four points lie on a plane in three dimensions. You can also use them to investigate whether three lines or four planes have a common point. But this process is more complicated because you must take into account the possibility of parallel lines and planes. The three-dimensional theory is developed fully here. The plane analogy is summarized in theorem 12; you can supply the details.

Set up a fixed coordinate system with origin O. A plane ε has linear equation $P \cdot X = b$ for some vector $P \neq O$ and some scalar b. If $b = 0$, then ε passes through the origin O. Otherwise, equations $P \cdot X = 0$ and $P \cdot X = b$ have no common solution, so the former represents the plane $\delta \parallel \varepsilon$ through O. Moreover, $\vec{OP}$ is the perpendicular to δ and ε through O. Since there's only one such line, it follows that equations $P \cdot X = b$ and $Q \cdot X = c$ represent equal or parallel planes just when $P = sQ$ for some scalar s.

You can write any linear equation $P \cdot X = b$ with a dot product $P^* \cdot X^*$ of *four* dimensional vectors P^* and X^* formed by appending fourth entries $p_4 = -b$ and $x_4 = 1$ to P and X:

$$P \cdot X = [p_1 \;\; p_2 \;\; p_3] \begin{bmatrix} x_1 \\ x_2 \\ x_3 \end{bmatrix} = b \leftrightarrow P^* \cdot X^* = [p_1 \;\; p_2 \;\; p_3 \;\; p_4] \begin{bmatrix} x_1 \\ x_2 \\ x_3 \\ 1 \end{bmatrix} = 0.$$

Now consider four planes ε_i with equations

$$P_i^* \cdot X^* = p_{i1}x_1 + p_{i2}x_2 + p_{i3}x_3 + p_{i4} = 0$$

for $i = 1$ to 4. The rows P_i^{*t} form a 4×4 matrix P^*. If the planes have a common point X, then $P^*X^* = O$ because for each i, its ith entry is the matrix product of (the ith row of P^*) times X^*. Since $X^* \neq O$, P^* is singular.

P^* is also singular when the planes ε_i are all parallel to a single line. In that case the planes $\delta_i \parallel \varepsilon_i$ through O all contain a single line g through O. Take any point $X \neq O$ on g. Then

$$P^* \begin{bmatrix} x_1 \\ x_2 \\ x_3 \\ 0 \end{bmatrix} = 0$$

because for each i, $p_{i1}x_1 + p_{i2}x_2 + p_{i3}x_3 = 0$ is an equation for δ_i.

Theorem 11. $\det P^* = 0$ if and only if planes ε_1 to ε_4 have a common point or are all parallel to a single line.

Proof. The previous discussion established the *if* clause. Conversely, suppose P^* is singular. Then $P^*W = 0$ for some four-dimensional vector $W \neq 0$. If $w_4 \neq 0$, then $X = \langle w_1/w_4, w_2/w_4, w_3/w_4 \rangle$ is a point on each ε_i because $P^*X^* = P^*(W(1/w_4)) = (P^*W)(1/w_4) = 0$. ♦

The plane analog of theorem 11 is simpler. Set up a fixed coordinate system in a plane ε, with origin O. A line g in ε has linear equation $P \cdot X = b$ for some vector $P \neq O$ and some scalar b. Equation $P \cdot X = 0$ represents the line $h \parallel g$ through O. Equations $P \cdot X = b$ and $Q \cdot X = c$ represent equal or parallel lines just when $P = sQ$ for some scalar s. You can rewrite equation $P \cdot X = b$ as $P^* \cdot X^* = 0$ with a dot product of *three*-dimensional vectors P^* and X^* formed by appending third entries $p_3 = -b$ and $x_3 = 1$ to P and X. Now consider three lines g_i with equations

$$P_i^* \cdot X^* = p_{i1}x_1 + p_{i2}x_2 + p_{i3} = 0$$

for $i = 1$ to 3. The rows P_i^t form a 3×3 matrix P^*.

Theorem 12. $\det P^* = 0$ if and only if lines g_1 to g_3 have a common point or are all parallel.

5.7 Centroid

Concepts

- Notation for centroid and medians
- Centroid as average of the vertices
- Centroid of any finite point set
- Minimizing the sum of squares of the distances from a point to three vertices

This section presents a few pretty and representative properties of the centroid G and medians of a triangle ΔABC, shown in figure 5.7.1. Theorems 1 to 3 present some properties that you might call "balancing" or "averaging." You can supply the simple proof of theorem 1.

Theorem 1. Each median divides ΔABC into two triangles with equal areas. The three medians divide it into six triangles with equal areas. The segments between the vertices and the centroid divide it into three triangles with equal areas. If d_a, d_b, and d_c are the distances from the centroid to edge lines a to c, then $ad_a = bd_b = cd_c$.

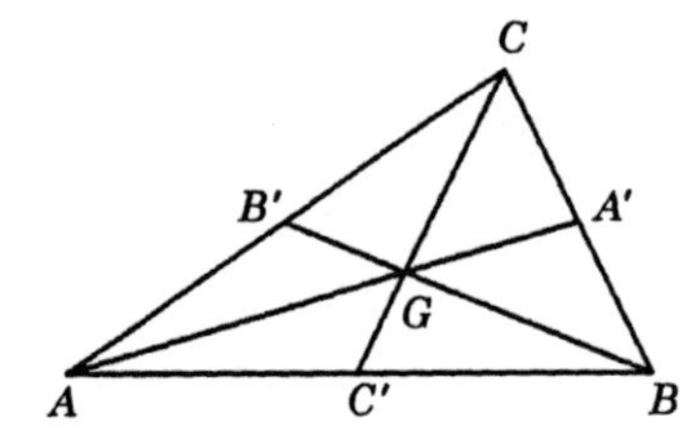

Figure 5.7.1
The edge midpoints A', B', and C'; the medians; and the centroid G of ΔABC

Theorem 2. If you set up a coordinate system and regard A to C and G as coordinate vectors, then

$$G = \tfrac{1}{3}(A + B + C) \qquad (A - G) + (B - G) + (C - G) = 0.$$

Proof. $A' = \tfrac{1}{2}(B + C)$ and by the medians theorem, $G = \tfrac{1}{3}(A + 2A')$. ♦

Theorem 2 suggests a generalization of the centroid concept to arbitrary finite sets of points P_1 to P_n: Set up a coordinate system, regard the points as coordinate vectors, and define their centroid G as the average $(P_1 + \cdots + P_n)/n$. For this to make sense, however, you need to show that you get the same G no matter what coordinate system you use. That problem is addressed in sections 6.1 and 6.2 under the **Invariance** headings.

Theorem 3. Consider a line g in plane ABC, and call one of its sides positive. Let e and e_a to e_c denote + or - the distances to g from G and A to C. Use + just on the positive side of g. Then $e = \tfrac{1}{3}(e_a + e_b + e_c)$.

Proof. Let e_a' denote + or - the distance from A' to g, with the sign chosen the same way. Then $e_a' = \tfrac{1}{2}e_b + \tfrac{1}{2}e_c$ and $e = \tfrac{1}{3}e_a + \tfrac{2}{3}e_a'$. (See figure 5.7.2.) ♦

The remaining properties considered in this section involve a pretty formula for the sum of the squares of the lengths m_a, m_b, and m_c of the medians.

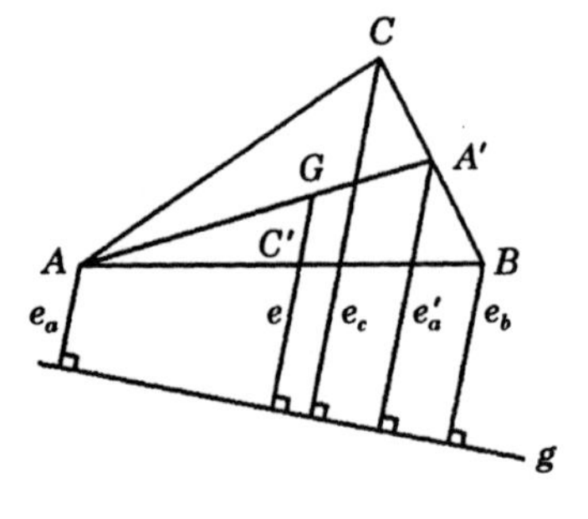

Figure 5.7.2
Theorem 3

The formula is the first of several in this chapter, selected mainly for their elegance. In turn, these are just a small sample of a great wealth of such formulas that you can find in the literature.

Theorem 4.

$$\begin{aligned} m_a^2 &= -\tfrac{1}{4} a^2 + \tfrac{1}{2} b^2 + \tfrac{1}{2} c^2 \\ m_b^2 &= \tfrac{1}{2} a^2 - \tfrac{1}{4} b^2 + \tfrac{1}{2} c^2 \\ m_c^2 &= \tfrac{1}{2} a^2 + \tfrac{1}{2} b^2 - \tfrac{1}{4} c^2 \\ m_a^2 + m_b^2 + m_c^2 &= \tfrac{3}{4}(a^2 + b^2 + c^2). \end{aligned}$$

Proof. Apply the law of cosines to ΔABC and $\Delta ABA'$ as in figure 5.7.1:

$$\begin{aligned} b^2 &= c^2 + a^2 - 2ca \cos \mathrm{m}\angle B \\ m_a^2 &= (\tfrac{1}{2} a)^2 + c^2 - 2(\tfrac{1}{2} a)c \cos \mathrm{m}\angle B. \end{aligned}$$

Derive $m_a^2 = -\tfrac{1}{4} a^2 + \tfrac{1}{2} b^2 + \tfrac{1}{2} c^2$ by subtracting twice the second equation from the first. You can supply the remainder of the proof. ♦

Few of the elegant triangle geometry formulas such as theorem 4 find immediate application. This one, however, leads to an interesting minimal property of the centroid: G is the point to which the sum of the squares of the distances from the vertices is least. Proving that requires a preliminary computation:

Lemma 5. $(PA)^2 + (PB)^2 + (PC)^2 = 3(PG)^2 + \tfrac{1}{3}(a^2 + b^2 + c^2)$ for any point P.

Proof. If $P = G$, this equation is just theorem 4. Therefore, let $P \neq G$. Construct the line $g \perp \overleftrightarrow{PG}$ through G, and regard as positive the side of g on which P lies. As in theorem 3, let e_a to e_c be + or − the distances from g to the corresponding vertices. Use + on the positive side. By theorem 3, $e_a + e_b + e_c = 0$ because G lies on g. Now consider the case in figure 5.7.3, where P isn't collinear with A and G. By the law of cosines,

$$\begin{aligned} (PA)^2 &= (PG)^2 + (GA)^2 - 2(PG)(GA) \cos \mathrm{m}\angle PGA \\ &= (PG)^2 + (GA)^2 - 2(PG)\, e_a. \end{aligned}$$

This equation holds even if P, G, and A fall on a line, because then $PA = \pm(PG - e_a)$. Similar equations hold for the other vertices. Adding all three, you get

$$\begin{aligned} &(PA)^2 + (PB)^2 + (PC)^2 \\ &\quad = 3(PG)^2 + (GA)^2 + (GB)^2 + (GC)^2 - 2(PG)(e_a + e_b + e_c) \\ &\quad = 3(PG)^2 + (GA)^2 + (GB)^2 + (GC)^2 \\ &\quad = 3(PG)^2 + (\tfrac{2}{3} m_a)^2 + (\tfrac{2}{3} m_b)^2 + (\tfrac{2}{3} m_c)^2 \\ &\quad = 3(PG)^2 + \tfrac{4}{9}(m_a^2 + m_b^2 + m_c^2) \\ &\quad = 3(PG)^2 + \tfrac{4}{9} \cdot \tfrac{3}{4}(a^2 + b^2 + c^2). \; ♦ \end{aligned}$$

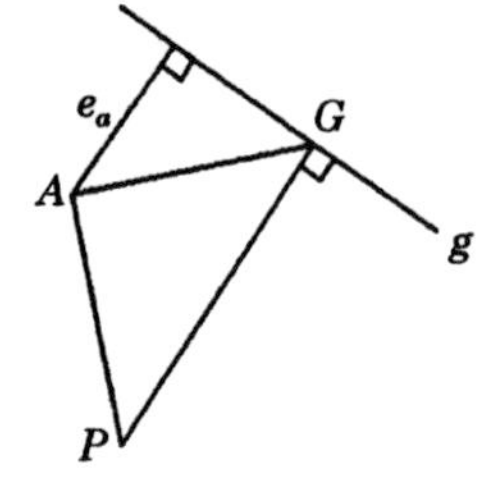

Figure 5.7.3
Proof of lemma 5

From lemma 5 you can see that $(PA)^2 + (PB)^2 + (PC)^2$ achieves its minimum value when $PG = 0$, which yields

Theorem 6. The centroid is the point to which the sum of the squares of the distances from the vertices is least. This minimum value is $^3/_4(a^2 + b^2 + c^2)$.

5.8 Orthocenter

Concepts
Notation for the feet of the altitudes of a triangle
Orthic triangle
Orthocentric quadruples

The feet of the altitudes of ΔABC through vertices A, B, and C are usually denoted by D, E, and F. When ΔABC is oblique, they form ΔDEF, the *orthic* triangle of ΔABC. You'll see it play an important role in the rest of this chapter. Its list of fascinating properties continues even further. For example, if ΔABC is acute, the orthic triangle has the shortest perimeter of any triangle whose vertices lie on each of the edges of ΔABC. Proving that requires methods best introduced later, in section 6.3, so it's presented as exercise 6.11.7.

Theorem 1. The altitudes and orthic triangle of an oblique ΔABC form angles as shown in figure 5.8.1. Their complements and sums satisfy the indicated equations. In particular, the altitudes of ΔABC bisect the angles of the orthic triangle, and $\Delta ABC \sim \Delta AEF \sim \Delta DBF \sim \Delta DEC$.

Proof. Because $\angle AEH$ and $\angle AFH$ are right, E and F lie on circle Γ_a with diameter $\overline{AH}$, and $m\angle AEF + m\angle FEH = 90° = m\angle AFE + m\angle EFH$. Figure 5.8.2 displays Γ_a and the corresponding part of figure 5.8.1. You can verify every angle congruence indicated there by observing that

$$m\measuredangle = 90° - m\measuredangle = m\measuredangle + m\measuredangle$$

$$m\measuredangle = 90° - m\measuredangle = m\measuredangle + m\measuredangle$$

$$m\measuredangle = 90° - m\measuredangle = m\measuredangle + m\measuredangle$$

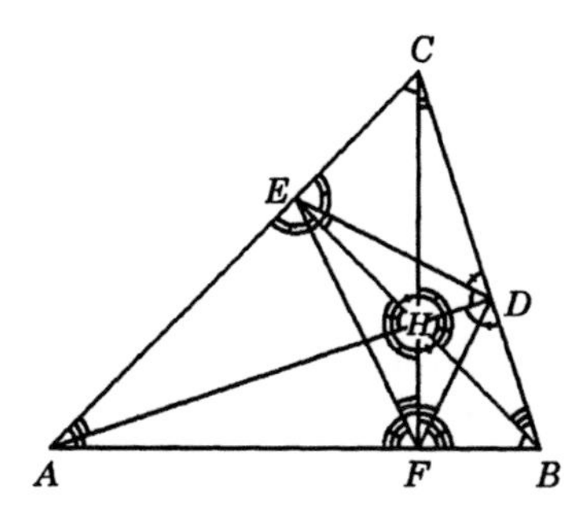

Figure 5.8.1
Theorem 1

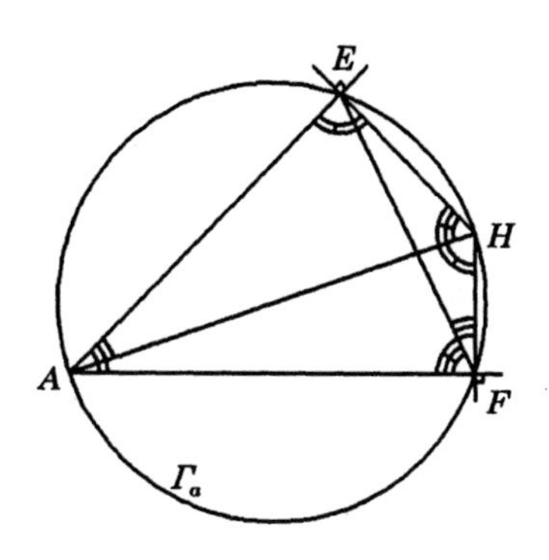

Figure 5.8.2
Proving theorem 1

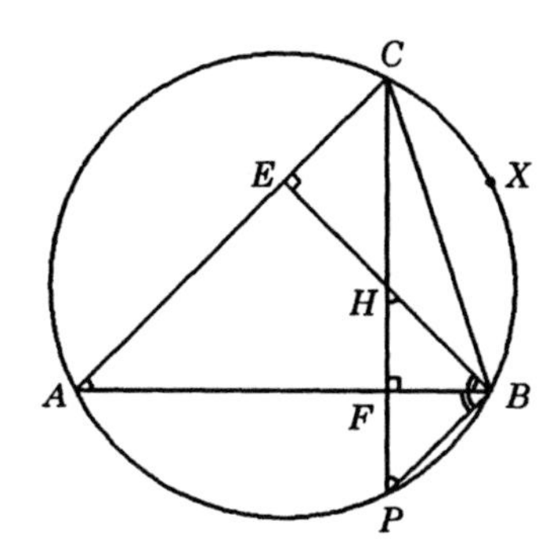

Figure 5.8.3
Theorem 2

the angles are inscribed in the same arc of Γ_a. Now, notice that $\angle AHE \cong \angle BHD$, move to circle Γ_b with diameter $\overline{BH}$, and verify the corresponding angle congruences there. In the same way, you can handle circle Γ_c with diameter $\overline{CH}$, which contains all the remaining angle congruences of figure 5.8.1. It remains to verify the rightmost equations in that figure, which involve sums of angle measures. They merely reflect the sums of the angle measures in right triangles ΔAFC, ΔBDA, and ΔCEB. ♦

The next result isn't particularly interesting in itself, but it finds many applications in triangle and circle geometry.

Theorem 2. If altitude line $\overleftrightarrow{CF}$ intersects the circumcircle of ΔABC at $P \neq C$, then $HF = FP$.

Proof. Suppose $\angle A$ and $\angle B$ are acute, as in figure 5.8.3. By the triangle sum theorem, $m\angle FHB = 90° - m\angle ABE = m\angle A$. Moreover, $\angle A \cong \angle HPB$

because they're inscribed in the circumcircle and subtend the same arc $\widehat{BXC}$. By the ASA theorem, $\Delta HFB \cong \Delta PFB$, so $HF = FP$.

You can provide a similar proof for a triangle with $\angle A$ or $\angle B$ obtuse. If $\angle A$ or $\angle B$ is right, then $H = F = P$. ♦

A quadruple $<A, B, C, H>$ of points is called *orthocentric* if A, B, and C form an oblique triangle and H is its orthocenter. Theorem 3, which is apparent from figure 5.8.1, says that you could record the four points of an orthocentric quadruple in any order.

Theorem 3. Any three points of an orthocentric quadruple form an oblique triangle whose orthocenter is the fourth point.

Theorem 4. The four triangles formed by an orthocentric quadruple have the same circumradius. Conversely, if three coplanar circles with radius R pass through a point H and intersect by twos at three distinct points A, B, and C, then $<A, B, C, H>$ is an orthocentric quadruple and A, B, and C lie on a circle with radius R.

Proof. Suppose an orthocentric quadruple $<A, B, C, H>$ forms four triangles as in figure 5.8.4. Consider the point P in figure 5.8.3. By theorem 2, $\Delta HBA \cong \Delta PBA$, so they have the same circumradius. But ΔPBA and ΔABC have the same circumcircle, so ΔHBA and ΔABC have the same circumradius.

Conversely, suppose circles with centers O_a, O_b, and O_c and radius R intersect as in figure 5.8.5. Then O_aHO_bC, O_bHO_cA, and O_cHO_aB are rhombi, so

$$\overleftrightarrow{AO_b} \parallel \overleftrightarrow{HO_c} \parallel \overleftrightarrow{BO_a} \qquad \overleftrightarrow{BO_c} \parallel \overleftrightarrow{HO_a} \parallel \overleftrightarrow{CO_b} \qquad \overleftrightarrow{CO_a} \parallel \overleftrightarrow{HO_b} \parallel \overleftrightarrow{AO_c}.$$

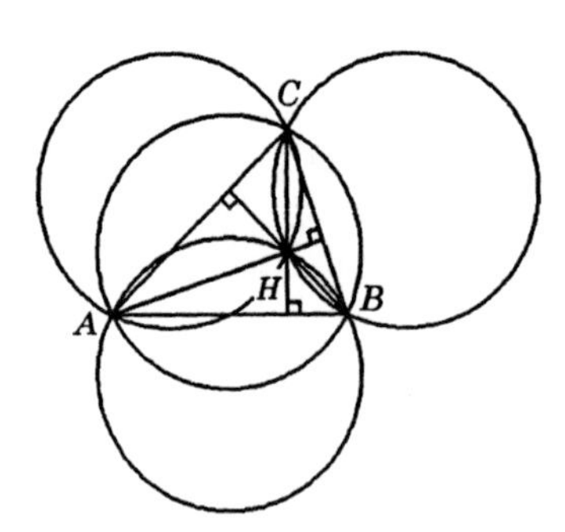

Figure 5.8.4
Theorem 4

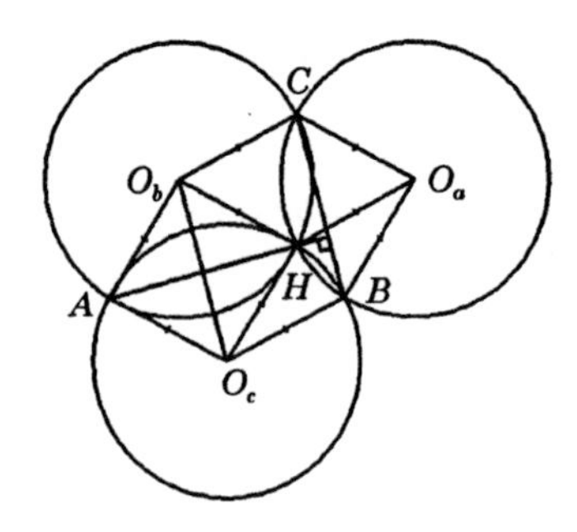

Figure 5.8.5
Proving theorem 4

By the parallel case of the second form of Desargues' theorem, $\overleftrightarrow{BC} \parallel \overleftrightarrow{O_bO_c}$. Consider the line $g \perp \overleftrightarrow{BC}$ through H. It's also perpendicular to $\overleftrightarrow{O_bO_c}$, so it must pass through A. That is, H lies on the altitude of ΔABC through A. Similarly, it lies on the other altitudes. ♦

5.9 Incenter and excenters

Concepts

Notation for the incircle and excircles of a triangle
Gergonne and Nagel points
Area formulas involving the inradius and exradii
Harmonic quadruples of points

The first results in this section complete the discussion of angle bisectors in section 5.1. You can prove theorem 1 by extending the techniques used there to derive the angle bisectors theorem. The earlier discussion of the *bisector* of an angle $\angle ABC$ —the line g containing all points equidistant from $\overleftrightarrow{AB}$ and $\overleftrightarrow{BC}$ and interior to $\angle ABC$ —led to the notions of incenter and incircle. The line $h \perp g$ through B in plane ABC also consists of points equidistant from $\overleftrightarrow{AB}$ and $\overleftrightarrow{BC}$. It's called an *external angle bisector* of ΔABC.

Theorem 1. As shown in figure 5.9.1, the interior and exterior angle bisectors of ΔABC meet by threes at the incenter I and three points E_a, E_b, E_c that are also equidistant from the triangle's edge lines. These four points form an orthocentric quadruple. ΔABC is the orthic triangle of $\Delta E_aE_bE_c$.

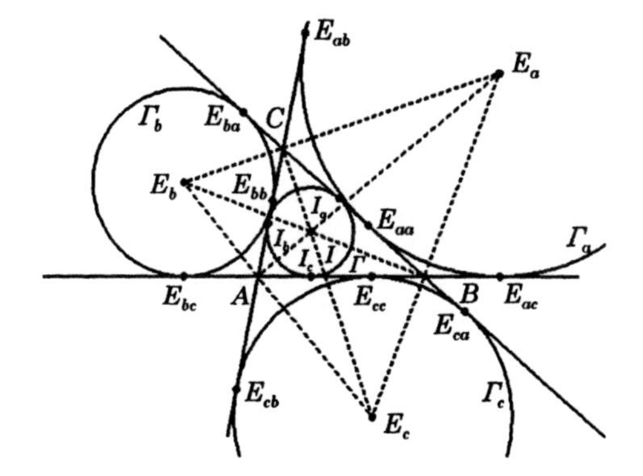

Figure 5.9.1
Theorem 1

Points E_a, E_b, E_c are called the *excenters* of ΔABC. Each is equidistant from the edge lines; these distances are called the *exradii* r_a, r_b, r_c of ΔABC. In figure 5.9.1, the circle Γ with center I and tangent to ΔABC at points I_a, I_b, I_c is the incircle. Its radius is r, the inradius. The circles $\Gamma_a, \Gamma_b, \Gamma_c$ whose centers and radii are the corresponding excenters and exradii are called the *excircles* of ΔABC. They touch the edge lines at points $E_{aa}, E_{ab}, \ldots$ as indicated in the figure. The next theorem—a tedious prerequisite for some surprising results—gives formulas for many of the distances in figure 5.9.1, in terms of the semiperimeter s of ΔABC.

Theorem 2.

$$\begin{aligned} s &= AE_{ab} = AE_{ac} = BE_{ba} = BE_{bc} = CE_{ca} = CE_{cb} \\ s - a &= AI_b = AI_c = BE_{ca} = BE_{cc} = CE_{ba} = CE_{bb} \\ s - b &= AE_{cb} = AE_{cc} = BI_a = BI_c = CE_{aa} = CE_{ab} \\ s - c &= AE_{bb} = AE_{bc} = BE_{aa} = BE_{ac} = CI_a = CI_b. \end{aligned}$$

Proof. These formulas all stem from the fact that any point P outside a circle Γ is equidistant from the points of contact of the two tangents through P. For example,

$$\begin{aligned} AE_{ab} &= b + CE_{ab} \qquad\qquad = b + CE_{aa} \\ &= b + a - BE_{aa} \qquad\quad = b + a - BE_{ac} \\ &= b + a - (AE_{ac} - c) = a + b + c - AE_{ac} = 2s - AE_{ab} \\ 2AE_{ab} &= 2s. \end{aligned}$$

(This method of reasoning is useful for exercise 4.6.6.) ♦

You can use these formulas with Ceva's theorem to derive several concurrence results. Two examples are shown in figure 5.9.2:

Corollary 3. Segments $\overline{AI_a}$, $\overline{BI_b}$, and $\overline{CI_c}$ are concurrent, as are segments $\overline{AE_{aa}}$, $\overline{BE_{bb}}$, and $\overline{CE_{cc}}$.

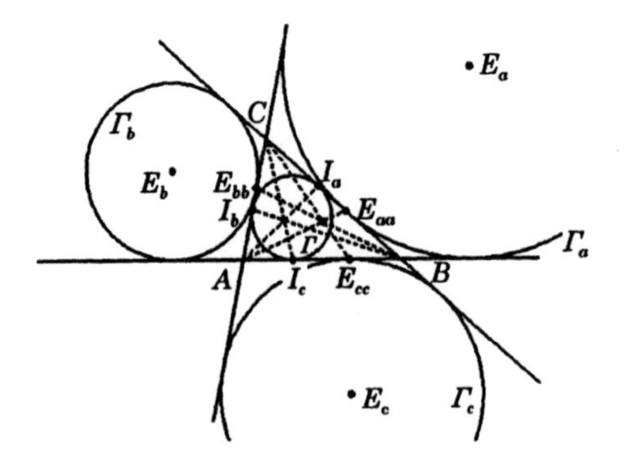

Figure 5.9.2
Constructing the Gergonne and Nagel points

The intersections of these triples of concurrent segments are called the triangle's *Gergonne* and *Nagel points*.

Theorem 4. The area $\text{a}\Delta ABC = rs = r_a(s - a) = r_b(s - b) = r_c(s - c)$.

Proof. $\text{a}\Delta ABC = \text{a}\Delta IAB + \text{a}\Delta IBC + \text{a}\Delta ICA = \frac{1}{2}\,cr + \frac{1}{2}\,ar + \frac{1}{2}\,br = rs$. To derive the second formula, note that $\Delta II_bA \sim \Delta E_aE_{ab}B$ in figure 5.9.1, so that by theorem 1, $r/r_a = II_b/E_aE_{ab} = I_bA/E_{ab}B = (s - a)/s$. ♦

You can use theorem 4 with the area formulas in section 5.5 to derive some spectacular equations about distances and angles related to a triangle. Exercises 5.11.25 and 5.11.26 provide samples.

The last part of this section features some very special properties of figure 5.9.1. Through the material in the next section, they lead to the final topic of this chapter, Feuerbach's theorem. That amazing result relates the incircle and excircles to another triangle and circle that you've already started to analyze. Part of the argument leading to Feuerbach's theorem is presented here in order to streamline section 5.10.

The next result involves a technique that has been studied deeply and used in many different geometric investigations: harmonic quadruples of points. Since Feuerbach's theorem is the only one of those applications included in this book, the theory of harmonic points is not fully developed here. Only the necessary features are mentioned. You may want to pursue this attractive subject further in another text.[14] A quadruple $<A, B, C, D>$ of points on a line g to which you've assigned a scale is called *harmonic* if

$$(A \text{ to } B)(C \text{ to } D) = (A \text{ to } D)(B \text{ to } C).$$

Clearly, the validity of this equation is independent of your choice of scale. Take care when you encounter this notion in the literature; you'll find many equivalent forms of the equation, with the letters permuted, or using ratios instead of products. The first result about harmonic quadruples provides an alternative equation that's simpler for some algebraic manipulations. The next ones show that certain constructions—in particular part of figure 5.9.1—yield harmonic quadruples.

Theorem 5. Let A, B, C, and D be collinear points and O be the midpoint of $\overline{BD}$. Then $<A, B, C, D>$ is a harmonic quadruple if and only if $(OB)^2 = (O \text{ to } A)(O \text{ to } C)$.

Proof. Choose a scale so that O has coordinate 0 and points A, B, C, and D have coordinates a, b, c, and $-b$. Then $<A, B, C, D>$ is harmonic

[14] For example, Davis 1949, chapter 6, or Eves 1963–1965, chapter 2.

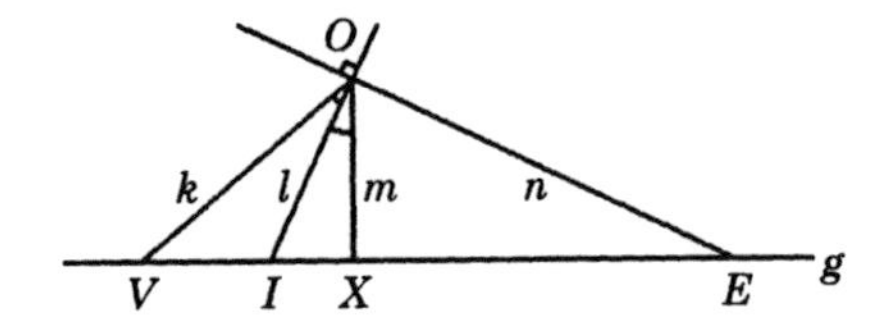

Figure 5.9.3
Theorem 6

$$\Leftrightarrow (A \text{ to } B)(C \text{ to } D) = (A \text{ to } D)(B \text{ to } C)$$
$$\Leftrightarrow (b - a)(-b - c) = (-b - a)(c - b)$$
$$\Leftrightarrow b^2 = ac$$
$$\Leftrightarrow (OB)^2 = (O \text{ to } A)(O \text{ to } C). \ \blacklozenge$$

Theorem 6. Suppose O is a point not on a line g; suppose lines k, l, m, and n pass through O and intersect g at points V, I, X, and E in the order V-I-X-E; and suppose $l \perp n$, as in figure 5.9.3. Then $I \neq E$, and $<V, I, X, E>$ is harmonic just in case $\angle VOI \cong \angle IOX$.

Proof. Clearly, $I \neq E$. Choose a scale for g so that $(V \text{ to } I)$ is positive. Then by the law of sines, $<V, I, X, E>$ is harmonic

$$\Leftrightarrow (VI)(XE) = (VE)(IX) \Leftrightarrow \frac{VI}{OI}\frac{XE}{OE} = \frac{VE}{OE}\frac{IX}{OI}$$

$$\Leftrightarrow \frac{\sin m\angle VOI}{\sin m\angle OVI}\frac{\sin m\angle XOE}{\sin m\angle OXE} = \frac{\sin m\angle VOE}{\sin m\angle OVE}\frac{\sin m\angle IOX}{\sin m\angle OXI}.$$

Since $\angle OVI \cong \angle OVE$ and $m\angle OXE = 180° - m\angle OXI$, it follows that $<V, I, X, E>$ is harmonic

$$\Leftrightarrow \sin m\angle VOI \sin m\angle XOE = \sin m\angle VOE \sin m\angle IOX$$
$$\Leftrightarrow \sin m\angle VOI \cos m\angle IOX = \cos m\angle VOI \sin m\angle IOX.$$

Since $\cos m\angle VOI \neq 0 \neq \cos m\angle IOX$, it follows that $<V, I, X, E>$ is harmonic

$$\Leftrightarrow \tan m\angle VOI = \frac{\sin m\angle VOI}{\cos m\angle VOI} = \frac{\sin m\angle IOX}{\cos m\angle IOX} = \tan m\angle IOX$$

$$\Leftrightarrow \angle VOI \cong \angle IOX. \ \blacklozenge$$

Corollary 7. In figure 5.9.1, let X denote the intersection of angle bisector $\overleftrightarrow{AI}$ with edge a. Then $<A, I, X, E_a>$ is a harmonic quadruple. Analogous results hold for the other bisectors.

Proof. This part of figure 5.9.1 is detailed in figure 5.9.4. Let C, A, and E_a play the roles of O, V, and E in theorem 6. ♦

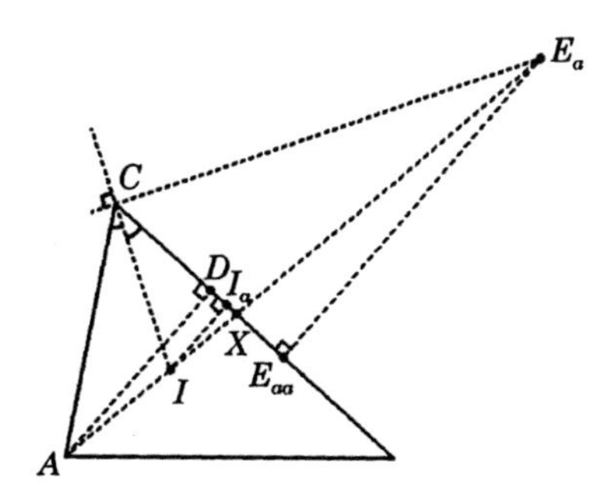

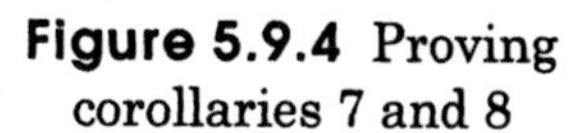
Figure 5.9.4 Proving corollaries 7 and 8

Corollary 8. In figure 5.9.1, let X and D denote the intersections of edge $\overline{BC}$ with angle bisector $\overleftrightarrow{AI}$ and the altitude through A. Then $<D, I_a, X, E_{aa}>$ is harmonic.

Proof. This part of figure 5.9.1 is detailed in figure 5.9.4. The altitude and radial lines $\overleftrightarrow{II_a}$ and $\overleftrightarrow{E_aE_{aa}}$ are all perpendicular to the edge line, hence they're parallel. By corollary 7, $(AI)(XE_a) = (AE_a)(IX)$, hence

$$\frac{AI}{IX} = \frac{AE_a}{XE_a}.$$

But distance ratios are invariant under parallel projection, so

$$\frac{AI}{IX} = \frac{DI_a}{I_aX} \qquad \frac{AE_a}{XE_a} = \frac{DE_{aa}}{XE_{aa}}.$$

Therefore, $(DI_a)(XE_{aa}) = (DE_{aa})(I_aX)$. ♦

Corollary 9. In figure 5.9.1, let X denote the intersection of edge a with the opposite angle's bisector $\overleftrightarrow{AI}$. Then edge midpoint A' is the midpoint of $\overline{I_aE_{aa}}$ and $(A'I_a)^2 = (A'D)(A'X)$. Analogous results hold for the other corresponding bisectors and edges.

Proof. $BI_a = s - b = CE_{aa}$ by theorem 2. Thus A' is the midpoint of $\overline{I_aE_{aa}}$. Now apply corollary 8 and theorem 5. ♦

Theorem 10. Let X denote the intersection of edge a of ΔABC with angle bisector $\overleftrightarrow{AI}$ and suppose $\angle B \not\cong \angle C$. Then $X \neq I_a$ and X lies on two tangents to the incircle Γ. One is the edge line and the other touches Γ at a point P. $\overleftrightarrow{AI}$ also bisects $\angle PXI_a$. Moreover,

- if $m\angle B > m\angle C$ as in figure 5.9.5, then B-I_a-X and $m\angle CXP = m\angle B - m\angle C$;
- if $m\angle B < m\angle C$ as in figure 5.9.1, then B-X-I_a and $m\angle CXP = m\angle B - m\angle C + 180°$.

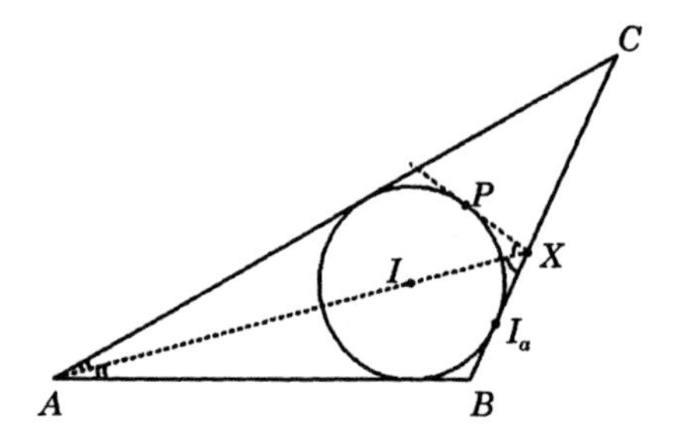

Figure 5.9.5 Theorem 10 in case $m\angle B > m\angle C$

Proof. By the SSS congruence theorem, $\Delta IXP \cong \Delta IXI_a$, hence $\angle IXP \cong \angle IXI_a$. Either $B\text{-}I_a\text{-}X$ and P lies in the interior of $\angle AXC$, or $B\text{-}X\text{-}I_a$ and P lies in the interior of $\angle AXB$. Since $\angle AXP \cong \angle AXI_a$, P must lie interior to whichever of $\angle AXB$ and $\angle AXC$ has the larger measure. Since $m\angle AXB = 180° - \tfrac{1}{2}\, m\angle A - m\angle B$ and $m\angle AXC = 180° - \tfrac{1}{2}\, m\angle A - m\angle C$ by the triangle sum theorem, P lies interior to $\angle AXC$ if $m\angle B > m\angle C$, hence $B\text{-}I_a\text{-}X$ in that case, and $B\text{-}X\text{-}I_a$ in the other. In case $m\angle B > m\angle C$,

$$\begin{aligned} m\angle CXP &= 180° - 2\, m\angle BXA \\ &= 180° - 2\,(180° - m\angle B - m\angle BAX) \\ &= -180° + 2\, m\angle B + 2\, m\angle BAX \\ &= m\angle A + 2\, m\angle B - 180° \\ &= m\angle A + 2\, m\angle B - (m\angle A + m\angle B + m\angle C) \\ &= m\angle B - m\angle C. \end{aligned}$$

You can verify the analogous equation for the other case. ♦

5.10 Euler line and Feuerbach circle

Concepts

Euler line
Feuerbach, or nine point circle
Power of a point relative to a circle.

If ΔABC isn't equilateral, then one of its vertices isn't equidistant from the other two, hence isn't on the perpendicular bisector g of the opposite edge. The altitude line h perpendicular to that edge doesn't intersect g. Since the circumcenter and orthocenter O and H lie on g and h, respectively, these points are different. $\overleftrightarrow{OH}$ is called the *Euler line* of ΔABC.

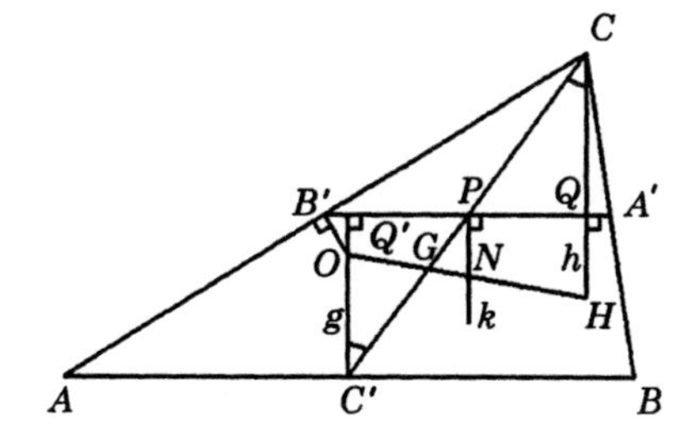

Figure 5.10.1
Theorems 1 and 3

Theorem 1. The centroid G of a nonequilateral triangle is on the Euler line, two thirds of the way from H to O.

Proof. If $\angle C$ is right, then $C = H$, $C' = O$, and the result is just the medians theorem. Assume $\angle C$ isn't right, as in figure 5.10.1. Then O is the orthocenter of the medial triangle $\Delta A'B'C'$. Since that's similar to ΔABC with ratio ½, $OC' = ½\,HC$. Further, $\angle OC'G \cong \angle HCG$ because $\overleftrightarrow{OC'} \parallel \overleftrightarrow{HC}$. By the medians theorem, $C'G = ½\,CG$. Therefore, $\Delta OC'G \sim \Delta HCG$ with ratio ½ by the SAS similarity theorem. It follows that $\angle C'GO \cong \angle CGH$. Therefore C, G, and H are collinear. ♦

Corollary 2. A nonequilateral triangle and its medial triangle have the same Euler line.

Proof. They have the same centroid. ♦

The circumcircle of the medial triangle $\Delta A'B'C'$ is called the *Feuerbach circle* of ΔABC.

Theorem 3. The midpoint N of segment $\overline{OH}$ is the center of the Feuerbach circle. Its radius is half the circumradius R of ΔABC.

Proof. Segments $\overline{CC'}$ and $\overline{A'B'}$ have the same midpoint P. Let Q and Q' be the feet of the perpendiculars to $\overleftrightarrow{A'B'}$ through C and C'. Then $PQ = PQ'$. The perpendicular bisector k of $\overline{A'B'}$ is parallel to $\overleftrightarrow{OC'}$ and intersects $\overline{OH}$ at a point N'. Since distance ratios are invariant under parallel projection, $N'H = N'O$. That is, $N' = N$. By analogous arguments, N also lies on the perpendicular bisectors of the other two edges of the medial triangle. ♦

The Feuerbach circle is often called the *nine point* circle of ΔABC. Theorem 4 shows why. It was discovered and published independently by Charles-

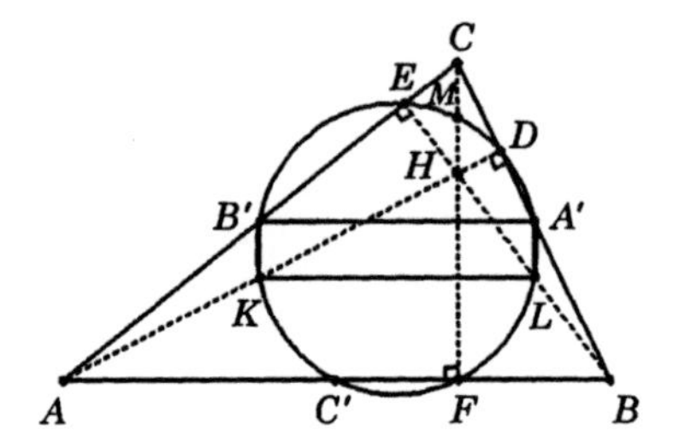

Figure 5.10.2
Theorem 4

Julien Brianchon and Victor Poncelet in 1821 and by Karl Wilhelm Feuerbach in 1822.[15]

Theorem 4 **(*Nine point theorem*).** The Feuerbach circle passes through the feet of the altitudes, so it's also the circumcircle of the orthic triangle. Moreover, it passes through the midpoints K, L, and N of segments $\overline{AH}$, $\overline{BH}$, and $\overline{CH}$.

Proof. By theorem 3.9.2, $\overleftrightarrow{LA'} \parallel \overleftrightarrow{HC} \parallel \overleftrightarrow{KB'}$ and $\overleftrightarrow{KL} \parallel \overleftrightarrow{AB} \parallel \overleftrightarrow{A'B'}$, as shown in figure 5.10.2. Moreover, $\overleftrightarrow{HC} \perp \overleftrightarrow{AB}$, so $A'B'KL$ is a rectangle, so $\overline{KA'}$ and $\overline{LB'}$ are diameters of a circle Φ. Similarly, $B'C'LM$ is a rectangle, so $\overline{LB'}$ and $\overline{MC'}$ are diameters of a circle, which must coincide with Φ. Therefore Φ is the Feuerbach circle and K, L, and M lie on it. Since $\angle KDA'$ is right, D lies on Φ. By analogous arguments, E and F lie on Φ. ♦

Corollary 5. The four triangles determined by an orthocentric quadruple have the same Feuerbach circle.

Corollary 6. The Feuerbach circle of the orthocentric quadruple consisting of the incenter and excenters of a triangle is its circumcircle.

The rest of the section is devoted to Feuerbach's theorem, the most involved result in this chapter: The Feuerbach circle of a triangle is *tangent* to the incircle and excircles. Several methods of proof are available. Feuerbach's original derivation depended heavily on distance computations using the law of cosines, similar to those near the end of section 5.5. They involved stupendous algebraic manipulations. The proof presented here, selected for its simplicity, was published by Roger A. Johnson in 1929.[16]

---

[15] The original publications are available in translation: Brianchon and Poncelet 1820 and Feuerbach 1822.

[16] Johnson [1929] 1960, chapter XI.

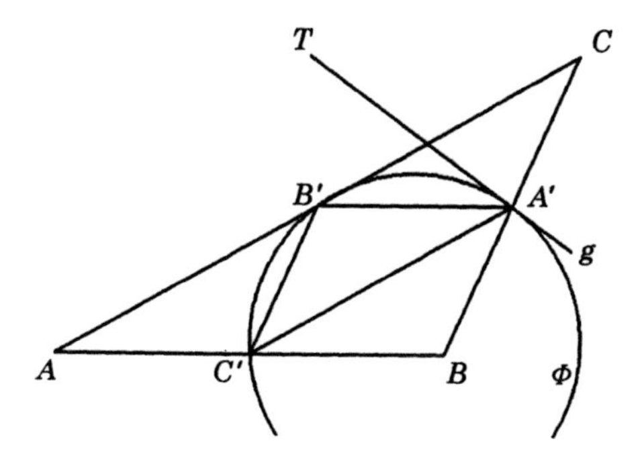

Figure 5.10.3
Lemma 7

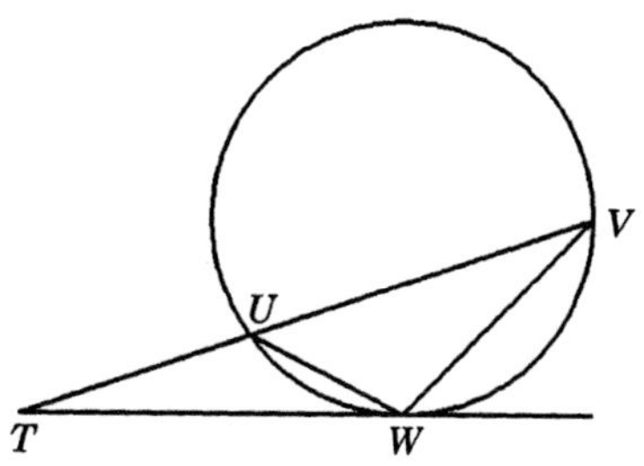

Figure 5.10.4
Lemma 8

If $\angle B \cong \angle C$, then clearly the Feuerbach circle, the incircle, and the excircle opposite A are all tangent at A'. Thus, to prove the theorem, you may assume $\text{m}\angle B > \text{m}\angle C$. (You can exchange B and C in the following argument to construct an analogous one for the opposite case.) Lemma 7 is the first step.

Lemma 7. Suppose $\text{m}\angle B > \text{m}\angle C$, as in figure 5.10.3, so that edge $\overleftrightarrow{BC}$ isn't tangent to the Feuerbach circle Φ. Let g be the tangent to Φ at A', and select a point T on g interior to $\angle CA'B'$. Then $\text{m}\angle CA'T = \text{m}\angle B - \text{m}\angle C$.

Proof. First, $\text{m}\angle CA'T = \text{m}\angle CA'B' - \text{m}\angle TA'B'$. Now use corollary 3.13.6 to show $\angle TA'B' \cong \angle A'C'B'$. (See exercise 4.6.5.) Clearly, $\angle CA'B' \cong \angle B$ and $\angle A'C'B' \cong \angle C$. ♦

The next step is a simple general result about intersecting tangent and secant lines.

Lemma 8. Let points U, V, and W lie on a circle Γ as in figure 5.10.4. Consider a point $T \neq W$ on the tangent to Γ at W, and suppose T-U-V. Then $(TU)(TV) = (TW)^2$.

Proof. Use corollary 3.13.6 as in the proof of lemma 7 to show $\angle TWU \cong \angle TVW$. Then $\Delta TWU \sim \Delta TVW$ by the AA theorem, hence $TU/TW = TW/TV$. ♦

Lemma 8 implies that for any points U and V on Γ collinear with T, the product $(TU)(TV)$ has the same value; it's called the *power* of T relative to Γ. This notion underlies extensive theories of coaxial systems of circles and circular inversion.[17] But the power concept is used only once in this book, in lemma 9, the next step en route to Feuerbach's theorem. You'll find in the literature some proofs of the theorem that make much more sophisticated use of those theories.

Reconsider figures 5.9.1 and 5.9.5, in particular the intersection X of edge line $\overleftrightarrow{BC}$ with bisector $\overleftrightarrow{AI}$, and the point P not on $\overleftrightarrow{BC}$ where a tangent from X touches the incircle. Notice $X \neq A'$, the midpoint of $\overline{BC}$. Thus line $\overleftrightarrow{PA'}$ isn't tangent to the incircle; they intersect at a point $Q \neq P$, shown in figure 5.10.5. Moreover, $A' \neq D$, the foot of the altitude through A; hence A', Q, and D are noncollinear.

Lemma 9. If $m\angle B > m\angle C$, then $m\angle A'QD = m\angle B - m\angle C$.

Proof. P, Q, and X aren't collinear, so they lie on a circle Δ, shown in figure 5.10.5. If $\overleftrightarrow{BC}$ and Δ aren't tangent, they intersect at a point $D' \neq X$; otherwise let D' denote X. Clearly, A'-P-Q; this implies that D' and X lie on the same side of $\overleftrightarrow{PA'}$. By corollary 5.9.9,

$$(A'X)(A'D) = (A'I_a)^2,$$

the power of A' relative to the incircle Γ. (See figure 5.10.6.) By lemma 8,

$$(A'I_a)^2 = (A'P)(A'Q),$$

which is also the power of A' relative to Δ. (See figure 5.10.5.) By lemma 8 again,

$$(A'P)(A'Q) = (A'X)(A'D').$$

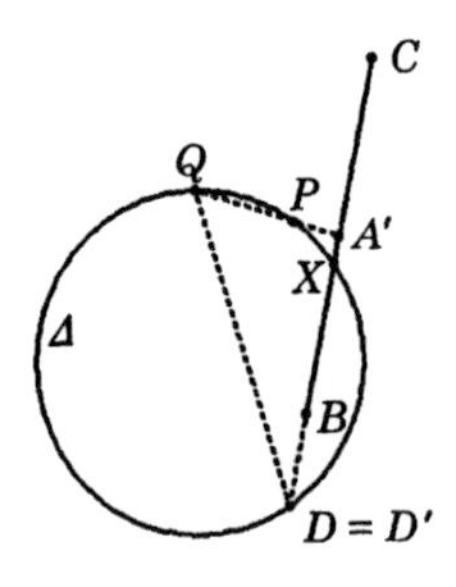

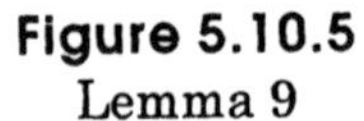
Figure 5.10.5
Lemma 9

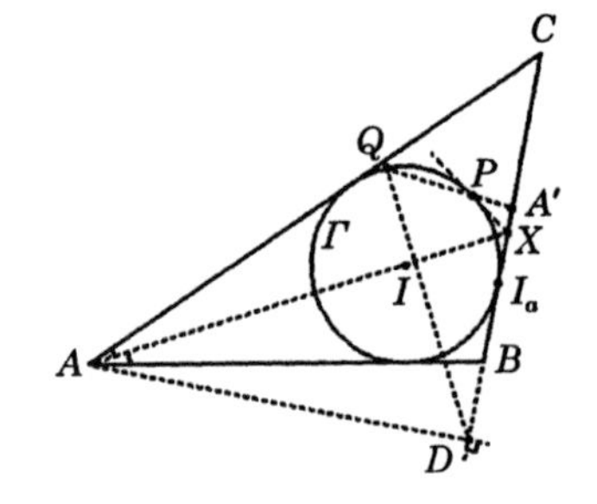

Figure 5.10.6
Proving lemma 9

[17] See Johnson [1929] 1960, chapter III; Coxeter and Greitzer 1967, chapter 5; or Davis 1949, chapters 8 and 9.

Thus $(A'X)(A'D) = (A'X)(A'D')$, hence $A'D' = A'D$, hence $D' = D$, because these points lie on $\overleftrightarrow{BC}$ on the same side of A'.[18] That is, P, Q, X, and D all lie on Δ, as in figure 5.10.5. Now you can compute the desired angle measure:

$$\begin{aligned} m\angle A'QD &= m\angle PQD = \tfrac{1}{2}\, m\widehat{PXD} \\ &= \tfrac{1}{2}(360° - m\widehat{PQD}) = 180° - \tfrac{1}{2}\, m\widehat{PQD} \\ &= 180° - m\angle PXD = m\angle CXP = m\angle B - m\angle C. \end{aligned}$$

The last equation in this chain was proved earlier, as theorem 5.9.10. ♦

Lemma 10. Under the assumption $m\angle B > m\angle C$, the Feuerbach circle is internally tangent to the incircle at point Q. (See figure 5.10.7.)

Proof. Clearly, $Q \neq D$, the foot of the altitude through A. Moreover, $\overleftrightarrow{DQ}$ isn't tangent to the Feuerbach circle Φ, because A' and the midpoint L of $\overline{AH}$ lie on opposite sides of $\overleftrightarrow{DQ}$, yet both fall on Φ. Therefore, $\overleftrightarrow{DQ}$ and Φ intersect at a point $Q' \neq D$. (See figure 5.10.8.) Let g be the tangent to Φ at A', and select a point T on g interior to $\angle CA'Q'$. Then

$$m\angle A'Q'D = \tfrac{1}{2}(360° - m\widehat{A'Q'D}) = 180° - \tfrac{1}{2}\, m\widehat{A'Q'D}.$$

Use corollary 3.13.6 as in the proof of lemma 7 to show

$$180° - \tfrac{1}{2}\, m\widehat{A'Q'D} = 180° - m\angle DA'T = m\angle CA'T.$$

By lemmas 7 and 9,

$$m\angle CA'T = m\angle B - m\angle C = m\angle A'QD,$$

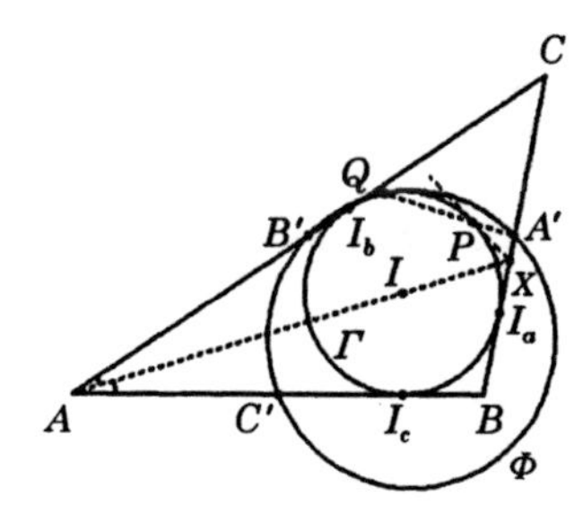

Figure 5.10.7
The Feuerbach circle Φ
and the incircle Γ of ΔABC

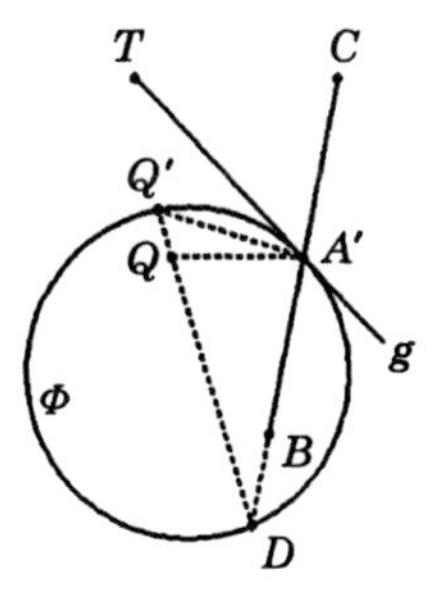

Figure 5.10.8
Proving lemma 10

[18] This implies $D' \neq X$, so $\overleftrightarrow{BC}$ and Δ can't be tangent—a case left open in the second sentence of this proof.

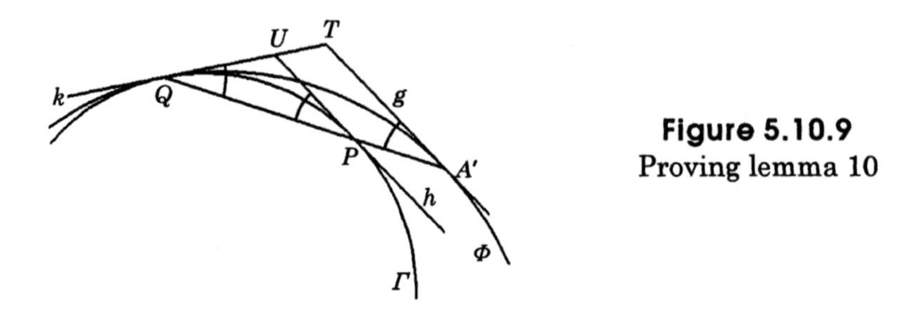

Figure 5.10.9
Proving lemma 10

hence $Q = Q'$, so Q lies on both the incircle and the Feuerbach circle.

By lemma 7 and theorem 5.9.10 the tangent g to the Feuerbach circle Φ at A' is parallel to the tangent h to the incircle Γ at P. The tangent k to Φ at Q intersects g and h at points T and U, shown in figure 5.10.9. By theorem 3.12.10 and the parallelism, $\angle UQP = \angle TQA' \cong \angle QA'T \cong \angle QPU$. By theorem 3.12.10 again, k is also tangent to Γ. ♦

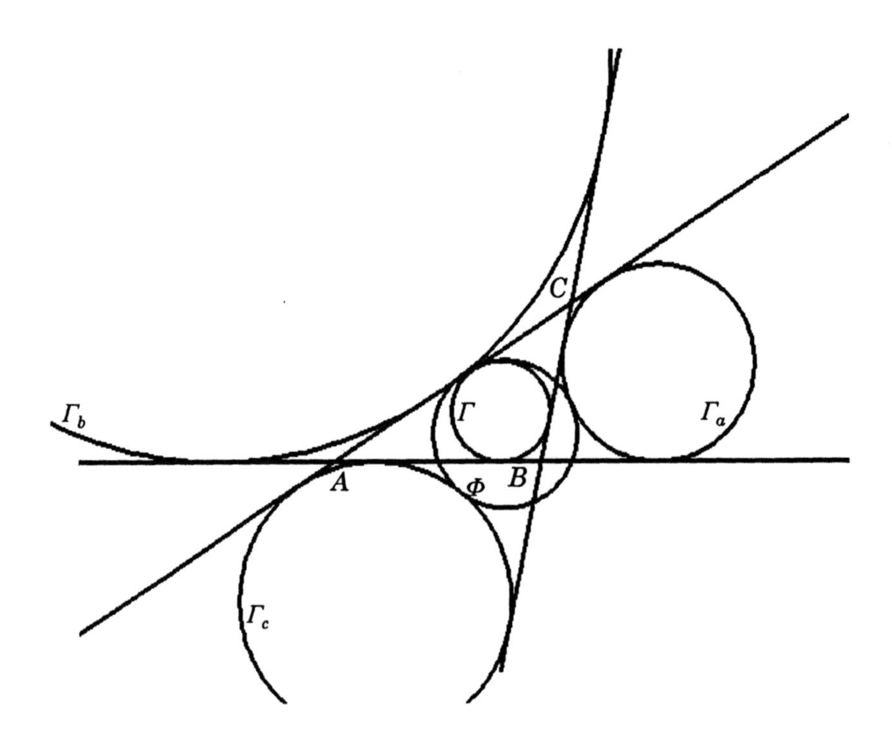

Figure 5.10.10 Feuerbach's theorem:

The Feuerbach circle Φ of ΔABC is tangent internally to the incircle Γ and externally to the excircles Γ_a, Γ_b, and Γ_c opposite A, B, and C.

Karl Wilhelm FEUERBACH was born in Jena in 1800. His father was a famous professor of law. His brothers included an archaeologist, a law professor, a philosopher, and an orientalist. In 1822 Feuerbach earned the doctorate from Jena and took a position at the gymnasium in Erlangen. He wrote three papers in mathematics, which included his famous theorem on the nine point circle and codiscovery of homogeneous coordinates. In 1827 Feuerbach became ill, and lived as a recluse until he died in 1834.

By exchanging B and C in the discussion from lemma 7 to this point, you can prove that the Feuerbach circle and the incircle are also tangent in case $m\angle B < m\angle C$. These two versions of lemma 10 constitute just half of Feuerbach's theorem; the other half has to do with the excircle. You're invited to prove it, as exercise 5.11.19, following the same pattern.

Theorem 11 (Feuerbach's theorem). The Feuerbach circle is tangent internally to the incircle and externally to the three excircles. (See figure 5.10.10.)

5.11 Exercises

Concepts

Analytic proofs of this chapter's major theorems
Circumcenter, incenter, circumsphere, insphere, and centroid of a tetrahedron
Orthocenters of rhombic tetrahedra
Homothetic centers of two circles
Finding triangle measurements from SAS, SSS, ASA, and SSA data
Tangent sum, difference, double angle, and half angle formulas
Chebyshev polynomials
Approximating sine values
Survey of spherical trigonometry
Using cross products and determinants in analytic geometry
Exercises that can utilize algebra software

This section contains thirty-eight exercises related to the earlier parts of the chapter. They're gathered together here because the material covered in chapter 5 is so tightly interrelated. Even routine problems such as exercise 1, intended to provide computational experience with the concepts in section 5.1, benefit from application of formulas derived much later in the chapter. Often a problem that seems to be about one subject yields most easily to methods introduced to study another. You'll find here a few straightforward applications of the major results in the text, and of the methods used in deriving them. But many of these exercises carry on the tradition started in chapter 4; they complete the development of theories and techniques started in the text, or extend them to new subjects.

For example, exercises 7 to 9 introduce the theory of tetrahedra: the three-dimensional analog of this whole chapter. Exercises 21 to 24 complete the development of plane trigonometry begun in chapters 3 and 4 and pursued hotly in section 5.5. With these exercises, the book contains the entire geometric content of the familiar trigonometry course, seen from an advanced viewpoint. Another family of exercises, 28 to 32, carries out a parallel development of spherical trigonometry, a neglected but highly practical branch of mathematics whose repertoire of applications is increasing.

Exercises in section 4.10 encouraged you to assemble an analytic geometry tool kit for use in a wide variety of applications. Analytic techniques were developed and extended in sections 3.11 and 5.6, using vector algebra notation and concepts summarized in appendix C. Several exercises in the present section apply and enhance those techniques. Most of the straightforward computations in exercises 1 and 2 are probably familiar to you from more elementary courses. But these quickly give way to much more complicated examples. Analytic proofs of some of the major theorems of triangle geometry require algebraic computations that touch or exceed the limit of human capability. Joint efforts of computer scientists and mathematicians have produced algebra software that can help with such work. You're invited to use it to carry out some of the computations in the last two exercises, which feature an analytic proof of Feuerbach's theorem, the climax of this chapter.

Related exercises are gathered under headings. Under a particular heading, the simpler exercises usually precede the deeper, more complex ones.

Basic exercises

The first few exercises in this section illustrate its most basic ideas. This arrangement shows why the exercises are gathered at the end of the chapter. An efficient solution of exercise 1, for example, will probably use concepts from sections 5.1, 5.5, and 5.9. Other basic exercises involve material covered in 5.7 and 5.8. Exercise 1 could be stated simply, *illustrate with a particular triangle as many as possible of the concepts introduced in chapter 5.*

Exercise 1. Consider in detail ΔACB shown with a coordinate system in figure 5.11.1. Calculate the following scalars and vectors:

centroid	G	inradius	r
circumcenter ...	O	Nagel point	N
circumradius ...	R	orthocenter	H
incenter	I		

Write all scalars in lowest terms with integer denominators. *Suggestion*: The Nagel point has y coordinate ≈ 0.27.

Exercise 1 used a particularly simple triangle to limit computational complexity. That restricts somewhat the repertoire of concepts it can illustrate. You may want to pose and solve similar exercises with less special triangles, and calculate more features. For example, try an isosceles triangle that's not right, and show that the feet of its altitudes do all lie on its Feuerbach circle.

The next exercise uses the same analytic methods in a general context. It's really three related exercises for the analytic geometry tool kit that you assembled in section 4.10. For any triangle ΔABC there's a Cartesian coordinate system with the arrangement shown in figure 5.11.2.

Exercise 2. Prove the edge bisectors, altitudes, and medians theorems analytically, using ΔABC in figure 5.11.2.

The angle bisectors theorem doesn't yield easily to these methods, mainly because the formulas for the slopes of the bisectors are too complicated. You'll find an exercise on that theorem at the end of this section, with a suggestion to use mathematical software to help with the algebra.

Exercise 3. Prove that the quadrilateral in figure 5.11.3, formed from the intersections of adjacent quadrisectors of the angles of a rhombus, is a square.

The next exercise has a particularly nice figure. Draw it!

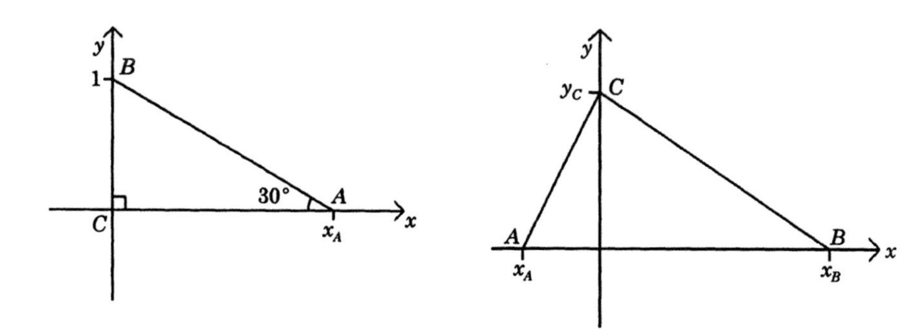

Figure 5.11.1
Exercise 1

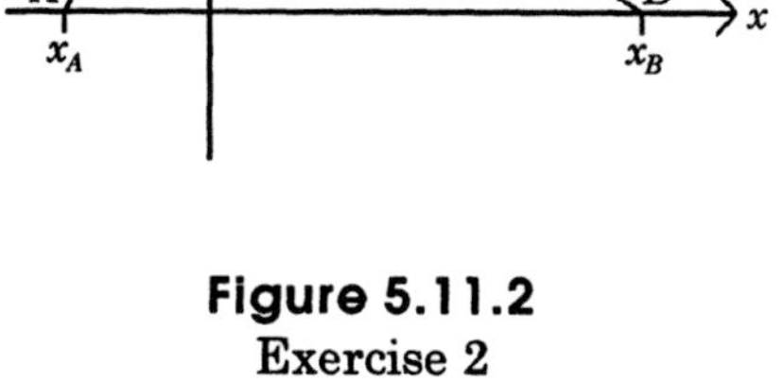

Figure 5.11.2
Exercise 2

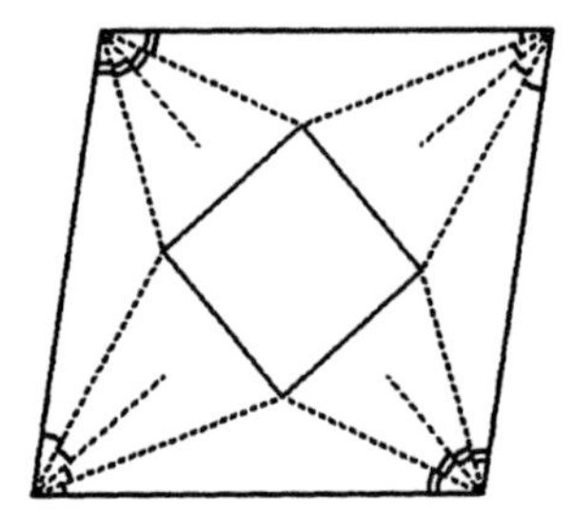

Figure 5.11.3
Exercise 3

Exercise 4. The altitude lines of an oblique ΔABC through vertices A, B, and C meet its circumcircle at points $X \neq A$, $Y \neq B$, and $Z \neq C$. Show that the edge lines of ΔXYZ are parallel to those of the orthic triangle, and these triangles are similar, with ratio 2.

Exercise 5. Show that points B, C, E, and F in theorem 5.8.1 all fall on a circle Γ that's *orthogonal* to Γ_a —that is, the tangents to Γ and Γ_a at their intersections are perpendicular. *Suggestion*: Where does the tangent g to Γ_a at F intersect $\overleftrightarrow{BC}$?

Exercise 6 is one of many inequalities that involve triangle features introduced in this chapter. To solve it, you need apply only the triangle inequality and the most basic properties of the centroid.

Exercise 6. Let m_a, m_b, and m_c be the lengths of the medians of a triangle with perimeter p. Show that $\frac{3}{4}p < m_a + m_b + m_c < p$.

Exercises on tetrahedra

Most of the plane geometric results covered in this chapter have three-dimensional analogs. You can derive some of them, such as the three in exercise 7, exactly as in plane geometry. Understanding the three-dimensional situations strengthens your hold on the basic plane theory. In other cases, such as exercise 8, the analogy isn't perfect; you must adjust the statement to get an interesting result. These two exercises will seem somewhat more difficult than the basic ones included under the previous heading, but that probably stems only from the effort necessary to visualize unfamiliar three-dimensional configurations. Exercise 9, the last on tetrahedra, is considerably more tedious. Its solution uses methods you've encountered already in the proof of theorem 5.9.2 and in exercise 4.6.6.

Exercise 7. Invent and prove three-dimensional analogs of the edge bisectors, angle bisectors, and medians theorems and related results in section 5.1. Use tetrahedra and spheres in place of triangles and circles. Relate the centroid of a tetrahedron to its vertices, the midpoints of its edges, and the centroids of its faces.

Exercise 8. This exercise develops a three-dimensional altitudes theorem. Since the altitude lines of a tetrahedron don't necessarily have a common point, the problem is to describe when that happens. First, you need some preliminary results relating tetrahedra and parallelepipeds.

Part 1. From a parallelepiped $\mathscr{P}$ with bases $AB'CD'$ and $C'DA'B$ and directrix $\overleftrightarrow{AC'}$, construct *twin* tetrahedra $ABCD$ and $A'B'C'D'$ as suggested by figure 5.11.4. Show that they're congruent, that $BCD \parallel B'C'D'$, and that every tetrahedron $ABCD$ can be constructed this way from some parallelepiped.

Part 2. $ABCD$ is called *rhombic* if the faces of $\mathscr{P}$ are all rhombic. Show that $ABCD$ is rhombic if and only if each edge lies in a plane perpendicular to the opposite edge line.

Part 3. Show that the altitude lines of a rhombic tetrahedron intersect at its twin's circumcenter.

Part 4. Show that if any two pairs of altitude lines of a tetrahedron intersect, then it's rhombic.

Part 5. The intersection of the four altitude lines of a rhombic tetrahedron is called its *orthocenter*. Show that the common perpendiculars of the three pairs of opposite edges of a rhombic tetrahedron intersect at its orthocenter.

Exercise 9. Demonstrate the equivalence of the following three properties of a tetrahedron $\mathscr{T}$:

(a) The incircles of the faces of $\mathscr{T}$ are tangent in pairs.
(b) There's a sphere tangent to all edges of $\mathscr{T}$.
(c) The sums of the lengths of the three pairs of opposite sides of $\mathscr{T}$ are equal.

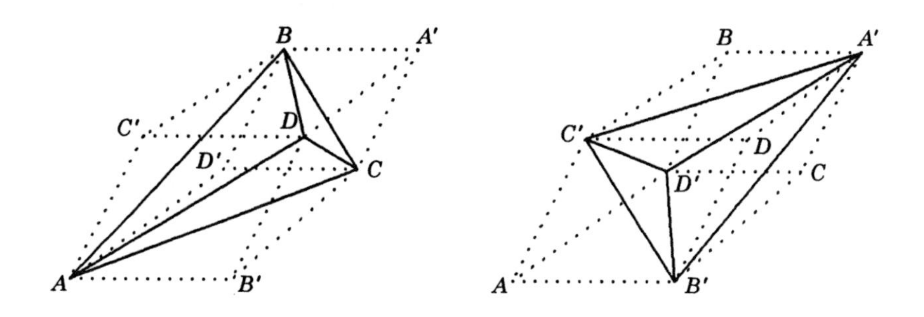

Figure 5.11.4 Twin tetrahedra

Exercises on Menelaus', Desargues', Ceva's, and Feuerbach's theorems

Exercises under this heading include applications of major theorems of this section, additional uses of methods employed to derive them, and an alternate proof. Analytic proofs are included under a later heading. Exercise 10 applies Menelaus' theorem to quadrilaterals. Can you generalize it to handle polygons with more than four vertices?

Exercise 10. Assign scales to the lines in figure 5.11.5, in order to use directed distances. Prove that

$$\frac{A \text{ to } E'}{E' \text{ to } C'}\,\frac{C' \text{ to } D'}{D' \text{ to } B}\,\frac{B \text{ to } D}{D \text{ to } C}\,\frac{C \text{ to } E}{E \text{ to } A} = 1.$$

Exercise 11. An intersection of two distinct common tangents to two distinct circles is called their *external homothetic center* if the circles lie on the same side of both tangents, and their *internal homothetic center* if they're on different sides of both tangents. Consider three circles, each two of which have an external homothetic center. Show that those three homothetic centers are collinear. What about internal homothetic centers? Combinations of internal and external homothetic centers?

Exercise 12. Consider points $D, E, F \neq A, B, C$ on the edge lines opposite these vertices of ΔABC. Suppose lines $\overleftrightarrow{AD}$, $\overleftrightarrow{BE}$, and $\overleftrightarrow{CF}$ are copencilar; let $\overleftrightarrow{BC}$ meet $\overleftrightarrow{EF}$ at A^*, $\overleftrightarrow{CA}$ meet $\overleftrightarrow{FD}$ at B^*, and $\overleftrightarrow{AB}$ meet $\overleftrightarrow{DE}$ at C^*. Show that A^*, B^*, and C^* are collinear.

Exercise 13. Draw the scene in figure 1.1.3 on a vertical panel placed obliquely, so that no tile edge is parallel or perpendicular to the image plane. Start with a perspective image of a single tile, analogous to figure 1.1.6. Then construct the image of the rest of the tiling. Imitate the section 1.1 discussion, using Desargues' theorem to show that your technique is consistent.

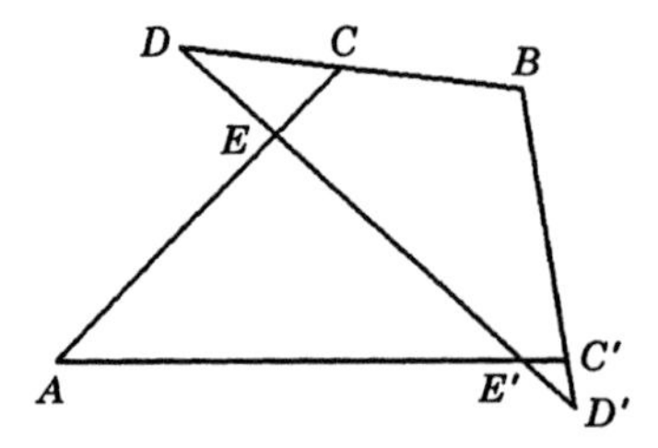

Figure 5.11.5
Exercise 10

Figure 5.11.6
$(O \text{ to } A)(O \text{ to } B) = (O \text{ to } C)$
if $(O \text{ to } P) = 1$

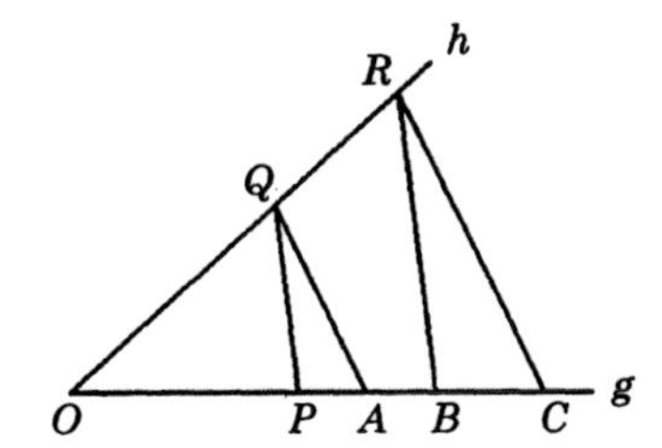

Exercise 14. Figure 5.11.6 displays a geometric algorithm for multiplying coordinates of points on a line g, relative to a particular scale. Let O and P be the points on g with coordinates 0 and 1, and select any line $h \neq g$ through O and any point $Q \neq O$ on h. Given any points A and B on g, find R on h and C on g so that $\overleftrightarrow{BR} \parallel \overleftrightarrow{PQ}$ and $\overleftrightarrow{RC} \parallel \overleftrightarrow{QA}$.

Part 1. Show that

$$(O \text{ to } A)(O \text{ to } B) = (O \text{ to } C).$$

This equation of course implies that you get the same point C regardless of your earlier choices of h and Q.

Part 2. Now suppose you knew Desargues' theorem but not the equation in part (1). Choose a point $Q' \neq Q, O$ on h. Construct C' according to the algorithm, using Q' in place of Q. Show how Desargues' theorem implies $C = C'$.

Part 3. Same as part (2), but now choose a line $h' \neq g, h$ through O, select $Q' \neq O$ on h', and construct C' according to the algorithm, using Q' and h' in place of Q and h.

In his 1899 book *Foundations of geometry*, David Hilbert showed how to formulate geometric axioms without mentioning scalars, how to *define* scalars as the points on a particular line g, and how to use Desargues' theorem as in the preceding exercise to define the product of two scalars. He handled addition similarly, and derived all the standard algebraic properties of these operations from his geometric axioms.

Exercise 15 demonstrates another way to prove Ceva's theorem, in addition to the two proofs in section 5.4. In exercise 16, you can apply the same method to derive some related formulas.

Exercise 15. Prove statement (i) of Ceva's theorem by considering ratios of areas of the six triangles that make up ΔABC in figure 5.4.1. Can you extend that method to apply to any of statements (ii) to (vi)?

Exercise 16. Prove that in figure 5.4.1,

$$\frac{WX}{AX} + \frac{WY}{BY} + \frac{WZ}{CZ} = 1 \qquad \frac{AW}{WX} = \frac{AY}{CY} + \frac{AZ}{ZB}.$$

Exercise 17 is a converse of part of theorem 5.7.1. To solve it, you may need to analyze when positive scalars x, y, and z have the same geometric and arithmetic means—that is, when $\sqrt[3]{xyz} = \frac{1}{3}(x + y + z)$.

Exercise 17. Show that if ΔAZW, ΔBXW, and ΔCYW in figure 5.4.1 have equal areas, then W is the centroid of ΔABC.

Exercise 18. Using concepts and techniques introduced in sections 5.1 and 5.4, describe the center of gravity of the union of the edges of a triangle. You may assume that the center of gravity of a segment is its midpoint.

Exercise 19. Finish the proof of Feuerbach's theorem begun in section 5.10 by showing that the Feuerbach circle of any triangle is tangent externally to its excircles.

Plane trigonometry

Several exercises under this heading complete the treatment of plane trigonometry started in section 3.13 and continued in 5.5. The last two exercises augment this chapter's list of arcane and entertaining formulas involving various triangle geometry concepts. They're just a sample of the huge stock to choose from. Exercise 20 connects the trigonometry in section 5.5 with Menelaus' theorem.

Exercise 20. Write Menelaus' product in terms of trigonometric functions of angles $\angle ACZ$, $\angle ZCB$, $\angle BAX$, $\angle XAC$, $\angle CBY$, and $\angle YBA$ in figure 5.4.1. How should you treat these angles consistently as positive or negative, so that the sign of the product is correct in all cases?

The next two problems summarize and provide examples for the part of a plane trigonometry course commonly called "solving triangles." They show how to determine the remaining side(s) or angle(s) of a triangle when you're given ASA, SSS, SAS, or SSA data. The last of these, often called the "ambiguous" case, is segregated from the others because SSA data sometimes don't determine the triangle uniquely. You made a preliminary study of that case in exercise 4.6.2.

Exercise 21, Part 1. Design an algorithm for computing the remaining angle measure and edge lengths of ΔABC given ASA data $m\angle A$, $b =$

CA, and $m\angle C$. Clearly, any valid input must satisfy the conditions $0° < m\angle A,\ m\angle C < 180°$, and $0 < b$. Is any further condition necessary? If so, and you don't check it at the onset, how might your algorithm fail? Demonstrate your algorithm for the data sets

$m\angle A$	b	$m\angle C$
55°	65	75°
85°	95	105°

Part 2. Same problem for SSS data $a = BC$, b, and $c = BA$. Must valid input satisfy any conditions beyond the obvious $0 < a, b, c$? Try your algorithm on the data sets

a	b	c
70	65	82
70	152	82
70	153	82
148	65	82

Part 3. Same problem for SAS data a, $m\angle B$, and c. Must valid input satisfy any conditions beyond the obvious $0 < a, c$ and $0° < m\angle B < 180°$? Try your algorithm on the data set

a	$m\angle B$	c
70	50°	82

Exercise 22. Design an algorithm for computing all possible values of the remaining angle measure and edge lengths of ΔABC given SSA data $a = BC$, $b = CA$, and $m\angle A$. According to exercise 4.6.2 there are at most two possible solution sets $m\angle B$, $m\angle C$, and $c = AB$. Clearly, any valid input must satisfy the conditions $0 < a, b$ and $0° < m\angle A < 180°$. Is any further condition necessary? If so, and you don't check it at the onset, how might your algorithm fail? Demonstrate your algorithm for the data sets

a	b	$m\angle A$
80	90	100°
90	80	100°
80	80	80°
30	60	30°
30	60	28°
28	60	28°

Exercise 23. Derive algebraically the formulas for $\tan(\alpha \pm \beta)$ in terms of $\tan\alpha$ and $\tan\beta$, and the formula for $\tan 2\alpha$ in terms of $\tan\alpha$. Derive a formula for $\tan \frac{1}{2}\alpha$ in terms of $\sin\alpha$ and $\cos\alpha$ that doesn't require a $\pm$ sign.

Exercise 24. Show that for $n = 0, 1, 2, \ldots$ there's a polynomial T_n of degree n —called a *Chebyshev* polynomial—such that for all x, $\cos nx = T_n(\cos x)$. Derive the *recursion* formula $T_{n+1}(y) = 2yT_n(y) - T_{n-1}(y)$. Write out the formulas for T_0 to T_5, sketch their graphs, and describe their roots, local extrema, and symmetry. What happens if you try something like this for $\sin nx$?

Exercise 25. Let r_a, r_b, and r_c be the exradii of ΔABC; r and R, its inradius and circumradius; Δ its area; s its semiperimeter; and h_a, h_b, and h_c be the lengths of its altitudes. Prove that

$$\frac{1}{r_a} + \frac{1}{r_b} + \frac{1}{r_c} = \frac{1}{r} = \frac{1}{h_a} + \frac{1}{h_b} + \frac{1}{h_c} \qquad r\,r_a r_b r_c = \Delta^2$$

$$r_a + r_b + r_c = 4R + r \qquad r_a r_b + r_b r_c + r_c r_a = s^2.$$

Exercise 26. Prove that the sum of the cosines of the measures of the angles of a triangle equals 1 plus the ratio of its inradius to its circumradius.

Exercise 27 shows how to approximate $\sin 1°$ accurately, using theorem 5.5.12 and values of $\sin 1\tfrac{1}{2}°$ and $\sin \tfrac{3}{4}°$ computed via subtraction and half angle formulas. Van der Waerden reports[19] that Aristarchus of Samos first used this technique around 280 B.C. to approximate $\sin 3°$. About A.D. 150 the Alexandrian astronomer Claudius Ptolemy employed it systematically to compute what amounts to a table of values $\sin\theta$ for $\theta = 1°, 2°, 3°, \ldots$. The method should remind you of real analysis arguments. That isn't surprising, since theorem 5.5.12 is an important step in justifying the formula for the derivative of the sine function.

Exercise 27, Part 1. Derive algebraic formulas for $\sin 1\tfrac{1}{2}°$ and $\sin \tfrac{3}{4}°$ and check them with your calculator.

Part 2. Show that $\sin\alpha - \sin\beta = 2\cos\mu \sin\nu$, where $\mu = \tfrac{1}{2}(\alpha + \beta)$ and $\nu = \tfrac{1}{2}(\alpha - \beta)$.

Part 3. Suppose $0° < \beta < \alpha < 90°$. Prove that

$$2\cos\mu \sin\nu < \frac{\alpha - \beta}{\beta}\sin\beta.$$

Part 4. Combine parts (2) and (3) to show that when $0° < \gamma < \beta < \alpha < 90°$,

$$\frac{\beta}{\alpha}\sin\alpha < \sin\beta < \frac{\beta}{\gamma}\sin\gamma.$$

[19] 1963, Chapter VII.

Part 5. Use $\alpha, \beta, \gamma = 1\frac{1}{2}°, 1°, \frac{3}{4}°$ in part (4), with the values you calculated in part (1), to estimate $\sin 1°$. What's the maximum possible error?

Spherical trigonometry

The exercises under this heading complete the survey of spherical trigonometry begun in the section 4.9 exercises. For a century, this material has been a mathematical backwater. In most geometry texts, you could only find it briefly mentioned as a contrast to the familiar plane Euclidean geometry. Its practical aspects were confined to applications texts: astronomy, geodesy, navigation. Its applications now reach into broader areas as we learn to write software to recognize objects in a three-dimensional environment and to control machines that move them about and/or display their images. These exercises generally follow the treatment in Marcel Berger's terse but penetrating monograph.[20]

Exercise 28, Part 1. Prove the law of sines for a spherical triangle ΔABC:

$$\frac{\sin \mathrm{m}\angle A}{\sin \mathrm{m}\widehat{BC}} = \frac{\sin \mathrm{m}\angle B}{\sin \mathrm{m}\widehat{CA}} = \frac{\sin \mathrm{m}\angle C}{\sin \mathrm{m}\widehat{AB}}.$$

Part 2. Prove Menelaus' theorem for a spherical ΔABC: If X, Y, Z are points on the great circles containing $\widehat{BC}, \widehat{CA}, \widehat{AB}$, and X, Y, Z lie on a great circle, then

$$\frac{\sin \mathrm{m}\widehat{AZ}}{\sin \mathrm{m}\widehat{ZB}} \frac{\sin \mathrm{m}\widehat{BX}}{\sin \mathrm{m}\widehat{XC}} \frac{\sin \mathrm{m}\widehat{CY}}{\sin \mathrm{m}\widehat{YA}} = 1.$$

Exercise 28 suggests some questions that would make a good project of larger scope. This form of Menelaus' theorem doesn't use "directed" arcs as its plane form uses directed distances. All terms in the Menelaus product are sines, so it comes out $+1$ instead of -1 as in the plane. Is it possible to make such distinctions consistently on the sphere? Is there a converse of this form of Menelaus' theorem? Is there a spherical form of Ceva's theorem?

Exercise 29 is mainly a problem in visualization. Parts 1 and 2 require careful reasoning about the sides of various planes. In part 2 you should concentrate on the plane ε through O perpendicular to $\vec{OB}$, and its intersections with planes OAB and OBC. The only computation you need for those parts is angle addition.

---

[20] Berger 1987, section 18.6.

Exercise 29, Part 1. Consider a spherical triangle ΔABC on a sphere Σ with center O. The perpendicular to OBC through O intersects Σ at two points. One lies on the same side of OBC as A; call it A'. Define B' and C' similarly. Show that A', B', and C' aren't coplanar with O. Thus they form a spherical $\Delta A'B'C'$ called the *polar* triangle of ΔABC. Show that ΔABC is in turn the polar triangle of $\Delta A'B'C'$.

Part 2. Let a', b', and c' be the measures of the arcs of the polar triangle opposite A', B', and C'. Show that

$$a' = 180° - \text{m}\angle A \qquad \text{m}\angle A' = 180° - a$$

and derive similar formulas for the measures of its other arcs and angles.

Part 3. Let a, b, and c be the measures of the arcs of ΔABC opposite A, B, and C. Show that

$$\begin{array}{ll} a < b + c & \text{m}\angle A + \text{m}\angle B < \text{m}\angle C + 180° \\ b < c + a & \text{m}\angle B + \text{m}\angle C < \text{m}\angle A + 180° \\ c < a + b & \text{m}\angle C + \text{m}\angle A < \text{m}\angle B + 180°. \\ a + b + c < 360° & \end{array}$$

Part 4. Consider the following problem in more detail: For a vertex V of the spherical triangle, which intersection of Σ with $\overleftrightarrow{OV'}$ should be called V'? Do parts (1) to (3) fail if you make one or more of these choices differently? What if you set up a coordinate system with origin O, made the sphere's radius 1, and chose $A' = B \times C$, etc.?

Exercise 30. Prove the *cosine laws* for a spherical triangle ΔABC: If a, b, and c are the measures of the arcs of ΔABC opposite A, B, and C, then

$$\cos a = \cos b \cos c + \sin b \sin c \cos \text{m}\angle A,$$
$$\cos \text{m}\angle A = -\cos \text{m}\angle B \cos \text{m}\angle C + \sin \text{m}\angle B \sin \text{m}\angle C \cos a.$$

Suggestion: Prove the first equation geometrically. You may have to consider the case $c = 90°$ separately. For the second, apply exercise 29.

Exercise 31. Imitate exercises 21 and 22 for spherical triangles. Include discussions of SAA and AAA data, too.

Exercise 32, Part 1. From the following data, find the airline distances from San Francisco to Bucharest and Christchurch. Assume that the earth is a sphere with radius $r = 3959$ miles.

City	*Latitude*	*Longitude*
San Francisco	37°46′ N	122°25′ W
Bucharest	44°25′ N	26°07′ E
Christchurch	43°33′ S	172°40′ E

How far have *you* been from home?

Part 2. Find the fraction of the earth's area covered by the spherical triangle formed by these cities. Describe the triangle by listing a few places on or near each of its edges.

Analytic geometry

In the section 4.10 exercises, you constructed a tool kit of analytic formulas and methods corresponding to many synthetic techniques developed in chapter 3. You were asked to avoid vector cross products, but to use solutions of linear systems instead.

Exercise 33. Restudy those tools now in the light of section 5.6, and employ cross products and determinants where appropriate to streamline them. (This is the main use of cross products: given vectors X and Y, to find a vector Z such that $X \cdot Z = 0 = Y \cdot Z$.)

Exercise 34. Extend as follows the tool kit that you revised in exercise 33:

Task:

Given	*Determine or find*
1. lines g and h, ...	their angle;
2. planes δ and ε, ...	their dihedral angle;
3. points P, Q not on plane ε, . . .	whether P and Q lie on the same side of ε;
4. points P, Q coplanar with . . . . . a line g but not on it,	whether P and Q lie on the same side of g in their plane;
5. noncollinear points P, Q, R, . . .	area of ΔPQR;
6. noncoplanar points P, Q, R, S, .	volume of tetrahedron $PQRS$.

The next two exercises continue the program begun in exercise 2 to prove the major theorems of this chapter analytically. Desargues' theorem is harder to handle than Menelaus' or Ceva's, even though it doesn't involve directed distances. Thus it's placed by itself. The suggested analytic methods don't favor any particular choice of coordinate system; their heavy reliance on parametric equations and vector algebra makes that nearly irrelevant.

Exercise 35. Prove Menelaus' and Ceva's theorems analytically. First, show that if $\vec{XY}$ is a directed line, $T = tX + (1 - t)Y$, and $t \neq 0$, then

$$\frac{X \text{ to } T}{T \text{ to } Y} = \frac{1-t}{t}.$$

Use fractions like this to formulate Menelaus' product, and use the determinant conditions in section 5.6 to indicate when three points are collinear or three lines have a common point.

Exercise 36. Prove Desargues' theorem analytically.

In exercise 2, you worked through analytic proofs of the edge bisectors, altitudes, and medians theorems, but not the angle bisectors theorem. It doesn't yield easily to analytic methods, mainly because the formulas for the slopes of the bisectors are too complicated. The problem is posed here, with the suggestion that you use algebra software to carry out, or at least verify, the tedious calculations. For this work and exercise 38, the author used the *weakest* such package available at the time.[21] It was able to carry out the computations, but only barely. Learning how to control it, though, led to a deeper understanding of the algebra. You have to know algebra very well in order to use algebra software effectively.

Exercise 37. Prove the angle bisectors theorem analytically, by the method of exercise 2.

Exercise 38. This final exercise leads you through an analytic proof that the incircle of ΔABC is internally tangent to the Feuerbach circle. It proceeds somewhat like Feuerbach's published proof (1822). But you're encouraged to employ algebra software to confront the calculations directly, instead of Feuerbach's devices to limit their complexity.

Suggestion: Use the coordinate system in figure 5.11.2, with

$$A = \langle x_A, 0\rangle \qquad B = \langle x_B, 0\rangle \qquad C = \langle 0, y_C\rangle$$
$$x_A < 0 \qquad x_B > 0 \qquad y_C > 0.$$

Use the standard abbreviations a for $BC = \sqrt{x_B^2 + y_C^2}$ and b for $CA = \sqrt{y_C^2 + x_A^2}$ in your computations, and replace a^2 and b^2 by $x_B^2 + y_C^2$ and $y_C^2 + x_A^2$ as appropriate. By introducing variables a and b in addition to x_A, x_B, and y_C, you minimize the role of square roots. You can apply polynomial algebra, and your polynomials will be linear in a and b.

First, compute the inradius r of ΔABC and the coordinates of its incenter I as in exercises 1, 2, and 37. Second, find the circumradius R of ΔABC via one of the equations derived in this chapter. Since the medial triangle $\Delta A'B'C'$ is similar to ΔABC with ratio ½, its circumradius $R' = ½ R$. Third, compute the coordinates of the circumcenter O' of the medial triangle. Fourth, show that Feuerbach's result is implied by the equation $(IO')^2 \doteq (r - R')^2$. Finally, rewrite this equation as $p(x_A, x_B, y_C, a, b) = 0$, where p is a sixth-degree homogeneous polynomial. You may then use algebra

[21] *Derive* 1989–.

software to compute the coefficients of p —the author's polynomial had 168 terms—and verify that they're all zero.

You may want to continue exercise 38 as a project: Prove that the excircles are externally tangent to the Feuerbach circle, and investigate the devices Feuerbach used to manage the algebra.

Chapter

6

Plane isometries and similarities

With this chapter you leave classical geometry, which concentrated on properties of figures constructed from points, lines, planes, circles, and spheres. You enter transformational geometry, whose fundamental objects are points and transformations. A transformation is a function that relates all points to others in a one-to-one way; it induces a correspondence between figures as well. You can use transformations to describe correspondences between various figures of classical geometry, to study positions of figures before and after motions, and to analyze the relationships between corresponding parts of symmetric figures. You can adapt transformational geometry to study aspects of a single figure that may differ when you use different coordinate systems. You'll investigate how figures change under various transformations; and —perhaps more important—you'll study figures and properties that *don't* change.

The transformations you'll study here—isometries and similarities—are closely related to the classical geometric notions of congruence and similarity. Isometries are transformations that preserve distance between points.[1] An isometry can change a triangle's location or its orientation in a plane. By the SSS congruence principle, though, an isometry cannot change a triangle's size or shape. In fact, two triangles are congruent *just in case* they're related by some isometry. This result doesn't lead to much new in the theory of *triangle* congruence, but does provide an easy way to extend the theory to figures of *any* type. You don't have to invent different congruence notions for each type, but can just call figures *congruent* if they're related by some isometry. The connection between the theory of similar figures and the study

[1] The word stems from the Greek words *isos* and *metron* for *equal* and *measure*.

of similarity transformations, which multiply all distances by a constant ratio, is analogous.

Formulation of a general theory of congruent figures is the first of many uses of transformations in the foundations of geometry. They provide a tool for determining what's common between the geometric theories that result from various alternative axiom systems—for example, from assuming the negation of Euclid's parallel axiom. Those studies lie beyond the scope of this text, but not far![2]

The correspondence between locations of objects before and after a motion is clearly an isometry. This connection is so suggestive that many authors use the term *motion* instead of *isometry*. But isometry doesn't encompass everything involved in motion; it ignores what happens *during* the event. That requires additional apparatus not included in transformational geometry, so *motion* is a misnomer here.

The connection between transformations and symmetry is perhaps not so obvious. We exhibit *bilateral* symmetry because there's a close correspondence between points on our left and right sides. Some features are preserved; we can easily recognize an acquaintance's mirror image. Most familiar animals appear bilaterally symmetric. About a thousand years ago, an artist in the Mimbres region of New Mexico exploited that when she portrayed the dancer in figure 6.0.1, costumed to evoke simultaneously images of a deer, a man, and a bat. But many features of life are not symmetric. For example, it's hard for us to recognize the mirror image of text in figure 6.0.2. In fact, symmetry is defined in terms of isometry: a figure is symmetric if it's unchanged under some isometry. A plane figure is bilaterally symmetric if it's unchanged when you reverse the two sides of some line.

The symmetric frieze ornament in figure 6.0.3, a pottery design from the San Ildefonso Pueblo in New Mexico, is designed to be repeated as required, and remains unchanged as your gaze steps along it. A rotationally symmetric design remains the same if you turn it through a particular angle. For example, the *dharmachakra*, the 24-spoked wheel of right conduct on the flag of India (figure 6.0.4) symbolizes righteousness, truth, and recurrence.[3]

You've tasted transformational geometry if you've studied coordinate transformations in analytic geometry. That application needs a different approach; you're not interested in corresponding points, but rather in corresponding coordinate pairs or triples assigned to a single point by different coordinate systems. This chapter's theory can be adapted to that end. But that's not pursued, because the book's overall emphasis isn't on analytic

[2] For an introduction, consult Behnke et al. [1960] 1974, chapters 4 and 5.

[3] According to an official Indian government publication (Sivaramamurti 1966, 8, 14), the *chakra* represents "the Wheel of Law . . . whose message of righteousness was binding even on the greatest monarch. . . . The wheel of life is inexorable . . . . [It] moves on into a cycle of births and deaths, culminating in final liberation . . . . The wheel is thus the symbol of creation."

Figure 6.0.1
Bilateral symmetry[4]

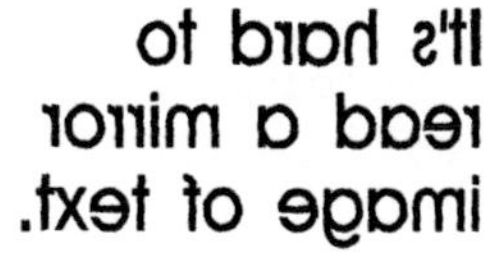

It's hard to
read a mirror
image of text.

Figure 6.0.2
Bilateral asymmetry

Figure 6.0.3
Frieze pattern[5]

Figure 6.0.4
India's flag

---

[4] From Brody, Scott, and LeBlanc 1983, plate 5. Where did the artist intentionally *break* the overall symmetry? Like most surviving Mimbres pottery, this example was recovered from a grave. Mimbres funeral ceremonies evidently included punching a hole in the bottom of each bowl to be buried.

[5] Kenneth M. Chapman 1970, plate 124p. This example originated not far from the Mimbres, but is probably less than a century old.

methods. Coordinate transformation techniques are demonstrated in exercise 6.11.31.

What methods does this chapter use to study transformations? It must depart from the classical methods of earlier chapters. Many practical applications involve computation with coordinates. That's so cumbersome that it's generally avoided in formulating the theory. The computations involve vectors and matrices, so matrix algebra affords some simplification—you can handle matrices as whole entities instead of manipulating their entries. But even greater simplification results from employing the algebra of compositions and inverses of functions. These are the substance of much of modern algebra, particularly group theory. The more deeply you study transformational geometry, particularly its applications to symmetry, the more benefit you gain by using group-theoretic methods. This chapter is an introduction to those techniques. The theory is developed with minimal use of matrix algebra. The matrix computations typical in applications are demonstrated in the exercises that conclude the chapter.

The chapter begins by collecting in section 6.1 the relevant general information about functions, composition, and inverses. Since this material is fundamental to several areas of higher mathematics, you may have encountered much of it already in other courses. Section 6.2 introduces three types of plane isometries—translations, rotations, and reflections—and some of their properties. Sections 6.3 to 6.5 discuss them in greater detail. The chapter's principal result is the structure theorem in 6.6. That enables you to classify every plane isometry as one of several types. You'll see that the earlier introductory treatment omitted one type: glide reflections, which are handled in section 6.7. The connection with matrix algebra is established in section 6.8, and used in 6.9 to refine the classification.

This chapter overview has only once mentioned the other kind of transformation this book studies: similarities. They're introduced in section 6.10. Their study requires only one new technique. Otherwise, methods already introduced for isometries work as well for similarities; there's just more detail.

The chapter concludes with exercises. These pursue further most of the ideas introduced in the previous sections, and will give you much more experience with computation.

For simplicity, chapter 6 is limited to *plane* transformational geometry. But *all* of it extends to three dimensions. In fact, most applications of plane isometries and similarities were originally handled by specialized methods that were less algebraic, more classical, in flavor. The corresponding three-dimensional results are just enough more complicated that a new apparatus was required to handle them. Logically, the three-dimensional material should be integrated into this chapter, because we really live in three-dimensional space, and its complexities make more apparent the need for the new methods. But that would make the chapter too bulky, it wouldn't be so easy for you to discern its simple structure, and the text would become less flexible for instructors and students. So the three-dimensional results are presented together in chapter 7, with a separate set of exercises. Much of the theory in that chapter is identical to the corresponding plane theory, once the definitions are properly formulated. So chapter 7 is even more streamlined, with many details left to you.

Chapter 2 and occasional passages in other previous chapters have offered some details of the history of the material presented there. It's not too hard to find sources for that, because it was often controversial. Researchers reported the background for their ideas, and cited where they agreed or differed with their predecessors. With transformational geometry, the task isn't so simple. Its techniques evolved over a century and a half, beginning in the late 1700s. They were introduced to solve hard applications problems. Mathematicians concentrated on their successes and on new problems. They rarely dwelled on philosophical, organizational, or pedagogical aspects of their work. The order of development of the material from this chapter on was often opposite to its logical order. Many of the complex techniques for applying the theory were introduced early in this period. Leonhard Euler, in 1775, would have found familiar most of the computational exercises in these chapters.[6] Determinants came into use during the early 1800s, and transformation groups and matrices in the mid-1800s. The last part of this material to fall into place is the first part presented; the language and techniques for dealing with functions, their compositions, inverses, and image sets were refined and cast in the form of section 6.1 only in the early 1900s. The history of transformational geometry is thus blurred. When you try to find the origin of a specific concept or result, you often find traces of it in works all the way back to Euler's time, even though it wasn't stated explicitly until much later. You can't readily use history pedagogically, as punctuation for the presentation of the theory. So there are fewer historical references in these later chapters. Don't let their absence give you the impression that the theory was revealed all at once, or that its history is bland! It's involved and intriguing—a challenging field of study.

---

[6] For example, Euler ([1775] 1968) formulated equations for isometries, even though the matrix algebra used for that purpose in theorems 6.8.2 and 7.1.7 wasn't developed until the late 1800s.

6.1 Transformations

Concepts

- Functions; domain, value, image, and range
- Linear functions
- Preserved and invariant properties of figures
- Function composition
- Associativity and noncommutativity
- Surjections, injections, and bijections
- Inverse functions
- Transformations
- Symmetric group, transformation groups
- Geometric transformations

This section summarizes elementary material about functions, on which the rest of this chapter is based. Together with appendices A and C, it forms the algebraic foundation for the entire book. You may have met much of this material already in precalculus and calculus courses. But its application to transformational geometry has a flavor so different from those subjects that the fundamentals are presented here for emphasis. You can regard transformational geometry as a very sophisticated development of these few elementary ideas and some others from chapter 3. For that reason it's important that you understand in complete and precise detail these definitions, examples, and first results. The first four exercises in section 6.11 will help you.

As you study this material in other geometry texts or other application areas, you'll encounter conflicting terminology and notation. Its applications spread over all mathematics, and what's convenient and graceful for one area can appear somewhat inappropriate for another. Moreover, this mathematics is younger than most of that in earlier chapters, and its terminology and notation are still evolving. This book's conventions result from many compromises. Several footnotes indicate its differences from other usage, and some of the reasons for the divergence.

This section concludes by mentioning some topics to which this chapter naturally leads, but which lie just beyond its scope.

Functions

The function concept pervades mathematics. In this book, for example, it has appeared among the undefined concepts and axioms:

- The *distance function* assigns to each pair of points X, Y the distance XY.

- A *scale* for a line g assigns to each point P on g a coordinate $c(P)$.
- The *area* function assigns to each polygonal region Σ its area $a\Sigma$.

Functions also occur in later developments, either implicitly or explicitly. For example,

- A function assigns to each pair P,Q of distinct points the *line* $\overleftrightarrow{PQ}$.
- Each *coordinate system* has a function that assigns to each point X a triple $<x_1,x_2,x_3>$ of coordinates.
- An *angular scale* at a point O in a plane ε assigns to each real number a ray in ε with origin O.

Section 3.13 introduced angular scales to help define $\sin x$ and $\cos x$ as real-valued functions of real numbers x. You've met many functions of that type in algebra. The following examples of *exponential*, *power*, *polynomial*, and *linear* functions are considered in detail later in this section:

$$2^x \quad x^3 \quad x^{1/3} \quad x(x^2-1) \quad ax+b \text{ for specified constants } a, b.$$

Also, in two or three dimensions, given matrix and vector constants A and B, you've worked with the *linear* function that assigns to each vector X the vector value $AX+B$.[7]

Here's the formal definition that unifies these examples. A *function* f assigns to each element x of a set D a unique *value* $f(x)$ in a set S. This sentence is often abbreviated $f:D \to S$ and read *f maps D into S*.[8] The uniquely determined set D of x to which f applies is called the *domain* of f. But the set S is *not* uniquely determined; if $f:D \to S$, then f also maps D into *any* set that contains S. Often it's more convenient to refer to a formula such as 2^x for the value $f(x)$ than to refer to f by name. In a case like that, you can refer to *the function* $x \to 2^x$ *that maps* x *to* 2^x. Since that doesn't mention the set D or S, you must determine them from the context. In this example, D could be the set $\mathbb{R}$ of all real numbers x, and S could be any set that contains all $x > 0$. When there's no doubt which letter is the bound variable, you can use just the formula and refer to *the function* 2^x. Be careful, though—the phrase *function* $ax+b$ is ambiguous unless you know that a and b are constants.[9]

---

[7] Even though they still call $ax+b$ a linear function, most authors of linear algebra texts reserve the term *linear* in higher dimensions to functions of the form AX. They'd call $AX+B$ an *affine* function.

[8] Authors often use the terms *function* and *mapping* interchangeably.

[9] Many older books that apply advanced algebra to geometry use *postfix* functional notation; instead of $f(x)$ they write $(x)f$. Because it clashes with calculus notation, that practice is disappearing. In this text it survives in usages like *the function* $x \to x'$. But here, the symbol $'$ is not a function name, just part of a formula for the function value.

Two functions f and g are regarded as *equal* if they have the same domain D and $f(x) = g(x)$ for every $x \in D$.

If $f : D \to S$, then to each subset $A \subseteq D$ corresponds the *image set* $f[A] = \{f(x) : x \in A\}$. The image $f[D]$ of the domain is called the *range* of f.

Preserved and invariant properties

In many applications, a function maps its domain D into itself. For example, all the trigonometric and algebraic functions mentioned under the previous heading map the domain $D = \mathbb{R}$ into itself. A function $f : D \to D$ is said to *preserve* a property $\mathcal{P}$ of members of D if $f(x)$ has property $\mathcal{P}$ whenever x does. You can use an image set to describe that situation—if $A = \{x \in D : x \text{ has } \mathcal{P}\}$, then f preserves $\mathcal{P}$ just when $f[A] \subseteq A$. For example, $x \to 2^x$ preserves the property of being an integer, but not that of being an odd integer. $\mathcal{P}$ is called *invariant* under f if $f(x)$ has $\mathcal{P}$ *if and only if* x does: $\mathcal{P}$ is invariant under f just when $f[A] = A$. For example, being positive is invariant under $x \to x^3$ but not under $x \to 2^x$.

You can consider properties of pairs, or sequences $<x_1, x_2, \ldots>$ of any number of elements of D the same way. For example, 2^x and x^3 are called *strictly increasing* functions because they leave *order* invariant:

$$x_1 \le x_2 \;\Leftrightarrow\; 2^{x_1} \le 2^{x_2} \;\Leftrightarrow\; x_1^3 \le x_2^3 .$$

On the other hand, $\sin x$ doesn't preserve order; you can have $x_1 \le x_2$ but $\sin x_1 > \sin x_2$.

You can consider properties of subsets or sequences of subsets of D the same way. For example, the property *being a bounded open interval* is invariant under $f(x) = 2^x$; if $a < b$, then

$$f[\{x : a < x < b\}] = \{y : 2^a < y < 2^b\}$$

because $a < x < b$ if and only if $2^a < 2^x < 2^b$. But that property isn't preserved by $\sin x$; for example,

$$\sin[\{x : 0° < x < 360°\}] = \{y : -1 \le y \le 1\},$$

which is a bounded *closed* interval, not an open one.

Function composition

If $f : S \to T$ and $g : D \to S$, then the *composition* of f with g is the function $f \circ g : D \to T$ such that $(f \circ g)(x) = f(g(x))$ for all $x \in D$.[10] For example,

[10] Many authors used to define composition backward. They called $g \circ f : x \to f(g(x))$ the composition of $g : D \to S$ with $f : S \to T$, letting the left-to-right order emphasize that you perform g first, then f. As you see, that collides with prefix notation for function values. (continued...)

if $D = S = T = \mathbb{R}$, $f : x \to 2^x$, and $g : x \to x^3$, then $f \circ g : x \to 2^{x^3}$.

You can check that if $A \subseteq D$, then $(f \circ g)[A] = f[g[A]]$. If $D = S = T$ and f and g both preserve some property $\mathscr{P}$ of elements, subsets, or sequences of elements or subsets of D, then so does $f \circ g$. And if $\mathscr{P}$ is invariant under both f and g, then it's invariant under $f \circ g$.

Functions $f : S \to T$ and $g : D \to S$ are said to *commute* if $f \circ g = g \circ f$.[11] This equation is *almost never true*. It doesn't make sense unless $D = S = T$, and even then it's usually false. Consider the example given two paragraphs earlier: $f \circ g \neq g \circ f$ because

$$f(g(x)) = 2^{x^3} \qquad g(f(x)) = (2^x)^3 = 2^{3x}.$$

As a second example, consider linear functions $h(x) = ax + b$ and $j(x) = cx + d$. *Compositions of linear functions are linear*:

$$\begin{aligned} h(j(x)) &= a(cx + d) + b = (ac)x + (ad + b) \\ j(h(x)) &= c(ax + b) + d = (ac)x + (bc + d), \end{aligned}$$

but h and j commute only when $ad + b = bc + d$.

Although the commutative law is generally not true in this context, function composition does have *some* features reminiscent of elementary algebra. Fundamental to everything in this chapter, for example, is the *associative law*: If $e : T \to U$, $f : S \to T$, and $g : D \to S$, then $(e \circ f) \circ g = e \circ (f \circ g)$, because

$$\begin{aligned} ((e \circ f) \circ g)(x) &= (e \circ f)(g(x)) = e(f(g(x)) \\ &= e((f \circ g)(x)) = (e \circ (f \circ g))(x) \end{aligned}$$

for each $x \in D$. The associative law lets you omit parentheses in such repeated compositions. Often that makes them more readable and intuitive. In calculus, for example, if you need to differentiate the function $2^{\sin^3 x}$, you regard it as the composition of 2^x with x^3 then $\sin x$; you don't have to ask whether it's the composition of 2^{x^3} with $\sin x$ or of 2^x with $\sin^3 x$.

Long—sometimes *very* long—compositions occur in this book. It's tedious to write the $\circ$ symbol so much. Therefore, compositions will usually be abbreviated by juxtaposition; for example, fg will stand for $f \circ g$, and

$$fg(x) = f(g(x)) \quad fg[A] = f[g[A]] \quad fgh = f(gh) = (fg)h.$$

Be careful not to confuse this with algebraic notation for multiplication of numbers: $2^x \sin x$ still means the product of 2^x and $\sin x$, not the composition! But remember that in this book, if f and g denote functions, then

---

[10] (...continued)
So those authors often used postfix notation, with $(x)(g \circ f) = ((x)g)f$. As mentioned in the previous footnote, that practice is disappearing.

[11] The word *commute* stems from the Latin prefix *con-* and verb *muto*, which mean *with* and *change*.

fg *usually* means their composition, and the equation $fg = gf$ is probably false!

Surjections, injections, and bijections

A function $f : D \to S$ is said to map D *onto* S *surjectively* if $S = f[D]$, the range of f.[12] For this phrase to make sense, you *must* specify S. For example, when $D = S = \mathbb{R}$, functions x^3 and $x(x^2 - 1)$ are surjective but 2^x and $\sin x$ are not. Should you change S to $\{x : x > 0\}$ or $\{x : -1 \le x \le 1\}$, then 2^x or $\sin x$, respectively, would become surjective.

f is said to map D into S *injectively* if $f(x_1) \ne f(x_2)$ whenever $x_1 \ne x_2$.[13] Functions 2^x and x^3 are injective but $x(x^2 - 1)$ and $\sin x$ are not. These functions display all four possible combinations of surjectivity and injectivity!

f is said to map D to S *bijectively* if it's both surjective and injective. The function x^3 is bijective; so is the linear function $ax + b$ unless $a = 0$. A bijective function $f : D \to S$ has an *inverse* $f^{-1} : S \to D$, which maps each $y \in S$ to the unique $x \in D$ for which $f(x) = y$. (Surjectivity means there's at *least* one such x; injectivity means there's at *most* one.) For these examples,

$$y = f(x) = x^3 \Rightarrow x = y^{1/3} \Rightarrow f^{-1}(y) = y^{1/3} \Rightarrow f^{-1}(x) = x^{1/3}$$

$$\begin{aligned} y = g(x) = ax + b &\Rightarrow x = a^{-1}y - a^{-1}b \\ &\Rightarrow f^{-1}(y) = a^{-1}y - a^{-1}b \Rightarrow f^{-1}(x) = a^{-1}x - a^{-1}b. \end{aligned}$$

You derive the formula for the inverse function value by solving equation $y = f(x)$ for x in terms of y, then reversing x and y. Notice that *the inverse of a nonconstant linear function is linear*!

You should verify that the inverse of a bijection $f : D \to S$ is a bijection $f^{-1} : S \to D$, and *its* inverse is the original function f.[14]

---

[12] Authors often write merely *f maps D onto S*, omitting the adverb *surjectively*. Then the subtle distinction between *into* and *onto* becomes too important. Moreover, since no convenient adjective or noun matches the preposition *onto*, they must refer awkwardly to *onto* functions or to the property of *onto-ness*. *Surjective* and *surjectivity* are more graceful. The adjectives *surjective*, *injective*, and *bijective* were coined by a pseudonymous group of French mathematicians for use in Bourbaki 1939, the first installment of a multivolume encyclopedia that reformulated advanced mathematics. *Sur* means *on* in French, and the root *-ject-* comes from the past participle of the Latin verb *iacio*, which means *throw*.

[13] Authors often refer to *one-to-one functions*, employing an adverb where an adjective would be appropriate. Moreover, since no convenient noun matches *one-to-one*, they must use the awkward term *one-to-one-ness*. *Injective* and *injectivity* are more graceful.

[14] Many authors use the term *one-to-one correspondence* for *bijection*. The term *correspondence* emphasizes the symmetric nature of such a relation, whereas *bijection* stresses its asymmetric left-to-right nature. The left-to-right convention became predominant in this context after authors found that symmetric terminology obscures the distinction between a bijection and its inverse. For an example of this awkwardness, consult the research papers by

(continued...)

It's also easy to verify the following propositions about compositions:

- If $f: S \to T$ and $g: D \to S$ are both surjective, then so is $fg: D \to T$.
- If $f: S \to T$ and $g: D \to S$ are both injective, then so is $fg: D \to T$.
- If $f: S \to T$ and $g: D \to S$ are both bijective, then so is $fg: D \to T$, and $(fg)^{-1} = g^{-1}f^{-1}: T \to D$.

Be sure you understand the reversed order of the functions in the previous sentence. If $z = f(g(x))$, then $z = f(y)$, where $y = g(x)$, so that $y = f^{-1}(z)$ and $x = g^{-1}(y) = g^{-1}(f^{-1}(z))$. In general, to undo the result of two or more successive operations, you must undo them all *in reverse order*. Think of donning your stockings, shoes, and overshoes, then removing them!

Exercise 6.11.2 provides an interesting sidelight on the previous paragraph, as follows. Suppose $f: S \to T$ and $g: D \to S$. If fg is surjective, then f must be surjective, but not necessarily g. If fg is injective, then g must be injective, but not necessarily f. If fg is bijective, then f is surjective and g injective, but neither is necessarily bijective. To ensure that you understand the concepts under this heading, complete that exercise, proving the positive clauses and finding counterexamples to justify the others.

Suppose $f: D \to D$ bijectively and $\mathscr{P}$ is a property of elements, subsets, or sequences of elements or subsets of D. It's possible for f but not f^{-1} to preserve $\mathscr{P}$. For example, consider $f(x) = x^3$, $f^{-1}(x) = x^{1/3}$ and the property $x > 2$; that inequality implies $x^3 > 2$ but not $x^{1/3} > 2$. But if $\mathscr{P}$ is *invariant* under f, then it's also invariant under f^{-1}. For example, the property $x_1 > x_2$ is invariant under both x^3 and $x^{1/3}$.

Transformations; identity functions

A bijection $f: D \to D$ is called a *transformation* of D.[15] One bijection, in particular, is ubiquitous: the *identity* function $\iota_D: x \to x$. Note that *the identity function for* $D = \mathbb{R}$ *is linear*: $\iota_{\mathbb{R}}(x) = x = 1x + 0$. It's easy to verify the following general propositions about identity functions:

- If $f: D \to S$, then $f\iota_D = f = \iota_S f$. If f is bijective, then $f^{-1}f = \iota_D$ and $ff^{-1} = \iota_S$. If f is a transformation of D, then $f^{-1}f = \iota_D = ff^{-1}$.
- Suppose $f: D \to S$ bijectively and $g: D \to S$ bijectively. If $fg = \iota_D$ or $gf = \iota_S$, then $g = f^{-1}$ and $f = g^{-1}$.
- *Every* property of elements, subsets, or sequences of elements or subsets of D is invariant under ι_D.

---

[14] (...continued)
Hermann Wiener (1890–1893), which introduced the general approach taken in this chapter.

[15] Authors often use the term *permutation* instead of *transformation*, particularly when the domain is finite.

The symmetric group; transformation groups

The family $\mathcal{S}$ of *all* transformations of D is called the *symmetric group* of D. This chapter is concerned with subfamilies $\mathcal{T}$ of $\mathcal{S}$ that satisfy three requirements:

- the identity ι_D must belong to $\mathcal{T}$;
- $\mathcal{T}$ must contain the inverse of each of its members; and
- $\mathcal{T}$ must contain the composition of any two of its members.

If $\mathcal{T}$ satisfies these conditions, it's called a *transformation group on* D. Thus a subfamily of the symmetric group is a transformation group just in case it contains the identity and is *closed* under inversion and composition. D always has at least two transformation groups: the symmetric group itself, and the *trivial* group that has only one member ι_D. Each clearly satisfies the conditions.

If a transformation f belongs to a transformation group $\mathcal{T}$, so do any n-term compositions $f \circ \cdots \circ f$ and $f^{-1} \circ \cdots \circ f^{-1}$; they're called *powers* of f and written f^n and f^{-n}. f^1 and f^0 stand for f and the identity. You can verify that some laws analogous to elementary algebra hold for this concept. For example, $f^{m+n} = f^m f^n$, $f^{mn} = (f^m)^n$, and $(f^n)^{-1} = f^{-n}$.

When one transformation group is a subset of another, it's called a *subgroup*. For example, the trivial group is a subgroup of every transformation group, and every group is a subgroup of itself. A subgroup of a group $\mathcal{G}$ that's different from $\mathcal{G}$ is called a *proper* subgroup.

Transformation groups are often characterized by invariants: If $\mathcal{P}$ is a property of elements, subsets, or sequences of elements or subsets of D, then the family of all permutations of D under which $\mathcal{P}$ is invariant forms a transformation group on D. This family clearly contains the identity and is closed under inversion and composition.

Members of the symmetric group of two- or three-dimensional space $\mathbb{R}^2$ or $\mathbb{R}^3$ are called *geometric transformations*.[16]

Linear transformations

Among the most useful geometric transformations are the two- and three-dimensional linear functions $X \to AX + B$ for specified matrix and vector constants A and B with invertible A. *The identity function is linear*: $X \to X = IX + O$. Just as in the one-dimensional case, *the composition of linear maps is linear*: If $\varphi(X) = AX + B$ and $\chi(X) = CX + D$, then

[16] This term is also used more broadly, for analogous concepts in projective, non-Euclidean, and higher-dimensional geometry, and sometimes for permutations of sets of lines or other figures instead of just points.

$$\varphi\chi(X) = A(CX + D) + B = (AC)X + (AD + B)$$

and $(AC)^{-1} = C^{-1}A^{-1}$. Moreover, *the inverse of a linear function is linear*:

$$\varphi^{-1}(X) = A^{-1}X - A^{-1}B$$

and $(A^{-1})^{-1} = A$. Notice how matrix multiplication and inversion correspond to function composition and inversion! The case $B = D = O$ shows that the noncommutativity of matrix multiplication and of linear function composition are two aspects of the same phenomenon.

Invariance of betweenness

Betweenness is invariant under linear transformations. To verify that, consider a transformation $X \to X' = AX + B$ and points P, Q, and R. You must show that P-Q-R if and only if P'-Q'-R'. First, P-Q-R implies that $Q = tP + (1 - t)R$ for some scalar t such that $0 \le t \le 1$, hence

$$\begin{aligned} Q' &= AQ + B = A(tP + (1 - t)R) + B \\ &= t(AP + B) + (1 - t)(AR + B) = tP' + (1 - t)R', \end{aligned}$$

hence P'-Q'-R'. Thus $X \to X'$ preserves betweenness. Its inverse is also linear, hence preserves betweenness, so P'-Q'-R' implies P-Q-R also.

It follows that any property you can define in terms of betweenness is invariant under linear transformations $X \to X'$ (regardless of whether it was originally defined that way). For example, collinearity is invariant: For any points P, Q, and R,

P, Q, and R are collinear
if and only if P-Q-R or Q-R-P or R-P-Q, hence
if and only if P'-Q'-R' or Q'-R'-P' or R'-P'-Q', hence
if and only if P', Q', and R' are collinear.

Similarly, all these properties are definable in terms of betweenness:

coplanarity	being	being
	a segment	a half plane
convexity	a ray	a plane
	a line	a half space
being	being the interior of	
an angle	an angle	
a triangle	a triangle	
a tetrahedron	a tetrahedron	

Therefore, they're all invariant under linear transformations.

The argument in the previous paragraph is more general than its context. It actually shows that if betweenness is invariant under *any* transformation

φ, then all properties definable in terms of betweenness are also invariant under φ.

Are linear transformations the *only* ones under which betweenness is invariant? The answer is *yes*, but its proof requires analysis methods beyond the scope of this book.[17]

Linear transformations don't necessarily preserve distance or even similarity. For example, under the transformation

$$X \to X' = \begin{bmatrix} 1 & 2 \\ 0 & 1 \end{bmatrix} X,$$

the image of ΔOUV is $\Delta O'U'V'$, where

$$O = <0,0> = O' \qquad V = <0,1>$$
$$U = <1,0> = U' \qquad V' = <2,1>.$$

Notice that $OV \neq O'V'$; and these triangles aren't similar.

The rest of this chapter is a detailed, fairly complete study of those geometric transformations that do preserve distance—called *isometries*—and those that preserve similarity but not necessarily distance—called *similarities*. All isometries and similarities are linear, but not vice-versa. Detailed study of linear transformations that aren't isometries lies beyond the scope of this book.[11]

6.2 Isometries

Concepts

- Definition
- Physical motions and time are not geometric concepts
- Translations
- Rotations
- Reflections
- The isometry group
- Noncommutativity
- Invariance
- Rigidity theorem
- Uniqueness theorem

[17] Chapter 15 of Martin 1982 is an introduction to that theory in two dimensions. Virtually all the results in *Methods of geometry* about isometries and similarities have both two- and three-dimensional analogues in the theory of general linear transformations. But those are often cast in projective geometric language, and hard to find in the literature. You could start a search with the research paper Ellers 1979 and its references.

This section introduces plane isometries and some related concepts. It presents common examples, but gives few details. Those come later, in sections 6.3 to 6.5 and 6.7, which study in depth four types of isometries. This section and the next are arranged to provide a quick route to the classification theorem, which shows that *every* isometry is one of those four types.

The rest of chapter 6 is concerned with points in a single plane ε. It's always the same plane, so it's never again mentioned explicitly. For example, in the rest of this chapter, *point* means *point in* ε.

A plane isometry $\varphi: P \to P'$ is a transformation of the plane that preserves distance:

(1) to each point P corresponds another, its *image* P';
(2) distinct points $P \neq Q$ have distinct images $P' \neq Q'$;
(3) every point X is the image P' of some point P;
(4) the distance between any points P and Q is the same as that between their images: $PQ = P'Q'$.

Condition (2) is redundant: If $P \neq Q$, then $PQ \neq 0$, so $P'Q' \neq 0$ by (4), hence $P' \neq Q'$. Nevertheless, the definition is usually phrased this way to emphasize that an isometry (1) is a function, (2) is injective, (3) is surjective, and (4) preserves distance.

Many authors use the term *motion* in place of *isometry*, suggesting physical action. That idea isn't formally incorporated into this text because it involves *time*—the situations before, during, and after the action—and time isn't ordinarily regarded as a geometric concept at this level. Instead, this text emphasizes the bijective correspondence between points P and P'. It alludes to physical motion only occasionally, as an informal guide to your intuition.

Translations

A familiar isometry $\tau: X \to X'$, the *translation* shown in figure 6.2.1, can be described by its effect on a single point O. Once you know $V = O'$, you can define at once the image of any point X not on $g = \overleftrightarrow{OV}$; it's the point X' that makes $OVX'X$ a parallelogram. Determining the image of a point W *on* g takes two parallelograms, as shown, or you can just refer to a scale on g. Phrasing the definition like this is awkward. It's easier just to say

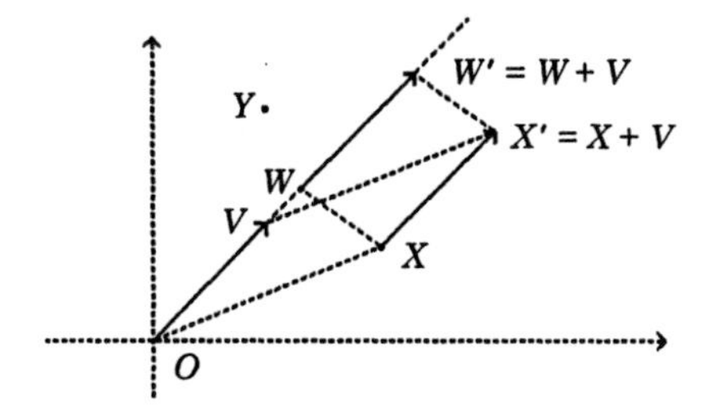

Figure 6.2.1
Translation
$\tau: X \to X' = X + V$

that if you introduce a coordinate system with origin O, then $X' = X + V$ for any X. By the parallelogram law, this gives the same result as the original definition, so it makes no difference which coordinate axes you use. You can regard the identity transformation as the translation τ with $V = O$.

Theorem 1. The translation $\tau : X \to X + V$ is an isometry.

Proof. It's surjective—any point Y is the image of $Y - V$. Find $Y - V$ in figure 6.2.1! And it preserves distance: $P' - Q' = (P + V) - (Q + V) = P - Q$, so $P'Q' = PQ$ by theorem 3.11.2. ♦

Using the word *translation* for this concept isn't as obscure as it might seem. The Latin prefix *trans-* means *across*, and *latus* is the past participle of the irregular verb *fero*, which means *carry*. So to *translate* X to X', you carry X across the parallelogram, just as you carried O to V.[18]

Rotations

You can conceive of a *rotation*, another kind of isometry, in much the same way. You've an intuition of rotation as a physical action. Hold this book in a vertical plane in front of you, your thumbs pressing on opposite spots on its covers, its spine toward you. Rotate it in that plane through $20°$ with your fingers, keeping your thumbs stationary. Which direction? Suppose friends on your right and left request the positive, counterclockwise direction. You try to comply, but one or the other always disagrees. Facing each other, they have opposite senses of direction. What one perceives as $20°$ the other sees as $-20°$. Perhaps intuition isn't entirely adequate! Your problem is to describe the rotation $X \to X'$ so that everyone agrees, and so that you can specify its effect on every point in all cases.

One solution is to specify not the center O and angle measure, but a particular angle, designating one ray as *initial* and the other as *terminal*. Actually, it's handier to mention just the rays; that lets you specify opposite rays for $180°$ and coincident rays for $0°$ or $360°$ rotations. To define the rotation $X \to X'$ about O that carries the initial ray $\vec{OA}$ to the terminal ray $\vec{OB}$, use an angular scale at O as in section 3.13. Suppose the rays correspond to angle parameters α and β, and let $\theta = \beta - \alpha$. Set $X' = O$ if $X = O$; otherwise find an angle parameter ξ for ray $\vec{OX}$, and let X' be the point such that $OX = OX'$ on the ray with angle parameter $\xi + \theta$. Figure 6.2.2 displays an example. This definition doesn't specify α, β, or ξ completely, but any other parameters α_1, β_1, and ξ_1 for the same rays would differ from them by multiples of $360°$, so $\xi_1 + (\beta_1 - \alpha_1)$ would correspond to the same ray $\vec{OX'}$.

[18] *Translate* sometimes has this sense in English religious parlance: St. Mark's body was *translated* from Alexandria to Venice in A.D. 828. *Transfer* has the same etymology.

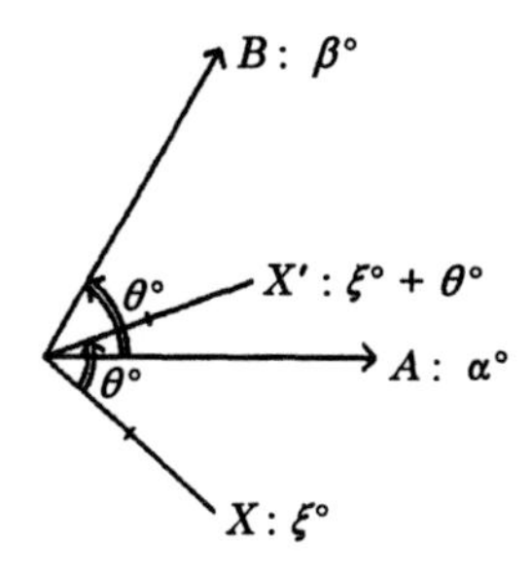

Figure 6.2.2 The rotation ρ_θ about O that maps $\vec{OA}$ to $\vec{OB}$

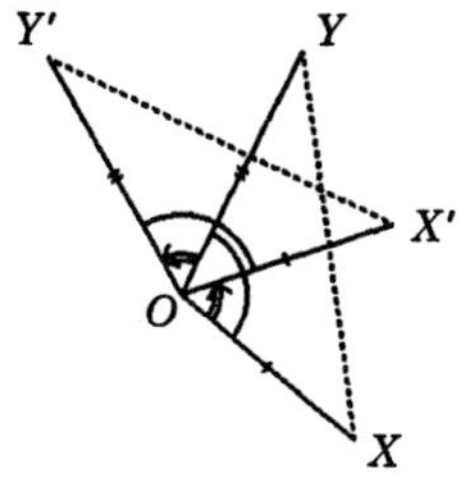

Figure 6.2.3 Rotation preserves distance: $XY = X'Y'$

Because the angle parameter θ figures so directly in the definition, the rotation is often denoted by ρ_θ. Since adding a multiple of $360°$ has no effect on ray $\vec{OX'}$, nor on the intuitive geometric notion of rotation, you don't have to restrict θ to the interval $0° < \theta < 180°$; it can take any value at all! If $\theta = 0°$, then $X' = X$ and ρ_θ is the identity. If $\theta = 180°$, ρ_θ is called the *half turn* σ_O about O.[19]

Theorem 2. For any θ, the rotation ρ_θ is an isometry.

Proof. Consider points X and Y noncollinear with O, with angle parameters ξ and η, as in figure 6.2.3. Then

$$m\angle XOY = |(\xi - \eta) \bmod 360°|.$$

The images X' and Y' have angle parameters $\xi' = \xi + \theta$ and $\eta' = \eta + \theta$, so

$$m\angle XOY = |(\xi' - \eta') \bmod 360°| = m\angle X'OY'.$$

Triangles ΔXOY and $\Delta X'OY'$ are thus congruent, and $XY = X'Y'$. You should also prove this equation for the remaining case, where X and Y are collinear with O. ♦

Reflections

The reflections constitute a third type of isometry. They're perhaps less familiar, but much easier to describe. The *reflection* $X' = \sigma_g(X)$ of a point X across a line g is X itself if X is on g, else it's the point on the

[19] Some authors call σ_O a *point reflection*.

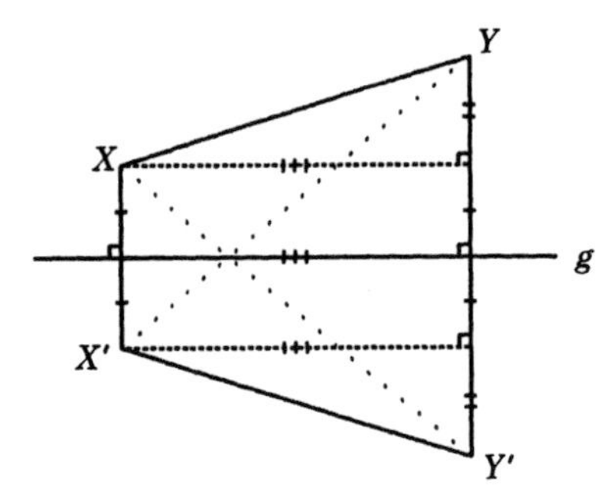

Figure 6.2.4 Reflection across g preserves distance: $XY = X'Y'$.

opposite side of g such that g is the perpendicular bisector of $\overline{XX'}$ —see figure 6.2.4.

Theorem 3. The reflection σ_g across line g is an isometry.

Proof. It's surjective because $X = \sigma_g(X') = \sigma_g(\sigma_g(X))$. Figure 6.2.4 sketches a proof that σ_g preserves the distance between points X and Y when they lie on the same side of g —that is, $XY = X'Y'$. If you show as well that $XY' = X'Y$, then interchange the labels Y and Y', you'll get a proof for X and Y on opposite sides of g. Finally, you can supply the even simpler argument that $XY = X'Y'$ when X and/or Y lies on g. ♦

The isometry group

There are other plane isometries besides the examples already given, but their details are postponed until section 6.7. The rest of this section concentrates on concepts that relate to *all* isometries.

Theorem 4. The isometries form a subgroup of the symmetric group of the plane.

Proof. This means that the identity transformation ι is an isometry, the inverse of an isometry φ is an isometry, and the composition of two isometries φ and χ is an isometry. That is, ι, φ^{-1}, and $\psi = \varphi\chi$ preserve distance. The first is trivial: ι preserves *everything*. To see that φ^{-1} preserves distance, consider two points P and Q. Let $X = \varphi^{-1}(P)$ and $Y = \varphi^{-1}(Q)$. Then $\varphi(X) = P$ and $\varphi(Y) = Q$. Moreover, $XY = \varphi(X)\varphi(Y)$ because φ is an isometry, so

$$\varphi^{-1}(P)\varphi^{-1}(Q) = XY = \varphi(X)\varphi(Y) = PQ.$$

Finally, to verify that $\psi(P)\psi(Q) = PQ$, note first that $\chi(P)\chi(Q) = PQ$ because χ is an isometry. Let $X = \chi(P)$ and $Y = \chi(Q)$. As before, $\varphi(X)\varphi(Y) = XY$, so

$$\psi(P)\psi(Q) = \varphi\chi(P)\varphi\chi(Q) = \varphi(X)\varphi(Y)$$
$$= XY = \chi(P)\chi(Q) = PQ. \blacklozenge$$

Theorem 4 is this book's first result of the form *the elements of a transformation group $\mathcal{T}$ that preserve ... form a subgroup of $\mathcal{T}$.* Here, $\mathcal{T}$ is the group of all transformations of the plane—its symmetric group—and the dots ... represent *distance.* You'll see many theorems that fit this pattern. They're all proved similarly.

Isometries don't necessarily commute. Figure 6.2.5 shows that $\varphi\chi \neq \chi\varphi$ for the reflections $\varphi = \sigma_g$ and $\chi = \sigma_h$ across selected lines g and h. You should investigate various other pairs φ and χ of isometries to get an idea of when their compositions $\varphi\chi$ and $\chi\varphi$ are or are not equal. Under the last heading of this section you'll see that any two *translations* commute. You'll study that question systematically later.

Invariance

A property of sets $\mathcal{X}$ or sequences $<X_1, X_2, \ldots>$ of points is called *invariant* under a transformation $X \to X'$ if $\mathcal{X}$ or $<X_1, X_2, \ldots>$ has that property when and only when its image $\{X' : X \in \mathcal{X}\}$ or $<X_1', X_2', \ldots>$ does. Distance, viewed as a property of pairs of points, is invariant under isometries by definition. Many other properties are invariant under isometries because you can define them in terms of distance. Theorem 5 presents a basic example.

Theorem 5. Betweenness is invariant under any isometry $X \to X'$.

Proof. If X_1, X_2, and X_3 are points, then

$$X_1\text{-}X_2\text{-}X_3 \leftrightarrow X_1X_2 + X_2X_3 = X_1X_3$$
$$\leftrightarrow X_1'X_2' + X_2'X_3' = X_1'X_3' \leftrightarrow X_1'\text{-}X_2'\text{-}X_3'. \blacklozenge$$

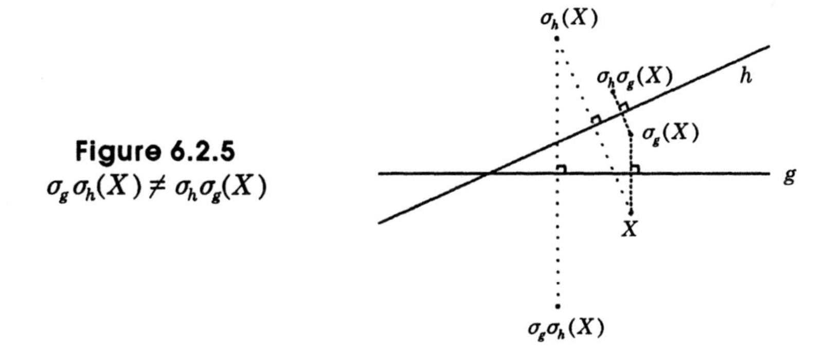

Figure 6.2.5
$\sigma_g\sigma_h(X) \neq \sigma_h\sigma_g(X)$

The critical idea of this proof is that you can define betweenness in terms of distance: X_1-X_2-X_3 $\Leftrightarrow$ $X_1X_2 + X_2X_3 = X_1X_3$. You can use similar arguments to show that *any* property definable in terms of distance is invariant under isometries. At the end of section 6.1 you saw a long list of frequently used properties that are definable in terms of betweenness, hence in terms of distance. Invariance is frequently used implicitly in the rest of this chapter. For example, whenever you use a compound concept like the area of an isometric image of a triangle, you're alluding to two facts:

- The image of a triangle $\mathcal{T}$ is a triangle. (*Being a triangle* is in the list of properties definable in terms of betweenness.)
- It has the same area as $\mathcal{T}$. (You can define *area of a triangle* in terms of distance via Hero's formula, theorem 5.5.10.)

You can supply such arguments when they're needed.

Fixpoints; rigidity and uniqueness theorems

A *fixpoint* of a transformation φ is a point X that it leaves fixed: $\varphi(X) = X$. Often it's easier to analyze a transformation by considering its fixpoints than by studying how it changes other points. Translations, except for the identity, have *no* fixpoints. Rotations, except for the identity, have exactly *one*. The reflection in a line g has a whole *line* of fixpoints. And the fixpoints of the identity transformation constitute the whole *plane*. Are these the only possibilities for the fixpoint set of an isometry? The next two results—the *rigidity theorems*—answer that question.

Theorem 6. If an isometry $X \to X'$ has fixpoints $P \neq Q$, then every point X on $\overleftrightarrow{PQ}$ is fixed.

Proof. There are three cases: P-Q-X, Q-X-P, and X-P-Q. In the first case, where P-Q-X, theorem 5 implies P'-Q'-X', i.e. P-Q-X'. Thus X and X' are on the same side of Q in $\overleftrightarrow{PQ}$ and $QX = Q'X' = QX'$, so $X = X'$. You can prove the remaining cases similarly. ♦

Theorem 7. The only isometry with three noncollinear fixpoints is the identity.

Proof. Suppose an isometry $\varphi : X \to X'$ has noncollinear fixpoints A, B, and C. Then all edge points of ΔABC are fixed. For every point X you can find two distinct edge points P and Q with X on $\overleftrightarrow{PQ}$ —see figure 6.2.6. By theorem 6, X is fixed. Thus φ leaves every point fixed. ♦

Corollary 8 **(*Uniqueness theorem*).** If χ and ψ are isometries, and

$$\chi(A) = \psi(A) \qquad \chi(B) = \psi(B) \qquad \chi(C) = \psi(C)$$

for noncollinear points A, B, and C, then $\chi = \psi$.

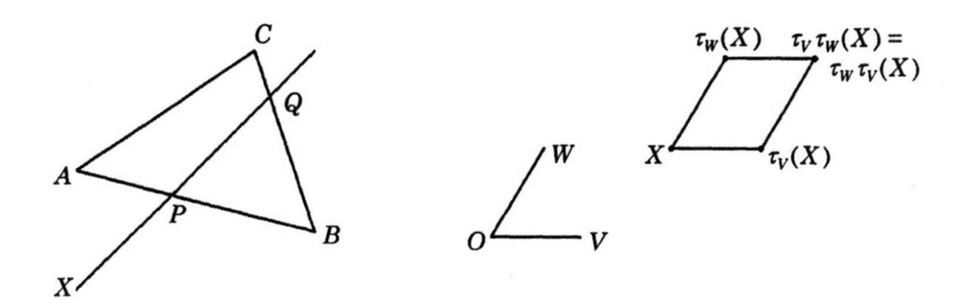

Figure 6.2.6
Proving theorem 7

Figure 6.2.7
Translations commute

Proof. A, B, and C are fixpoints of $\varphi = \chi\psi^{-1}$, so φ is the identity. ♦

The uniqueness theorem is used often to prove that two isometries defined different ways are actually the same. Here's a typical example, which you can check by another method. Set up a coordinate system with origin O, and consider translations $\tau_V : X \to X + V$ and $\tau_W : X \to X + W$. Figure 6.2.7 shows that $\tau_V\tau_W(X) = \tau_W\tau_V(X)$ for the indicated point X. Often, such figures are valid only for a restricted set $\mathscr{X}$ of points X. But if $\mathscr{X}$ includes three noncollinear points, no matter how near each other, then by the uniqueness theorem, the equation holds for *all* points X. In this case, you can certainly pick three noncollinear points X for which the equation holds, so $\mathscr{X}$ is the entire plane and $\tau_V\tau_W = \tau_W\tau_V$. That is, any two translations commute. Of course, you could prove the same result algebraically:

$$\begin{aligned}\tau_V\tau_W(X) &= \tau_W(X) + V = (X + W) + V = (X + V) + W \\ &= \tau_V(X) + W = \tau_W\tau_V(X).\end{aligned}$$

But the uniqueness theorem is especially helpful when there are different cases to consider and no algebraic solution.

6.3 Reflections

Concepts
- Mirror images
- Triangle orientation
- Reflections reverse orientation
- Reflections in perpendicular lines
- Half turns

Section 6.2 introduced the reflection $X' = \sigma_g(X)$ of a point X across a line g: If X lies on g, then $X' = X$, else $X' \neq X$ and g is the perpendicular bisector of $\overline{XX'}$. You saw that σ_g is a plane isometry. It's self-inverse: $\sigma_g^{-1} = \sigma_g$ because $\sigma_g\sigma_g(X) = X$ for each X. Why is it called a *reflection*? Stand with your eye at point Y on the same side of a mirror g as point X in figure 6.3.1. The mirror image of X appears to be at X', because light from X reflects from the mirror to Y so that the incidence and reflection angles $\angle XPO$ and $\angle YPQ$ are equal.[20]

The light ray reflected from X to Y strikes mirror g at the point P for which the path length $XP + PY$ is as short as possible. Figure 6.3.2 shows such a path for an arbitrary point P on g. Since $XP = X'P$, the shortest $XP + PY$ is the smallest sum $X'P + PY$, which occurs when P lies on $\overline{YX'}$ as in figure 6.3.1.

As a child you learned that a mirror reverses orientation; it reflects your right-hand gesture as one of your image's left hand. To learn some technological skills such as driving, you must develop intuition to process these reversed perceptions efficiently. This situation has a mathematical counterpart. Figure 6.3.3 should reinforce your intuition that if a figure $\mathscr{X}$ is oriented, then its image $\mathscr{X}' = \{X' : X \in \mathscr{X}\}$ under a reflection $\sigma_g : X \to X'$ has the opposite orientation.

What *is* an oriented figure? That concept is too general to formulate easily. In fact, it's more general than necessary; you need only consider triangles. Section 3.3 defined ΔABC as the *ordered* triple of its vertices. Thus, for example, ΔABC and ΔCBA are different, even though we use the same figure to represent them. Using figure 6.3.3, you can classify the six triangles with these vertices according to your intuitive notion of orientation:

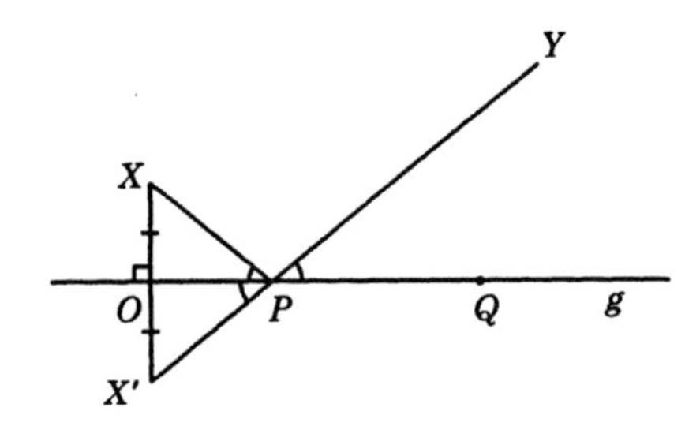

Figure 6.3.1 Mirror image X' of X as seen from Y

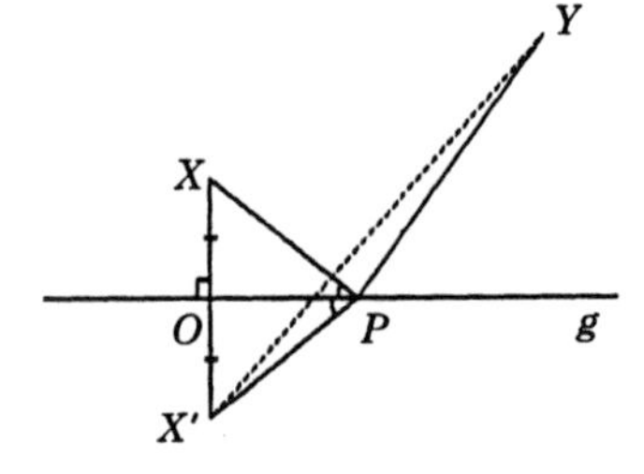

Figure 6.3.2 Path from X to Y via another point P on g

[20] The word *reflection* stems from the past participle of the Latin verb *reflexo*, which means *bend back*; the Greek σ in the traditional notation refers to the German word *Spiegel* for *mirror*.

Figure 6.3.3
Reflection across g
reverses orientation.

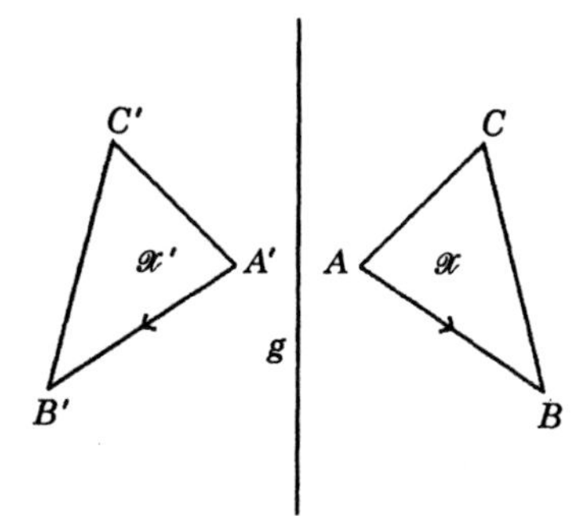

Class (+):	ΔABC	ΔBCA	ΔCAB
Class (-):	ΔCBA	ΔACB	ΔBAC.

How can you distinguish these classes mathematically, without relying on figure 6.3.3 and your intuition? The determinant formula for triangle area derived in section 5.6 provides the tool. Choose a coordinate system, and regard the vertices as coordinate vectors. The triangles all have the same area; according to theorem 5.6.5, it's $\pm \frac{1}{2}$ times each of the determinants

$$\det[A - B, C - B] \quad \det[B - C, A - C] \quad \det[C - A, B - A]$$
$$\det[C - B, A - B] \quad \det[A - C, B - C] \quad \det[B - A, C - A].$$

Those in the first row are all equal. For example, by subtracting the right-hand column from the left, multiplying a column by -1, and interchanging columns, you get

$$\begin{aligned} \det[A - B, C - B] &= \det[(A - B) - (C - B), C - B] \\ &= \det[A - C, C - B] \\ &= -\det[A - C, B - C] \\ &= \det[B - C, A - C]. \end{aligned}$$

Moreover, each determinant in the second row is the negative of the one above it. Thus, all of the triangles in one row have the same *positive* determinant—they constitute class (+). The others have the same *negative* determinant—they constitute class (-). You can then call the triangles *positively* or *negatively* oriented, with respect to the chosen coordinate system. (The coordinate system for figure 6.3.3 evidently oriented ΔABC positively.)

To show that a reflection $\sigma_g : X \to X'$ reverses orientation, you could find equations for the coordinates of X' in terms of those of X, then show that $\det[A' - B', C' - B']$ and $\det[A - B, C - B]$ always differ in sign, so that $\Delta A'B'C'$ and ΔABC have opposite orientation. (In fact, these determinants always have the same magnitude; they differ *only* in sign!) Deriving the required equations for the reflection across an arbitrary line g is tedious. (See exercise 6.11.24.) Moreover, it's unnecessary, because a cleaner argument in section 6.9 will yield this general result. But you can check it easily for some special lines g. For example, if g is the first axis, then σ_g has equations $x_1' = x_1$, $x_2' = -x_2$, and you get $\det[A' - B', C' - B']$ from $\det[A - B, C - B]$ by changing the signs of the entries in its second row. If g is the second axis, change signs in the first row. If g is the diagonal line $x_1 = x_2$, then σ_g has equations $x_1' = x_2$, $x_2' = x_1$, and you get $\det[A' - B', C' - B']$ from $\det[A - B, C - B]$ by interchanging its rows. In each of these examples, the determinant merely changes sign.

The remainder of this section is concerned with the commutativity of reflections σ_g and σ_h across lines g and h. When does that occur? That is, for what lines g and h is $\sigma_g \sigma_h = \sigma_h \sigma_g$? Clearly, this equation holds if $g = h$. Figure 6.3.4 shows that if $g \perp h$, then $\sigma_g \sigma_h(P) = \sigma_h \sigma_g(P)$ for points P not on g or h. You can verify that it also holds for points P *on* the lines. Thus the reflections commute if the lines are equal or perpendicular. Conversely, suppose the reflections commute, but $g \neq h$. Find a point X on g but not h. Then

$$\sigma_g \sigma_h(X) = \sigma_h \sigma_g(X) = \sigma_h(X),$$

so $Y = \sigma_h(X)$ is a fixpoint of σ_g, and Y lies on g. Moreover, $X \neq Y$ because X isn't on h, so h is the perpendicular bisector of $\overline{XY}$ and $g = \overleftrightarrow{XY}$, hence $g \perp h$. This argument, and further inspection of figure 6.3.4, demonstrates

Theorem 1. σ_g and σ_h commute just when $g = h$ or $g \perp h$. If $g \perp h$, then $\sigma_g \sigma_h = \sigma_h \sigma_g$ is the half turn σ_O about the intersection $O = g \cap h$.

Corollary 2. A half turn σ_O commutes with a line reflection σ_h just when O lies on h. In that case, $\sigma_O \sigma_h = \sigma_h \sigma_O = \sigma_g$, the reflection across the line $g \perp h$ through O.

Proof. If O lies on h, then

$$\sigma_O \sigma_h = \sigma_g \sigma_h \sigma_h = \sigma_g = \sigma_h \sigma_h \sigma_g = \sigma_h \sigma_O$$

by theorem 1. Conversely,

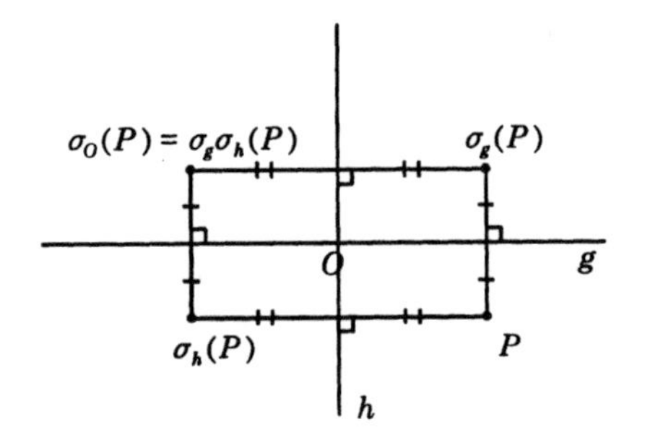

Figure 6.3.4 σ_g and σ_h commute if $g \perp h$.

$$\begin{aligned}\sigma_O\sigma_h = \sigma_h\sigma_O &\\ &\Rightarrow \sigma_O\sigma_h(O) = \sigma_h\sigma_O(O) = \sigma_h(O)\\ &\Rightarrow \sigma_h(O) \text{ is a fixpoint of } \sigma_O\\ &\Rightarrow \sigma_h(O) = O \text{ —the } \textit{only} \text{ fixpoint of } \sigma_O \text{ —}\\ &\Rightarrow O \text{ is a fixpoint of } \sigma_h\\ &\Rightarrow O \text{ lies on } h. \blacklozenge\end{aligned}$$

You can complete this discussion of commutativity and reflections by verifying that two half turns σ_O and σ_P *never* commute unless $O = P$. In the next section, you'll see what kind of isometry $\sigma_O\sigma_P$ is in general.

6.4 Translations

Concepts

- Definition of a translation
- Equations for a translation
- Compositions of translations and vector addition
- Translation group
- Compositions of reflections in parallel lines
- Compositions of half turns
- Translations preserve orientation

Section 6.2 presented two definitions of a translation: a purely geometric one using parallelograms and an equivalent formulation in terms of vector algebra. The latter is more convenient, but requires you to choose a coordinate system at the start of the discussion. Each translation is determined by the image V of the origin O; you can define the translation τ_V with vector V by setting $\tau_V(X) = X + V$ for each point X (regarded as a coordinate vector). This includes the case $\tau_V(O) = O + V = V$. As shown in figure 6.2.1, you can construct the point $X + V$ from V, O, and X using parallelograms, with no reference to coordinates. Therefore, the definition of τ_V depends only on V and O, not on other details of the coordinate system. Theorem 1 displays the *equations* for τ_V in terms of the coordinate system.

Theorem 1. The components of $X' = \tau_V(X)$ are

$$\begin{cases} x_1' = x_1 + v_1 \\ x_2' = x_2 + v_2 \,. \end{cases}$$

Not only do translations τ_V correspond one to one with their vectors V, but composition of translations corresponds to vector addition. Further, this analogy relates the inverse of a translation to the negative of its vector, and the identity, or *trivial*, translation corresponds to the zero vector:

Theorem 2. For any vectors V and W, $\tau_V\tau_W = \tau_{V+W} = \tau_W\tau_V$ and $\tau_V^{-1} = \tau_{-V}$. Moreover, τ_O is the identity transformation ι.

Proof. For all points X,

$$\begin{aligned}\tau_V\tau_W(X) &= \tau_V(X+W) = (X+W)+V = X+(W+V)\\ &= X+(V+W) = \tau_{V+W}(X)\\ \tau_O(X) &= X+O = X = \iota(X).\end{aligned}$$

Moreover,

$$\begin{aligned}&\tau_W\tau_V = \tau_{W+V} = \tau_{V+W} = \tau_V\tau_W\\ &\tau_V\tau_{-V} = \tau_{V+(-V)} = \tau_O = \iota = \tau_{(-V)+V} = \tau_{-V}\tau_V. \blacklozenge\end{aligned}$$

Corollary 3. The translations form a transformation group on the plane.

Proof. The composition $\tau_V\tau_W$ of two translations is the translation τ_{V+W}. The inverse τ_V^{-1} of a translation is the translation τ_{-V}. Finally, the identity transformation is a translation: $\iota = \tau_O$. ♦

The next results show how to build translations from reflections or half turns.

Theorem 4. The composition $\sigma_g\sigma_h$ of the reflections across equal or parallel lines g and h is a translation τ. With respect to any coordinate system, τ has vector $V = 2(P - Q)$, where P and Q are the intersections of g and h with a common perpendicular k.

Proof. If $g = h$, then V is the zero vector and $\sigma_g\sigma_h = \iota = \tau_V$. If $g \neq h$, as in figure 6.4.1, then

$$V = 2(P - Q) = \sigma_g\sigma_h(X) - X,$$

so $\tau_V(X) = \sigma_g\sigma_h(X)$ for any point X on the side of h opposite g. This equation holds for three noncollinear points—pick them near X in figure 6.4.1—so by the uniqueness theorem, $\sigma_g\sigma_h = \tau_V$. ♦

Corollary 5. The composition $\sigma_P\sigma_Q$ of half turns about points P and Q is a translation τ. With respect to any coordinate system, τ has vector $V = 2(P - Q)$.

Proof. Let g and h be the perpendiculars to $k = \overleftrightarrow{PQ}$ through P and Q. Then $\sigma_P\sigma_Q = \sigma_g\sigma_k\sigma_k\sigma_h = \sigma_g\sigma_h$. Now apply theorem 4. ♦

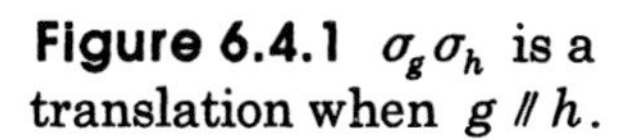
Figure 6.4.1 $\sigma_g \sigma_h$ is a translation when $g \parallel h$.

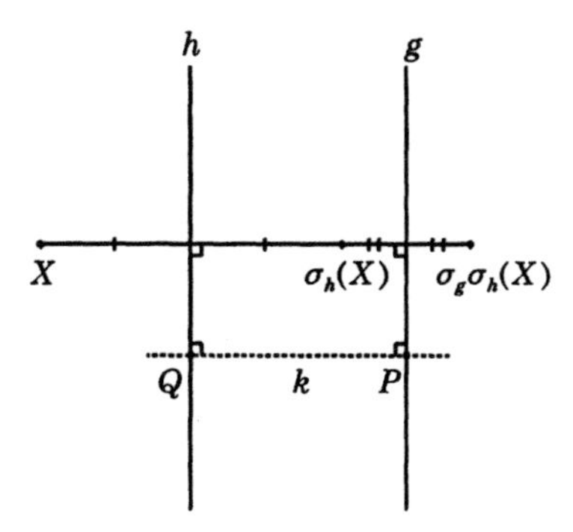

Given vector V and point P you can find a point Q, or given V and Q you can find P, such that $V = 2(P - Q)$. That yields

Corollary 6. Every translation τ is the composition $\sigma_P \sigma_Q$ of two half turns. You can choose P or Q arbitrarily; the other then depends on τ. If $P \neq Q$, then τ is the composition $\sigma_g \sigma_h$ of the reflections across the perpendiculars g and h to $\overleftrightarrow{PQ}$ through P and Q.

In section 6.5 you'll need to analyze all compositions of *three* reflections. The next result provides a start for that study.

Corollary 7. A composition $\sigma_g \sigma_h \sigma_k$ of reflections across parallel or equal lines g, h, and k is the reflection across a fourth line copencilar with these three.

Proof. A line l perpendicular to g, h, and k intersects them at points P, Q, and S. Choose a coordinate system, and find the point R on l such that $P - Q = R - S$. Let $V = 2(P - Q) = 2(R - S)$ and j be the perpendicular to l through R. By theorem 4, $\sigma_g \sigma_h = \tau_V = \sigma_j \sigma_k$, so $\tau_V(X) = \sigma_g \sigma_h(X)$ for any point X. ♦

Transformations correspond so closely to vectors that properties of vectors are often attributed to translations. For example, we refer to parallel translations or to the angle between them. By the *length* $|\tau|$ of a translation we mean the length of its vector: $|\tau| = P\tau(P)$ for any point P.

The last idea considered in this section is the effect of translation on orientation. Figure 6.4.2 should reinforce your intuition that any translation preserves orientation. You could deduce that from theorem 4 or corollary 5 if you'd shown already that every reflection reverses orientation or every half turn preserves it. However, neither of those results has been proved yet. Instead, you can use coordinates and determinants to construct the following argument, similar to the one devised for some special reflections in section 6.3.

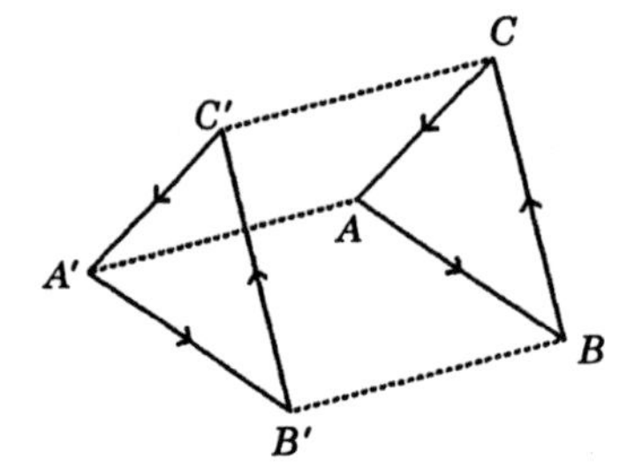

Figure 6.4.2 Translation preserves orientation.

Theorem 8. Suppose a translation τ_V maps ΔABC to $\Delta A'B'C'$. Then

$$\det[A - B, C - B] = \det[A' - B', C' - B'],$$

so τ_V preserves orientation.

Proof.
$$\begin{aligned}\det[A' - B', C' - B'] &= \det[(A+V) - (B+V), (C+V) - (B+V)] \\ &= \det[A - B, C - B]. \blacklozenge\end{aligned}$$

6.5 Rotations

Concepts
- Definition of a rotation
- Equation of a rotation about the origin
- Composition of rotations and angle addition
- Group of rotations about O
- Compositions of reflections in intersecting lines
- Compositions of rotations and translations

Section 6.2 defined the rotation $X \to X'$ about a point O that carries initial ray $\vec{OA}$ to terminal ray $\vec{OB}$ using an angular scale p at O as follows. Suppose p assigns those rays angle parameters α and β, and let $\theta = \alpha - \beta$. For any point $X \neq O$, find an angle parameter ξ corresponding to ray $\vec{OX}$, and let $X' = \rho_\theta(X)$ be the point on the ray with angle parameter $\xi + \theta$, such that $OX' = OX$. When you study a rotation in a broader context, you're often not free to choose the scale; some other aspect of your problem may have specified it already. Does it make any difference which angular scale you use? The answer is no. According to theorem 3.13.3, if p' is another scale at O, then there are constants ζ and n such that $n = \pm 1$ and for all ξ, $p(\xi) = p'(\zeta + n\xi)$. If you used scale p' to describe ρ, you'd find that the initial and terminal rays and ray $\vec{OX}$ correspond to parameters

$\alpha' = \zeta + n\alpha$, $\beta' = \zeta + n\beta$, and $\xi' = \zeta + n\xi$. To determine $\rho_\theta(X)$, you'd compute

$$\begin{aligned}\xi' + (\beta' - \alpha') &= (\zeta + n\xi) + (\zeta + n\beta) - (\zeta + n\alpha)\\ &= \zeta + n(\xi + \beta - \alpha) = \zeta + n(\xi + \theta),\end{aligned}$$

and you'd find that the corresponding ray is $p'(\zeta + n(\xi + \theta)) = p(\xi + \theta) = \vec{OX}'$, the same one you got with scale p.

Now choose a coordinate system with origin O. Use any angular scale at O. For any real number θ consider the rotation $\rho_\theta : X \rightarrow X'$ that carries the ray with parameter $0°$ to the one with parameter θ. Theorem 1 displays the *equations* for ρ_θ in terms of the coordinate system. Later, section 6.8 will derive equations for a rotation about a point different from the origin.

Theorem 1. The components of $X' = \rho_\theta(X)$ are

$$\begin{cases} x_1' = x_1 \cos\theta - x_2 \sin\theta \\ x_2' = x_1 \sin\theta + x_2 \cos\theta\ . \end{cases}$$

Proof. Suppose X has angle parameter ξ. Then $OX' = OX$ and X' has parameter $\xi + \theta$. Using the cosine and sine addition formulas (corollary 5.5.5) you get

$$\begin{aligned}x_1 &= OX \cos\xi\\ x_2 &= OX \sin\xi\\ x_1' &= OX' \cos(\xi + \theta)\\ &= OX[\cos\xi \cos\theta - \sin\xi \sin\theta] = x_1 \cos\theta - x_2 \sin\theta\\ x_2' &= OX' \sin(\xi + \theta)\\ &= OX[\cos\xi \sin\theta + \sin\xi \cos\theta] = x_1 \sin\theta + x_2 \cos\theta. \ \blacklozenge\end{aligned}$$

Not only do rotations ρ_θ about O correspond to angle parameters θ, but composition of these rotations corresponds to angle addition. Further, this analogy relates the inverse of a rotation to the negative of its parameter, and the identity rotation corresponds to parameter zero:

Theorem 2. $\rho_\eta \rho_\theta = \rho_{\eta+\theta} = \rho_\theta \rho_\eta$ and $\rho_\theta^{-1} = \rho_{-\theta}$ for any angle parameters η and θ. Moreover, $\rho_{0°}$ is the identity transformation ι.

Proof. First, $\rho_\eta \rho_\theta(O) = \rho_\eta(O) = O$. Consider a point $X \neq O$ for which ray $\vec{OX}$ has angle parameter ξ; let $\rho_\eta(X) = X'$ and $\rho_\theta(X') = X''$. Then $\vec{OX}'$ and $\vec{OX}''$ have parameters $\xi + \eta$ and $(\xi + \eta) + \theta = \xi + (\eta + \theta)$, so $\rho_{\eta+\theta}(X) = X''$. It follows that $\rho_\eta \rho_\theta = \rho_{\eta+\theta}$. Similarly, $\rho_\theta \rho_\eta = \rho_{\theta+\eta} = \rho_{\eta+\theta}$. Clearly, $\rho_{0°} = \iota$. Finally, $\rho_\theta^{-1} = \rho_{-\theta}$ because $\rho_{-\theta}\rho_\theta = \rho_{-\theta+\theta} = \rho_{0°} = \iota$. ♦

According to theorem 2, the composition of rotations about O, the inverse of any such rotation, and the identity transformation are all rotations about O:

Corollary 3. The rotations about O form a transformation group on the plane.

In section 6.4 you saw that a composition of reflections across parallel lines is a translation. The next result shows that if the lines intersect, the composition is a rotation. In fact, *every* rotation is such a composition of two reflections.

Theorem 4. If rays $\vec{OA}$ and $\vec{OB}$ have angle parameters α and β, then $\rho_{2(\alpha-\beta)}$ is the composition $\sigma_a\sigma_b$ of the reflections across lines $a = \overleftrightarrow{OA}$ and $b = \overleftrightarrow{OB}$.

Proof. If $a = b$, then $\rho_{2(\alpha-\beta)} = \iota = \sigma_a\sigma_b$. If $a \perp b$, the result is contained in theorem 6.3.1.

Now assume $\angle AOB$ is acute, as in figure 6.5.1. Select a point X on the side of b opposite A but on the same side of a as B, with $m\angle XOB < m\angle BOA$ as shown. Let $X' = \sigma_b(X)$ and $X'' = \sigma_a(X') = \sigma_a\sigma_b(X)$. Then rays $\vec{OX}$, $\vec{OB}$, $\vec{OX'}$, $\vec{OA}$, and $\vec{OX''}$ all fall in a single half plane as shown, and $m\angle XOX'' = 2m\angle BOA$, so $X'' = \rho_{2(\alpha-\beta)}(X)$. You can select three noncollinear points X that satisfy the same conditions, so $\sigma_a\sigma_b = \rho_{2(\alpha-\beta)}$ by the rigidity theorem.

If $\angle AOB$ is obtuse, replace A by a point $A' \neq O$ such that A-O-A'. Then $\angle A'OB$ is acute, $\overleftrightarrow{OA'} = a$, and $\vec{OA'}$ has parameter $\alpha + 180°$. By the previous paragraph, $\sigma_a\sigma_b = \rho_{2(\alpha+180°-\beta)} = \rho_{2(\alpha-\beta)+360°} = \rho_{2(\alpha-\beta)}$. ♦

Given an angle parameter θ and a ray $\vec{OA}$ with parameter α you can find a ray $\vec{OB}$ with parameter β, or given θ and $\vec{OB}$ you can find $\vec{OA}$, such that $\theta = 2(\alpha - \beta)$. Setting $a = \overleftrightarrow{OA}$ and $b = \overleftrightarrow{OB}$ yields

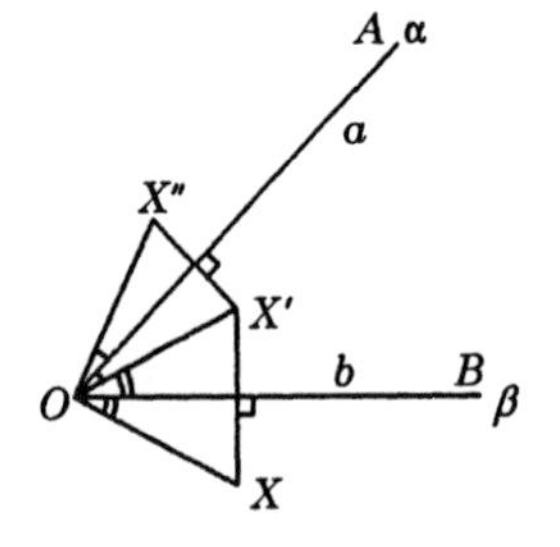

Figure 6.5.1
Proof of theorem 4

Corollary 5. Every rotation ρ about O is the composition $\sigma_a\sigma_b$ of two reflections across lines through O. You can choose either line a or b arbitrarily; the other then depends on ρ.

The next result continues the analysis of compositions of *three* reflections that you started with corollary 6.4.

Corollary 6. A composition $\sigma_g\sigma_h\sigma_k$ of reflections across lines through O is the reflection across a fourth line copencilar with these three.

Proof. Find points A, B, and $D \neq O$ on g, h, and k. Suppose rays $\vec{OA}$, $\vec{OB}$, and $\vec{OD}$ correspond to angle parameters α, β, and δ. Find the ray $\vec{OC}$ that corresponds to the parameter γ for which $\alpha - \beta = \gamma - \delta$, and let $j = \vec{OC}$. Then $\sigma_g\sigma_h = \rho_{2(\alpha-\beta)} = \rho_{2(\gamma-\delta)} = \sigma_j\sigma_k$, so $\sigma_g\sigma_h\sigma_k = \sigma_j$. ♦

Theorems 6.4.2 and 6.5.2 showed that the composition of two translations or of two rotations about the same center is a translation or rotation, respectively. Theorems 6.4.4 and 6.5.4 showed that the composition of two reflections is a translation or a rotation. The final two results in this section describe the composition of a rotation and a translation, and of rotations about different centers.

Corollary 7. The composition of a rotation $\rho \neq \iota$ and any translation τ is a rotation about some center.

Proof. By corollary 6.4.6, $\tau = \sigma_g\sigma_h$ for some line g through the center O of ρ and some line h parallel to g. By corollary 5, $\rho = \sigma_f\sigma_g$ for some line $f \neq g$ through O. Therefore $\rho\tau = \sigma_f\sigma_g\sigma_g\sigma_h = \sigma_f\sigma_h$, a rotation about the intersection P of f and h. You can supply a similar argument for $\tau\rho$. ♦

Corollary 8. Any composition of two rotations is a rotation or a translation.

Proof. By theorem 2 it's enough to consider rotations ρ and ρ' about distinct centers O and O'. Let $h = \overleftrightarrow{OO'}$. By corollary 5, $\rho = \sigma_g\sigma_h$ and $\rho' = \sigma_h\sigma_{g'}$ for some lines g and g' through O and O'. Therefore $\rho\rho' = \sigma_g\sigma_h\sigma_h\sigma_{g'} = \sigma_g\sigma_{g'}$. By theorem 6.4.4 or theorem 4, this composition is a translation or a rotation. ♦

Corollary 7 doesn't mention the relationship of the angle parameters of the constituent rotations to that of the composition. Corollary 8 doesn't show how the parameters of the constituent rotations determine whether the composition φ is a rotation or a translation, and in the former case how they are related to the angle parameter of φ. You can work out the connections in exercise 6.11.10 by analyzing the lines used in the proofs, or you can refer ahead to theorem 6.9.3, which derives these results through matrix algebra.

6.6 Structure theorem

Concepts
- Congruent figures
- Structure theorem
- Even and odd isometries
- Group of even isometries

Previous sections introduced the concept of a plane isometry, and described three familiar types in detail: reflections, translations, and rotations. You saw that all translations and rotations are compositions of two reflections. This section's main result is the structure theorem. It shows that *every* isometry is the identity transformation, or a reflection, or the composition of two reflections, *or the composition of three*. It's the fundamental tool for classifying plane isometries. Section 6.7 investigates the only type not yet introduced: the composition of three reflections. In section 6.8 the structure theorem plays a central role in connecting the study of isometries with matrix algebra.

The proof of the structure theorem proceeds via three preliminary results, which are arranged to make the argument as concise as possible. Lemma 2 and theorem 3 are hard to illustrate cleanly because each describes two or three different cases. You should work out separate figures for all cases.

Lemma 1. Suppose ΔABC and ΔABZ are congruent triangles. Then $C = Z$ or $\sigma_{\overleftrightarrow{AB}}(Z)$.

Proof. By theorem 3.12.5 case (2), the circles with centers A and B and radii $AC = AZ$ and $BC = BZ$ intersect at exactly two points Z and Z', and $\overleftrightarrow{AB}$ is the perpendicular bisector of $\overline{ZZ'}$. See figure 6.6.1. ♦

Lemma 2. Suppose ΔABC and ΔAYZ are congruent triangles. Then there exists an isometry χ such that $B = \chi(Y)$, $C = \chi(Z)$, and χ is the

(0) identity transformation,
(1) reflection across a line through A, or
(2) composition of the reflections across two lines through A.

Figure 6.3.1
Proving lemma 1

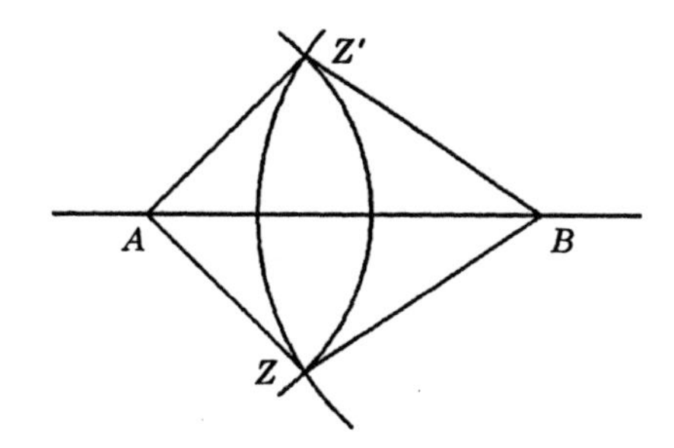

Proof. If $B = Y$, lemma 1 yields the result with alternative (0) or (1). Otherwise, the perpendicular bisector h of $\overline{BY}$ passes through A because $AB = AY$. Then $B = \sigma_h(Y)$; let $Z' = \sigma_h(Z)$. By the SSS congruence principle, $\Delta ABZ' \cong \Delta AYZ \cong \Delta ABC$. By lemma 1 with Z' in place of Z, either

(1) $C = Z'$, or
(2) $C = \sigma_g(Z')$, where $g = \overleftrightarrow{AB}$.

In case (1), let $\chi = \sigma_h$; in case (2), $\chi = \sigma_g \sigma_h$. ♦

Theorem 3. Suppose ΔABC and ΔXYZ are congruent triangles. Then there exists an isometry φ such that $A = \varphi(X)$, $B = \varphi(Y)$, $C = \varphi(Z)$, and φ is

(0) the identity transformation,
(1) a reflection,
(2) the composition of reflections across two lines, or
(3) the composition of reflections across three lines.

Proof. If $A = X$, lemma 2 yields the result with alternative (0), (1), or (2). Otherwise, let k be the perpendicular bisector of $\overline{AX}$. Then $A = \sigma_k(X)$; let $Y' = \sigma_k(Y)$ and $Z' = \sigma_k(Z)$. By the SSS principle, $\Delta AY'Z' \cong \Delta XYZ \cong \Delta ABC$. By lemma 2 with Y' and Z' in place of Y and Z, there exists an isometry χ such that $B = \chi(Y')$, $C = \chi(Z')$, and χ is the

(1) identity transformation,
(2) reflection across a line through A, or
(3) composition of the reflections across two lines through A.

Let $\varphi = \chi\sigma_k$. ♦

According to theorem 3, congruent triangles are always related by some isometry. By the SSS congruence principle, triangles related by an isometry are always congruent. Thus, you should be able to *define* triangle congruence in terms of isometry in section 3.5. Some axiomatizations of geometry do exactly that. But this idea has more practical consequences, too. You can now easily *extend* the notion of congruence: Two figures $\mathcal{F}$ and $\mathcal{F}'$ are called *congruent*—in symbols, $\mathscr{F} \cong \mathscr{F}'$ —if $\mathscr{F}' = \varphi[\mathscr{F}]$ for some isometry φ. That's much more graceful than defining congruence separately for each interesting class of figure. You should check that this new definition is consistent with the section 3.5 definitions for segment and angle congruence, and with any other special cases you may have encountered earlier.

Theorem 4 (Structure theorem). The group of plane isometries consists of the identity, reflections, and compositions of two or three reflections.

Proof. Consider any isometry ψ, select any triangle $\triangle XYZ$, and let $A = \psi(X)$, $B = \psi(Y)$, and $C = \psi(Z)$. By theorem 3, there exists an isometry φ such that $A = \varphi(X)$, $B = \varphi(Y)$, $C = \varphi(Z)$, and φ is

(0) the identity transformation,
(1) a reflection,
(2) the composition of reflections across two lines, or
(3) the composition of reflections across three lines.

By the uniqueness theorem, $\varphi = \psi$. ♦

To what extent do these isometry types overlap? The identity transformation ι is a composition of two reflections: $\iota = \sigma_g \sigma_g$ for any line g. Similarly, every reflection is a composition of three: $\sigma_g = \sigma_g \sigma_g \sigma_g$. You could write the structure theorem more tersely: Every plane isometry is a composition of two or three reflections. Is there any overlap between *these* two classes? The equation $\sigma_g \sigma_h = \sigma_j \sigma_k \sigma_l$ for lines g, h, j, k, and l leads to contradiction, as follows. Let $\varphi = \sigma_g \sigma_h$ and $\chi = \sigma_l \sigma_k$ so that $\varphi\chi = \sigma_j$. By theorems 6.4.4 and 6.5.4 and corollaries 6.5.7 and 6.5.8, each of φ and χ, and hence their composition, is a rotation or a translation. But the fixpoint set of a rotation or a translation is empty, a single point, or the entire plane, whereas that for a reflection is a line. Therefore, $\varphi\chi \neq \sigma_j$ —contradiction! *No* composition of two reflections is a composition of three.

Compositions of two reflections are called *even* isometries; compositions of three are called *odd*. By the previous paragraph, every isometry is even or odd, but none is both. The inverse of an even isometry—a translation or rotation—is also a translation or rotation, hence it's even. Summarizing these results about even isometries, here is

Theorem 5. The even isometries—the translations and rotations—constitute a group.

The notions of even and odd isometry are closely connected with that of orientation, mentioned several times in earlier sections. Since orientation was defined via coordinates and a determinant, that aspect of the theory is presented later, in section 6.8.

You've considered the reflections and the even isometries already in considerable detail. The next section investigates the remaining type: compositions of three reflections.

6.7 Glide reflections

> **Concepts**
> When do a translation and a reflection commute?
> Glide reflections

In sections 6.4 and 6.5 you saw that the composition φ of reflections across three copencilar lines g, h, and k is a reflection across another line of the same pencil. This section investigates φ when g, h, and k aren't copencilar.

The first consideration seems unrelated, but it will provide the answer. When does a translation commute with a reflection?

Theorem 1. Let $\tau : P \rightarrow P'$ be a translation that's not the identity. Then τ and a reflection σ_g commute if and only if $\overleftrightarrow{PP'} \parallel g$ for some point P.

Proof. Figure 6.7.1 depicts the *if* argument. To verify the converse, suppose $\overleftrightarrow{PP'}$ and g intersect at a single point O. You'll find that $\tau\sigma_g(O)$ and $\sigma_g\tau(O)$ lie on different sides of g. ♦

When τ and σ_g commute, their composition φ is called a *glide reflection*. In that case, $\varphi\varphi = \tau\sigma_g\sigma_g\tau = \tau\tau$. This equation lets you determine τ if you know φ; for any point P, let $P'' = \varphi\varphi(P)$. Then $\tau(P)$ is the midpoint of $\overline{PP''}$. You can also determine g, since $\sigma_g = \varphi\tau^{-1}$. Thus you can call τ *the* corresponding translation and g *the* axis of φ. The vector corresponding to τ is often called the vector of φ.

When its translation is the identity, a glide reflection φ is just a reflection. Otherwise, you can see from figure 6.7.1 that φ has no fixpoint. In that case, φ is an odd motion, but not a reflection. Therefore, it's the composition of the reflections in three noncopencilar lines. Theorem 2 is the converse of this statement.

Theorem 2. If lines k, h, and g are not copencilar, then $\sigma_k\sigma_h\sigma_g$ is a glide reflection, but not a reflection.

Proof, case 1: k and h intersect at a point O, which may not lie on g. Find the line $h' \perp g$ through O; it intersects g at a point P. (See figure

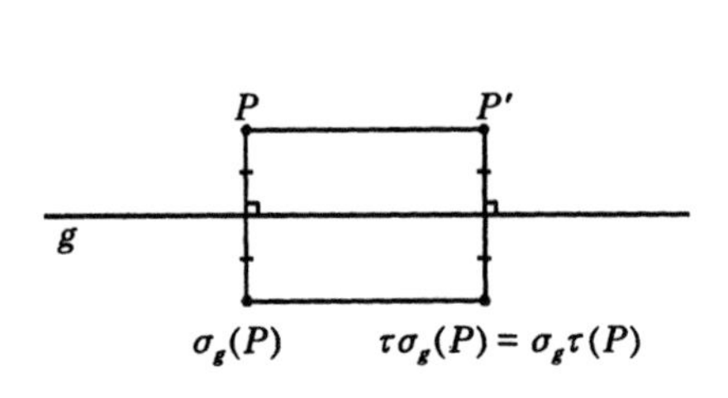

Figure 6.7.1
A commuting translation τ and reflection σ_g

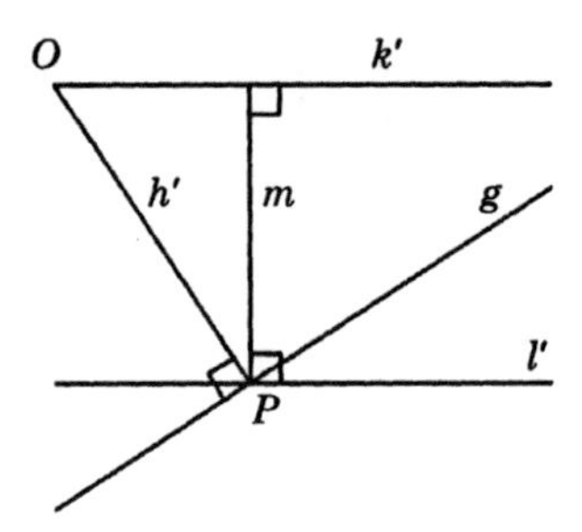

Figure 6.7.2
Case (1) of the proof of theorem 2

6.7.2.) Let k' be the image of h' under the rotation $\sigma_k\sigma_h$, so that $\sigma_k\sigma_h = \sigma_{k'}\sigma_{h'}$. Find the line $m \perp k'$ through P, and the line $l' \perp m$ through P. Then $\sigma_{h'}\sigma_g = \sigma_P = \sigma_{l'}\sigma_m$ and $\iota \neq \sigma_{k'}\sigma_{l'} = \tau$, a translation. If $P' = \tau(P)$, then P' lies on m, hence $\tau\sigma_m$ is a glide reflection, but not a reflection. Finally,

$$\sigma_k\sigma_h\sigma_g = \sigma_{k'}\sigma_{h'}\sigma_g = \sigma_{k'}\sigma_P = \sigma_{k'}\sigma_{l'}\sigma_m = \tau\sigma_m.$$

Case 2: $k \parallel h$. Then k and g intersect at a point. By case (1), $\sigma_g\sigma_h\sigma_k$ is a glide reflection $\tau\sigma_m$ with translation $\tau \neq \iota$ and axis m. Choose a point P, and set $O = \tau^{-1}(P)$ and $\tau(P) = Q$. Then $\vec{PO} = \vec{PQ}$, so τ^{-1} and σ_m commute and $\tau^{-1}\sigma_m$ is a glide reflection, but not a reflection. Finally,

$$\sigma_k\sigma_h\sigma_g = (\sigma_g\sigma_h\sigma_k)^{-1} = (\tau\sigma_m)^{-1} = \sigma_m^{-1}\tau^{-1} = \tau^{-1}\sigma_m. \blacklozenge$$

6.8 Isometries and orthogonal matrices

Concepts

Orthogonal matrices A
Inverses and products of orthogonal matrices are orthogonal
Equation of a rotation about the origin O
Equation of a reflection across a line through O
Equations $X' = AX + V$ always represent isometries
Every isometry has a unique equation $X' = AX + V$
Angle parameters for arbitrary rotations
Determinants of orthogonal matrices and parity of isometries

So far, your study of isometries has employed mostly synthetic methods. You've used coordinates only to formulate equations for translations, rotations about the origin, and a few special line reflections. And those have figured only in rudimentary discussions of orientation. It's time to use coordinates more seriously to derive equations for *all* plane isometries and to complete the discussion of orientation. Section 6.9 will then use equations to study various compositions of isometries.

For this section, use a fixed coordinate system with origin O, and the usual angular scale.

The most efficient way to formulate equations for arbitrary plane isometries uses matrix algebra. For example, you can restate as single matrix equations the pairs of equations derived in sections 6.5 and 6.3 for a rotation ρ_θ about O and a reflection σ_h across the first axis. For each of these isometries $X \to X'$ the components of X' are

$$\rho_\theta \begin{cases} x_1' = x_1 \cos\theta - x_2 \sin\theta \\ x_2' = x_1 \sin\theta + x_2 \cos\theta \end{cases} \qquad \sigma_h \begin{cases} x_1' = x_1 \\ x_2' = -x_2 \end{cases}$$

$$X' = A_\theta X \qquad\qquad X' = HX$$

$$A_\theta = \begin{bmatrix} \cos\theta & -\sin\theta \\ \sin\theta & \cos\theta \end{bmatrix} \qquad H = \begin{bmatrix} 1 & 0 \\ 0 & -1 \end{bmatrix}.$$

Orthogonal matrices

In this process, *orthogonal* matrices play a central role: matrices whose transposes are their inverses. That is, a square matrix A is orthogonal if A^tA is the identity matrix I. For example, the matrices A_θ and H are orthogonal:

$$A_\theta^t A_\theta = \begin{bmatrix} \cos\theta & \sin\theta \\ -\sin\theta & \cos\theta \end{bmatrix} \begin{bmatrix} \cos\theta & -\sin\theta \\ \sin\theta & \cos\theta \end{bmatrix}$$

$$= \begin{bmatrix} \cos^2\theta + \sin^2\theta & 0 \\ 0 & \sin^2\theta + \cos^2\theta \end{bmatrix} = I$$

$$H^t H = \begin{bmatrix} 1 & 0 \\ 0 & -1 \end{bmatrix} \begin{bmatrix} 1 & 0 \\ 0 & -1 \end{bmatrix} = \begin{bmatrix} 1 & 0 \\ 0 & (-1)^2 \end{bmatrix} = I.$$

These examples show that *some* isometries that fix the origin O correspond to orthogonal matrices. In fact, *every* orthogonal matrix A represents an isometry $X \to X' = AX$ that fixes O. To prove that, note that $X \to X'$

bijectively because A is invertible. Moreover, $O' = AO = O$ and for any points P and Q,

$$\begin{aligned}(P'Q')^2 &= (P' - Q')\cdot(P' - Q')\\ &= (P' - Q')^t\,(P' - Q')\\ &= (AP - AQ)^t\,(AP - AQ)\\ &= \big(A(P - Q)\big)^t A(P - Q)\\ &= (P - Q)^t A^t A(P - Q)\\ &= (P - Q)^t\,(P - Q) = (PQ)^2.\end{aligned}$$

Thus $P'Q' = PQ$; the transformation $X \to X'$ preserves distance.

For any vector V, the transformation $X \to X'' = X' + V = AX + B$ is also an isometry:

$$\begin{aligned}(P''Q'')^2 &= (P'' - Q'')\cdot(P'' - Q'')\\ &= \big((P' + V) - (Q' + V)\big)\cdot\big((P' + V) - (Q' + V)\big)\\ &= (P' - Q')\cdot(P' - Q') = (P'Q')^2 = (PQ)^2.\end{aligned}$$

You'll soon see that you can represent *every* plane isometry $X \to X''$ this way; there's always an orthogonal matrix A and a vector V such that $X'' = AX + V$ for every point X.

First, here are some simple and useful facts about orthogonal matrices. Clearly, the identity matrix I is orthogonal. The inverse of an orthogonal matrix A is orthogonal, for if $A^tA = I$, then $A^t = A^{-1}$ and

$$(A^{-1})^t A^{-1} = (A^{-1})^t A^t = (AA^{-1})^t = I^t = I.$$

Moreover, the product of orthogonal matrices A and B is orthogonal:

$$(AB)^t(AB) = B^tA^tAB = B^tB = I.$$ [21]

Now consider a line $g = \overleftrightarrow{OP}$, where ray $\overrightarrow{OP}$ has angle parameter θ. Find the ray $\overrightarrow{OB}$ with parameter $\frac{1}{2}\theta$, and let $b = \overleftrightarrow{OB}$, as in figure 6.8.1. Consider the rotation $\rho_\theta = \sigma_b\sigma_h = \sigma_g\sigma_b$. This equation implies

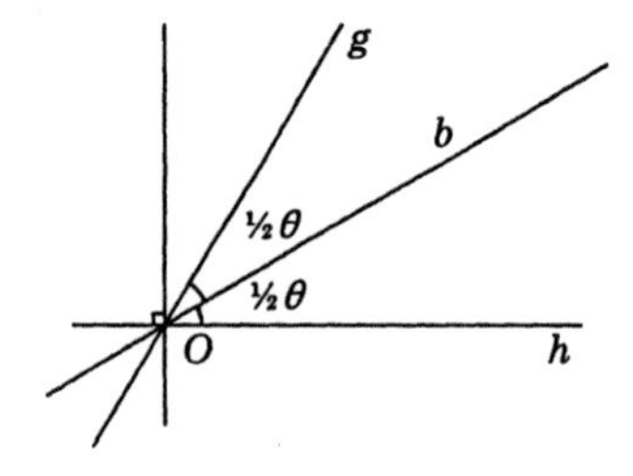

Figure 6.8.1
$\sigma_h\sigma_b = \sigma_b\sigma_g$

---

[21] This paragraph closely parallels the definition of a transformation group in section 6.1. Had this book defined the notion *matrix group*, it could speak in terms of the *group* of orthogonal matrices. But it avoids the more abstract group concepts because they're not so useful in this context.

$$\sigma_g = \sigma_b\sigma_h\sigma_b = \sigma_b\sigma_h\sigma_h\sigma_h\sigma_b = \rho_\theta\sigma_h\rho_\theta^{-1}.$$

Isometries ρ_θ and σ_h correspond to orthogonal matrices A_θ and H, so σ_g corresponds to the orthogonal matrix

$$\begin{aligned}
A_\theta H A_\theta^{-1} &= A_\theta H A_{-\theta} \\
&= \begin{bmatrix} \cos\theta & -\sin\theta \\ \sin\theta & \cos\theta \end{bmatrix} \begin{bmatrix} 1 & 0 \\ 0 & -1 \end{bmatrix} \begin{bmatrix} \cos(-\theta) & -\sin(-\theta) \\ \sin(-\theta) & \cos(-\theta) \end{bmatrix} \\
&= \begin{bmatrix} \cos\theta & \sin\theta \\ \sin\theta & -\cos\theta \end{bmatrix} \begin{bmatrix} \cos\theta & \sin\theta \\ -\sin\theta & \cos\theta \end{bmatrix} \\
&= \begin{bmatrix} \cos^2\theta - \sin^2\theta & 2\sin\theta\cos\theta \\ 2\cos\theta\sin\theta & -(\cos^2\theta - \sin^2\theta) \end{bmatrix} = \begin{bmatrix} \cos 2\theta & \sin 2\theta \\ \sin 2\theta & -\cos 2\theta \end{bmatrix}.
\end{aligned}$$

Theorem 1. Any isometry φ that fixes the origin has equation $\varphi(X) = AX$, for some orthogonal matrix A. Every orthogonal matrix A has the form $A = A_\theta$ or $A = A_\theta H A_\theta^{-1}$ for some angle parameter θ.

Proof. You've already seen that for any orthogonal matrix A the transformation $X \to AX$ is an isometry that fixes O. Now consider an arbitrary isometry φ that fixes O. According to sections 6.6 and 6.7, φ must be

(0) the identity,
(1) a reflection σ_g across a line g,
(2) a composition of two reflections, or
(3) a glide reflection.

A glide reflection with a fixpoint is a reflection, so case (3) is included in case (1). By sections 6.4 and 6.5, a composition φ of two reflections is

(2a) a translation, or
(2b) a rotation ρ.

A translation with a fixpoint is the identity, so case (2a) is included in case (0). Since ρ fixes only its center and σ_g fixes only the points on g, just these cases remain:

(0) φ is the identity, with orthogonal matrix I, or
(1) φ is a reflection σ_g across a line g through the origin, with orthogonal matrix $A_\theta H A_\theta^{-1}$, or
(2b) φ is a rotation ρ about the origin, with orthogonal matrix A_θ. ♦

Finally, consider an *arbitrary* isometry ψ. Let $B = \psi(O)$ and τ be the translation with vector $-B$. The isometry $\varphi = \tau\psi$ fixes the origin:

$$\varphi(O) = \tau\psi(O) = \tau(B) = B - B = O.$$

By the previous paragraph, there's an orthogonal matrix A such that $\varphi(X) = AX$ for every point X. Since τ^{-1} is the translation with vector $-(-B) = B$,

$$\psi(X) = \tau^{-1}\varphi(X) = \tau^{-1}(AX) = AX + B.$$

The transformation determines A and B uniquely; for if $AX + B = CX + D$ for every point X, then substituting $X = O$ yields $B = C$, and substituting the kth unit vector shows that the kth columns of A and C are equal. This discussion has proved

Theorem 2. The plane isometries are the transformations $\varphi : X \to X' = AX + B$ where A is an orthogonal matrix and B is an arbitrary vector. A and B are determined uniquely by φ.

A more concise argument for theorem 2 is presented in the proof of theorem 7.1.7. It doesn't refer to matrices A_θ and H, but uses the uniqueness theorem and some simpler matrix algebra. But that proof conveys little geometric insight; it's reserved until you've gained more experience.

A rotation $\varphi : X \to X'$ about a center $P \neq O$ has equation $X' = AX + B$ for some orthogonal matrix A. Let τ be the translation with vector $-B$. Then $\tau\varphi : X \to AX$ is an even isometry that fixes O —that is, a rotation ρ_θ about O with angle parameter θ. Thus $A = A_\theta$. Rotations with *any* center correspond to angle parameters just like the rotations about O!

Parity

If A is an orthogonal matrix, then

$$1 = \det I = \det A^t A = \det A^t \det A = (\det A)^2,$$

hence $\det A = \pm 1$. If φ is an isometry with matrix A, this determinant is called its *parity*, par φ. A rotation ρ_θ about O —an even isometry— has parity $\det A_\theta = \cos^2\theta + \sin^2\theta = 1$. In contrast, a reflection σ_g across a line g through O —an odd isometry—has parity $\det(A_\theta H A_\theta^{-1}) = (\det A_\theta)(\det H)(\det A_\theta)^{-1} = \det H = -1$. Theorem 3 shows that this pattern persists; parity provides an easy algebraic way to tell when φ is even or odd.

Theorem 3. φ is even if par $\varphi = +1$, and odd if par $\varphi = -1$.

Proof. A transformation $\varphi : X \to X' = AX + B$ preserves orientation of a triangle ΔPQR just in case $\det[P' - R', Q' - R']$ and $\det[P - R, Q - R]$ have the same sign. But

$$\begin{aligned}
&\det[P' - R', Q' - R'] \\
&\quad = \det[(AP + B) - (AR + B), (AQ + B) - (AR + B)] \\
&\quad = \det[AP - AR, AQ - AR] \\
&\quad = \det[A(P - R), A(Q - R)] \\
&\quad = \det A[P - R, Q - R] \\
&\quad = \det A \det[P - R, Q - R] \\
&\quad = \pm\det[P - R, Q - R]. \blacklozenge
\end{aligned}$$

Theorem 4. par $\varphi\chi$ = par φ par χ and par φ^{-1} = par φ for any isometries φ and χ.

6.9 Classifying isometries

Concepts

Compositions of reflections
Identity, reflections, translations, rotations, and glide reflections
Groups of even isometries
Rotations about arbitrary centers
Compositions of rotations and translations
Sets of fixpoints and fixed lines
Conjugacy
Transforming coordinate systems
Conjugacy classes

In previous sections you've seen that all plane isometries have simple matrix equations, and you've studied some particular isometries, their matrices, and their relationships in considerable detail. Parts of these analyses aren't yet complete. For example, although you've seen that the sets of all translations and of all rotations about a point O form groups, you haven't studied in detail the compositions of translations and rotations, nor of rotations about different centers. This kind of work often involves analyzing an isometry with a complicated definition, to fit it into the pattern of this chapter: Is it the identity, a translation, rotation, reflection, or glide reflection? What is its vector or center or axis? You've already started using the appropriate tools, without much attention to them. This section organizes them. They are employed to solve the example problems just mentioned, and to study in detail another kind of relationship—conjugacy—that's used to analyze symmetries in chapter 8.

Classification

The structure theorem in section 6.6 displayed a preliminary classification. A plane isometry is

the identity, or
a reflection, or
a composition of reflections across two lines g and h, or
a composition of reflections across three lines g, h, and k.

These classes aren't disjoint, since $\iota = \sigma_g \sigma_g$ and $\sigma_g = \sigma_g \sigma_g \sigma_g$ for any g. Clearly, $\sigma_g \sigma_h$ is the identity only when $g = h$. And by corollaries 6.4.7 and 6.5.6 and theorem 6.7.2, $\sigma_g \sigma_h \sigma_k$ is a reflection just when g, h, and k are copencilar. Therefore, a slight modification of the language just displayed yields a classification with disjoint classes:

Theorem 1. A plane isometry is

(0) the identity, or
(1) a reflection, or
(2) a composition of reflections across two distinct lines, or
(3) a composition of reflections across three noncopencilar lines.

For the rest of this heading, use a fixed coordinate system with origin O. Theorem 1 permits further refinement:

Corollary 2. A plane isometry is

(0) the identity ι, or
(1) a reflection σ_g across a line g, or
(2a) a translation τ_V with vector $V \neq O$, or
(2b) a rotation ρ_θ with center O and $\theta \bmod 360° \neq 0°$, or
(3) a glide reflection $\sigma_g \tau_V$ with $V \neq O$.

You can often use fixpoint information effectively to identify an isometry. This table shows the fixpoint sets of the isometry classes in corollary 2:

Class	*Isometry*		*Fixpoint set*
(0)	ι		the entire plane
(1)	σ_g		the line g
(2a)	τ_V	with $V \neq O$	empty
(2b)	ρ_θ	with $\theta \bmod 360° \neq 0°$	the center O
(3)	$\sigma_g \tau_V$	with $V \neq O$	empty

In exercise 6.11.12 you can work out an analogous description of the sets of fixed lines. You'll find several examples of the use of fixpoint information later in this section.

Classes (0) and (2) together form the group of *even* isometries. Those in (1) and (3) are called *odd*. By theorem 6.8.3, you can tell whether an isometry is even or odd by ascertaining whether its *parity*—the determinant of its matrix—is +1 or −1.

Rotations and translations

The group of even isometries has some interesting subgroups. Corollary 6.4.3 showed that the translations constitute a subgroup. In chapter 8 a yet smaller one is used to analyze ornamental patterns: the group of all translations τ_{cV}, where V is a specified nonzero vector. This set forms a group because it's closed under composition and inversion, and contains the identity:

$$\tau_{cV}\tau_{dV} = \tau_{(c+d)V} \qquad \tau_{cV}^{-1} = \tau_{-(cV)} = \tau_{(-c)V} \qquad \iota = \tau_O = \tau_{0V}.$$

You can allow c and d to be any scalars, or restrict them to be integers. You get two different groups, depending on your choice.

Theorem 6.5.3 showed that the rotations about a specified center also form a subgroup. For those about the origin, you've seen that composition corresponds to addition of angle parameters: $\rho_\eta\rho_\theta = \rho_{\eta+\theta}$. This is also true for other centers. If $\varphi : X \to A_\eta X + B$ and $\chi : X \to A_\theta X + C$ are rotations about a center P with angle parameters η and θ, then

$$\begin{aligned}\varphi\chi(X) &= \varphi(A_\theta X + C) = A_\eta(A_\theta X + C) + B \\ &= A_\eta A_\theta X + A_\eta C + B = A_{\eta+\theta} X + A_\eta C + B,\end{aligned}$$

so $\varphi\chi$ has angle parameter $\eta + \theta$. It follows that φ^{-1} has parameter $-\eta$.

You can use the matrix representation of a rotation $\varphi : X \to A_\eta X + B$ with angle parameter η to analyze compositions of φ with other isometries. For example, if $\tau : X \to X + V$ is any translation, then

$$\begin{aligned}\varphi\tau(X) &= \varphi(X + V) = A_\eta(X + V) + B = A_\eta X + A_\eta V + B \\ \tau\varphi(X) &= \tau(A_\eta X + B) = A_\eta X + B + V,\end{aligned}$$

so $\varphi\tau$ and $\tau\varphi$ are both rotations with the same angle parameter η, unless φ was the identity. For another rotation $\chi : X \to A_\theta X + C$ you get

$$\varphi\chi(X) = \varphi(A_\theta X + C) = A_\eta(A_\theta X + C) + B = A_{\eta+\theta} X + A_\eta C + B.$$

As in the previous paragraph, $\varphi\chi$ is a rotation with parameter $\eta + \theta$, unless $A_{\eta+\theta}$ is the identity matrix, in which case $\varphi\chi$ is a translation. This proves

Theorem 3. The composition of a rotation that's not the identity with any translation is a rotation with the same angle parameter. The composition of rotations with parameters χ and ψ is a rotation with parameter $\eta + \theta$ unless $(\eta + \theta) \bmod 360° = 0°$. In that case, it's a translation.

You can use theorem 3 to derive the equation of a rotation ψ with angle parameter θ but unknown center P if you can find $B = \psi(O)$. Consider $\rho = \tau_{-B}\psi$; since $\rho(O) = O$, ρ is a rotation about O with angle parameter θ. You know its equation, and from $\psi = \tau_B \rho$ you can derive the equation for ψ.

Conjugacy

In section 6.8, the isometry $\rho_\theta \sigma_h \rho_\theta^{-1}$ played a major role in deriving the matrix equation for a reflection across an arbitrary line g through O. This pattern occurs so commonly, particularly in studying symmetry, that it's studied in detail. For any transformations μ and φ, the composition $\varphi \mu \varphi^{-1}$ is called the *conjugate of* μ *by* φ. When λ is a conjugate of μ, we write $\lambda \sim \mu$. The next two theorems present some basic properties of conjugacy. They're used frequently, without mention, in later discussions.

Theorem 4. Conjugacy is an equivalence relation. That is, for any λ, μ, and ν,

$$\begin{array}{ll} \mu \sim \mu & \text{(reflexivity)} \\ \lambda \sim \mu \Leftrightarrow \mu \sim \lambda & \text{(symmetry)} \\ \lambda \sim \mu \ \& \ \mu \sim \nu \Rightarrow \lambda \sim \nu & \text{(transitivity).} \end{array}$$

Conjugate isometries have the same parity.

Proof. μ is its own conjugate by ι. If λ is the conjugate of μ by φ, then μ is the conjugate of λ by φ^{-1}. If λ is the conjugate of μ by φ and μ is the conjugate of ν by χ, then λ is the conjugate of ν by $\varphi\chi$:

$$\lambda = \varphi \mu \varphi^{-1} = \varphi \chi \nu \chi^{-1} \varphi^{-1} = (\varphi\chi)\nu(\varphi\chi)^{-1}.$$

Finally, if $\lambda = \varphi \mu \varphi^{-1}$, then $\operatorname{par}\lambda = \operatorname{par}\varphi \operatorname{par}\mu \operatorname{par}(\varphi^{-1}) = (\operatorname{par}\varphi)^2 \operatorname{par}\mu = \operatorname{par}\mu$, by corollary 6.8.4. ♦

The equivalence classes generated by the conjugacy relation are called *conjugacy classes.*

Theorem 5. The only transformation that's conjugate with the identity ι is ι itself. Now consider any transformations φ, χ, and ψ, and let χ' and ψ' denote the conjugates of χ and ψ by φ. Then

(a) $\chi'\psi'$ is the conjugate of $\chi\psi$ by φ,
(b) $(\chi')^{-1}$ is the conjugate of χ^{-1} by φ, and
(c) χ' and ψ' commute just in case χ and ψ commute.

Proof. First, $\varphi\iota\varphi^{-1} = \varphi\varphi^{-1} = \iota$. Moreover,

(a) $\varphi\chi\psi\varphi^{-1} = \varphi\chi\iota\psi\varphi^{-1} = \varphi\chi\varphi^{-1}\varphi\psi\varphi^{-1}$.

(b) $(\chi')^{-1} = (\varphi\chi\varphi^{-1})^{-1} = (\varphi^{-1})^{-1}\chi^{-1}\varphi^{-1} = \varphi\chi^{-1}\varphi^{-1}$.

(c) $\chi'\psi' = \psi'\chi' \Leftrightarrow \varphi\chi\varphi^{-1}\varphi\psi\varphi^{-1} = \varphi\psi\varphi^{-1}\varphi\chi\varphi^{-1}$
$\Leftrightarrow \varphi\chi\psi\varphi^{-1} = \varphi\psi\chi\varphi^{-1}$
$\Leftrightarrow \chi\psi = \varphi^{-1}\varphi\psi\chi\varphi^{-1}\varphi$
$\Leftrightarrow \chi\psi = \psi\chi$. ♦

Transforming coordinate systems

The conjugacy concept is even more closely connected to the equations of isometries than the section 6.8 examples suggest. The connection is based on the idea of transforming a coordinate system $\mathscr{C}$. All you need to describe $\mathscr{C}$ is the isosceles right triangle ΔOU_1U_2 in figure 6.9.1 that consists of the origin O and the axis points U_1 and U_2 with coordinate 1. An isometry φ transforms $\mathscr{C}$ into another system $\mathscr{D}$ defined by the image ΔPV_1V_2 of ΔOU_1U_2. What if you use *both* coordinate systems—for instance, to study different aspects of some geometric figure? How are a point's $\mathscr{C}$ and $\mathscr{D}$ coordinates related? Theorem 6 answers that question, and theorem 7 connects these considerations with conjugacy.

Theorem 6. If an isometry φ transforms coordinate system $\mathscr{C}$ into system $\mathscr{D}$, then the $\mathscr{C}$ coordinates of each point X are the same as the $\mathscr{D}$ coordinates of $Y = \varphi(X)$.

Proof. Consider figure 6.9.1. The figure formed by ΔOU_1U_2 and X is congruent to that formed by ΔPV_1V_2 and Y. ♦

Theorem 7. Isometries χ and ψ are conjugate if and only if there are coordinate systems under which they have the same matrix equation.

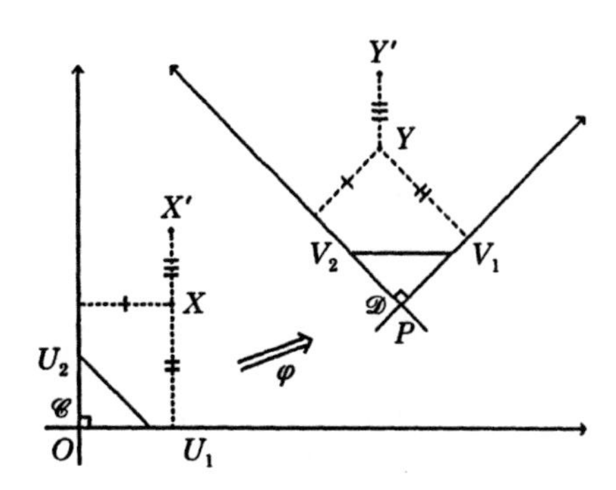

Figure 6.9.1
Transforming a coordinate system

Proof. Consider coordinate systems $\mathscr{C}$ and $\mathscr{D}$, points X and Y, and an isometry φ as in theorem 6. Suppose $\chi : X \to X'$ has equation $X' = AX + B$ with respect to $\mathscr{C}$. *Important: This equation treats X and X' as vectors of $\mathscr{C}$ coordinates!* By theorem 6, the $\mathscr{C}$ coordinates of X' are the same as the $\mathscr{D}$ coordinates of $Y' = \varphi(X')$. Thus

$$\begin{aligned} X' = AX + B &\leftrightarrow X' = \chi(X) \\ &\leftrightarrow \varphi^{-1}(Y') = \chi\varphi^{-1}(Y) \\ &\leftrightarrow Y' = \varphi\chi\varphi^{-1}(Y). \end{aligned}$$

If $\psi = \varphi\chi\varphi^{-1}$, then $X' = AX + B \leftrightarrow Y' = \psi(Y)$. Since the $\mathscr{D}$ coordinates of Y and Y' are the same as the $\mathscr{C}$ coordinates of X and X', you can conclude that

$$Y' = AY + B \leftrightarrow Y' = \psi(Y)$$

when you regard Y and Y' in the left-hand equation as vectors of $\mathscr{D}$ coordinates.

Conversely, suppose χ and ψ have the same equation with respect to coordinate systems $\mathscr{C}$ and $\mathscr{D}$. By theorem 6.6.3, $\mathscr{C}$ and $\mathscr{D}$ are related by an isometry φ as in figure 6.9.1. Since the $\mathscr{C}$ coordinates of X and X' are the same as the $\mathscr{D}$ coordinates of Y and Y',

$X' = AX + B$	$\leftrightarrow$	$Y' = AY + B$	$\leftrightarrow$	$Y' = \psi(Y)$.
with X and X' interpreted as vectors of $\mathscr{C}$ coordinates		with Y and Y' interpreted as vectors of $\mathscr{D}$ coordinates		

But the first paragraph of this proof guaranteed $X' = AX + B \leftrightarrow Y' = \varphi\chi\varphi^{-1}(Y)$. Therefore $\psi(Y) = \varphi\chi\varphi^{-1}(Y)$ for all Y; having the same equation implies conjugacy. ♦

You now have *two* proofs of theorem 4; the word *same* in theorem 7 makes it *obvious* that conjugacy is an equivalence relation!

Conjugacy classes

By theorem 7, the half turns about any two points O and P are conjugate; with respect to coordinate systems with origins O and P, each has matrix form $X \to X' = -X$. By theorem 6.8.2, no other isometries have equations of this form, so the half turns constitute a single conjugacy class. But this result doesn't give complete information about the conjugate $\chi = \varphi\sigma_P\varphi^{-1}$ of σ_P by an isometry φ. Even without theorem 7, you can conclude that it's a half turn; χ is even and self-inverse. You can classify it completely by fixpoint analysis:

$$\begin{aligned}
&X \text{ is a fixpoint of } \chi \\
&\leftrightarrow \varphi\sigma_P\varphi^{-1}(X) = X \\
&\leftrightarrow \sigma_P\varphi^{-1}(X) = \varphi^{-1}(X) \\
&\leftrightarrow \varphi^{-1}(X) \text{ is a fixpoint of } \sigma_P \\
&\leftrightarrow \varphi^{-1}(X) = P \\
&\leftrightarrow X = \varphi(P).
\end{aligned}$$

Thus $\chi = \varphi\sigma_P\varphi^{-1} = \sigma_{\varphi(P)}$ because χ has exactly one fixpoint, $\varphi(P)$.

You can just as easily analyze reflections. By theorem 7, reflections across any two lines g and h are conjugate; with respect to coordinate systems with first axes g and h, each has the matrix form

$$X \to X' = \begin{bmatrix} 1 & 0 \\ 0 & -1 \end{bmatrix} X.$$

By theorem 6.8.2, no other isometries have equations of this form, so the reflections constitute a single conjugacy class. Again, this result doesn't give complete information about the conjugate $\psi = \varphi\sigma_g\varphi^{-1}$ of σ_g by an isometry φ. You can classify it completely by fixpoint analysis:

$$\begin{aligned}
&X \text{ is a fixpoint of } \psi \\
&\leftrightarrow \varphi\sigma_g\varphi^{-1}(X) = X \\
&\leftrightarrow \sigma_g\varphi^{-1}(X) = \varphi^{-1}(X) \\
&\leftrightarrow \varphi^{-1}(X) \text{ is a fixpoint of } \sigma_g \\
&\leftrightarrow \varphi^{-1}(X) \text{ lies on } g \\
&\leftrightarrow X \text{ lies on the image set } \varphi[g].
\end{aligned}$$

The fixpoints of ψ constitute the line $\varphi[g]$, so ψ must be the reflection across that line. That is, $\varphi\sigma_g\varphi^{-1} = \sigma_{\varphi[g]}$. This paragraph and the previous one have proved

Theorem 8. The half turns constitute a single conjugacy class. Moreover, $\varphi\sigma_P\varphi^{-1} = \sigma_{\varphi(P)}$ for any point P and any isometry φ. Similarly, the reflections constitute a single conjugacy class, and $\varphi\sigma_g\varphi^{-1} = \sigma_{\varphi[g]}$ for any line g and any isometry φ.

Analyzing translations is a little more complicated. Consider a point O and translations τ and υ; let

$$\begin{aligned}
T &= \tau(O) \qquad & t &= OT \\
U &= \upsilon(O) \qquad & u &= OU.
\end{aligned}$$

By theorem 7, τ and υ are conjugate just in case $t = u$. With respect to coordinate systems with origin O and with T and U on the positive rays of the first axes, they have matrix forms

$$X \to X' = X + \begin{bmatrix} t \\ 0 \end{bmatrix} \qquad X \to X' = X + \begin{bmatrix} u \\ 0 \end{bmatrix},$$

and these are the same just in case $t = u$. Theorem 9 provides complete information about the conjugate $\upsilon = \varphi\tau\varphi^{-1}$ of τ by an isometry φ:

Theorem 9. The conjugacy class of a translation τ whose vector has length t consists of all translations with vectors of that length. If τ and an isometry φ have equations $\tau: X \to X' = X + T$ and $\varphi: X \to X' = AX + B$ with respect to a single coordinate system, then $\upsilon = \varphi\tau\varphi^{-1}$ is the translation with vector AT.

Proof. The second sentence remains to be proved. Since $X = A^{-1}X' - A^{-1}B$, it follows that φ^{-1} has matrix form $X \to A^{-1}X - A^{-1}B$ and

$$\begin{aligned}\upsilon(X) &= \varphi\tau\varphi^{-1}(X) = \varphi\tau(A^{-1}X - A^{-1}B) = \varphi(A^{-1}X - A^{-1}B + T) \\ &= A(A^{-1}X - A^{-1}B + T) + B = X + AT. \blacklozenge\end{aligned}$$

The final results of this section provide analogous information for glide reflections and rotations. Their proofs differ. The argument for theorem 10 can use the concise techniques of function composition. But theorem 11 uses matrix computation, because that's convenient for handling angle parameters.

Theorem 10. Choose a coordinate system. The conjugacy class of a glide reflection χ with axis g and vector V consists of all glide reflections with vectors whose length is the same as that of V. The conjugate of χ by an isometry $\varphi: X \to X' = AX + B$ is the glide reflection with axis $\varphi[g]$ and vector AV.

Proof. Find points O and P on g with $V = P - O$, so that $\tau_V = \sigma_O\sigma_P$ and $\chi = \sigma_g\tau_V$. If φ is an isometry, then by theorems 8 and 9,

$$\varphi\chi\varphi^{-1} = \varphi\sigma_g\varphi^{-1}\varphi\tau_V\varphi^{-1} \qquad \varphi\sigma_g\varphi^{-1} = \sigma_{\varphi[g]} \qquad \varphi\tau_V\varphi^{-1} = \tau_{AV}.$$

Define $O' = \varphi(O)$, $P' = \varphi(P)$, and $V' = P' - O'$, so that O' and P' lie on $\varphi[g]$ and

$$\varphi\tau_V\varphi^{-1} = \varphi\sigma_O\varphi^{-1}\varphi\sigma_P\varphi^{-1} = \sigma_{O'}\sigma_{P'} = \tau_{V'}.$$

Thus $\varphi\chi\varphi^{-1}$ is the glide reflection with axis $\varphi[g]$ and vector V'.

Conversely, assuming that ψ is a glide reflection with axis g' whose vector V' has the same length as V, find points O' and P' on g' with $V' = P' - O'$, so that $\tau_{V'} = \sigma_{O'}\sigma_{P'}$ and $\psi = \sigma_{g'}\tau_{V'}$. By theorem 6.6.3, there's an isometry φ such that $\varphi(O) = O'$ and $\varphi(P) = P'$. It follows that $\varphi[g] = g'$. Then

$$\sigma_{g'} = \sigma_{\varphi[g]} = \varphi\sigma_g\varphi^{-1}$$
$$\tau_{V'} = \sigma_{\varphi(O)}\sigma_{\varphi(P)} = \varphi\sigma_O\varphi^{-1}\varphi\sigma_P\varphi^{-1} = \varphi\sigma_O\sigma_P\varphi^{-1} = \varphi\tau_V\varphi^{-1}$$
$$\psi = \varphi\sigma_g\varphi^{-1}\varphi\tau_V\varphi^{-1} = \varphi\sigma_g\tau_V\varphi^{-1} = \varphi\chi\varphi^{-1}. \blacklozenge$$

Theorem 11. Suppose χ is a rotation, not the identity, with angle parameter θ and center P. Its conjugate by an isometry φ is a rotation ψ with center $\varphi(P)$. Moreover, ψ has angle parameter θ or $-\theta$, depending on whether φ is even or odd.

Proof. $\psi = \varphi\chi\varphi^{-1}$ is a rotation with center $\varphi(P)$ because that's its only fixpoint:

$$\varphi\chi\varphi^{-1}(X) = X \leftrightarrow \chi\varphi^{-1}(X) = \varphi^{-1}(X) \leftrightarrow \varphi^{-1}(X) = P \leftrightarrow X = \varphi(P).$$

Suppose φ is even. From the matrix representations $\varphi : X \to A_\eta X + B$ and $\chi : X \to A_\theta X + C$, you get

$$\begin{aligned}\varphi\chi\varphi^{-1}(X) &= A_\eta A_\theta A_\eta^{-1} X + \text{a constant vector}\\ &= A_\eta A_\theta A_{-\eta} X + \cdots\\ &= A_{\eta+\theta-\eta} X + \cdots\\ &= A_\theta X + \cdots,\end{aligned}$$

so $\varphi\chi\varphi^{-1}$ is a rotation with parameter θ.

Now suppose ω is an odd isometry. Consider the even isometry $\varphi = \sigma_h\omega$, where h is the first axis. Then $\varphi\chi\varphi^{-1}$ is a rotation ρ with parameter θ:

$$\rho = (\sigma_h\omega)\chi(\sigma_h\omega)^{-1} = \sigma_h\omega\chi\omega^{-1}\sigma_h^{-1} = \sigma_h(\omega\chi\omega^{-1})\sigma_h$$
$$\sigma_h\rho\sigma_h = \omega\chi\omega^{-1}.$$

From the matrix equation $\rho : X \to A_\theta X + B$, you see that

$$\omega\chi\omega^{-1}(X) = HA_\theta HX + \text{a constant vector},$$

where H is the matrix corresponding to the reflection σ_h. Thus

$$H = \begin{bmatrix} 1 & 0 \\ 0 & -1 \end{bmatrix}$$

$$\begin{aligned}HA_\theta H &= \begin{bmatrix} 1 & 0 \\ 0 & -1 \end{bmatrix}\begin{bmatrix} \cos\theta & -\sin\theta \\ \sin\theta & \cos\theta \end{bmatrix}\begin{bmatrix} 1 & 0 \\ 0 & -1 \end{bmatrix}\\ &= \begin{bmatrix} \cos\theta & -\sin\theta \\ -\sin\theta & -\cos\theta \end{bmatrix}\begin{bmatrix} 1 & 0 \\ 0 & -1 \end{bmatrix} = \begin{bmatrix} \cos\theta & \sin\theta \\ -\sin\theta & \cos\theta \end{bmatrix}\\ &= \begin{bmatrix} \cos(-\theta) & -\sin(-\theta) \\ \sin(-\theta) & \cos(-\theta) \end{bmatrix} = A_{-\theta}.\end{aligned}$$

Therefore, $\omega\chi\omega^{-1}$ is a rotation with parameter $-\eta$. ♦

Corollary 12. Rotations with angle parameters η and θ are conjugate just in case $|\eta \bmod 360°| = |\theta \bmod 360°|$.

You can use theorem 11 to derive the equation of a rotation ψ with center P, as follows. Let ρ be the rotation about the origin with the same angle parameter, and let $\varphi = \tau_P$, so that $\psi = \varphi\rho\varphi^{-1}$. The equations of φ, ρ, and φ^{-1} are easy to find, and from those you derive the equation for ψ.

6.10 Similarities

Concepts
- Similarities and their ratios
- Contractions and dilations
- Invariants
- Similarity group
- Uniqueness theorem
- Classification theorem
- Equation $X' = rAX + B$ of a similarity
- Fixpoint theorem
- Homothetic rotations and reflections
- Even and odd similarities

Studying geometric transformations more general than isometries usually requires methods beyond those developed in this chapter. But one broader family of transformations, the similarities, needs little new. Let r be a positive real number. A *plane similarity with ratio* r is a plane transformation $\psi : P \rightarrow P'$ such that for all points P and Q,

$$P'Q' = r\,(PQ).$$

A similarity can't have two *different* ratios because $r\,(PQ) = r'\,(PQ)$ and $P \neq Q$ imply $r = r'$; so it makes sense to call r *the* ratio of ψ.

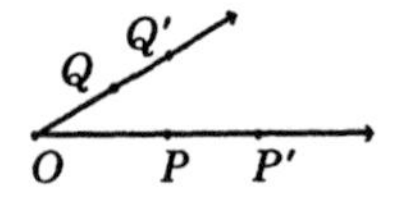

Figure 6.10.1 Dilation about O with ratio r

$r = 1.7 \quad OP' = r\,(OP)$
$\qquad\qquad OQ' = r\,(OQ)$

Evidently, every isometry is a similarity with ratio 1. As another example, select any point O and consider the transformation $\varphi_{r,O} : P \to P'$ such that $O' = O$ and for $P \neq O$, P' is the point on $\vec{OP}$ with $OP' = r\,(OP)$. (See figure 6.10.1.) If O is the origin of a Cartesian coordinate system, so that you can regard points as vectors, then $P' = rP$ for every point P. If $r <$ 1 or $r > 1$, then $\varphi_{r,O}$ is called a *contraction* or *dilation about* O.[22]

Theorem 1. $\varphi_{r,O}$ is a similarity with ratio r.

Proof. By theorem 3.11.2,

$$\begin{aligned}(P'Q')^2 &= (P' - Q')\cdot(P' - Q') = (rP - rQ)\cdot(rP - rQ)\\ &= r^2(P - Q)\cdot(P - Q) = r^2(PQ)^2. \blacklozenge\end{aligned}$$

You'll see soon, in the classification theorem, that *every* similarity is an isometry or the composition of an isometry with a dilation or contraction. That's why there's not much new in this section. First, though, theorems 2 and 3 record some general properties of similarities that you'll need for this analysis.

Theorem 2. A similarity $\psi : X \to X'$ with ratio r leaves angle measure and betweenness invariant. Thus any geometric property that you can define in terms of betweenness is invariant. In particular, ψ is a collineation. The image of a circle with radius t is a circle with radius rt. The image of a triangle with area s is a triangle with area $r^2 s$.

Proof. The equation $m\angle XYZ = m\angle X'Y'Z'$ follows from the SSS similarity principle. For betweenness, proceed as in the proof of theorem 6.2.5, taking into account ratio r:

$$\begin{aligned}X\text{-}Y\text{-}Z &\Leftrightarrow XY + YZ = XZ\\ &\Leftrightarrow r\,(XY) + r\,(YZ) = r\,(XZ)\\ &\Leftrightarrow X'Y' + Y'Z' = X'Z'\\ &\Leftrightarrow X'\text{-}Y'\text{-}Z'.\end{aligned}$$

If O is the center of circle Γ with radius t, then a point X lies on Γ if and only if $OX = t$. This condition is equivalent to $O'X' = rt$, so X lies on Γ if and only if X' lies on the circle with center O' and radius rt. The area result follows from theorem 3.9.7. ♦

Theorem 3. A composition of similarities with ratios r and s is a similarity with ratio rs. The inverse of a similarity with ratio r is a similarity

[22] Many authors also use the term *dilation* when $r < 1$. That jars, so this text uses *contraction* in that case. Don't confuse *dilation* with *dilatation*; the latter term has a different meaning —see the heading **Dilatations** in section 6.11.

with ratio r^{-1}. Therefore, the similarities form a transformation group that contains all isometries.

Proof. If $P \to P'$ and $P' \to P''$ are similarities with ratios s and r, then for all points P and Q, $P''Q'' = r(P'Q') = rs(PQ)$, so the composition $P \to P''$ is a similarity with ratio rs. Moreover, $PQ = r^{-1}r(PQ) = r^{-1}(P'Q')$, so the inverse transformation $P' \to P$ is a similarity with ratio r^{-1}. ♦

***Corollary 4* (*Uniqueness theorem*).** Plane similarities χ and ψ coincide if $\chi(X) = \psi(X)$ for three noncollinear points X.

Proof. Suppose χ and ψ have ratios r and s, points X_1 and X_2 are distinct, $\chi(X_1) = \psi(X_1)$, and $\chi(X_2) = \psi(X_2)$. Then $r(X_1X_2) = \chi(X_1)\chi(X_2) = \psi(X_1)\psi(X_2) = s(X_1X_2)$, hence $r = s$. Since its ratio is $r/s = 1$, the similarity $\varphi = \chi\psi^{-1}$ is an isometry. Since φ fixes three noncollinear points, it's the identity, by theorem 6.2.7. ♦

Lemma 5 is a preliminary version of theorem 8. It's introduced now, to facilitate use of the equation of a similarity in the final version of theorem 8.

***Lemma 5*.** Given any point O and any similarity ψ with ratio r, there's an isometry χ such that $\psi = \chi\varphi_{r,O}$.

Proof. The composition $\chi = \psi\varphi_{r,O}^{-1}$ is a similarity with ratio $rr^{-1} = 1$, hence it's an isometry. Moreover, $\psi = \chi\varphi_{r,O}$. ♦

***Theorem 6*.** The similarities are the transformations $X \to X' = rAX + B$ where $r > 0$, A is an orthogonal matrix, and B an arbitrary vector. The transformation determines r, A, and B uniquely; r is the similarity ratio.

Proof. Every such equation defines a similarity: the composition of the isometry $X \to AX + B$ with $\varphi_{r,O}$. On the other hand, given a similarity $\psi: X \to X'$, lemma 5 provides an isometry χ such that $\psi = \chi\varphi_{r,O}$. By theorem 6.8.2 there's an orthogonal matrix A and a vector B such that $\chi(X) = AX + B$ for all X, so that ψ has the desired equation. If equations $X' = rA_1X + B_1$ and $X' = sA_2X + B_2$ represented the same transformation, it would be a similarity with ratio $r = s$. But then the transformation $X \to X'' = r^{-1}X'$ would be an isometry with equations $X'' = A_1X + r^{-1}B_1$ and $X'' = A_2X + r^{-1}B_2$, so $A_1 = A_2$ and $r^{-1}B_1 = r^{-1}B_2$ by theorem 6.8.2, hence $B_1 = B_2$. ♦

***Theorem 7* (*Fixpoint theorem*).** If $\psi: X \to X'$ is a similarity with ratio r, then $r = 1$ or ψ has exactly one fixpoint.

Proof. First consider the case $r < 1$. Select any point X_0 and define points $X_1, X_2, \ldots$ by the rule $X_{m+1} = X'_m$. Then

$$X_1X_2 = X_0'X_1' = r\,(X_0X_1),$$
$$X_2X_3 = X_1'X_2' = r\,(X_1X_2) = r^2\,(X_0X_1),$$
$$\vdots$$

so in general,

$$X_mX_{m+1} = r^m\,(X_0X_1).$$

Now let $m < n$ and consider the ith coordinates x_{mi} and x_{ni} of points X_m and X_n:

$$\begin{aligned}|x_{mi} - x_{ni}| &\le X_mX_n \le X_mX_{m+1} + X_{m+1}X_{m+2} + \cdots + X_{n-1}X_n \\ &= r^m\,(X_0X_1) + r^{m+1}\,(X_0X_1) + \cdots + r^n\,(X_0X_1) \\ &= r^m\,(1 + r + \cdots + r^{n-m})(X_0X_1).\end{aligned}$$

The sum in parentheses is the initial part of a convergent geometric series, so

$$|x_{mi} - x_{ni}| \le r^m\left(\sum_{n=0}^{\infty} r^n\right)(X_0X_1) = \frac{r^m\,(X_0X_1)}{1-r}.$$

Since $\lim_m r^m = 0$, you can make $|x_{mi} - x_{ni}|$ arbitrarily small by taking m sufficiently large and $n > m$. That is, $x_{0i}, x_{1i}, x_{2i}, \ldots$ is a Cauchy sequence; it has a limit x_i. Let X be the vector whose ith component, for each i, is x_i. Then X is a fixpoint. To see that, use the equation $X' = rAX + B$ provided by theorem 6:

$$\begin{aligned}&i\text{th component of } X' = i\text{th component of } rAX + B \\ &= (\Sigma_j\, ra_{ij}\,x_j) + b_i \\ &= (\Sigma_j\, ra_{ij} \lim_n x_{nj}) + b_i \\ &= (\lim_n \Sigma_j\, ra_{ij}\,x_{nj}) + b_i \\ &= \lim_n((\Sigma_j\, ra_{ij}\,x_{nj}) + b_i) \\ &= \lim_n(i\text{th component of } rAX_n + B) \\ &= \lim_n(i\text{th component of } X_n') \\ &= \lim_n(i\text{th component of } X_{n+1}) \\ &= \lim_n x_{n+1,i} = x_i = i\text{th component of } X.\end{aligned}$$

Thus $X' = X$, as claimed.

Now consider the case $r > 1$. Since ψ^{-1} is a similarity with ratio < 1, it has a fixpoint X by the preceding paragraph. But then X is a fixpoint of ψ, too.

Finally, to prove uniqueness, suppose $r \ne 1$ and ψ has fixpoints X and Y. Then $X'Y' = r\,(XY) = r\,(X'Y')$, so $X'Y' = 0$, $X' = Y'$, and $X = Y$. ♦

The Cauchy sequence technique for proving the fixpoint theorem has a flavor different from that of the rest of this book. The only other part based on limits is the discussion of π in section 3.14. Cauchy sequences occur frequently in real analysis. In fact, similarities with ratio $r < 1$ are examples of the *contraction mappings* that you study in analysis: mappings such that $X'Y' \le r\,(XY)$ for all X and Y. And the proof of the fixpoint theorem is just an adaptation of a familiar proof of the contraction mapping theorem. That result[23] —*every* contraction mapping has a fixpoint—has several major applications. For example, it establishes the convergence of various iterative approximation methods such as the Newton–Raphson technique, and it implies the existence of solutions to first-order ordinary differential equations satisfying Lipschitz conditions. You can prove the fixpoint theorem without using analysis methods, but that proof will be less succinct, and may require different arguments for two and three dimensions.[24]

Compositions $\chi\varphi_{r,O}$ where χ is a rotation about O or a reflection across a line through O are called *homothetic rotations* or *reflections*.[25]

***Theorem 8* (*Classification theorem*).** A plane similarity is an isometry, a homothetic rotation, or a homothetic reflection.

Proof. By the fixpoint theorem, if a similarity ψ isn't an isometry, then it has a fixpoint O. Let r be its ratio, and define $\chi = \psi\varphi_{r^{-1},O}$, so that $\psi = \chi\varphi_{r,O}$. Then χ is an isometry with fixpoint O'. By the discussion following corollary 6.9.2, it's a rotation about O' or a reflection across a line g through O'. ♦

A similarity $\psi : X \to X' = rAX + B$ with $r > 0$ preserves orientation just in case the isometry $\chi : X \to AX + B$ does—that is, just in case $\det A = 1$. To see this, note that for any points P, Q, and R,

$$\begin{aligned}
&\det[P' - R', Q' - R'] \\
&\quad = \det[(rAP + B) - (rAR + B), (rAQ + B) - (rAR + B)] \\
&\quad = \det[rA(P - R), rA(Q - R)] \\
&\quad = (\det rA)\det[P - R, Q - R] \\
&\quad = r^2(\det A)\det[P - R, Q - R].
\end{aligned}$$

ψ is called *even* if orientation is preserved; *odd* if it's reversed.

---

[23] See Bartle 1964, 170 for the contraction mapping theorem.

[24] See Exercises 6.11.41–44 and Martin 1982, 141.

[25] *Homothetic* stems from the Greek prefix *homo-* and verb *tithenai*, which mean *same* and *put*. It's the opposite of *antithetic*.

6.11 Exercises

Concepts

Curved mirrors
Classifying isometries defined as compositions or by equations
Constructions with isometries and similarities
Billiards and dual billiards problems
Equations of reflections
Determining equations of isometries and similarities
Coordinate transformations
Alias and *alibi* interpretations of equations
Reflection calculus
Distance-preserving functions
Dilatations

This section contains forty-five exercises related to earlier parts of the chapter. Like other exercises in this book, they're essential components of its development. A few—for example, exercises 1, 5, and 12—provide more experience with reasoning techniques introduced earlier. Others, particularly those on the classification and the equations of isometries, show you how to do the computations required to use the theorems derived in the text. The theory is streamlined into an efficient outline of the results important for applications. But those computations seem unwieldy at first. That would obscure the main points of the theory, so the computations are clustered here. Some may require awkward matrix arithmetic. There was no effort to avoid that. If you find a calculation too tedious, use mathematical software.[26]

This chapter's theory is often applied to analytic geometry problems. That work usually has a different flavor because it uses *coordinate* transformations. Those apply to the numbers used to locate points, not to the points themselves, like geometric transformations here. Exercise 31 shows how to make the transition. A few exercises introduce topics that could be developed much further. Those on reflection calculus, and on dilatations, for example, lead into deep studies in foundations of geometry. Chapter 7 extends the theory of isometries and similarities to three dimensions. Most of the exercises in this section can be extended as well. You'll see that in section 7.7.

Basic exercises

Exercises 1 and 2 present some important properties that apply to transformations in general, not just those in geometry. For instance, exercise 2, part 1, asks what you can say about functions φ and χ whose composition is

[26] *X(Plore)* (Meredith 1993) is a good choice.

surjective, and what you can *not* say, in general. To show that you can *not* claim that φ or χ has some property $\mathscr{P}$, you must provide a *counterexample*: specific functions whose composition is surjective, but for which $\mathscr{P}$ is not true. Searching for counterexamples for exercise 2, you should concentrate on functions whose domains and ranges are *very small* finite sets. You won't need more than that to make the required properties fail, and you'll avoid confusion by other phenomena that are really irrelevant. That kind of reasoning may seem nongeometric to you, but no—those tiny counterexamples are always contained in the larger ones that may seem closer to geometry, and they're much easier to understand.

Exercise 1. Prove that two self-inverse transformations of a set D commute if and only if their composition is self-inverse.

Exercise 2. Consider functions $\varphi : S \to T$ and $\chi : D \to S$, so that $\varphi\chi : D \to T$.

Part 1. Prove that if $\varphi\chi$ is surjective, then φ must be surjective. Find an example where $\varphi\chi$ is surjective, but χ is neither surjective nor injective.

Part 2. Prove that if $\varphi\chi$ is injective, then χ must be injective. Find an example where $\varphi\chi$ is injective, but φ is neither injective nor surjective.

Part 3. Prove that if $\varphi\chi$ is bijective, then φ must be surjective and χ injective. Find an example where $\varphi\chi$ is bijective, but neither φ nor χ is.

Exercises 3 and 4 present some fundamental facts about linear functions. It wasn't necessary to state them in the text of chapter 6 or 7 because all linear functions considered there are isometries or similarities, and for those function classes these facts are evident. Had chapter 6 introduced isometries through their equations, however, the results in these exercises would have been incorporated in the text before that point. To prove them, use appendix C material about Gauss elimination, matrix invertibility, and determinants.

Exercise 3. Consider a matrix A, a vector B, and the linear function $X \to AX + B$ that maps $\mathbb{R}^n$ to itself, where $n = 2$ or 3. Show that these conditions are equivalent:

(a) A is invertible,
(b) $X \to AX + B$ is bijective,
(c) $X \to AX + B$ is injective,
(d) $X \to AX + B$ is surjective.

Exercise 4. Let $n = 2$ or 3. Show that the range of a linear function $X \to AX + B$ that maps $\mathbb{R}^n$ to itself is a point, line, plane, or the entire space. In which cases is φ bijective? *Suggestion*: First consider the function $\varphi : X \to AX$.

Theorem 6.6.3 shows that you can map any triangle onto any congruent one with an isometry; the uniqueness theorem, corollary 6.2.8, says that you can do that in only one way. The latter statement requires you to regard a triangle as an *ordered* triple of vertices. You'd lose the uniqueness if you defined a triangle as a mere *set* of vertices or as its boundary—the union of its edges—for there are several isometries that map the set of vertices and the boundary of an equilateral triangle onto themselves. You'll see much more detail on these considerations in chapter 8. In exercise 5 you can study the same question for a simpler example, a segment. As defined in section 3.2, a segment is the *set* of points between its vertices, *not* an ordered pair of vertices.

Exercise 5. Show that if Σ and Υ are segments of the same length, then there's a rotation ρ such that $\rho[\Sigma] = \Upsilon$. Could there be *another* isometry φ such that $\varphi[\Sigma] = \Upsilon$? Could there be another *rotation*?

At the beginning of section 6.3 you saw how to find the shortest path from one point to another via a point on a mirror line g by reflecting one of the points across g. And you saw how this relationship corresponds to the equality of incidence and reflection angles. Exercise 6 shows that the relationship holds for a smoothly curved mirror as well. You need differential calculus to work with the curve.

Exercise 6. Consider a smooth curve Γ and points A and B not on Γ. (See figure 6.11.1.) Suppose point O lies on Γ and

$$AO + OB \le AX + XB$$

for every point X on Γ. Let g be the tangent to Γ at O and select points U and V on g such that U-O-V. Prove that $\angle UOA \cong \angle VOB$. *Suggestion*: Use a coordinate system with first axis g and origin O. Since Γ is smooth, there's a differentiable function f on an open interval $\mathcal{I}$ containing 0 such that for any $x_1 \in \mathcal{I}$, a point $X = \langle x_1, x_2 \rangle$ lies on Γ just in case $x_2 = f(x_1)$.

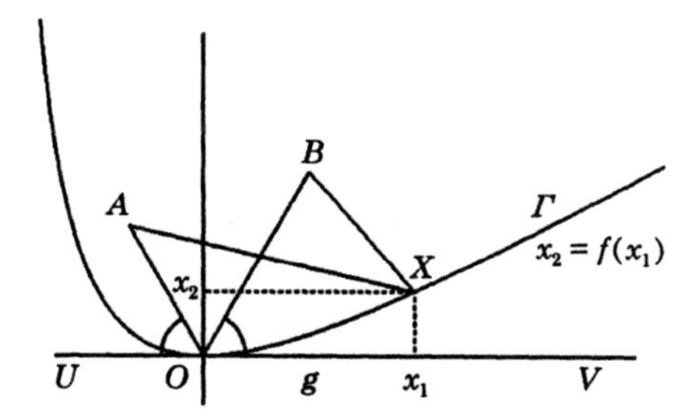

Figure 6.11.1
Exercise 6

The next exercise invites you to use a technique from section 6.3 to derive Fagnano's theorem, a morsel of triangle geometry mentioned in section 5.8.

Exercise 7. Show that of all triangles with vertices X, Y, and Z on the edges of an acute ΔABC, the orthic triangle has the shortest perimeter. *Suggestion*: Show first that for a given X on $\overline{BC}$ the shortest $XY + YZ + ZX$ is $\sigma_{\overleftrightarrow{AB}}(X)\,\sigma_{\overleftrightarrow{CA}}(X) = 2(AX)\sin m\angle A$.

Classifying isometries

Most exercises under this heading ask you to classify an isometry. That means decide whether it's the identity or one of these types of isometries, and provide the information in the right-hand column:

For a	*provide*
reflection	axis
translation	vector
rotation	center and an angle parameter
glide reflection	axis and vector.

Exercises 8 and 9 are aperitifs—you'll use this classification to verify some properties of isometries in general.

Exercise 8. Show that an isometry φ is even if and only if $\varphi = \chi\chi$ for some isometry χ.

Exercise 9. Show that every plane isometry is the composition of two self-inverse isometries.

Exercise 10. Prove a result like theorem 6.9.3 by continuing the proofs of corollaries 6.5.7 and 6.5.8 and analyzing the angles formed by the lines mentioned there, without referring to any equations of transformations. *Suggestion*: Be careful about the *signs* of the angle parameters. You don't need to get precisely the same result as the theorem. It was postponed from section 6.5 to section 6.9 because consistent assignment of angle parameters for different centers is so hard to describe.

Exercise 11 presents a nearly obvious result that you may use in several later exercises to find the axis of a glide reflection.

Exercise 11. Show that the axis of a glide reflection $\varphi : X \to X'$ consists of the midpoints of all segments $\overline{XX'}$.

Exercise 12. Construct a table analogous to the one following corollary 6.9.2, to show the possible sets of lines that are fixed by the various isometries. A fixed line might or might not consist entirely of fixed points. Lines that do are called *pointwise* fixed. What isometries have pointwise fixed lines? What isometries have fixed lines, *no* point of which is fixed?

The remaining exercises under this heading and some later ones require you to classify various isometries according to the scheme presented earlier. The main tools available are their fixpoints and parity—discussed in earlier sections—and exercises 11 and 12. *Use those tools!*

Exercise 13. Let $V_1V_2\cdots V_n$ be a convex polygon and $\theta_1, \theta_2, \ldots, \theta_n$ be the measures of the corresponding vertex angles. At each vertex V_i draw arc $\widehat{A_iB_i}$ with measure θ_i and radius $V_iA_i = V_iB_i$ small enough that no two of these arcs intersect. (See figure 6.11.2 for an example with $n = 4$.) Consider translations

τ_1 with vector $\overrightarrow{B_1A_2}$

τ_2 with vector $\overrightarrow{B_2A_3}$

$\vdots \qquad \vdots$

τ_n with vector $\overrightarrow{B_nA_1}$

and for $i = 1, \ldots, n$ consider rotation ρ_i with center V_i and angle parameter θ_i. Classify the isometry $\tau_n\rho_n\tau_{n-1}\rho_{n-1}\cdots\tau_1\rho_1$.[27]

Exercise 14. Classify the composition of glide reflections with perpendicular axes g and h and vectors V and W.

Exercise 15. Classify the compositions of the reflection across the first coordinate axis and the translation with vector $<1,2>$, in both orders.

Exercise 16. Let a, b, and c denote the lines containing the edges opposite vertices A, B, and C of ΔABC. Let D, E, and F denote the feet of

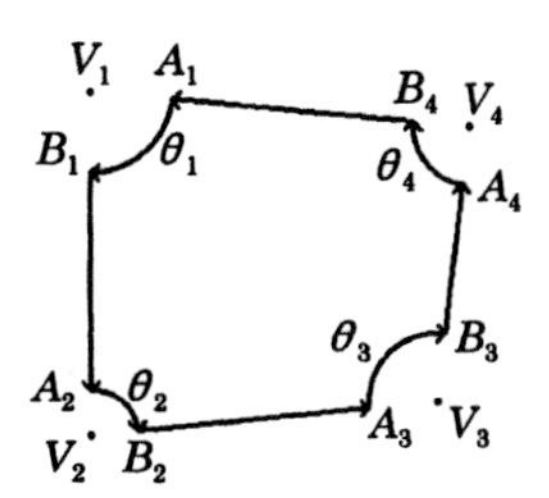

Figure 6.11.2
Exercise 13

[27] This exercise is due to Lawrence Levy (1970, 32).

the corresponding altitudes. Show that D and F lie on the axis of $\varphi = \sigma_a \sigma_b \sigma_c$. Classify φ when

(a) $m\angle A = m\angle B = 60°$ (c) $m\angle A = m\angle C = 45°$
(b) $m\angle A = m\angle B = 45°$ (d) $m\angle A = 30°$ and $m\angle C = 90°$.

Constructions

Following a more traditional approach to geometry, you'd have paid more attention to constructing the objects under study—probably with tools limited to straightedge and compasses. In fact, Euclid's axiomatization is sometimes viewed as a study of just those constructions, rather than of the space in which we live. This text has avoided them because they're no longer in the mathematical mainstream. The next seven exercises, however, merge transformational geometry into the older stream.[28] Each can be solved by traditional methods with no reference to transformations, but the isometries and similarities you've studied in this chapter shed more light on them. And if you seek the *simplest* constructions with straightedge and compasses, transformations may point the way.

Exercise 17. Suppose you want to carom ball A off the back, right, and front walls g, h, and k of a billiards table, to strike ball B, as in figure 6.11.3. How do you find the point P on g at which you must aim? *Suggestion*: Try more modest coups first.

Exercise 18. Consider nonparallel lines $g \neq h$ and three points P, Q, and R. Find a three-segment path beginning on g and ending on h, such that P, Q, and R are the midpoints of its segments. Can there be more than one such path? What if $g = h$ or $g \parallel h$?

Did your solution of exercise 17 require that the table be rectangular? Probably not. Could you extend it to apply to a convex polygonal table with any number of edges? Probably so—you can buy or build novelty billiards games like that. Exercise 18 didn't place any limitations on points

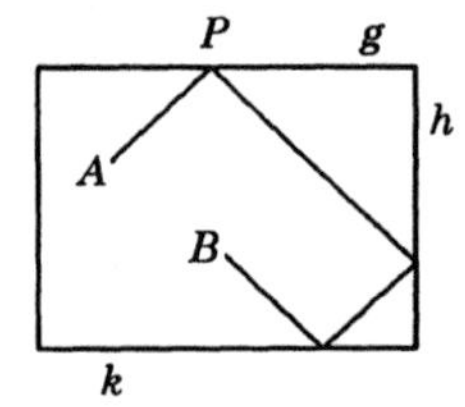

Figure 6.11.3
Exercise 17

[28] Several exercises under this heading are due to Max Jeger [1964] 1966.

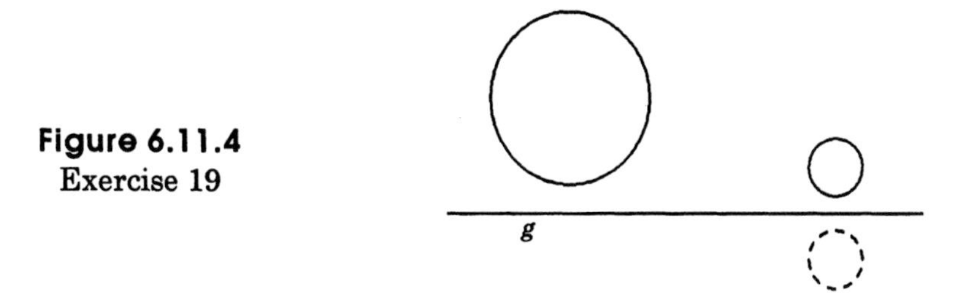

Figure 6.11.4
Exercise 19

P, Q, and R. Can you extend it to apply to any number of points? Exercises 34 and 35 consider *closed* polygonal paths of this sort.

Exercise 19. Consider two circles on the same side of line g. Find a point P on g such that tangents through P to the circles make equal angles with g. How many such points are there? *Suggestion*: Reflect one of the circles across g, as in figure 6.11.4.

Exercise 20. Consider two points A and B on the same side of line g. Find points O, P, and Q on g such that $2\text{m}\angle OPA = \text{m}\angle QPB$. *Suggestion*: Consider exercise 19.

Exercise 21. Consider a point P and concentric circles Γ and Δ. Look for points Q and R on Γ and Δ such that ΔPQR is equilateral. Under what condition on P, Γ, and Δ is a solution possible? How many solutions can there be? *Suggestion*: On which *two* circles must Q lie?

Exercise 22. Given ΔABC with $\text{m}\angle A < 90°$, find points C_1 and C_2 on the edge opposite C and points A_1 and B_1 on the edges opposite A and B such that $A_1B_1C_1C_2$ is a rectangle and $C_1C_2 = 2(B_1C_1)$.

Exercise 23. Consider a point B, two circles Γ and Δ, and any number $t > 0$. Find points C and D on Γ and Δ such that C-B-D and $BC/BD = t$. How many solutions could there be? Describe, as completely as you can, the set $\mathscr{B}$ of points B for which a solution is possible. *Suggestions*: On which *two* circles must C lie? Describing $\mathscr{B}$ is difficult; you may want to plot the points B corresponding to several thousand pairs of points C and D on Γ and Δ. The details of your description may vary depending on the situation of the circles.

Exercises on equations

Each exercise under this heading is concerned with the equation $X' = AX + B$ of a transformation $\varphi : X \to X'$. In exercises 24 to 26 the problem is simply to find the equation—that is, the entries of A and B.

Exercise 24. Two rotations about the point $Z = <1, 2>$ each map the line $2x_1 - x_2 = 0$ onto the line $x_1 - x_2 = -1$. Determine their equations, accurate to four decimal places.

Exercise 25. Consider a point $P = <p_1, p_2> \neq O$ and the line $g \perp \vec{OP}$ through O.

Part 1. Show that σ_g has matrix

$$A = I - \frac{2}{P^t P} K, \text{ where } K = PP^t = \begin{bmatrix} p_1^2 & p_1 p_2 \\ p_2 p_1 & p_2^2 \end{bmatrix}.$$

Part 2. Consider the line $h \perp \vec{OP}$ through a point Q. Show that $\sigma_h : X \to X'$ has equation

$$X' = AX + \frac{2}{P^t P} KQ.$$

Part 3. Find an equation for the reflection across the line h with equation $x_1 + 2x_2 = 3$.

Part 4. Find an equation for the glide reflection with axis h and vector $<2, -1>$.

Exercise 26. Consider points $A = <1, 0>$, $B = <2, 0>$, and $C = <0, 2>$, and the feet D and F of the altitudes of ΔABC through A and C. According to theorem 5.8.1, $\Delta ABC \sim \Delta DBF$.

Part 1. Find the equations of the similarity $\varphi : X \to X'$ that maps ΔABC onto ΔDBF.

Part 2. What's its ratio?

Part 3. Express φ as a composition of some or all of these: reflection, translation, rotation, glide reflection, dilation, and contraction.

Part 4. Find its fixpoint.

Suggestion: The entries of the matrix and vector for φ form the solution of a system of six linear equations in six unknowns. You might want to solve that for part (1), then work on (3). Or you could solve (3) first geometrically, then do (1).

Exercise 27 asks you to classify four transformations. First, show that they're isometries. Then you have the same tools at hand as for earlier classification exercises. Use them! The logic is the same, but now you compute with coordinates.

Exercise 27. Show that each of the following systems of equations represents an isometry φ, and classify it.

(a) $\begin{cases} x_1' = \frac{3}{5}x_1 + \frac{4}{5}x_2 \\ x_2' = -\frac{4}{5}x_1 + \frac{3}{5}x_2 \end{cases}$ (c) $\begin{cases} x_1' = \frac{3}{5}x_1 + \frac{4}{5}x_2 + 1 \\ x_2' = -\frac{4}{5}x_1 + \frac{3}{5}x_2 - 1 \end{cases}$

(b) $\begin{cases} x_1' = \frac{3}{5}x_1 - \frac{4}{5}x_2 \\ x_2' = -\frac{4}{5}x_1 - \frac{3}{5}x_2 \end{cases}$ (d) $\begin{cases} x_1' = \frac{3}{5}x_1 - \frac{4}{5}x_2 + 1 \\ x_2' = -\frac{4}{5}x_1 - \frac{3}{5}x_2 - 1 \end{cases}$

The next three exercises could have been placed under the earlier **Classifying isometries** heading. To solve them you don't really need to work with their equations. But that might be the simplest way.

Exercise 28. Classify the compositions of the $1°$ rotation about point $<2,3>$ and the translation with vector $<4,5>$, in both orders.

Exercise 29. Classify the compositions of the $1°$ rotation about point $<2,3>$ and the $4°$ rotation about point $<5,6>$, in both orders.

Exercise 30. Classify the compositions of the $10°$ rotation about the origin and the reflection across the line with equation $x_1 + x_2 = 1$, in both orders.

A geometric transformation φ associates with points of a figure those of a corresponding figure—its image—perhaps in another place. You could call this the *alibi* version of the concept—*alibi* is the Latin word for *elsewhere*. There's another version, often used in analytic geometry. Suppose the figure is a coordinate system; then so is its image. A point P with coordinates $<x_1, x_2>$ relative to the first system has coordinates $<x_1', x_2'>$ relative to the second. Associated with φ is a *coordinate transformation* that maps the coordinate pair $<x_1, x_2>$ to $<x_1', x_2'>$. Some texts regard *that* as the basic geometric transformation concept: Points of a figure Φ have two corresponding coordinate pairs. You could call it the *alias* version—*alias* is the Latin word for *otherwise*, to which we often add *known as*. Suppose you want to use two coordinate systems to model different aspects of some phenomenon. It's usually easy to find the isometry φ that transforms one into the other; it's part of the problem framework. Then you can derive an equation for φ. How do you use that to find equations that relate the corresponding coordinate pairs $<x_1, x_2>$ and $<x_1', x_2'>$ of a point P? (These considerations were introduced, with slightly different notation, under the heading **Transforming coordinate systems** in section 6.9.) Figures Φ in your model may be described by equations that relate the two coordinates of points P of Φ. How do you convert an equation that relates x_1 and x_2 to one for x_1' and x_2'? Exercise 31 provides some examples of this process. You'll have to decide when and how to use the equation of φ and that of its inverse. You'll see that how you transform equations that describe figures depends on whether they're single equations or pairs of parametric equations.

Exercise 31. Suppose that, in addition to your standard Cartesian coordinate system with origin O, you're using an auxiliary one with

- origin $O' = <3, 2>$,
- unit point U' for the first axis located below and left of O' on the line through O' with slope 1, and
- unit point V' for the second axis located above and left of O'.

A point $X = <x_1, x_2>$ has coordinates x_1' and x_2' relative to this coordinate system.

Part 1. Find formulas for x_1' and x_2' in terms of x_1 and x_2.

Part 2. Find the equation of the unit circle Γ with center O in terms of the auxiliary coordinate system: X must lie on Γ if and only if x_1' and x_2' satisfy the equation.

Part 3. $x_2 = x_1^2$ is the equation of a parabola Π. Find the equation of Π in terms of the auxiliary coordinate system.

Part 4. $x_1 = \cos t$ and $x_2 = 2\sin t$ for $0 \le t < 2\pi$ are parametric equations for an ellipse Λ. Find parametric equations for Λ in terms of the auxiliary coordinate system.

Reflection calculus

During the 1920s Gerhard Thomsen discovered how to formulate many familiar properties of plane figures in terms of equations involving compositions of point and line reflections. The first few exercises under this heading introduce this method. You can solve them by manipulating equations involving reflections, half turns, translations, and occasionally vector algebra. You never have to use coordinates explicitly. You can call this method a *reflection calculus*.

Exercise 32. Show that for any points O, P, and Q, the composition $\sigma_O \sigma_P \sigma_Q$ is a half turn about point $R = O - P + Q$, and $OPRQ$ is a (possibly degenerate) parallelogram.

Exercise 33. Describe the figure formed by points A, B, C, and G for which $\sigma_A \sigma_G \sigma_B \sigma_G \sigma_C \sigma_G = \iota$. *Suggestion*: Reason with vectors.

Exercise 34. Construct a pentagon with given edge midpoints M_1 to M_5. Is it unique? Generalize your solution to apply to polygons with more edges.

Gerhard THOMSEN was born in Hamburg in 1899, the son of a physician. He served a year in World War I, then became one of the first students of the new university at Hamburg. Thomsen completed the Ph.D. in 1923 with a dissertation on differential geometry. He served as assistant in Karlsruhe and Hamburg, studied a year with Tullio Levi-Civita in Rome, then presented his Habilitationsschrift in Hamburg, on a problem in gravitational physics. In 1929 he became Ausserordentlicher Professor at Rostock. There he continued work on differential geometry, collaborating with Wilhelm Blaschke, and published as well in mathematical physics and foundations of geometry. In 1933, reacting to the Nazification turmoil in his university, he published an inflammatory lecture that seemed to support some Nazi aims but attacked Nazi suppression of education in the sciences. This evidently attracted the attention of the secret police. Thomsen died, an apparent suicide, on a Rostock railroad track in 1934.

Exercise 35. When can you construct a quadrilateral with given edge midpoints? When is it unique? How about a hexagon? *Suggestion*: What kind of figure is formed by the midpoints of the edges of a quadrilateral? Of a hexagon?

Exercises 34 and 35 are closely related to exercise 17 on billiards and to exercise 18. Can you imagine the pentagon and hexagon as *dual billiards* tables? You play on the outside, grazing the corners instead of rebounding from the sides!

A development of the reflection calculus and results obtained with it were published in Thomsen 1933a. Exercises 36 to 39 are selected from section 4 of that book.

Exercise 36. Consider points P and Q, lines g and h, and the isometry $\varphi = \sigma_g \sigma_P \sigma_h \sigma_Q$. Under what condition on the points and lines is $\varphi = \iota$? *Suggestion*: First, classify $\sigma_g \sigma_P$ and $\sigma_h \sigma_Q$.

Exercise 37. Prove that these conditions on lines g and h are equivalent:

(a) $g \parallel h$;
(b) for *every* point P, $\sigma_g \sigma_h \sigma_P \sigma_g \sigma_h \sigma_P = \iota$ but $\sigma_g \sigma_h \sigma_P \neq \iota$;
(c) for *some* point P, $\sigma_g \sigma_h \sigma_P \sigma_g \sigma_h \sigma_P = \iota$ but $\sigma_g \sigma_h \sigma_P \neq \iota$.

Exercise 38. Consider some lines a, b, and c and the isometry

$$\varphi = \sigma_c\sigma_a\sigma_b\sigma_a\sigma_b\sigma_a\sigma_c\sigma_b\sigma_a\sigma_c\sigma_a\sigma_c\sigma_a\sigma_b.$$

Prove that φ is the identity if and only if a, b, and c are copencilar or form an isosceles triangle with congruent edges on b and c. *Suggestion*: It's easy to show that copencilarity implies $\varphi = \iota$. When there's a triangle, consider the line g through the feet B^* and C^* of the altitudes perpendicular to b and c. When is $g \parallel a$? What isometry is $\gamma = \sigma_b\sigma_a\sigma_c$?

Exercise 39. Show that for any lines a, b, and c,

$$\sigma_a\sigma_b\sigma_c\sigma_a\sigma_b\sigma_c\sigma_b\sigma_c\sigma_a\sigma_b\sigma_c\sigma_a\sigma_c\sigma_b\sigma_a\sigma_c\sigma_b\sigma_c\sigma_b\sigma_a\sigma_c\sigma_b = \iota.$$ [29]

Exercises on the definitions of isometry and similarity

The next two exercises are concerned with functions that map the plane to itself but satisfy (seemingly) weaker conditions than isometries or similarities.

Exercise 40. Show that any function $\varphi : X \to X'$ from the plane to itself that preserves distance is an isometry. That is, from the hypothesis that $X'Y' = XY$ for all X and Y, deduce that φ is injective and surjective. *Suggestion*: For surjectivity, given Y, you must find X such that $Y = X'$; select two points P and Q and observe the relationship of Y to P' and Q'.

Exercise 41. Let r be any positive number and φ be a function from the plane to itself such that $X'Y' = r\,(XY)$ for all X and Y. Prove that φ is a similarity with ratio r. *Suggestion*: Use exercise 40.

It's possible to weaken the definitions of isometry and similarity even further than these exercises suggest. For example, invariance of the property of separation by distance 1 —that is, the condition $XY = 1$ if and only if $X'Y' = 1$ —implies that φ is an isometry.[30]

[29] Responding to a question from Thomsen, Hellmuth Kneser (1931) showed that this equation is the shortest that's true for *all* lines a, b, and c, except for equations like $\sigma_a\sigma_a = \iota$ and others such as $\sigma_a\sigma_b\sigma_b\sigma_a = \iota$ that are derivable from it.

[30] Modenov and Parkhomenko [1961] 1965, appendix.

Dilatations

A *dilatation*[31] is a plane transformation $\varphi : X \to X'$ such that the image $g' = \varphi[g]$ of any line g is a line parallel or equal to g. The last four exercises of this chapter develop the theory of dilatations, which plays a major role in some axiomatizations of geometry.[32] You should solve them without using any limit concepts from analysis. Your solutions will provide a proof of the fixpoint theorem for similarities, theorem 6.10.8, that, unlike the proof in the text, requires no analysis. In your solutions of exercises 43 to 45, you're expected to use the results in the preceding exercises.

Exercise 42. Prove that the dilatations form a transformation group. Of the various types of isometries and similarities considered in this chapter, which are dilatations and which not?

Exercise 43. Show that

(a) the only dilatation with two distinct fixpoints is the identity, and
(b) if φ and χ are dilatations and $\varphi(X) = \chi(X)$ for two distinct points X, then $\varphi = \chi$.

Suggestions: For (a) show first that a dilatation $\varphi : X \to X'$ with distinct fixpoints P and Q fixes every point *not* on $\overleftrightarrow{PQ}$, then consider points *on* the line. Note that any line through a fixpoint is fixed and any point on two fixed lines is fixed.

Exercise 44. Show that every dilatation $\varphi : X \to X'$ that's not a translation has a fixpoint. *Suggestion*: Find P and Q such that $PP'Q'Q$ is a quadrilateral, then show that it can't be a parallelogram.

Exercise 45. Prove that every dilatation that's not a translation is a contraction, a dilation, or the composition of a contraction or dilation with the half turn about its center.

---

[31] Be careful not to confuse the terms *dilation* and *dilatation*! The first stems from the Latin prefix *dis-*, meaning *apart*, with the past participle *latus* of the irregular Latin verb *fero*, which means *carry*. The second is from the related regular verb *dilato*, which means *broaden*.

[32] For example, Artin 1957, chapter 2.

Chapter 7

Three-dimensional isometries and similarities

Chapter 6 introduced plane transformational geometry. You studied plane isometries and similarities in depth. New concepts entered gradually, so that each new idea was related as closely as possible to basic geometry, where our visual intuition provides a clear guide. With that background, you've studied the effects of plane motions, and plane Cartesian coordinate transformations. You're prepared to analyze symmetries of plane ornamentation undertaken later, in chapter 8. And you're prepared to study more general types of plane transformations that fall beyond the scope of this book.

But you haven't many techniques yet to handle motions and coordinate transformations in three-dimensional space, where you live. Nor are you ready yet to analyze the three-dimensional symmetries of polyhedra and crystals. For that, this chapter provides a complete coverage of three-dimensional isometries and similarities. You'll see that it's a straightforward extension of the two-dimensional theory. Of course, there are more isometry types in three dimensions, but the algebra techniques you used in two dimensions apply without change in this chapter.

Because you've already met, in simpler form, most of the ideas discussed in this chapter, it's more streamlined. Concepts are introduced and studied in the order that leads most quickly to the main results of the theory. To that end, the chapter relies more heavily on algebra. Sometimes this is "brutally" efficient. For example, section 7.1 derives the matrix equation $X' = AX + B$ for an isometry $X \to X'$ directly from the uniqueness and rigidity theorems, before considering in detail *even one* particular type of isometry.

That plan makes various general techniques available for use earlier. But following it in chapter 6 would have prevented you from applying your geometric experience effectively in learning new concepts.

The greater reliance on algebra is also due to the greater complexity of three dimensions. Many problems to which we now apply algebraic plane transformational geometry methods could be solved without that theory. Algebra reveals its usefulness as it helps reduce the complexity in three dimensions.

7.1 Isometries

Concepts

- Definition
- The isometry group
- Invariance
- Rigidity theorem
- Uniqueness theorem
- Orthogonal matrices
- Equations of isometries
- Conjugacy
- Parity

This section extends to three dimensions the notion of isometry that you studied in chapter 6. It also presents various properties shared by all such isometries: the fixpoint, rigidity, and uniqueness theorems, representation of isometries $X \to X'$ by equations $X' = AX + B$ with orthogonal matrices A, the relationship of these equations to conjugacy, and the distinction between even and odd isometries. You'll study particular types of isometries in later sections.

The equations require a Cartesian coordinate system. Introduce a "standard" system $\mathscr{C}$ now, with origin O and unit vectors U_1 to U_3, and refer to it as necessary.

In the three-dimensional theory, an *isometry* is a distance-preserving transformation of the set of *all* points—not just those in some plane. As in two dimensions, you can show that the identity transformation ι is an isometry, the inverse φ^{-1} of an isometry φ is an isometry, and the composition $\varphi\chi$ of two isometries φ and χ is an isometry:

Theorem 1. The isometries form a subgroup of the symmetric group of the set of all points.

There's a whole catalog of properties of three-dimensional figures that are invariant under isometries: the invariant properties of plane figures that you've already considered—especially betweenness—plus

measure of a dihedral angle
volume of a tetrahedron
perpendicularity
parallelism, etc.

being a
- plane
- half space
- dihedral angle
- tetrahedron
- sphere

This text won't pursue those details. As necessary, you can supply invariance proofs like that of theorem 6.2.5, because you can define all these concepts in terms of distance (using betweenness if appropriate).

Rigidity and uniqueness theorems

The uniqueness theorem is a major tool in the plane theory. To see that an isometry under analysis is really a familiar one, you need only show that they agree at three noncollinear points. An analogous tool performs the same function in three dimensions. The plane version was derived from the *rigidity theorems* 6.2.6 and 6.2.7; its three-dimensional form requires analogs of those results. The first one is already valid in three dimensions:

Theorem 2. If an isometry $X \to X'$ has distinct fixpoints P and Q, then it fixes every point X on $\overleftrightarrow{PQ}$.

You can check that the original proof is valid, too. For the next result, you must edit the statement of theorem 6.2.7; its proof remains valid.

Theorem 3. If an isometry $X \to X'$ has noncollinear fixpoints P, Q, and R, then it fixes every point X on the plane PQR.

Three-dimensional geometry enters with a new rigidity result. Its proof, exercise 7.9.1, is merely a three-dimensional version of the proof of theorem 6.2.7.

Theorem 4. The only isometry with four noncoplanar fixpoints is the identity.

With this foundation, you can easily state the three-dimensional *uniqueness theorem*. Its proof is the same as that of the two-dimensional version, corollary 6.2.8.

Corollary 5 **(*Uniqueness theorem*).** If χ and ψ are isometries and $\chi(P) = \psi(P)$ for four noncoplanar points P, then $\chi = \psi$.

Orthogonal matrices

The entire analytic treatment of isometries with matrix algebra works in three dimensions just as it did in two. Appendix C presents an outline of the elementary features of vector and matrix algebra, independent of dimension. The notion of orthogonality, introduced in section 6.8, applies to square matrices A of *any* dimension: A is *orthogonal* if $A^tA = I$, i.e., its transpose is its inverse. As noted there, the identity matrix I is orthogonal, the inverse of an orthogonal matrix is orthogonal, and the product of two orthogonal matrices is orthogonal. That section also showed that for *any* orthogonal matrix A and vector B, the mapping $X \to AX + B$ is an isometry.

These analogies suggest that *every* three-dimensional isometry φ should have an equation $\varphi(X) = AX + B$ for some orthogonal matrix A and some vector B. The argument for the corresponding plane result, theorem 6.8.2, required the equations for two special types of isometries, including some rotations. You *could* take that approach in three dimensions, deriving analogs of those special equations. But that's difficult because equations of three-dimensional rotations are generally more complicated. The argument in the next proof, however, doesn't depend on individual equations nor on the dimension; so the desired result holds in three dimensions as well. The first step is to show that you can map your preselected coordinate system O, U_1, U_2, U_3 into *any* other one by applying a transformation of the form $X \to AX + B$ with an orthogonal matrix A.

Theorem 6. Given points B, V_1, V_2, V_3 such that $BV_j = 1$ for $j = 1$ to 3 and $\vec{BV}_1$ to $\vec{BV}_3$ are perpendicular, there's an orthogonal matrix A such that $AU_j + B = V$ for $j = 1$ to 3.

Proof. Construct the matrix A whose jth column is $A_j = V_j - B$. Then $AU_j + B = A_j + B = V_j$, as desired. Moreover, $A_j \cdot A_j = (BV_j)^2 = 1$. Now suppose $j \neq k$. Since $\vec{BV}_j \perp \vec{BV}_k$, Pythagoras' theorem implies

$$(OA_j)^2 + (OA_k)^2 = (BV_j)^2 + (BV_k)^2 = (V_jV_k)^2.$$

But $V_j - V_k = A_j - A_k$, so

$$(V_jV_k)^2 = (V_j - V_k)\cdot(V_j - V_k) = (A_j - A_k)\cdot(A_j - A_k) = (A_jA_k)^2.$$

The converse part of Pythagoras' theorem now yields $\overleftrightarrow{OA}_j \perp \overleftrightarrow{OA}_k$, and $A_j \cdot A_k = 0$ by theorem 3.11.10. It follows that A is orthogonal: $A^tA = I$. To see that, note that the j,kth entry of A^tA is the jth row of A^t times the kth column of A —that is, $A_j \cdot A_k$. According to the equations just derived, that's 1 if $j = k$ or 0 if $j \neq k$. It's the j,kth entry of I.[1] ♦

---

[1] The word *orthogonal*, a synonym for *perpendicular*, stems from the Greek words *orthos* and *gonia* for *upright* and *angle*. Matrices like A were termed orthogonal because their column vectors are perpendicular, as noted in this proof.

Theorem 7. The three-dimensional isometries are the mappings $X \to \varphi(X) = AX + B$, where A is an orthogonal matrix and B is an arbitrary vector. The isometry φ determines A and B uniquely.

Proof. You've already seen that this equation always defines an isometry. It remains to show that an arbitrary isometry φ has such an equation, and only one. First, φ maps your standard coordinate system O, U_1, U_2, U_3 into another system B, V_1, V_2, V_3. By theorem 6, there's an orthogonal matrix A such that $AU_j + B = V_j$ for $j = 1$ to 3. The isometry $X \to AX + B$ agrees with φ at four noncoplanar points O and U_1 to U_3. By the uniqueness theorem, the two isometries coincide: $\varphi(X) = AX + B$ for every X, as required. Finally, you can use the last paragraph of the proof of theorem 6.8.1 to show that φ determines A and B uniquely. ♦

The short algebraic argument in the proof of theorem 6 could have been used in chapter 6 to avoid some matrix calculations. But it has so little visual content that the longer argument was used instead. It's difficult to determine just how this material was originally discovered. In essence, it's already present in Euler [1775] 1968.

Conjugacy

Chapter 6 used the notion of conjugacy to classify plane isometries. It plays the same role in three dimensions. Its definition in section 6.9 applies to *any* transformations μ and φ; the composition $\varphi\mu\varphi^{-1}$ is the conjugate of μ by φ. We write $\lambda \sim \mu$ to indicate that a transformation λ is a conjugate of μ. The basic theorems 6.9.4 and 6.9.5 hold for three-dimensional isometries, too. In particular, conjugacy is an equivalence relation, and the equivalence class of an isometry is called its conjugacy class. Sections 7.2 to 7.5 use conjugacy to classify isometries.

The section 6.9 discussion of coordinate transformations, equations of isometries, and their conjugacy is essentially valid for three dimensions, too. An isometry φ transforms your standard coordinate system $\mathscr{C}$ into another system $\mathscr{D}$ defined by $P = \varphi(O)$ and $V_j = \varphi(U_j)$ for $j = 1$ to 3 as in figure 7.1.1. Theorem 9, the analog of theorem 6.9.6, shows how a point's $\mathscr{C}$ and $\mathscr{D}$ coordinates are related.

Theorem 9. If an isometry φ transforms coordinate system $\mathscr{C}$ into system $\mathscr{D}$, then the $\mathscr{C}$ coordinates of each point X are the same as the $\mathscr{D}$ coordinates of $Y = \varphi(X)$.

Proof. Consider figure 7.1.1. The figure formed by O, U_1 to U_3, and X is congruent to that formed by P, V_1 to V_3, and Y. ♦

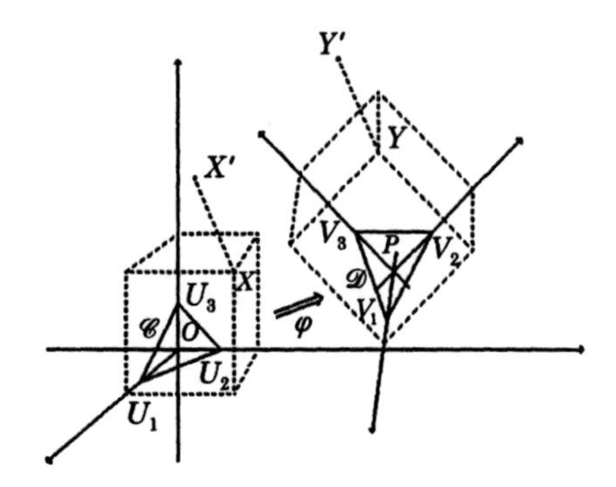

Figure 7.1.1
Transforming a coordinate system

Theorem 10, the main result about conjugacy, is the analog of theorem 6.9.7; you can use its proof unchanged, if you cite theorem 6 of this section in place of theorem 6.6.3.

***Theorem 10* (*Conjugacy theorem*).** Two isometries are conjugate if and only if there are coordinate systems under which they have the same matrix equation.

Parity

As in two dimensions, the determinant par φ of the orthogonal matrix A corresponding to an isometry φ is called its *parity*. Since equation $A^tA = I$ implies $(\det A)^2 = (\det A^t)(\det A) = \det(A^tA) = \det I = 1$, the parity is ± 1. Isometries with parity 1 are called *even*. For example, the identity is an even isometry. The others are called *odd*. This section concludes with some simple facts about parity that you've already met in the plane theory. The connection between parity and orientation is discussed in the next section.

***Theorem 11*.** $\text{par}\,\varphi\chi = \text{par}\,\varphi\ \text{par}\,\chi$ and $\text{par}\,\varphi^{-1} = \text{par}\,\varphi$ for any isometries φ and χ.

***Corollary 12*.** The even isometries constitute a subgroup of the isometry group.

***Corollary 13*.** Conjugate isometries have the same parity.

Leonhard EULER was born in Basel in 1707. He was taught first by his father, a protestant minister, who had attended Jakob Bernouilli's lectures at the university there. In 1720 Euler entered the university. He studied with Johann Bernouilli, and received a degree in philosophy in 1723 with a comparison of Newton's and Descartes' ideas. He started theological studies, but dropped those for mathematics. Although he published some papers on mechanics, Euler could find no job at Basel. In 1727, however, Peter the Great's reign was ending, the Academy of Science at St. Petersburg was being formed, and the intelligentsia there was becoming dominated by Germans. On Bernouilli's recommendation, they offered Euler a job; he became Professor there in 1731. The Academy had scientific, educational, and technological assignments, and provided a very rich atmosphere. Euler produced about 90 papers on mathematics there during the period 1727–1741, and a treatise on mechanics. He worked on cartography, shipbuilding, and navigation as well.

Euler's brother also had an Academy position, as an artist. In 1733 Euler married the daughter of another Swiss artist at the Academy. Two sons were born in St. Petersburg. Later, in Berlin, they had another son and two daughters; eight other children died in infancy. In 1738, Euler contracted a disease that left him blind in one eye.

Political turmoil in Russia, particularly resentment against German influence, arose during the late 1730s, simultaneous with Frederick the Great's reorganization of the Academy of Science at Berlin. After prolonged negotiations, Euler moved there in 1741. However, he retained many connections with the St. Petersburg Academy, editing its mathematical publications and publishing there his own texts on calculus and navigation. Euler stayed in Berlin for 25 years. During that period he produced about 380 papers in pure and applied mathematics, and was elected to the Academies in London and Paris as well. Euler was very influential in the development and administration of the Berlin Academy, actually running it for many years. But he and the King were incompatible; Frederick wouldn't appoint him President.

In 1766, Euler returned to St. Petersburg. The long, stable reign of Catherine the Great had begun. One of Euler's sons became professor of physics there; another, an artillery officer. An illness in 1769–1771 led to Euler's total blindness. But his scientific output actually increased! Besides many research papers, he published texts on algebra, calculus, navigation, and life insurance, and treatises on optics and lunar theory. His first wife died in 1773; in 1776 he married her half-sister. Euler died of a brain hemorrhage in the midst of heavy work in 1783.

7.2 Reflections

Concepts
- Reflections across planes
- Conjugate reflections
- Structure theorem
- Even and odd isometries
- Orientation

Reflections play a fundamental role in the theory of plane isometries. They provide components from which you can construct all other isometries, and they facilitate the distinction between even and odd isometries. Their three-dimensional counterparts are the *reflections across planes*. The reflection $X' = \sigma_\varepsilon(X)$ of a point X across a plane ε is X itself if X is on ε, else it's the point on the opposite side of ε such that ε is the perpendicular bisector of $\overline{XX'}$. The plane ε is called the *mirror* of σ_ε. To study these reflections, it's sometimes convenient to use the theory of plane isometries. Theorem 1 follows directly from the plane theory and the definition of σ_ε. (See figure 7.2.1.) You can supply the proof.

Theorem 1. σ_ε is self-inverse. Its fixpoints are the points in ε. It fixes any plane $\alpha \perp \varepsilon$, and its restriction to α is the (two-dimensional) reflection of the points on α across the intersection $g = \alpha \cap \varepsilon$.

Corollary 2. σ_ε is an isometry.

Proof. (See figure 7.2.1.) Any two points X and Y lie in some plane $\alpha \perp \varepsilon$, and σ_ε operates on X and Y just as the (two-dimensional) reflection of α across $g = \alpha \cap \varepsilon$. That preserves the distance XY. ♦

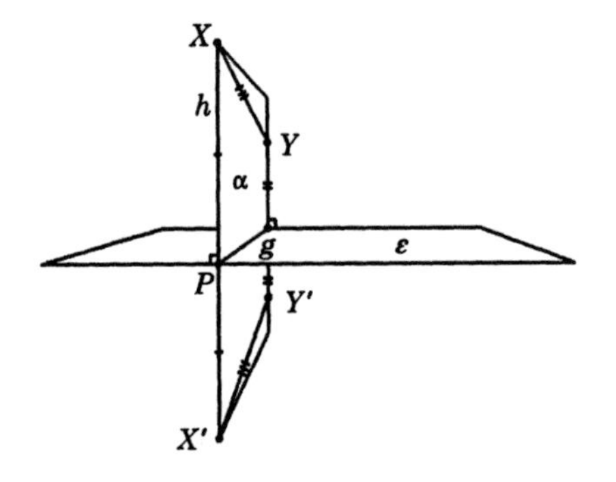

Figure 7.2.1
Reflection
$\sigma_\varepsilon : X \to X'$ across ε

Theorem 3. An isometry φ that leaves a plane ε pointwise fixed is the identity or σ_ε.

Proof. Select a point X not on ε, construct the line $h \perp \varepsilon$ through X, and let $P = h \cap \varepsilon$. (See figure 7.2.1.) Since φ fixes P and ε, it maps h to a line through P perpendicular to ε. That is, φ fixes h, so $\varphi(X)$ is one of the two points X and $\sigma_\varepsilon(X)$ on h whose distance from P is PX. Thus φ agrees either with ι or with σ_ε at four noncoplanar points: X and any three noncollinear points on ε. The uniqueness theorem yields $\varphi = \iota$ or σ_ε. ♦

Corollary 4. $\varphi \sigma_\varepsilon \varphi^{-1} = \sigma_{\varphi[\varepsilon]}$ for any isometry φ and any plane ε.

Proof. Use the argument under the section 6.9 heading **Conjugacy classes** to show that $\varphi \sigma_\varepsilon \varphi^{-1}$ leaves plane $\varphi[\varepsilon]$ pointwise fixed, but isn't the identity. ♦

The equation of the reflection $\sigma_\varepsilon : X \to X'$ across a plane ε is generally unwieldy—see exercise 7.9.20. But for some special planes it's very simple. For example, if ε is the plane through the first two coordinate axes, then σ_ε has equations

$$\begin{aligned} x_1' &= x_1 \\ x_2' &= x_2 \\ x_3' &= -x_3 \end{aligned} \qquad X' = AX \qquad A = \begin{bmatrix} 1 & 0 & 0 \\ 0 & 1 & 0 \\ 0 & 0 & -1 \end{bmatrix}.$$

This matrix A is orthogonal, with determinant -1.

Any plane ε contains the first two axes of some coordinate system, so any two reflections have the same equations with respect to suitably chosen coordinate systems. With the conjugacy theorem, corollary 4, and corollary 7.1.13, this implies

Theorem 5. The reflections constitute a single conjugacy class. All reflections are odd.

Structure theorem

According to the structure theorem in two-dimensional transformational geometry, every plane isometry is the identity, a reflection, or the composition of two or three reflections. That result is the basis for classifying plane isometries. An analogous one holds in the three-dimensional theory. You'll notice that its proof is organized just like that of theorem 6.6.4, the two-dimensional result.

Theorem 6 (Structure theorem). The three-dimensional isometry group consists of the identity, the reflections, and the compositions of two, three, and four reflections.

Proof. An isometry φ maps a tetrahedron Γ into a congruent tetrahedron Δ. If $\Gamma = \Delta$, then φ is the identity by the rigidity theorem. If Γ and Δ share three vertices but not four, then φ is the reflection across the plane δ through those vertices, by theorem 3.

If Γ and Δ share *two* vertices, then there's a plane γ through those vertices such that Γ and the tetrahedron $\sigma_\gamma[\Delta]$ share *three* vertices. (Make γ "bisect" a certain dihedral angle.) By the previous paragraph, $\sigma_\gamma\varphi$ is the identity or the reflection across a plane δ, so $\varphi = \sigma_\gamma$ or $\sigma_\gamma\sigma_\delta$.

If Γ and Δ share *one* vertex, then there's a plane β through it such that Γ and the tetrahedron $\sigma_\beta[\Delta]$ share *two*. By the previous paragraph, $\sigma_\beta\varphi = \sigma_\gamma$ or $\sigma_\gamma\sigma_\delta$ for some planes γ and δ, so $\varphi = \sigma_\beta\sigma_\gamma$ or $\sigma_\beta\sigma_\gamma\sigma_\delta$.

Even if Γ and Δ share *no* vertices, there's a plane α —the perpendicular bisector of any segment joining two corresponding vertices—such that Γ and the tetrahedron $\sigma_\alpha[\Delta]$ have at least *one* vertex in common. By the previous paragraph, $\sigma_\alpha\varphi = \sigma_\beta\sigma_\gamma$ or $\sigma_\beta\sigma_\gamma\sigma_\delta$ for planes β, γ, and δ, so φ is the composition of three or four reflections. ♦

Corollary 7. An isometry is even or odd depending on whether it's a composition of an even or odd number of reflections.

Proof. Every isometry is a composition of reflections, which are odd. ♦

Orientation

You can distinguish even from odd isometries in three dimensions, as in two, by their effects on oriented figures. Informally, you can consider left- and right-hand gestures. Formally, consider tetrahedra—section 3.10 defined them as *ordered* quadruples of noncoplanar points. In figure 7.2.2, the vertices of tetrahedron $PQRS$ are arranged as you might point with your right thumb, with P on the thumb, and ΔQRS oriented as your fingers are curled. The reflection across the plane maps them to points P' to S', arranged like the corresponding left-hand gesture. There are twenty-four tetrahedra with vertices P to S listed in various orders. Using the figure, you can classify them according to your intuitive notion of orientation:

Class $\mathscr{R}$: $PQRS$, $PRSQ$, and ten others;
Class $\mathscr{L}$: $PQSR$, $PRQS$, and ten others.

How can you distinguish these classes mathematically, without relying on the figure and your intuition? The determinant formula for tetrahedral

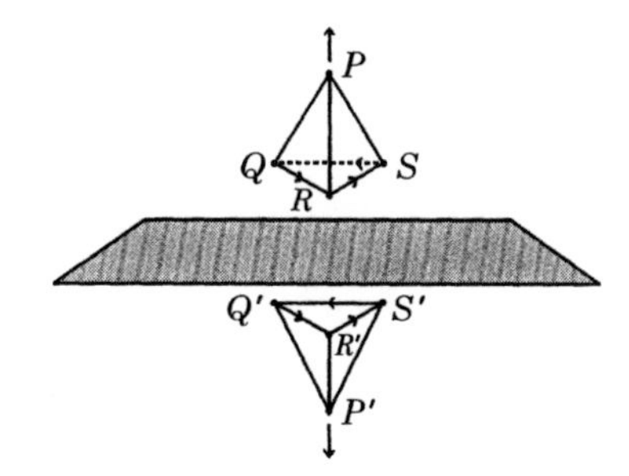

Figure 7.2.2
Reflection across a plane reverses orientation.

volume, theorem 5.6.10, provides the tool. These tetrahedra all have the same volume: $\pm\frac{1}{6}$ times each of the determinants

$$\mathscr{R}: \det[P-R, Q-R, S-R],\ \det[P-S, R-S, Q-S], \ldots$$
$$\mathscr{L}: \det[P-S, Q-S, R-S],\ \det[P-Q, R-Q, S-Q], \ldots.$$

You can show that the determinants in the first class are all equal to the same value; and all those in the second, to its negative. Thus all those in one class have the same *positive* determinant; the others have the same *negative* determinant. Accordingly, you can call the tetrahedra *positively* or *negatively* oriented. (*Which* class of tetrahedra is positive depends on the standard coordinate system that you chose at the onset of this chapter. If you reverse the scale on any one axis, you'll interchange the positive and negative classes.) The following computation shows that an even isometry $X \to X' = AX + B$ preserves a tetrahedron's orientation, whereas an odd one reverses it:

$$\begin{aligned}
&\det[P'-R', Q'-R', S'-R'] \\
&\quad = \det[(AP+B)-(AR+B), (AQ+B)-(AR+B), \\
&\qquad\quad (AS+B)-(AR+B)] \\
&\quad = \det[AP-AR,\ AQ-AR,\ AS-AR] \\
&\quad = \det[A(P-R), A(Q-R), A(S-R)] \\
&\quad = \det(A[P-R, Q-R, S-R]) \\
&\quad = (\det A)(\det[P-R, Q-R, S-R]) \\
&\quad = \pm\det[P-R, Q-R, S-R].
\end{aligned}$$

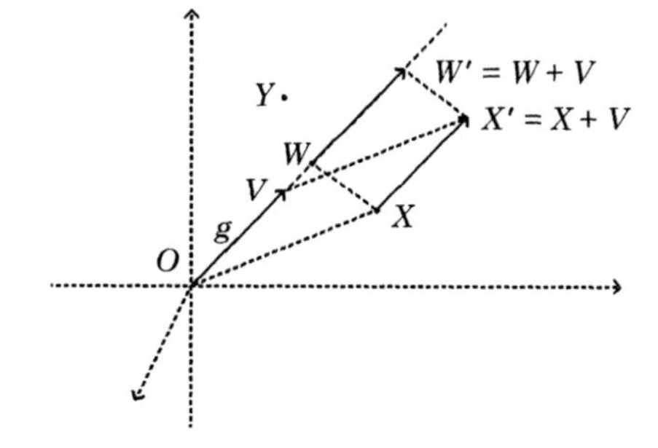

Figure 7.3.1
Translation
$\tau: X \to X' = X + V$

7.3 Translations and rotations

Concepts

- Translations
- Conjugate translations
- Rotations
- Angle parameters for rotations
- Conjugate rotations
- Commuting reflections
- Half turns

The previous section introduced three-dimensional reflections, and showed that each three-dimensional isometry is the identity, a reflection, or the composition of two, three, or four reflections. Section 7.2 studies the compositions of *two* reflections in detail. They're the translations and the rotations.

Translations

Translations are handled in three dimensions exactly as in the plane. The translation corresponding to a vector V is the transformation τ_V : $X \to X' = X + V$ displayed in figure 7.3.1 on the preceding page. (This is just the corresponding two-dimensional figure 6.2.1 with a third axis inserted.) The following theorem summarizes some main facts about translations. You can prove it just as you proved the analogous planar results in sections 6.2 and 6.4.

Theorem 1. Composition of translations corresponds to vector addition: For any vectors V and W, $\tau_V \tau_W = \tau_{V+W}$ and $\tau_V^{-1} = \tau_{-V}$. Moreover, $\tau_O = \iota$. Therefore, the translations form a subgroup of the isometry group.

Just as in two dimensions, you can construct translations from reflections.

Theorem 2. The composition $\sigma_\delta \sigma_\varepsilon$ of reflections across parallel planes δ and ε is the translation τ with vector $2(P - Q)$, where P and Q are the intersections of δ and ε with any line k perpendicular to δ and ε.

Proof. Verify that $\sigma_\delta \sigma_\varepsilon(X) = \tau(X)$ for four noncoplanar points X, as in figure 7.3.2, then apply the uniqueness theorem. ♦

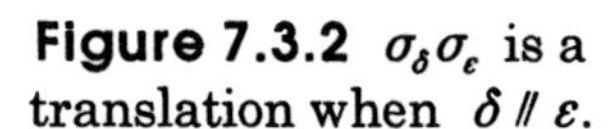

Figure 7.3.2 $\sigma_\delta\sigma_\varepsilon$ is a translation when $\delta \parallel \varepsilon$.

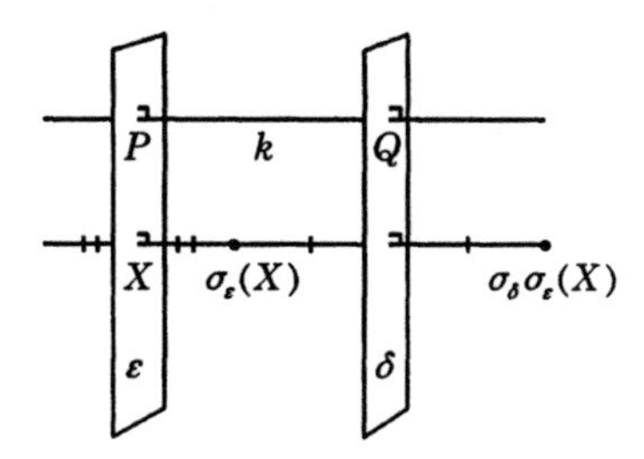

The conjugacy properties of three-dimensional translations are also analogous to the results for plane transformations. Theorem 3 is in fact *identical* to theorem 6.9.9.

Theorem 3. The conjugacy class of a translation τ whose vector has length t consists of all translations with vectors of length t. If τ and an isometry φ have equations $\tau : X \to X' = X + T$ and $\varphi : X \to X' = AX + B$ with respect to a single coordinate system, then $\varphi\tau\varphi^{-1}$ is the translation with vector AT.

Proof. The first sentence follows directly from the conjugacy theorem. The proof of the second is the same as that of theorem 6.9.9. ♦

Rotations

Pursuing the analogy with the theory of plane isometries, you'll next ask what's the composition $\sigma_\delta\sigma_\varepsilon$ of the reflections across two distinct *intersecting* planes δ and ε. It's called a *rotation* about its *axis* $g = \delta \cap \varepsilon$. The following theorem justifies this terminology. To prove it, apply theorem 7.2.1.

Theorem 4. Consider distinct planes δ and ε through a line g perpendicular to a plane α. (See figure 7.3.3.) Choose an angular scale on α with center $P = \alpha \cap g$. Choose half planes δ' and ε', each with edge g. If rays $\alpha \cap \delta'$ and $\alpha \cap \varepsilon'$ have angle parameters ζ and η, then the restriction of $\rho = \sigma_\delta\sigma_\varepsilon$ to α is the (two-dimensional) rotation of α about P with angle parameter $\theta = 2(\eta - \zeta)$. The fixpoints of ρ constitute its axis g.

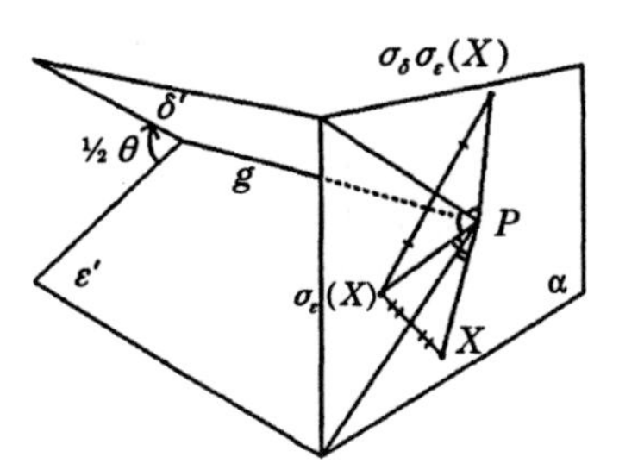

Figure 7.3.3 $\sigma_\delta\sigma_\varepsilon$ is a rotation when δ intersects ε.

In theorem 4 you can choose δ' and ε' in four ways. Another choice might add 180° to ζ and/or η. Since angle parameters are determined only mod 360°, your choice has no effect on θ. Since θ depends only on the *difference* of two parameters, it depends only on ρ and the *orientation* of the angular scale. You can specify orientation by designating a ray on g as positive. (Informally, align it with your thumb when the knuckles and fingertips of your right fist fall along rays with zero and positive angle parameters.)

For theoretical considerations it's best to describe a rotation as a specific composition of plane reflections. For practical computations, however, it's often helpful to specify a positive axial ray and an angle parameter. For example, a rotation with angle parameter θ about one of the three coordinate axes fixes that coordinate of each point X, and operates on its other coordinates like the corresponding two-dimensional rotation about the origin. For the first coordinate axis, the rotation has equations

$$\begin{aligned} x_1' &= x_1 \\ x_2' &= x_2 \cos\theta - x_3 \sin\theta \\ x_3' &= x_2 \sin\theta + x_3 \cos\theta \end{aligned} \qquad X' = AX \qquad A = \begin{bmatrix} 1 & 0 & 0 \\ 0 & \cos\theta & -\sin\theta \\ 0 & \sin\theta & \cos\theta \end{bmatrix}.$$

The analogous rotations about the second and third axes have matrices

$$\begin{bmatrix} \cos\theta & 0 & -\sin\theta \\ 0 & 1 & 0 \\ \sin\theta & 0 & \cos\theta \end{bmatrix} \qquad \begin{bmatrix} \cos\theta & -\sin\theta & 0 \\ \sin\theta & \cos\theta & 0 \\ 0 & 0 & 1 \end{bmatrix}.$$

For other axes, the matrices have no such obvious form. In practice, if you must compute with the matrix of a rotation, you generally choose one of the coordinate axes as the rotational axis.

It's harder to phrase results about conjugacy and rotations in three dimensions than it was in the plane theory, because the angles are harder to describe. It seems easiest to use the next theorem to refer angle questions back to the planes whose reflections combine to form a rotation.

Theorem 5. Suppose line g is the intersection of planes δ and ε. Consider the rotation $\rho = \sigma_\delta \sigma_\varepsilon$ and an isometry φ. Then $\varphi\rho\varphi^{-1}$ is the rotation $\sigma_{\varphi[\delta]}\sigma_{\varphi[\varepsilon]}$ with axis $\varphi[g]$. A rotation is conjugate with ρ just in case it's the composition of the reflections across planes ζ and η that form a dihedral angle whose measure is the same as that of one of those formed by δ and ε.

Proof. $\varphi\rho\varphi^{-1} = \varphi\sigma_\delta\sigma_\varepsilon\varphi^{-1} = \varphi\sigma_\delta\varphi^{-1}\varphi\sigma_\varepsilon\varphi^{-1} = \sigma_{\varphi[\delta]}\sigma_{\varphi[\varepsilon]}$. You can check that the stated condition on ζ and η is equivalent to the existence of an isometry φ such that $\varphi[\delta] = \zeta$ and $\varphi[\varepsilon] = \eta$. ♦

When do two reflections commute? What's their composition in that case? To answer those questions, you can prove the following corollary of theorems 2 and 4. It's similar to the solution of the corresponding two-dimensional problem.

Theorem 6. Reflections σ_δ and σ_ε commute just when planes δ and ε are equal or perpendicular. In the former case, $\sigma_\delta \sigma_\varepsilon$ is the identity. In the latter, it's self-inverse but not the identity. Its restriction to a plane perpendicular to its axis is a (two-dimensional) half turn.

A self-inverse three-dimensional rotation that's not the identity is called the *half turn* about its axis g, and it's denoted by σ_g.[2] Several exercises in section 7.7 examine half turns in detail. Exercise 21 derives the equation of σ_g in terms of the equation of g.

7.4 Glide and rotary reflections

Concepts

- Commuting translations and reflections
- Glide reflections
- Commuting rotations and reflections
- Rotary reflections
- Reflections across points

Previous sections have analyzed reflections, and compositions of two reflections—that is, translations and rotations. What kind of isometry is a composition of three reflections? It's either the composition of a translation with a reflection, or of a rotation with a reflection. The next two headings describe these possibilities in turn. Each one asks first, when is the order of these components significant?

Glide reflections

Theorem 1. A translation τ_V commutes with the reflection σ_ε across a plane ε just when $V = O$ or $\vec{OV} \parallel \varepsilon$.

Proof. You can easily verify that $\sigma_\varepsilon \tau_V$ and $\tau_V \sigma_\varepsilon$ coincide under either condition. To show that they differ when neither condition holds, consider what they do to a point on ε. ♦

---

[2] Some authors call σ_g the *reflection across* g.

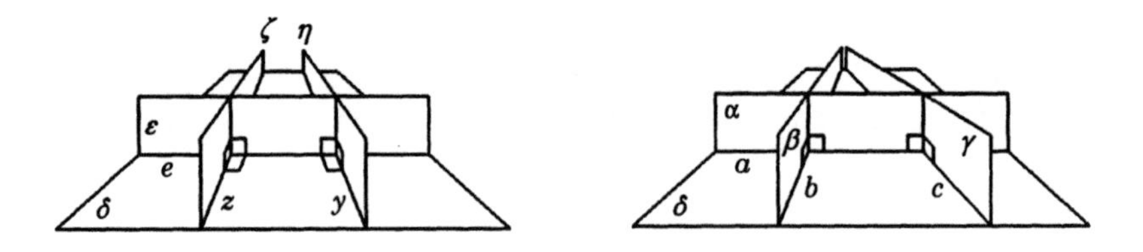

Figure 7.4.1 Proving theorem 2

When $V = O$ or $\vec{OV} \parallel \varepsilon$, the composition $\varphi = \sigma_\varepsilon \tau_V$ is called a *glide reflection*. The translational component is uniquely determined by φ because $\varphi\varphi = \tau_{2V}$. By decomposing the translational component into reflections across planes ζ and η, you can see that $\varphi = \sigma_\varepsilon \sigma_\zeta \sigma_\eta$ is a glide reflection when $\varepsilon \perp \zeta, \eta$ and $\zeta \parallel \eta$. (See the left-hand diagram of figure 7.4.1—ignore plane δ.) The reflectional component of φ is unique, too, because its mirror is the set of midpoints $\frac{1}{2}(X + \varphi(X))$ for all X. A glide reflection has no fixpoint unless it's a reflection.

Theorem 2. If planes α, β, and γ have a common perpendicular plane δ, then $\sigma_\alpha \sigma_\beta \sigma_\gamma$ is a glide reflection.

Proof. Let $a = \alpha \cap \delta$, $b = \beta \cap \delta$, and $c = \gamma \cap \delta$, as in the right-hand diagram of figure 7.4.1. By theorem 7.2.1, each of σ_α, σ_β, and σ_γ fixes δ, and they operate on a point X in δ in the same way as the two-dimensional reflections of δ across a, b, and c, respectively. Thus, $\varphi = \sigma_\alpha \sigma_\beta \sigma_\gamma$ affects X in the same way as the composition of three two-dimensional reflections. By the theory of plane isometries, that's a two-dimensional glide reflection. Therefore, there exist lines e, z, and y in δ such that $e \perp z$, $e \perp y$, and φ operates on points in δ in the same way as the composition of its two-dimensional reflections across e, z, and y. Construct planes ε, ζ, and η perpendicular to δ through e, z, and y, as in the left-hand diagram of figure 7.4.1. Each of σ_ε, σ_ζ, and σ_η fixes δ, and they affect points in δ in the same way as the two-dimensional reflections of δ across e, z, and y, respectively. That is, $\varphi(X) = \sigma_\varepsilon \sigma_\zeta \sigma_\eta (X)$ for all X in δ. Thus, the even isometry $\sigma_\eta \sigma_\zeta \sigma_\varepsilon \varphi$ fixes each point in δ. By theorem 7.2.3, it's the identity, so $\varphi = \sigma_\varepsilon \sigma_\zeta \sigma_\eta$. This is a glide reflection, since $\sigma_\zeta \sigma_\eta$ is a translation τ_V with $V = O$ or $\vec{OV} \parallel \varepsilon$. ♦

Rotary reflections and inversions

Next, consider compositions of *rotations* and reflections. When is the order of the components significant? The following theorem says when it's *not*. Its proof, similar to that of theorem 1, is exercise 7.9.2.

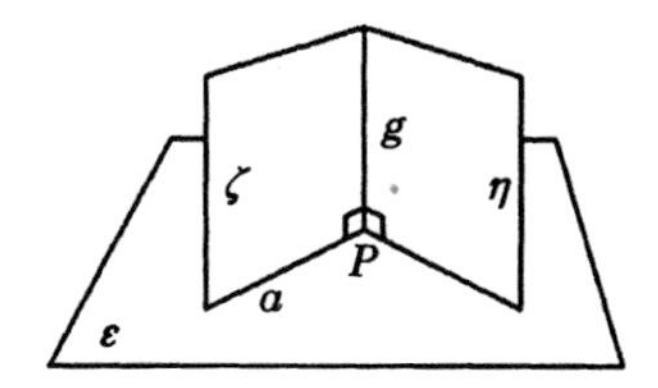

Figure 7.4.2
Proving theorem 4

Theorem 3. A rotation ρ commutes with a reflection σ_ε just when $\rho = \iota$, or its axis g is perpendicular to ε, or it's a half turn about a line in ε.

When $\rho = \iota$ or ρ is a rotation with axis perpendicular to ε, the composition $\varphi = \sigma_\varepsilon \rho$ is called a *rotary reflection*. By decomposing the rotational component into reflections across planes ζ and η, you can see that $\varphi = \sigma_\varepsilon \sigma_\zeta \sigma_\eta$ is a rotary reflection when $\varepsilon \perp \zeta, \eta$ and ζ intersects η, as in figure 7.4.2. Clearly, φ has either a plane ε of fixpoints, or exactly one fixpoint P, called the *center* of φ. In the first case, $\varphi = \sigma_\varepsilon$. In the second, if φ isn't self-inverse, it has exactly one fixed plane ε —its *mirror*—which consists of all points of the form $\frac{1}{2}(X + \varphi(X))$. In that case, φ also determines its rotational component uniquely, since $\rho = \sigma_\varepsilon \varphi$.

However, a self-inverse rotary reflection has no unique mirror or rotational component. This situation occurs when $\zeta \perp \eta$ in figure 7.4.2. Then, φ is a composition of reflections across three mutually perpendicular planes. It's called the *reflection* σ_P *across center* $P = g \cap \varepsilon$.[3] The next result justifies that terminology—the point reflection depends only on P, not on g or ε. You can supply the proof. See exercises 7.7.7, 7.7.8, and 7.7.10 for a more detailed study of point reflections.

Theorem 4. The reflection σ_P across a point P is self-inverse, and $P = \frac{1}{2}(X + \varphi(X))$ for every point X, so $\varphi(X) = 2P - X$. If ε is any plane through P and g is the line through P perpendicular to ε, then $\sigma_P = \sigma_\varepsilon \sigma_g$.

The next result shows that a composition of three reflections is a rotary reflection, unless theorem 2 has classified it already as a glide reflection.

Theorem 5. If planes α, β, and γ have no common perpendicular plane, then $\sigma_\alpha \sigma_\beta \sigma_\gamma$ is a rotary reflection.

Proof. You can show that $h = \beta \cap \gamma$ is a line. Erect plane $\delta \perp \alpha$ through h, then find plane η through h such that $\sigma_\beta \sigma_\gamma = \sigma_\delta \sigma_\eta$ (by making two

[3] Some authors call σ_P the *inversion* with center P, but that usage conflicts with the notion of inversion with respect to a circle or sphere.

dihedral angles have equal measure). Erect plane $\varepsilon \perp \eta$ through the line $a = \alpha \cap \delta$, then find plane $\zeta \perp \varepsilon$ through a. (See figure 7.4.2.) It follows that $\sigma_\alpha \sigma_\beta \sigma_\gamma = \sigma_\alpha \sigma_\delta \sigma_\eta = \sigma_a \sigma_\eta = \sigma_\varepsilon \sigma_\zeta \sigma_\eta$. This isometry is a rotary reflection because $\sigma_\zeta \sigma_\eta$ is a rotation with axis $g = \zeta \cap \eta$ perpendicular to ε. ♦

7.5 Classifying isometries

Concepts
Commuting translations and rotations
Screws
Classification

According to the structure theorem, each isometry is the identity, a reflection across a plane, or a composition of reflections across two, three, or four planes. In section 7.2 you studied reflections. Section 7.3 was devoted to compositions of two reflections: the translations and rotations. And section 7.4 described compositions of three: the glide and rotary reflections. This section covers the last case, compositions of four reflections. Then it summarizes the entire classification scheme.

Screws

A composition of four reflections could be a translation or a rotation. But there's one more type, whose description is facilitated by the next theorem. Its proof, which resembles those of theorems 7.4.2 and 7.4.4, is exercise 7.9.3.

Theorem 1. A translation τ_V commutes with a rotation ρ just when $V = O$, $\rho = \iota$, or $\overleftrightarrow{OV}$ is parallel to the axis g of ρ.

When τ_V commutes with ρ, the composition $\varphi = \rho\tau_V$ is called a *screw*. A screw determines its rotational and translational components ρ and τ_V uniquely. If it's neither a rotation nor a translation, then φ has no fixpoints but exactly one fixed line g, $V = \varphi(P) - P$ for any point P on g, and $\rho = \varphi\tau_V^{-1}$.

The next result seems like a special case, but Hermann Wiener showed in 1890[4] that it provides the key to identifying any composition of four reflections.

---

[4] Wiener 1890–1893, part II.

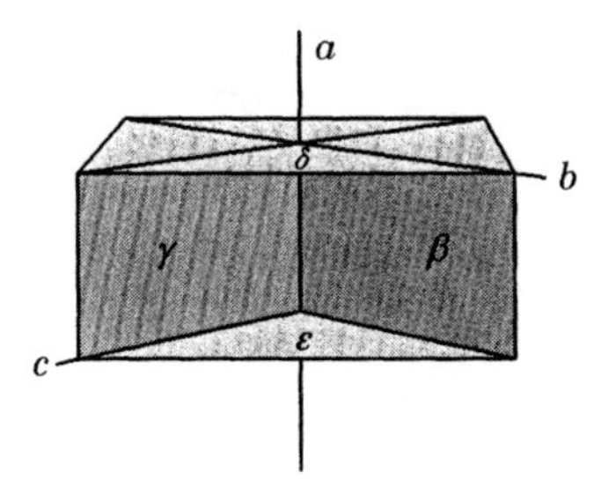

Figure 7.5.1
Proving lemma 2

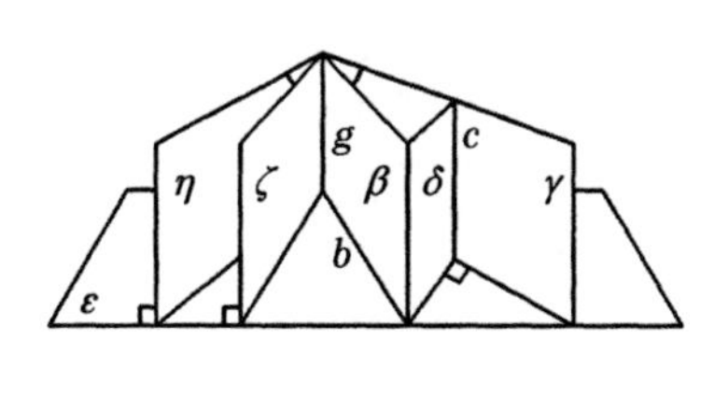

Figure 7.5.2
Proving theorem 3

Lemma 2. For any lines b and c, the composition $\varphi = \sigma_b \sigma_c$ is a screw.

Proof. This argument is designed to apply when b and c are skew (see figure 7.5.1). You can verify that it also works when they're coplanar. (You'll need additional figures for the parallel and intersecting cases.) First, find a line a perpendicular to b and c, plane β through a and b, and plane γ through a and c. Erect planes $\delta \perp \beta$ through b and $\varepsilon \perp \gamma$ through c. Then $\delta \perp \gamma$ also, so $\sigma_b \sigma_c = (\sigma_\beta \sigma_\delta)(\sigma_\gamma \sigma_\varepsilon) = (\sigma_\beta \sigma_\gamma)(\sigma_\delta \sigma_\varepsilon)$. This is a screw because $\sigma_\beta \sigma_\gamma$ is a rotation with axis a and $\sigma_\delta \sigma_\varepsilon$ is a translation parallel to a. ♦

Theorem 3. Every even isometry φ is a screw.

Proof. Suppose φ is neither a translation nor a rotation. Then there exist distinct points P and Q such that $\varphi(P) = Q$. Let δ be the perpendicular bisector of segment $\overline{PQ}$, so that $\varphi\sigma_\delta(Q) = Q$. Since $\varphi\sigma_\delta$ is odd, it's a composition of three reflections. Since it has a fixpoint, it's a rotary reflection, and $\varphi\sigma_\delta = \sigma_\varepsilon \sigma_\zeta \sigma_\eta$ for some planes ε, ζ, and η such that $\varepsilon \perp \zeta, \eta$. The intersection $g = \zeta \cap \eta$ is a line; otherwise, $\varphi\sigma_\delta = \sigma_\varepsilon$, and φ would equal $\sigma_\varepsilon \sigma_\delta$, a rotation or translation. Erect the plane $\gamma \perp \delta$ through g, as shown in figure 7.5.2. Find the plane β through g such that $\sigma_\zeta \sigma_\eta = \sigma_\gamma \sigma_\beta$ (by making two dihedral angles have equal measure). Then $\varepsilon \perp \beta$ also, hence $\varphi = \varphi\sigma_\delta\sigma_\delta = \sigma_\varepsilon \sigma_\zeta \sigma_\eta \sigma_\delta = \sigma_\varepsilon \sigma_\gamma \sigma_\beta \sigma_\delta = \sigma_c \sigma_b$, where $b = \beta \cap \delta$ and $c = \gamma \cap \varepsilon$. By lemma 2, this is a screw. ♦

Hermann WIENER was born in 1857 in Karlsruhe. His father Christian Wiener was Professor of Descriptive Geometry at the Technical Institute there. His younger brother Otto became a physicist, noted for work on the wave theory of light.

Wiener began university studies at Karlsruhe, but completed his Ph.D. at München in 1880 with a dissertation on plane curves. His training was influenced significantly by Felix Klein, a close collaborator at München with Wiener's cousin Alexander Brill. In 1880, Klein moved to Leipzig, and Wiener followed. After a year, though, he returned to Karlsruhe as assistant to his father. He continued research, and obtained a lecturer position in the university at Halle in 1885.

There Wiener wrote a series of research papers that clarified the relationships of several fundamental theorems and techniques in Euclidean and projective geometry, including Desargues' theorem and the use of self-inverse transformations. He made the first extensive use of calculations with reflections, like those in this and the preceding chapter. He regarded this calculus as an algebra of *geometric* objects, in contrast to the algebra of *numbers* used in coordinate geometry.

Wiener also pioneered the abstract axiomatic approach to geometry, emphasizing its usefulness, but he never built a complete theory. Hilbert's famous remark that he could regard geometry as a theory of tables, chairs, and bar stools just as well as one of points, lines, and planes stemmed from his excitement after hearing a lecture by Wiener in 1890. That took place in Halle at the first meeting of the German Mathematical Society. Wiener was one of the original members.

In 1894 Wiener became Professor of Mathematics at the technical university in Darmstadt, where he remained until he retired in 1927. Darmstadt lay outside the mainstream, and Wiener never continued his major research work. In 1911 he revised and marketed a set of geometric models originally developed by his father; they came into widespread use in secondary and technical school instruction. Wiener served his university well—he was elected Dean of his faculty for two terms. He died in 1939.

Wiener's results were extended and incorporated into major works on foundations of geometry for the next sixty years. In particular, crucial steps in the work of Hilbert, Thomsen, and the present author were established or proposed originally by Hermann Wiener.

Classification theorem

You've now completed the analysis of compositions of reflections. The following theorem refines and summarizes the classification begun with the structure theorem.

---

[5] Wiener 1890–1893.

[6] See Reid 1970, chapter 8.

Theorem 4 (Classification theorem). A three-dimensional isometry is

(0) the identity ι, or
(1) the reflection σ_ε across a plane ε, or
(2a) a translation τ_V with vector $V \neq O$, or
(2b) a rotation $\rho \neq \iota$, or
(3a) a glide reflection $\sigma_\varepsilon \tau_V$ with mirror ε and vector $V \neq O$, or
(3b) a reflection across a point P, or
(3c) a rotary reflection $\sigma_\varepsilon \rho$ with mirror ε and rotational component ρ that's not self-inverse, or
(4) a screw $\rho \tau_V$ with rotational component $\rho \neq \iota$ and vector $V \neq O$.

These classes are disjoint.

You might like to provide a table of fixed points, lines, and planes for each of the eight isometry types listed in the classification theorem, analogous to those after corollary 6.9.2 and in exercise 6.11.12.

Finding fixpoints, fixed lines, and fixed planes of an isometry $\varphi : X \to X'$ when you know its equation $X' = AX + B$ is a substantial exercise in solving possibly singular linear systems. You'll find examples under the heading **Exercises on equations** in section 7.7. The practical details of those computations tend to differ considerably from those of the theory presented so far. They're outlined in some paragraphs that precede the exercises.

Section 6.9 applied the classification of plane isometries to describe rotations and translations in more detail. It also pursued the notion of conjugacy beyond the level achieved by theorems 7.2.5, 7.3.3, and 7.3.5. For analogous three-dimensional studies, see the **Conjugacy exercises** heading in section 7.7.

7.6 Similarities

Concepts
Complete analogy with the two-dimensional theory
Hometheties and homothetic rotations
Classification theorem

You could really write this section on three-dimensional similarities yourself. Their theory is entirely analogous to its plane counterpart. Except for its main result, the classification theorem, this section just makes the necessary adjustments to the definitions and theorem statements. You can adapt every proof in section 6.10 to work in three dimensions.

A *similarity with ratio* r is a transformation $\psi: P \to P'$ of the set of all points, such that $P'Q' = r\,(PQ)$ for all P and Q. As an example, select any point Z and consider the transformation $\varphi_{r,Z}: P \to P'$ such that $Z' = Z$ and for $P \neq Z$, P' is the point on $\vec{ZP}$ with $ZP' = r\,(ZP)$. If $r = 1$, then $\varphi_{r,Z}$ is the identity; if $r < 1$ or $r > 1$, it's called a *contraction* or *dilation about* Z.

A similarity ψ: $X \to X'$ with ratio r leaves invariant angle measure, betweenness, and any geometric property that you can define in terms of them. In particular, ψ is a collineation, and the image of a plane is a plane. The image of a circle or sphere with radius t is a circle or sphere with radius rt. That of a triangle with area s is a similar triangle with area $r^2 s$. And the image of a tetrahedron with volume v is a similar tetrahedron with volume $r^3 v$.

A composition of similarities with ratios r and s is a similarity with ratio rs. The inverse of a similarity with ratio r is a similarity with ratio r^{-1}. Therefore, the similarities form a transformation group that contains the isometries.

The section 6.10 proof of the fixpoint theorem was chosen specifically to apply to the three-dimensional case as well: Unless a similarity ψ is an isometry, it has exactly one fixpoint.

The uniqueness theorem takes a slightly different form in three dimensions: Similarities χ and ψ coincide if $\chi(X) = \psi(X)$ for *four noncoplanar* points X. This follows from the fixpoint theorems, since $\chi\psi^{-1}$, a similarity with four noncoplanar fixpoints, must be an isometry, hence the identity.

A similar argument shows that every similarity $\psi: X \to X'$ is the composition of an isometry and a dilation or contraction with the same ratio r and arbitrarily chosen center, so that ψ has an equation $X' = rAX + B$ with an orthogonal matrix A. Moreover, ψ determines A and B uniquely, and every such equation defines a similarity. Since ψ preserves orientation just in case the isometry $X \to AX$ does, ψ is termed *even* or *odd* depending on whether $\det A = +1$ or -1.

The similarities $\varphi_{r,Z}$ and $\varphi_{r,Z}\,\sigma_Z$ are called *homotheties* with center Z. A composition of one of these with a rotation about an axis through Z is called a *homothetic rotation*. If ψ is a similarity but not an isometry, then it has a fixpoint Z. Define $\chi = \psi\varphi_{r^{-1},Z}$ or $\chi = \psi\sigma_Z\varphi_{r^{-1},Z}$ depending on whether ψ is even or odd, so that χ is an even isometry with fixpoint Z and $\psi = \chi\varphi_{r,Z}$ or $\chi\varphi_{r,Z}\sigma_Z$. By theorem 7.5.4, χ is a rotation about an axis through Z, so ψ is a homothetic rotation. This proves

***Theorem 1* (*Classification theorem*).** A similarity is an isometry or a homothetic rotation.

7.7 Exercises

Concepts

Distance-preserving functions
Dilatations
Compositions of commuting point and plane reflections and half turns
Conjugates
Compositions of reflections, translations, and rotations
Classifying isometries and similarities, given their equations
Finding fixpoints and fixed lines of isometries, given their equations
Reflection calculus
Three-reflections theorems

This section contains thirty-two exercises related to the material covered earlier in this chapter. The first three ask for some straightforward proofs omitted earlier. Several others, listed under the headings **Commuting reflections and half turns**, **Conjugacy exercises**, and **Classifying isometries**, extend the chapter's methods beyond the essentials covered earlier, and provide a nearly complete treatment of the interrelationships among reflections, translations, and rotations. The conjugacy exercises firmly establish results you'll often find used without mention in applications. The exercises under the heading **Exercises on equations** familiarize you with the algebraic and numerical computations often required to apply the theory. Some require tedious arithmetic to evaluate determinants and solve linear systems. There was no effort to avoid that. You're expected to use mathematical software for that work. Finally, exercises 5 and 6 and those under the heading **Reflection calculus** provide excursions into areas less central to the program of this book. You've already met their two-dimensional counterparts in chapter 6.

General exercises

Exercises 1 to 3 ask for several proofs omitted from earlier sections of this chapter.

Exercise 1. Prove theorem 7.1.4: The only isometry with four noncoplanar fixpoints is the identity. *Suggestion*: Imitate the proof of theorem 6.2.7.

Exercise 2. Prove theorem 7.4.3: A rotation ρ commutes with a reflection σ_ε just when $\rho = \iota$, or its axis g is perpendicular to the mirror ε, or it's a half turn about a line in ε. *Suggestion*: Imitate the proof of theorem 7.4.1.

Exercise 3. Prove theorem 7.5.1: A translation τ_V commutes with a rotation ρ just when $V = O$, $\rho = \iota$, or $\vec{OV}$ is parallel to the axis of ρ. *Suggestion*: Imitate the proofs of theorems 7.4.2 and 7.4.4.

Exercise 4 is the three-dimensional analog of exercise 6.11.8.

Exercise 4. Show that an isometry φ is even if and only if $\varphi = \chi\chi$ for some isometry χ.

Exercises 5 and 6 pursue excursions into less central themes, already begun in two-dimensional exercises under the section 6.11 headings **Exercises on the definitions of isometry and similarity** and **Dilatations**. If you've solved those in the most efficient way, you should need little to adapt your work to three dimensions.

Exercise 5. Formulate and prove three-dimensional versions of exercises 6.11.40–6.11.41 on the definitions of isometry and similarity.

Exercise 6. Formulate and prove three-dimensional analogs of exercises 6.11.42–6.11.45 on dilatations.

Commuting reflections and half turns

The conditions in plane geometry under which reflections and half turns commute—the nontrivial self-inverse isometries—were laid out at the end of section 6.3. Those ideas played essential roles in the rest of that chapter. In three-dimensional geometry there are *three* types of nontrivial self-inverse isometries. A complete treatment of their commutativity would have clogged the earlier sections of this chapter unacceptably, so they covered only a few aspects. In fact, the theory in this chapter was streamlined to allow that omission. But its *applications* require *all* the commutativity details, so they're presented here systematically, as exercises.

Exercise 7. Show that the composition $\sigma_P\sigma_Q$ of two point reflections is the translation with vector $2(P - Q)$, and they don't commute unless $P = Q$.

Exercise 8. Show that point and plane reflections σ_P and σ_ε commute just when P lies on ε, and in that case $\sigma_P\sigma_\varepsilon$ is the half turn about the line $g \perp \varepsilon$ through P.

Exercise 9. Show that

Part 1. a half turn σ_g commutes with a plane reflection σ_ε if and only if g lies in ε or $g \perp \varepsilon$;

Part 2. if g lies in ε, then $\sigma_g \sigma_\varepsilon$ is the reflection across the plane perpendicular to ε through g;

Part 3. if $g \perp \varepsilon$, then $\sigma_g \sigma_\varepsilon$ is the reflection across $g \cap \varepsilon$.

Exercise 10. Show that a point reflection σ_P and a half turn σ_g commute if and only if P lies on g, and in that case $\sigma_P \sigma_g$ is the reflection in the plane $\varepsilon \perp g$ through P.

Exercise 11. Show that half turns σ_g and σ_h about distinct lines g and h commute if and only if $g \perp h$, and in that case, $\sigma_g \sigma_h$ is the reflection across the intersection of g and h.

Conjugacy exercises

The conjugacy classes of two-dimensional isometries were described in complete detail under that heading in section 6.9. That information is important for applications, because authors often handle, for example, all 60° rotations interchangeably without mentioning explicitly the conjugacy relation that connects them. The same is true in three-dimensional applications. Earlier sections of chapter 7 contained the most essential conjugacy results: the general ideas near the end of section 7.1, and theorems 7.2.5, 7.3.3, and 7.3.5. Further results were omitted because their proofs are unwieldy and repetitive. They're offered here as exercises.

Exercise 12 shows that, like the plane reflections, the half turns and point reflections each constitute single conjugacy classes.

Exercise 12. Show that $\varphi \sigma_P \varphi^{-1} = \sigma_{\varphi(P)}$ and $\varphi \sigma_g \varphi^{-1} = \sigma_{\varphi[g]}$ for any point P, any line g, and any isometry φ. *Suggestion*: Consider fixpoints!

Exercise 13. Describe the conjugacy class of a glide reflection.

Exercise 14. Describe the conjugacy class of a rotary reflection.

Exercise 15. Describe the conjugacy class of a screw.

Classifying isometries

Earlier in this chapter, a glide reflection was defined as a composition of a translation and a commuting plane reflection, a rotary reflection as a composition of a rotation and a commuting plane reflection, and a screw as a composition of a translation and a commuting rotation. With these notions, you can classify *all* isometries. Exercises 16 to 18 give you some practice. What kinds of isometries are the analogous compositions whose components *don't* commute?

Exercise 16. What's the composition of a translation and a plane reflection when they don't commute?

Exercise 17. What's the composition of a rotation and a plane reflection when they don't commute?

Exercise 18. What's the composition of a translation and a rotation when they don't commute?

Exercises 16 to 18 don't cover the *whole* subject of commutativity. When do isometries of the various *other* types commute, and what are their compositions? Most of the combinations are too rare to warrant special treatment—except one: the composition of two rotations. In two dimensions its description is simple; see theorem 6.9.3. In three, that's not true unless the rotations commute. Exercise 19 investigates the situation.

Exercise 19. When do two rotations commute? What's their composition in that case and in general?

Exercises on equations

As in two-dimensional transformational geometry, the theory of isometries in three dimensions seems very different from the practice of computing with them. That's the effect of streamlining to present the essential details of the theory efficiently, in a form that's easy to remember. The exercises under this heading present the computational details. In the first two, closely related to two-dimensional exercise 6.11.25, you derive equations for plane reflections and half turns in terms of the equations of their mirrors and axes.

Exercise 20. Consider a point $P = \langle p_1, p_2, p_3 \rangle \neq O$ and the plane $\delta \perp \vec{OP}$ through O. Show that σ_δ has matrix

$$A = I - \frac{2}{P^t P} K,$$

where $K = PP^t$. Now consider the plane $\varepsilon \perp \vec{OP}$ through a point Q. Show that $\sigma_\varepsilon : X \to X'$ has equation

$$X' = AX + \frac{2}{P^t P} KQ.$$

Exercise 21. Find an equation for the half turn $X \to X'$ about the line g with parametric equations $X = tP + Q$. *Suggestion*: Use the matrix $K = PP^t$ introduced in exercise 20.

In the next two exercises, you start from geometric descriptions of certain isometries and find their equations. These involve very tedious computations. Use mathematical software wherever possible!

Exercise 22. Find equations for

(a) the reflection across the plane ε with equation $x_1 + 2x_2 + 3x_3 = 4$;

(b) the glide reflection with mirror ε and vector $<3, -2, 1/9>$;

(c) the half turn about the line through points $<1, 1, 2>$ and $<2, 1, 1>$.

Exercise 23. Consider the plane ε in exercise 22, its intersections P, Q, and R with the coordinate axes, and the line $g \perp \varepsilon$ through Q. Find equations for

(a) the rotation $\rho : X \to X'$ about g that maps $\vec{QR}$ to $\vec{QP}$, with coefficients accurate to four decimal places;

(b) the rotary reflection $\rho\sigma_\varepsilon$.

Suggestion: The equation has the form $X' = AX + B$ for some matrix A and some vector B. Find four points X, no three of which are coplanar, for which you know X'. From the four equations $X' = AX + B$ construct a system of twelve linear equations for the twelve entries of A and B, then solve it.

The final two exercises under this heading ask you to classify an isometry or similarity, given its equation. To *classify* an isometry,

- identify its class among those listed in the classification theorem;
- find an equation for its mirror—if it has one;
- find its vector—if it has one;
- find an equation for its axis—if it has one—and specify a vector to indicate a positive ray, then find a corresponding angle parameter.

To complete a classification, you may need to find fixpoints or fixed lines of an isometry $\varphi : X \to X' = AX + B$. For a fixpoint, you'll have to solve the linear system $X = X'$ —that is, $X = AX + B$, or $(I - A)X = B$. If $I - A$ is invertible, there's just one fixpoint X. Otherwise, reduce the system to an equivalent one for which you can identify the solutions readily. There are four cases. You might

- derive an *inconsistent* equation equivalent to $0 = 1$ —there's *no* fixpoint;
- reduce the system to three *trivial* equations, each equivalent to $0 = 0$ —*every* point is fixed;

- reduce it to two trivial equations and a nontrivial one that identifies a *plane* of fixpoints, with equation $c_1x_1 + c_2x_2 + c_3x_3 = d$;
- reduce it to one trivial equation and two equations

$$\begin{cases} x_i - c_i x_k = d_i \\ x_j - c_j x_k = d_j \,, \end{cases}$$

where $\{i, j, k\} = \{1, 2, 3\}$ —you can convert those to parametric equations for a *line* of fixpoints:

$$\begin{cases} x_i = t\,c_i + d_i \\ x_j = t\,c_j + d_j \\ x_k = t \cdot 1 + 0\,. \end{cases}$$

To find fixed lines g of an isometry φ, note that φ maps points $tP+Q$ with $P \neq O$ to points

$$A(tP+Q)+B = tP^* + Q',$$

where $P^* = AP$ and $Q' = AQ+B$. Thus φ maps the line $g \parallel \vec{OP}$ through Q onto a line $g' \parallel \vec{OP}^*$ through Q'. It follows that $g \parallel g'$ just in case P^* is a scalar multiple of P. But φ preserves length, so $g \parallel g'$ just in case $P^* = \pm P$. The first step in finding fixed lines is to find all vectors P such that $P^* = \pm P$. Solve the linear systems $AP = \pm P$ —that is, $(A \pm I)P = O$. Second, find the points Q for which $g = g'$. For each P, both Q and Q' must lie on g, so

$$Q' - Q = AQ + B - Q = (A - I)Q + B = tP$$

for some t. You must solve the system

$$(I - A)Q + tP = B$$

of three linear equations in four unknowns: t and the components of Q.

That general procedure for finding fixed lines may be more involved than what you really need. For these classification exercises, your main fixed-line problem is probably to find the axis g of a screw $\varphi: X \to X' = AX + B$ that has no fixpoint. Here's a neater alternative for that special situation. Since φ affects points X on g like a translation with vector V,

$$\begin{aligned} &X' - X = V = (X')' - X', \\ &AX + B - X = [A(AX+B) + B] - [AX + B] \\ &(A^2 - 2A + I)X = (I - A)B \\ &(I - A)^2 X = (I - A)B. \end{aligned}$$

You can find the points X on g by solving that singular linear system!

You can recognize a similarity $X \to AX + B$ with ratio r by computing $AA^t = r^2 I$; then classify the corresponding isometry $X \to r^{-1}AX + r^{-1}B$.

These exercises are stated in terms of exact fractions. To get *exact* solutions, you'll need mathematical software that operates with exact fractions.[7] You *won't* be able to carry out the computations by hand or by calculator with decimal arithmetic.

You *could* convert each fraction in an exercise to a decimal approximation, and use decimal arithmetic. Do some exercises that way, too! Carry as many decimal digits as your calculator permits. You'll have to make certain decisions in the classification procedure by testing whether some computed number $d=0$. With approximate decimal arithmetic, you'll never get $d=0$ *exactly*. So you'll have to decide how close to zero is close enough. In practice, that can be a critical decision. It's important enough in some areas of computer graphics that it's worth covering in more detail *there*. But *these* exercises provide exact fractions so you can clearly relate theory to computation without worrying about approximation errors that depend on your equipment.

Exercise 24. For these matrices A verify that the equation $X' = AX$ defines an isometry or similarity $\varphi: X \to X'$, and classify it.

$$\text{(a)}\quad \begin{bmatrix} \frac{6}{7} & -\frac{2}{7} & -\frac{3}{7} \\ -\frac{2}{7} & \frac{3}{7} & -\frac{6}{7} \\ -\frac{3}{7} & -\frac{6}{7} & -\frac{2}{7} \end{bmatrix} \qquad \text{(b)}\quad \begin{bmatrix} \frac{12}{49} & \frac{31}{49} & -\frac{36}{49} \\ \frac{24}{49} & -\frac{36}{49} & -\frac{23}{49} \\ -\frac{41}{49} & -\frac{12}{49} & -\frac{24}{49} \end{bmatrix}$$

$$\text{(c)}\quad \begin{bmatrix} -\frac{9518}{9555} & \frac{131}{1911} & \frac{526}{9555} \\ -\frac{142}{1911} & -\frac{1894}{1911} & -\frac{211}{1911} \\ -\frac{449}{9555} & \frac{218}{1911} & -\frac{9482}{9555} \end{bmatrix} \qquad \text{(d)}\quad \begin{bmatrix} -6 & 2 & 3 \\ 2 & -3 & 6 \\ 3 & 6 & 2 \end{bmatrix}.$$

Exercise 25. For these matrices A and vectors B verify that the equation $X' = AX + B$ defines an isometry or similarity $\varphi: X \to X'$, and classify it.

(a) Same A as exercise 24(a), $B = \langle 4/7, 8/7, 12/7 \rangle$;

(b) Same A as exercise 24(a), $B = \langle 5, 6, 7 \rangle$;

(c) Same A as exercise 24(b), $B = \langle 4/7, 8/7, 12/7 \rangle$;

(d) Same A as exercise 24(c), $B = \langle 968/195, 88/39, -616/195 \rangle$;

$$\text{(e)}\quad A = \begin{bmatrix} -\frac{9518}{9555} & -\frac{131}{1911} & \frac{526}{9555} \\ -\frac{142}{1911} & \frac{1894}{1911} & -\frac{211}{1911} \\ -\frac{449}{9555} & -\frac{218}{1911} & -\frac{9482}{9555} \end{bmatrix} \qquad B = \langle 968/195, 88/39, -616/195 \rangle;$$

[7] X(*Plore*) (Meredith 1993) is a good choice.

(f) $A = \begin{bmatrix} -9518 & -655 & 526 \\ -710 & 9470 & -1055 \\ -449 & -1090 & -9482 \end{bmatrix}$ $B = \langle 968, 440, -616 \rangle$.

Reflection calculus

Just as in two dimensions, geometers have developed approaches to three-dimensional geometry that rely very heavily on algebraic computations with compositions of self-inverse isometries—plane and point reflections and half turns. Because of the greater variety, these approaches are not so elegant as their plane counterparts. The *three-reflections theorems* provide basic tools in both two and three dimensions. The two-dimensional results are theorems 6.4.7, 6.5.6, and 6.7.2, and exercise 6.11.32. Their analogs haven't appeared yet in this more streamlined chapter. They're presented as exercises 26, 27, and 29. The intervening exercise 28 is an extension of lemma 7.5.2 that you'll need for exercise 29.

Exercise 26. Show that for any points P, Q, and R, the composition $\sigma_P \sigma_Q \sigma_R$ is the reflection across point $S = P - Q + R$, and $PQRS$ is a (possibly degenerate) parallelogram.

Exercise 27. Show that the composition φ of the reflections across three planes is self-inverse in just three cases:

(a) they're all perpendicular to a single line g, or
(b) their intersection contains a line g, or
(c) they're mutually perpendicular.

Also show that in case

(a) φ is the reflection across a plane perpendicular to g, and
(b) φ is the reflection across a plane through g, and
(c) φ is the reflection across the intersection of the three planes.

Exercise 28. According to lemma 7.5.2, for any lines b and c the composition $\sigma_b \sigma_c$ is a screw. Fill in the blanks and prove this extension of the lemma:

- $\sigma_b \sigma_c = \iota$ if and only if ...;
- it's a translation if and only if ...;
- it's a rotation if and only if ...;
- if it's not a translation, then its axis is the line that's perpendicular to b and c, and its angle parameter is ...;
- if it's not a rotation, then its vector is ....

Exercise 29. Show that the composition φ of the half turns about three lines is self-inverse in just three cases:

(a) the lines are mutually perpendicular, or
(b) some pair of them is parallel, or
(c) they have a common perpendicular line m.

Show also that in case

(a) $\varphi = \iota$,
(b) φ is the half turn about a line l,
(c) φ is the half turn about a line $l \perp m$.

Exercise 30 is the three-dimensional analog of exercise 6.11.9.

Exercise 30. Show that every isometry φ is the composition of two self-inverse isometries. *Suggestion*: Consider separately the various types of isometry.

Working in the 1920s and early 1930s on the two-dimensional reflection calculus—see that heading in section 6.11—Gerhard Thomsen evidently considered analogous three-dimensional questions. Thomsen's Ph.D. student H. Boldt developed that material into a full theory, published as his dissertation in 1934. It isn't as clean as its two-dimensional counterpart. That's understandable; three-dimensional theorems are generally more complicated. As you've seen in previous exercises on commuting reflections and on the three-reflections theorems, enlarging the class of nontrivial self-inverse isometries from two types to three brings minor complications to almost every elementary theorem. Consequently, even only slightly less elementary results become unwieldy when stated solely in terms of reflections and half turns. Exercises 31 and 32 are taken from Boldt 1934. They're closely related to the two-dimensional reflection calculus exercises in section 6.11.

Exercise 31. Describe the figure formed by points A, B, C, and G for which $\sigma_A \sigma_G \sigma_B \sigma_G \sigma_C \sigma_G = \iota$. Find an analogous result with points A, B, C, D, and G. *Suggestion*: Reason with vectors.

Exercise 32, Part 1. Under what conditions on lines g and h is $\sigma_g \sigma_h \sigma_g \sigma_h = \iota$?

Part 2. Under what conditions on a line g and points P and Q is $\sigma_g \sigma_P \sigma_Q \sigma_g \sigma_Q \sigma_P = \iota$?

Part 3. Under what conditions on a line g and points P and Q is $\sigma_g \sigma_P \sigma_Q \sigma_g \sigma_P \sigma_Q = \iota$?

Part 4. Under what conditions on lines g and h do there exist points P and Q such that $\sigma_g \sigma_h = \sigma_P \sigma_Q$?

Part 5. Under what conditions on lines g and h do there exist points P and Q such that $\sigma_g \sigma_h \sigma_g \sigma_h = \sigma_P \sigma_Q$?

Suggestion: Exercise 28 is helpful.

Chapter

8

Symmetry

Artists and artisans have always exploited symmetry in constructing attractive and efficient designs. This chapter's title, a self-referential calligraphic design by Scott Kim,[1] displays twofold rotational symmetry. Turn the book 180° and admire! The design in figure 8.0.1, from the Mimbres culture 1250 years ago in New Mexico, displays fourfold rotational symmetry. It adorns a medium with a practical aspect: the interior of a water bowl.[2] You saw a frieze pattern in figure 6.0.3 that displayed the same rotational symmetry as this chapter's title. Frieze patterns are designed to be repeated indefinitely to decorate edges and ribbons. You can also wind a frieze around an object, overlaying repeated parts appropriately, to get a pattern on the inside or outside of a circular band. Wound about a bowl, the figure 6.0.3 frieze would resemble figure 8.0.1.

The frieze pattern in figure 8.0.2 was used about two hundred years ago on pottery from Tesuque Pueblo, about 270 miles from the Mimbres region.[3] This one combines repetitive and reflectional symmetry; it appears the same upside down. But it has no rotational symmetry.

Figures 8.0.3 and 8.0.4 show Kin Kletso, one of the smaller villages in the Chaco Canyon area of New Mexico that were occupied around A.D. 1050–1175. Its architecture uses repetition in two perpendicular directions, perhaps like that of the office building or neighborhood you're working in.

---

1 From Kim 1981, 28. See also the signature concluding the preface to the present book.

2 From Brody, Scott, and LeBlanc 1983, plate 37. Most Mimbres pottery has been recovered from graves. Mimbres funeral ceremonies evidently included punching a hole in the bottom of each bowl to be buried.

3 From Chapman 1970, plate 66a. You'll see this design again as a circular band pattern on a jar in figure 8.5.6.

Figure 8.0.1 Mimbres bowl with fourfold rotational symmetry

Figure 8.0.2 Tesuque frieze ornament

Parts of Chaco structures once repeated this design in *three* perpendicular directions; grids of rooms were stacked floor upon floor. The numerals in figure 8.0.4 indicate the number of floors. You'll see reflectional and rotational aspects in this grid, as well as three circular *kivas* for rituals, placed in the grid with little regard to its symmetry. The overall design is at once utilitarian, ceremonial, and delightful and satisfying for the onlooker.

Nature doesn't require human intervention to create symmetry. Many animals and flowers display bilateral or rotational symmetry. Figure 8.0.5 shows the structure of Devils Postpile in California's Sierra Nevada. Its hexagonally symmetric basalt columns fit together precisely, not because some hand stacked them like cells in a beehive, but because they pulled apart as an ancient lava flow cooled and contracted. Gravity and exposure to cool air and substratum affect the lava flow's behavior in the vertical direction. But if its horizontal plane sections are perfectly homogeneous, they should pull apart into regular shapes that fit together to fill the planes: equilateral triangles, squares, or regular hexagons. Cooling tends to shrink their perimeters. Of those types of figures with a given area, the hexagon has the shortest perimeter. Figure 8.0.5 and the top view in figure 8.0.6 show many basalt columns that are approximately regular hexagonal prisms.

We like to stack other objects, too, such as the balls in figure 8.0.7. Can you visualize their repetition in three noncoplanar directions? Can you see the rotational and reflectional symmetry? Can you imagine each ball contained in a polyhedral cell, so that the cells are stacked face to face with no space between? Without our assistance, nature stacks molecules in crystals.

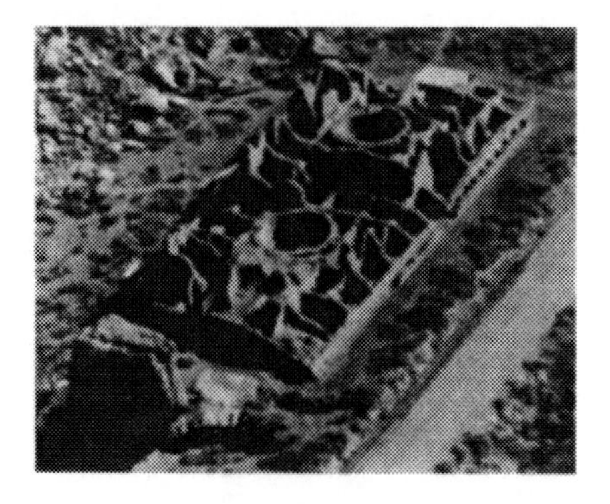

Figure 8.0.3 Aerial photograph of Kin Kletso[4]

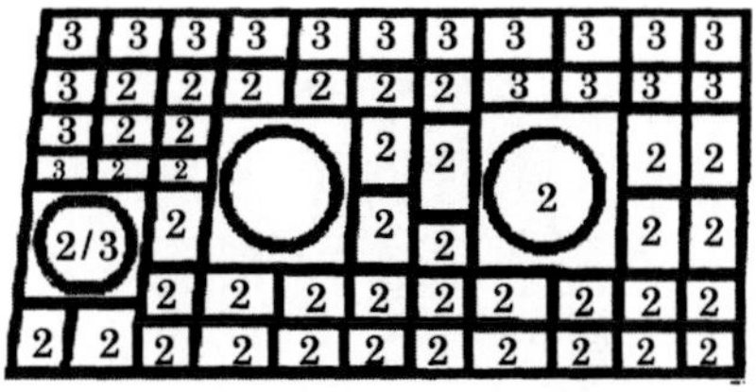

Figure 8.0.4 Kin Kletso floor plan[5]

Figure 8.0.5 Basalt columns at Devils Postpile[6]

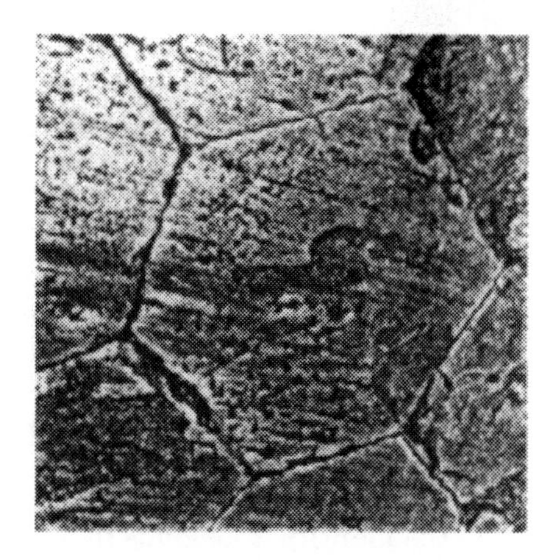

Figure 8.0.6 Top view of columns[6]

---

[4] Lekson [1984] 1986, figure 4.87.

[5] Ibid., figure 4.89. The numerals represent the number of floors of the building at each location.

[6] Photographs by Dr. Wymond Eckhardt.

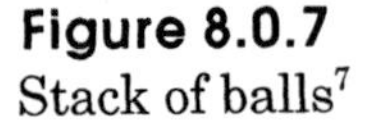

Figure 8.0.7
Stack of balls[7]

Figure 8.0.8 Contemporary Mexican lamp[8]

Polyhedral ornamentation is common; consider the elegant lamp in figure 8.0.8. Try to puzzle out its rotational and reflectional symmetry—the most complicated of all these examples.

This chapter is about geometry, art and design. It analyzes in detail the symmetries of examples like figures 8.0.1 to 8.0.8. The chapter 6 introduction called a figure *symmetric* if it's unchanged by some isometry. Two figures display the same symmetry if they're unchanged by the same isometries. To analyze the symmetry of a figure F, study the isometries that leave it unchanged—they're called its *symmetries*. This is the official definition of that word, and it's consistent with the usage in the sentence before last. An isometry φ is a symmetry of F if and only if $\varphi[F] = F$. According to the definition in section 6.1, the set $\mathcal{G}_F$ of symmetries of F is a subgroup of the group of all isometries:

- the identity ι belongs to $\mathcal{G}_F$ because $\iota[F] = F$;
- $\mathcal{G}_F$ contains the inverse of each of its members, because $\varphi[F] = F$ implies $\varphi^{-1}[F] = \varphi^{-1}\varphi[F] = F$;
- $\mathcal{G}_F$ contains the composition of any two of its members, because $\varphi[F] = F$ and $\chi[F] = F$ imply $\varphi\chi[F] = \varphi[F] = F$.

$\mathcal{G}_F$ is called the *symmetry group* of F.[9]

How do the symmetry groups of the objects in figures 8.0.1 to 8.0.8 differ? They have several distinguishing properties. Perhaps the most evident has

---

[7] Photograph by the present author.

[8] From a Sundance Catalog Company advertisement.

[9] Don't confuse this with the *symmetric* group, which consists of *all* transformations of F!

to do with translations belonging to the groups. Translations τ and υ are called *independent* if for some (hence any) point O the points O, $\tau(O)$, and $\upsilon(O)$ are noncollinear; a translation φ is *independent from* τ and υ if O, $\tau(O)$, $\upsilon(O)$, and $\varphi(O)$ are noncoplanar. The symmetry groups of this chapter's title, and of the Mimbres bowl and Mexican lamp in figures 8.0.1 and 8.0.8 contain no translation at all. The symmetry groups of the friezes in figures 6.0.3 and 8.0.2 contain translations, but no two are independent. A single layer of rectangular rooms, "tiles", or balls in figure 8.0.3, 8.0.6, or 8.0.7 has a symmetry group with two independent translations but not three—provided you regard them as extending indefinitely in two dimensions. The groups of the stacks of rooms or balls or of the Devils Postpile in figure 8.0.5 each contain three independent translations, provided you regard them as extending indefinitely in all three dimensions. You can regard these four classes of symmetry groups as zero-, one-, two-, or three-*dimensional*: They contain no translation, one translation on which all others depend, or two or three independent translations. If you study these examples and figures 6.0.1 and 6.0.4 very carefully, ignoring inconsequential irregularities, you can catalog all the symmetries in their groups. You'll conclude that the groups all differ.

Exactly what qualities do you consider in distinguishing symmetry groups? This isn't as simple a question as it might seem at first. Independence of translations is one. On the other hand, you wouldn't distinguish between the symmetry groups of figure 8.0.1 and that of the bowl it depicts, even though their centers differ. (One's in your hand, the other's in a museum.) Precise criteria are described in section 8.1 under the **Conjugacy** heading.

You know these are not the only ways to construct symmetric designs. Look about you for ornaments such as logos or balls with zero-dimensional groups, one-dimensional fences or ribbons, two-dimensional tiling or wallpaper patterns, and three-dimensional packing schemes that display different rotational and reflectional symmetries. How can you classify such designs? How many different classes are there? That's the subject of this chapter. To study it you'll use rather elaborate group-theoretic analyses.

This book is not the place to find a great wealth of examples of symmetry. Sections 8.1 to 8.4 are devoted to the details of the geometric theory. Barely more than a minimal number of figures illustrate its applications. Those from the art world are selected as much as possible from a single area, the Pueblo cultures of New Mexico. This shows how group-theoretic geometric analysis, a sophisticated product of Europe, can be used to study art from an almost entirely independent culture. The exercises in section 8.5 suggest numerous other applications. To gain familiarity with them, you'll need to consult the references cited later in this chapter introduction, particularly *their* bibliographies. *Need to* isn't the right way to say it! You'll probably become immersed in wonderful figures from the worlds of advertising, anthropology, architecture, art, botany, commercial design, crystallography, geology, molecular chemistry, music, and zoology. You'll begin to notice examples all about you, and you'll startle your friends by stopping frequently to admire beautiful symmetries they don't even notice!

Much of this chapter's mathematics is presented almost in the reverse order from that of its discovery. Artists acquire extensive intuitive knowledge of the possibilities for zero-, one-, and two-dimensional symmetry groups without recording any detailed conscious analysis. Some cultures tacitly identify which combinations they use, and regard others as foreign. Artists of some cultures have gathered together at one place and time descriptions of *all* possibilities in one or another of the dimensional classes, without explicitly proving that the possibilities were exhausted.

Since the Renaissance, European scientists have applied mathematical methods to this subject, and recorded their findings. The group-theoretic approach, which brings unity to their work, was developed only since the late 1800s, to solve the hardest problem of those mentioned so far: classifying crystal symmetries. A long train of research[10] led to simultaneous publications by E. S. Fedorov ([1891] 1971) and Arthur Schönflies (1891), enumerating and classifying the 230 types of crystal symmetry. Classifying 230 types of *anything* requires an elaborate methodology; the group theory was present in these works, though not in the present-day form. Fedorov, Director of Mines in the Urals, did this as a sideline, and wrote mostly in Russian, hence proceeded outside the main stream of science at that time.[11] Schönflies, a professor at Göttingen, worked very much *in* the mainstream of geometry. The two corresponded at length, and Schönflies granted Fedorov priority for the discovery.[12]

As you'll see in section 8.1, the study of plane figures with zero-dimensional symmetry group is simple enough that it needs only a small amount of organization beyond the basic geometry of regular polygons, which had been developed by Euclid's time. So it's reasonable to presume that the analysis of those symmetries was common knowledge by Renaissance times. Section 8.4 describes polyhedra with many different zero-dimensional groups. By the mid 1800s, the German and French mineralogists Hessel and Bravais had completed the investigation of symmetric polyhedra. They used little theory beyond Euclid's analysis of the regular polyhedra; however, there are so many cases that the present book does not present the details of their study. Fedorov completed the study of the seventeen two-dimensional sym-

---

[10] See the historical accounts Schneer 1983 and Burckhardt 1967/1968.

[11] Russian scientists were concerned about their linguistic isolation. At the 1900 International Congress of Philosophy, they were major participants in discussions of a possible international scientific language. Louis Couturat and Giuseppe Peano, the senior French and Italian mathematicians at the congress, soon developed specifications for two such languages, *Ido* and *Interlingua*. The Italian contributions to that congress played a major role in the foundations of geometry; see sections 2.8 and 2.9.

[12] Burckhardt 1967/1968, section 5.

metry groups along with his 1891 classification of crystals,[13] but he didn't mention the one-dimensional problem.

This field of geometry then passed from the spotlight for some years. Hilbert included a higher-dimensional generalization of the classification in his list (1900) of problems that would guide mathematics into the twentieth century. Several books on crystallography appeared during the next ten years. But significant renewed attention evidently had to await the development in Göttingen, during the 1910 decade, of modern higher algebra as a routine tool for organizing studies in all disciplines of mathematics.

Detailed study of three-dimensional symmetry groups remains a domain for specialists. However, popular interest in the one- and two-dimensional groups was rekindled in the early 1920s. First, Hilbert included them in his 1920/1921 lectures on visual geometry. Their translation *Geometry and the imagination* (Hilbert and Cohn-Vosson [1932] 1952) introduced the present author to this material and sparked his love for geometry. Next, three papers appeared in the *Zeitschrift für Kristallographie* in 1924 and 1926. The first was just a letter to the editor by the well known Zürich mathematician Georg Pólya. He pointed out that the crystallographic methods could be used to classify familiar ornamental patterns. He didn't publish the details:

> The proof of the completeness of the enumeration is an easy exercise for anyone who knows Schönflies' results and his text. I'm not including my proof here because it doesn't seem sufficiently rounded out.[14]

The very next paper in that journal, by its editor Paul Niggli, also a professor at Zürich, provided the proof (Niggli 1924). He first pointed out that Fedorov had already done so in principle, but no one had yet presented the proof clearly. Niggli then adapted some parts of his own book on crystallography to yield the classification of plane ornaments. Soon after, he published the analogous but simpler analysis of the one-dimensional symmetry groups (Niggli 1926). This Zürich activity stemmed from Schönflies' and Hilbert's work in Göttingen. Pólya, a Hungarian, had done postdoctoral research there in 1912/1913 before he accepted the professorship at Zürich. Andreas Speiser, by then already a professor at Zürich, had completed his Ph.D. research at Göttingen in 1909. In 1923, Speiser published the first group theory text incorporating the new Göttinger approach to algebra. Its second edition (Speiser 1927) featured the classifications by Fedorov, Pólya, and Niggli. That was the first appearance in book form of the group-theoretic analysis of ornamental symmetry. From then on, presentations and extensions of these methods appeared frequently.

---

[13] Fedorov 1891.

[14] Pólya 1924, 280, translated by the present author. Pólya was renowned for his mathematics and revered for his pedagogy. The style of his letter, with its pictures of seventeen different ornaments, made it famous.

The scope of symmetry analysis, its applications, and its literature are vast. This chapter only touches the surface. Some major areas are mentioned only in passing, if at all. For further information, you may want to consult some of the references mentioned next. See the bibliography for more detailed citations.

Sources that concentrate on art and scientific applications, with little or no attention to instruction in technical geometry	
Blackwell 1984	*Geometry in architecture*
Hargittai 1986	*Symmetry: Unifying human understanding*
Hargittai and Hargittai 1994	*Symmetry: A unifying concept*
Jones [1856] 1972	*The grammar of ornament: Illustrated by examples of various styles of ornament*
Schattschneider 1990	*Visions of symmetry: Notebooks, periodic drawings, and related work of M. C. Escher*
Senechal and Fleck 1977	*Patterns of symmetry*
Senechal and Fleck 1988	*Shaping space: A polyhedral approach*
Stevens 1974	*Patterns in nature*
Stewart and Golubitsky 1992	*Fearful symmetry: Is God a geometer?*
Washburn and Crowe 1988	*Symmetries of culture: Theory and practice of plane pattern analysis*
Weyl 1952	*Symmetry*

Sources with some emphasis on introductory instruction in geometry or group theory	
Beck et al. 1969	*Excursions into mathematics.* Chapter 1, "Euler's formula for polyhedra, and related topics."
Crowe 1986	*Symmetry, rigid motions, and patterns*
Farmer 1996	*Groups and symmetry: A guide to discovering mathematics*
Foster (no date)	*The Alhambra, past and present: A geometer's odyssey*
Gallian 1994	*Contemporary abstract algebra*
Hilton et al. 1997	*Mathematical reflections in a room with many mirrors*

Sources that pursue the study of symmetry beyond the scope of this book	
Armstrong 1988	*Groups and symmetry*
Cromwell 1997	*Polyhedra*
Fejes Tóth 1964	*Regular figures*
Grünbaum and Shephard 1987	*Tilings and patterns*
Hilbert and Cohn-Vosson [1932] 1952	*Geometry and the imagination*
Klemm 1982	*Symmetrien von Ornamenten und Kristallen*
Martin 1982	*Transformation geometry: An introduction to symmetry*
Senechal 1995	*Quasicrystals and geometry*
Shubnikov and Koptsik 1974	*Symmetry in science and art*
Yale [1968] 1988	*Geometry and symmetry*

Arthur Moritz SCHÖNFLIES was born in 1853 in Landsberg an der Warte. From 1870 to 1877 he studied at the University of Berlin, and earned the doctorate with a dissertation in synthetic geometry supervised by Kummer. From 1878 through 1880 he taught in gymnasiums in Berlin and at Colmar, in the Alsace. Schönflies returned to academic life, earned Habilitation in Göttingen in 1884, and was appointed Ausserordentlicher Professor there in 1892. During this period he pursued his own research in geometry, and edited the works of Plücker. Schönflies published a monograph on the general theory of rigid motions, and wrote the major article on that subject in the *Enzyklopädie der mathematischen Wissenschaften*. He applied this to crystallography, and with Fedorov characterized the 230 crystal types. Schönflies wrote the authoritative work on mathematical crystallography in 1891. In 1895, with W. Nernst, he produced a noted popular book on mathematics in science.

Schönflies moved to a full professorship at Königsberg in 1899, and at about fifty undertook study of a new area, set theory. He produced several monumental reports on the subject during 1900–1913, providing some of the first clean discussions and proofs of now-standard concepts and theorems about infinite sets. Schönflies moved again, in 1911, to Frankfurt am Main. There he continued his work on set theory and in 1923 produced a new edition of his crystallography book. In all, he produced no fewer than four major *Enzyklopädie* articles. Schönflies retired in 1923, and died in Frankfurt in 1928.

8.1 Polygonal symmetry

Concepts
- Simple closed polygons
- Regular polygons
- Conjugacy
- Symmetry groups of polygons
- Finite cyclic isometry groups
- Modular arithmetic
- Dihedral groups

We're all familiar with the use of polygons[15] in design, both for practical purposes and for ornamentation. For $n \geq 3$, a *simple closed polygon* with *vertices* V_0 to V_{n-1} is defined as the union of n *edges*

$$\Sigma_0 = \overline{V_0V_1},\ \Sigma_1 = \overline{V_1V_2},\ \ldots,\ \Sigma_{n-2} = \overline{V_{n-2}V_{n-1}},\ \Sigma_{n-1} = \overline{V_{n-1}V_0},$$

provided

- no pair of successive edges, nor Σ_n and Σ_0, are collinear;
- no edges intersect, except that Σ_k and Σ_{k+1} share V_{k+1} for $k = 0$ to $n - 1$, and Σ_n and Σ_0 share V_0.

Numerical prefixes *penta-*, *hexa-*, ..., *dodeca-* specify polygons with 5, 6, ..., 12 edges and vertices. Figure 8.1.1 displays a simple closed pentagon. Since *all* polygons in this chapter are simple and closed, those adjectives will be omitted.

The *regular* polygons are favorites, because all edges are congruent, as are the angles between intersecting edges. Exercise 8.5.5 invites you to survey the uses of these highly symmetric objects in practical and ornamental design.

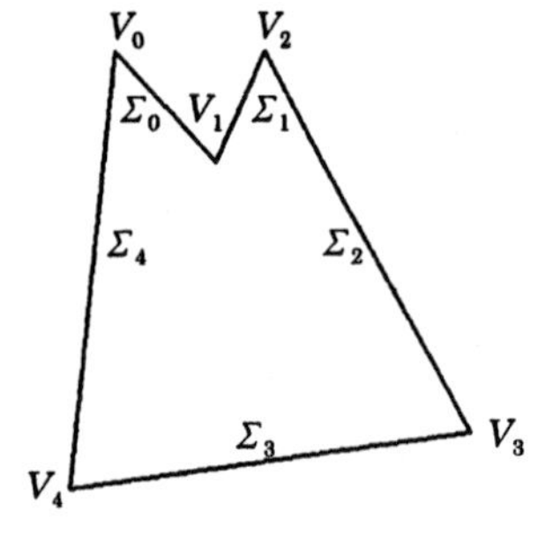

Figure 8.1.1
Simple closed pentagon

---

[15] The Greek prefix and root *poly-* and *gon* mean *many* and *angle*.

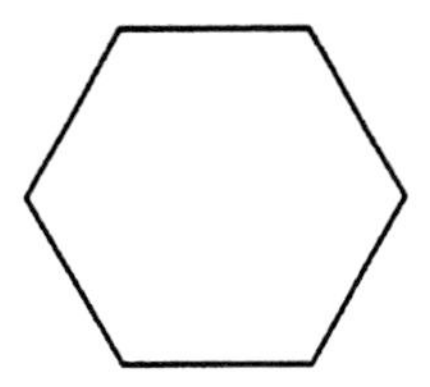

Figure 8.1.2 Regular hexagon

Figure 8.1.3 Decagonal stellation of a (dotted) regular pentagon

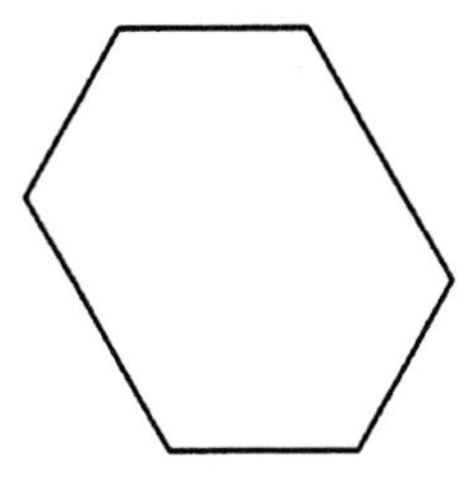

Figure 8.1.4 *Affine-regular* hexagon (opposite edges are parallel)

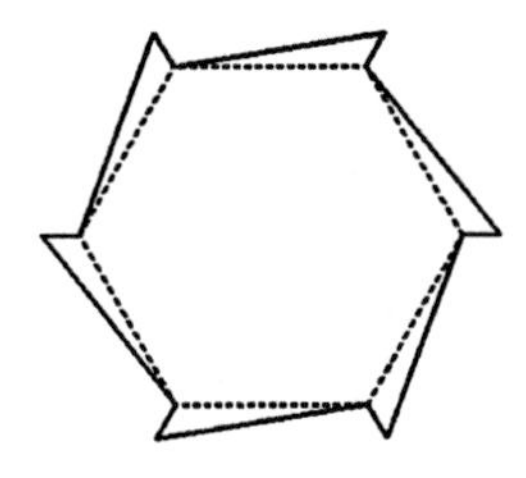

Figure 8.1.5 *Eared* dodecagon made from a (dotted) regular hexagon

You should consider somewhat less symmetric polygons, too. Figures 8.1.2 to 8.1.5 present some examples. (Figure 8.1.1 showed no symmetry at all.)

You can easily catalog, and thus distinguish, the symmetries of these examples. That amounts to listing the symmetry groups of the figures:

Figure	*Rotations*	*Reflections*
8.1.1	$0°$	none
8.1.2	$n \cdot 60°$	6 lines through the center
8.1.3	$n \cdot 72°$	5 lines through the center
8.1.4	$n \cdot 180°$	2 lines through the center
8.1.5	$n \cdot 60°$	none

(n represents any integer.) These figures all have different symmetry.

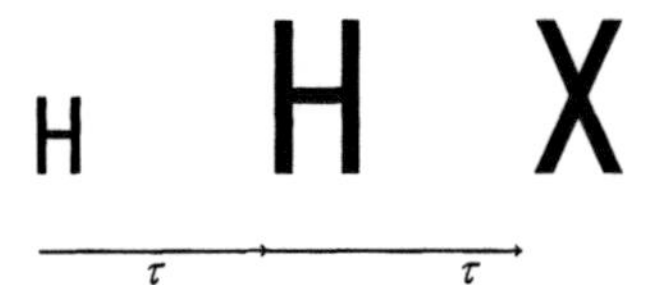

Figure 8.1.6 Letters with the same symmetry

Conjugacy

What criterion governs whether figures have different symmetry? It can't be that their symmetry groups differ, because we regard the letters **H** and **H** in figure 8.1.6 as having the same symmetry, even though the half turns and horizontal and vertical reflections in their symmetry groups $\mathscr{G}_{\mathsf{H}}$ and $\mathscr{G}_{\mathsf{H}}$ have different centers and axes. How are these groups related? The letters are similar. A similarity, the composition $\varphi = \delta\tau$ of a horizontal translation τ followed by a dilation δ with ratio $r \approx 2.2$, maps **H** onto **H**. You can use φ to relate the groups. To each symmetry ψ of the smaller **H** corresponds the conjugate similarity $\psi^\varphi = \varphi\psi\varphi^{-1}$. It's an isometry because its similarity ratio is $r \cdot 1 \cdot r^{-1} = 1$. Moreover, ψ^φ leaves the larger **H** unchanged: $\varphi^{-1}[\mathsf{H}]$ is the smaller **H**, which is unchanged by ψ, then mapped back onto the larger one by φ, so that $\varphi\psi\varphi^{-1}[\mathsf{H}] = \mathsf{H}$. Therefore ψ^φ is a symmetry of the larger **H**. The conjugation $\psi \to \psi^\varphi$ maps $\mathscr{G}_{\mathsf{H}}$ to $\mathscr{G}_{\mathsf{H}}$. This correspondence is bijective because it has an inverse, the conjugation $\omega \to \omega^{\varphi^{-1}}$, which maps $\mathscr{G}_{\mathsf{H}}$ to $\mathscr{G}_{\mathsf{H}}$. The close relationship between the groups doesn't depend on *all* details of the figures—just those that we commonly call symmetry properties. For example, the conjugation $\psi \to \psi^\chi$ defined by the similarity $\chi = \delta\tau^2$ maps the symmetry group $\mathscr{G}_{\mathsf{H}}$ of the smaller **H** in figure 8.1.6 bijectively onto the symmetry group $\mathscr{G}_{\mathsf{X}}$ of the large **X**. (Each group consists of the identity, the half turn about the letter's center, and the reflections across its horizontal and vertical axes.) When a conjugation $\psi \to \psi^\varphi$ maps one isometry group $\mathscr{G}$ onto another group $\mathscr{G}^\varphi = \psi^\varphi[\mathscr{G}]$, they're called *conjugate*. When their symmetry groups are conjugate, two figures are said to display the *same symmetry*. In particular, similar figures have conjugate symmetry groups, and display the same symmetry.

Theorem 1. The conjugation $\psi \to \psi^\varphi$ defined by a similarity χ maps an isometry group $\mathscr{G}$ bijectively onto a conjugate isometry group $\mathscr{G}^\varphi$. It leaves the identity fixed: $\iota^\varphi = \iota$. Moreover, for any isometries ψ and ω in $\mathscr{G}$, $\psi^\varphi\omega^\varphi = (\psi\omega)^\varphi$ and $(\psi^\varphi)^{-1} = (\psi^{-1})^\varphi$.

Proof. The previous paragraph established that conjugation is bijective and the elements of $\psi^\varphi[\mathscr{G}] = \mathscr{G}^\varphi$ are isometries. The equations

$$\iota^\varphi = \varphi\iota\varphi^{-1} = \iota$$
$$\psi^\varphi\omega^\varphi = \varphi\psi\varphi^{-1}\varphi\omega\varphi^{-1} = \varphi\psi\omega\varphi^{-1} = (\psi\omega)^\varphi$$

$$(\psi^{\varphi})^{-1} = (\varphi\psi\varphi^{-1})^{-1} = \varphi\psi^{-1}\varphi^{-1} = (\psi^{-1})^{\varphi}$$

imply that $\psi^{\varphi}[\mathcal{G}]$ contains the identity, the composition of any of its elements ψ^{χ} and ω^{χ}, and the inverse of any element ψ^{χ}, so it's an isometry group. ♦

The last statement in theorem 1 implies that you can perform a group operation in $\mathcal{H} = \mathcal{G}^{\chi}$ either directly—for example, $\psi^{\chi}\omega^{\chi}$ —or by operating on the corresponding isometries in $\mathcal{G}$, then mapping the result—$\psi\omega$ in this example—back to its counterpart $(\psi\omega)^{\chi}$ in $\mathcal{H}$. The groups are conjugate in the sense of the original Latin participle *conjugatus*: joined together as in friendship or wedlock.

When we use the word *same* in informal language, we assume that it refers to a relationship that enjoys the reflexive, symmetric, and transitive properties of an equivalence relation. (See appendix A.) For the vague notion of figures *displaying the same symmetry properties*, the previous paragraph and theorem 1 developed a precise counterpart: *possessing conjugate symmetry groups*. Whenever such a notion of sameness is formalized, the justification of its use should include verification that it is indeed an equivalence relation. The next theorem does that for conjugacy.

Theorem 2. The conjugation $\psi \to \psi^{\iota}$ defined by the identity ι leaves each isometry ψ fixed—that is, $\psi^{\iota} = \psi$ —hence it maps each isometry group $\mathcal{G}$ to itself: $\mathcal{G}^{\iota} = \mathcal{G}$. If a conjugation $\psi \to \psi^{\varphi}$ defined by a similarity φ maps an isometry group $\mathcal{G}$ to $\mathcal{H} = \mathcal{G}^{\varphi}$, then the inverse conjugation $\omega \to \omega^{\varphi^{-1}}$ maps $\mathcal{H}$ back to $\mathcal{G}$. Moreover, if a conjugation $\omega \to \omega^{\chi}$ maps $\mathcal{H}$ to a third isometry group $\mathcal{I}$, then the conjugation $\psi \to \psi^{\chi\varphi}$ maps $\mathcal{G}$ to $\mathcal{I}$.

The proof is straightforward; most of it is contained in the previous discussion. The three sentences of theorem 2 imply that conjugacy is an equivalence relation:

- every isometry group $\mathcal{G}$ is conjugate to itself; (*Reflexivity*)
- if an isometry group $\mathcal{G}$ is conjugate to an isometry group $\mathcal{H}$, then $\mathcal{H}$ is also conjugate to $\mathcal{G}$; (*Symmetry*)
- if $\mathcal{G}$ is conjugate to $\mathcal{H}$, and $\mathcal{H}$ is conjugate to an isometry group $\mathcal{I}$, then $\mathcal{G}$ is also conjugate to $\mathcal{I}$. (*Transitivity*)

Symmetry groups of polygons

You've probably already noticed the following properties of symmetry groups of polygons:

Theorem 3. The symmetry group of a polygon Π is finite. It contains no translation or glide reflection with nonzero vector.

Proof. Clearly, the symmetry group of a bounded figure can contain no translation or glide reflection with nonzero vector. Select any three consecutive vertices of Π. Any symmetry φ of Π must map them into three consecutive image vertices, and by the uniqueness theorem, corollary 6.2.8, these images determine φ completely. There are only finitely many possibilities for the image vertices, so there are only finitely many possibilities for φ. ♦

The remainder of this section considers finite symmetry groups in detail, and concludes with a converse of theorem 3: *Every finite group of plane isometries is the symmetry group of some polygon.*

Finite cyclic symmetry groups

The simplest finite isometry groups are those that contain only rotations, such as the symmetry group of the eared dodecagon in figure 8.1.5. These groups are called *cyclic.*

Theorem 4. Consider a nontrivial finite isometry group $\mathscr{G}$ consisting of m rotations. These must be $\rho^k_{O,360°/m}$ for $k = 0$ to $m - 1$, with a uniquely determined center O.

Proof. Suppose $\mathscr{G}$ contained nontrivial rotations $\rho_{N,\eta}$ and $\rho_{O,\theta}$ with $N \neq O$. Let $P = \rho_{N,\eta}(O)$. Then $O \neq P$ and $\mathscr{G}$ would contain $\rho_{N,\eta}\rho_{O,\theta}\rho^{-1}_{N,\eta} = \rho_{P,\theta}$ and the nontrivial translation $\rho_{O,\theta}\rho^{-1}_{P,\theta}$, contradicting theorem 3! Thus, all nontrivial rotations $\rho_{O,\theta}$ in $\mathscr{G}$ must have the same center O. Assign them angle parameters between $0°$ and $360°$, and let θ denote the smallest of these parameters. $\mathscr{G}$ contains the rotations $\rho^k_{O,\theta}$ for all integers k, so they can't all be distinct: $\rho^j_{O,\theta} = \rho^k_{O,\theta}$ with $j < k$ for some j and k, hence $\rho^{k-j}_{O,\theta} = \iota$. That is, *some* positive power of $\rho_{O,\theta}$ is the identity; let m be the *first* power for which $\rho^m_{O,\theta} = \iota$, so that $\theta = 360°/m$. If $\rho^j_{O,\theta} = \rho^k_{O,\theta}$ with $0 \le j < k < m$, then $\rho^{k-j}_{O,\theta} = \iota$ with $0 < k - j \le k < m$, contradiction! Thus, the powers $\rho^k_{O,\theta}$ for $k = 0$ to $m - 1$ are all distinct.

It remains to show that these powers constitute *all* the members of $\mathscr{G}$. Indeed, if $\rho_{O,\eta}$ is any member of $\mathscr{G}$, then you can divide η by θ to find an integer quotient p and a remainder λ such that $0° \le \lambda < \theta$ and $\eta =$

$p\theta + \lambda$. It follows that $\rho_{O,\lambda} = \rho_{O,\eta}\,\rho_{O,\theta}^{-p}$ belongs to $\mathcal{G}$, which contradicts the choice of θ unless $\lambda = 0°$; and that entails $\eta = p\theta$. Finally, divide p by m to get an integer quotient q and remainder k such that $0 \le k < m$ and $p = qm + k$. As desired,

$$\rho_{O,\eta} = \rho_{O,p\theta} = \rho_{O,\theta}^{p} = \rho_{O,\theta}^{qm+k} = (\rho_{O,\theta}^{m})^{q}\rho_{O,\theta}^{k} = \rho_{O,\theta}^{k}$$

Therefore, $\mathcal{G}$ has exactly m elements $\rho_{O,\theta}^{k}$ for $k = 0$ to $m - 1$. ♦

The proof of theorem 4 suggests a connection between a finite cyclic isometry group $\mathcal{G}$ with m elements and *arithmetic modulo* m.[16] If $\theta = 360°/m$ and $0 \le j, k < m$, then

$$\rho_{O,\theta}^{j}\rho_{O,\theta}^{k} = \rho_{O,j\theta}\rho_{O,k\theta} = \rho_{O,(j+k)\theta} = \rho_{O,\theta}^{j+k}.$$

It's possible that $j + k > m$, which is inconvenient if you want to consistently use angle parameters between $0°$ and $360°$. So you can divide $i = j + k$ by m to get integer quotient q and integer remainder l such that $0 \le l < m$ and $i = qm + l$. It follows that

$$\rho_{O,\theta}^{j+k} = \rho_{O,\theta}^{qm+l} = (\rho_{O,\theta}^{m})^{q}\rho_{O,\theta}^{l} = \rho_{O,\theta}^{l}$$

The remainder l is often written $i \bmod m$, so that the rule for composition of rotations in $\mathcal{G}$ can be stated

$$\rho_{O,\theta}^{j}\rho_{O,\theta}^{k} = \rho_{O,\theta}^{l} \text{ where } l = (j + k) \bmod m.$$

If $0 < k < m$, then $-k = (-1)m + (m - k)$ and $0 \le m - k < m$, hence $(-k) \bmod m = m - k$. The equations

$$(\rho_{O,\theta}^{k})^{-1} = (\rho_{O,\theta}^{k})^{-1}\rho_{O,\theta}^{m} = \rho_{O,\theta}^{m-k}$$

and $0 \bmod m = 0$ then yield the rule for inversion of a rotation in $\mathcal{G}$:

$$(\rho_{O,\theta}^{k})^{-1} = \rho_{O,\theta}^{l} \text{ where } l = (-k) \bmod m.$$

A cyclic isometry group with two elements contains just the identity and a half turn about the center O. It's the symmetry group of a parallelogram that's neither a rectangle nor a rhombus. For any larger cyclic isometry group $\mathcal{G}$ with center O and m elements you can construct a regular polygon with m edges and center O, then add ears as in figure 8.1.5 to get a polygon with $2m$ edges and symmetry group $\mathcal{G}$.

Theorem 5. Every finite cyclic isometry group is the symmetry group of some polygon. Any two finite cyclic isometry groups with the same number of elements are conjugate.

---

[16] Analogous to arithmetic modulo 360 as introduced in section 3.13.

Proof. The previous discussion established the first statement. Further, if you follow its algorithm, starting with two finite cyclic groups, each with n elements, you'll construct similar polygons, whose symmetry groups are conjugate by theorem 1. ♦

The second statement in theorem 5 suggests use of a common symbol $\mathscr{C}_m$ to designate all finite cyclic groups with m elements.

Dihedral groups

All regular polygons with m edges are similar, so their symmetry groups are conjugate, by theorem 1. They're called *dihedral* groups, and a common symbol $\mathscr{D}_m$ is used to designate them.[17] Study the regular pentagon and hexagon in figures 8.1.7 and 8.1.8, and list their symmetries. Do the same for an equilateral triangle and a square. You'll see a pattern, and conclude

Theorem 6. The symmetry group $\mathscr{D}_m$ of a regular polygon Π with m edges has $2m$ elements. It consists of the cyclic group $\mathscr{C}_m$ plus m reflections across lines through the center of Π. If m is odd, each of these lines passes through a vertex and the midpoint of the opposite edge. If m is even, half of them pass through two opposite vertices, and the other half, through two opposite midpoints.

It's also customary to consider dihedral groups with four and two elements. It's not too painful to consider a nontrivial line segment as a regular polygon

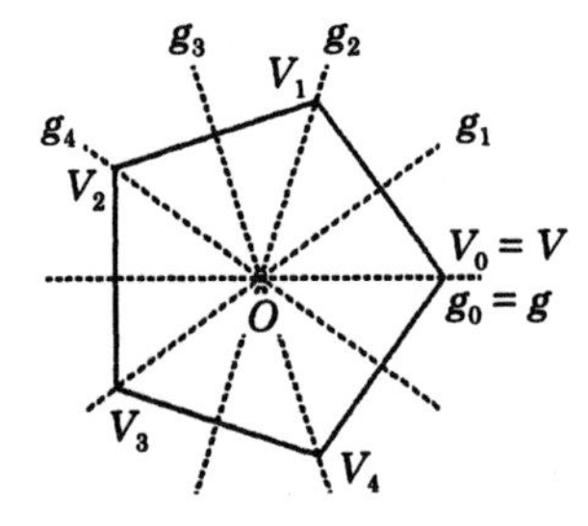

Figure 8.1.7 Regular pentagon: lines of symmetry

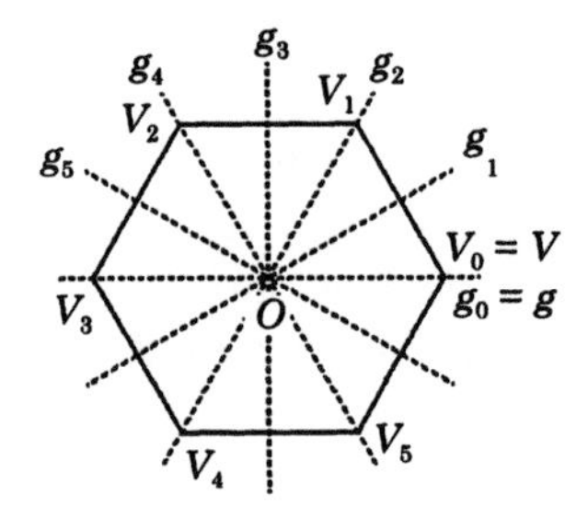

Figure 8.1.8 Regular hexagon: lines of symmetry

---

[17] The Greek prefix and root *di-* and *hedron* mean *two* and *face*. This terminology stems from the fact that symmetry groups were first studied in detail for three-dimensional figures. A polygon is the boundary of a polygonal region, which was regarded as a flat figure, almost a polyhedral region, but with only two faces.

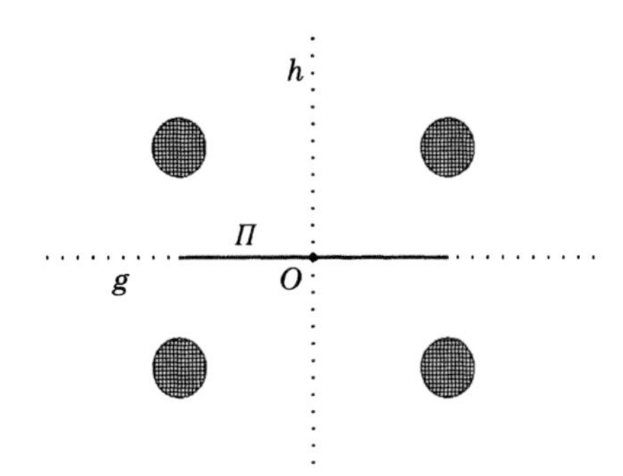

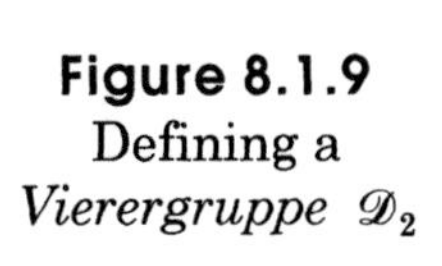
Figure 8.1.9 Defining a *Vierergruppe* $\mathscr{D}_2$

with *two* vertices. (See figure 8.1.9.) Its symmetry group contains four elements: the identity, the half turn about its midpoint O, the reflection across its line g, and the reflection across the line $h \perp g$ through O. Clearly, all such groups are conjugate; they're regarded as dihedral groups and designated by the symbol $\mathscr{D}_2$. Such a group is also commonly called a *Vierergruppe*.[18] Finally, a group that contains just the identity and a single line reflection is also called dihedral. All such groups are conjugate, and are designated by the symbol $\mathscr{D}_1$.

The final result of this section is a converse of theorem 3.

***Theorem** 7.* Every finite group of plane isometries is cyclic or dihedral.[19]

Proof. No finite isometry group $\mathscr{G}$ can contain any translation or glide reflection with nonzero vector. Let $\mathscr{C}$ be the cyclic subgroup consisting of all rotations in $\mathscr{G}$. If $\mathscr{C} = \mathscr{G}$, then $\mathscr{G}$ is cyclic. Thus you can assume that $\mathscr{G}$ contains the reflection across some line g. The mapping $\varphi \to \varphi\sigma_g$ is a bijection from $\mathscr{C}$ to the set of all reflections in $\mathscr{G}$ because

- $\varphi\sigma_g$ is odd, hence a reflection, for each φ in $\mathscr{C}$;
- $\varphi\sigma_g = \chi\sigma_g$ implies $\varphi = (\varphi\sigma_g)\sigma_g = (\chi\sigma_g)\sigma_g = \chi$ for each φ and χ in $\mathscr{C}$;
- $\psi\sigma_g$ is even, hence in $\mathscr{C}$, for each reflection ψ in $\mathscr{G}$, and $\psi = (\psi\sigma_g)\sigma_g$.

If $\mathscr{C}$ contains only the identity, then $\mathscr{G}$ contains only one reflection, hence $\mathscr{G}$ is a dihedral group $\mathscr{D}_1$. Thus you can assume that $\mathscr{C}$ is nontrivial, with

18 This term apparently stems from an early diagram that displayed some patterns like figure 8.1.9 in the four plane quadrants defined by g and h. They looked like prints of someone on *all fours*—*alle Vierer* in German.

19 Several authors call this result *Leonardo's theorem*, because Weyl (1952, 66, 99) wrote that Leonardo da Vinci, in notebooks dating from about 1500, had essentially tabulated the symmetries in the finite cyclic and dihedral groups. But Senechal (1990, 329) suggests that closer inspection of the notebooks doesn't support Weyl's claim.

center O. This point lies on the mirror of each reflection ψ in $\mathscr{G}$; otherwise, $\mathscr{G}$ would contain $\psi\rho_{O,360°/m}$, a glide reflection with nonzero vector. Define m distinct lines $g_k = \rho_{O,180°k/m}[g]$ for $k = 0, \ldots, m-1$, so that $g_0 = g$, $\sigma_{g_k}\sigma_g = \rho_{O,360°k/m}$, and $\sigma_{g_k} = \rho_{O,360°k/m}\sigma_g$ is in $\mathscr{G}$. In fact, $\mathscr{G}$ consists precisely of $\mathscr{C}_m$ plus σ_{g_k} for $k = 0, \ldots, m-1$. Choose any point $V \neq O$ on g, and define $V_k = \rho_{O,360°k/m}(V)$ for $k = 0, \ldots, m-1$ so that $V_0 = V$. If $m = 2$, then segment $\overline{V_0V_1}$ has symmetry group $\mathscr{G}$. If $m > 2$, then V_0 to V_{m-1} are the vertices of a regular polygon with symmetry group $\mathscr{G}$. That is, $\mathscr{G}$ is a dihedral group $\mathscr{D}_m$. (Figures 8.1.7 and 8.1.8 illustrate cases $m = 5$ and 6 of this construction.) ♦

8.2 Friezes

Concepts
- Frieze groups and friezes
- Lattices and grids
- Decision trees
- Classifying friezes
- Frieze group notation

The previous section considered bounded plane ornaments: designs derived from regular polygons. They possess rotational and reflectional symmetries in various combinations. Their symmetry groups are finite, and the discussion concluded with a detailed analysis of *all* finite groups of plane isometries.

The next more complicated symmetric figures display repetition as well. These patterns, called *friezes*, repeat indefinitely, or at least are always *extensible*, in two opposite directions, but *not* in any other direction. You've seen examples already in figures 6.0.3 and 8.0.2. *Repetition* means doing *something* over and over. So you must construct a frieze by making a fundamental cell first, then appending duplicate cells as many times as you wish in two opposite directions. There must be a *smallest* repeated cell. The continuous stripe in figure 8.2.1, for example, is *not* regarded as a frieze.

The examples so far have been bounded by parallel lines, but that's not really necessary. A frieze can extend indefinitely far in a different direction, as long as it doesn't *repeat* in that direction and its opposite. Figure 8.2.2 depicts an unbounded frieze; but the closely related pattern suggested by figure 8.2.3 is *not* a frieze because it repeats in two nonopposite directions.

As in the previous section, the task here is to *classify* frieze patterns. You might object that since the notion *frieze* is not yet totally clear, how can frieze *classes* be defined precisely? A single technique solves both problems. Friezes will be classified according to their symmetry groups. It's easy to

Figure 8.2.1
Not a frieze

Figure 8.2.2
An unbounded frieze

define *frieze group* without referring to the notion *frieze* itself. Then we can define a *frieze* as a figure whose symmetry group is a frieze group! Because the groups display less variety than the friezes, it's easier to classify the groups.

The symmetry groups of the examples in figures 8.0.2 and 8.2.2 contain line reflections as well as horizontal translations. That of figure 6.0.3 contains those translations and some half turns. In each case the translations in the group are the powers of a particular translation, with a shortest possible vector V, stretching from some cell to the corresponding point in its nearest replica. There are always two choices for V; it could point either right or left.

This suggests defining a *frieze group* to be a group $\mathscr{G}$ of plane isometries that contains all integral powers τ^n of a nontrivial *fundamental* translation τ, but no other translations. $\vec{\tau}$ will denote the corresponding vector, so that $\mathscr{G}$ contains the translations with vectors $n\vec{\tau}$ for $n = 0, \pm 1, \pm 2, \ldots$, but no others. The only other fundamental translation for $\mathscr{G}$ is τ^{-1}. As suggested earlier, a *frieze* is defined as a plane figure whose symmetry group is a frieze group.

It's useful to consider some terminology, notation, and preliminary results, to gain familiarity with the techniques before considering details of the classification. First, lines parallel and perpendicular to $\vec{\tau}$ are called *horizontal* and *vertical*. Next, a point O determines a *lattice* of points $O_n = \tau^n(O)$ for $n = 0, \pm 1, \pm 2, \ldots$; the midpoint of segment $\overline{O_n O_{n+1}}$ is denoted by $O_{n+\frac{1}{2}}$. Similarly, a vertical line v determines a *grid* of vertical lines

Figure 8.2.3
Not a frieze

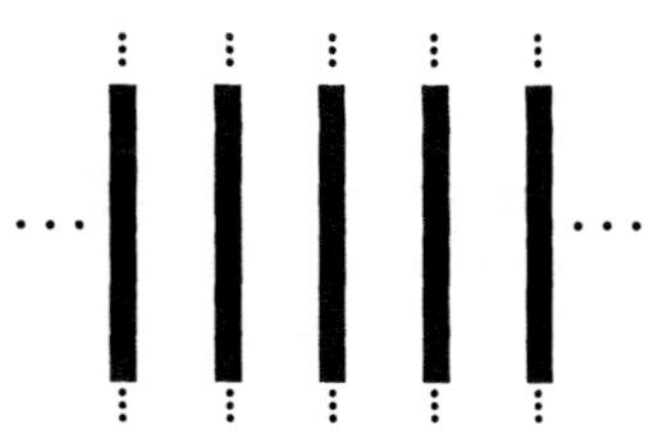

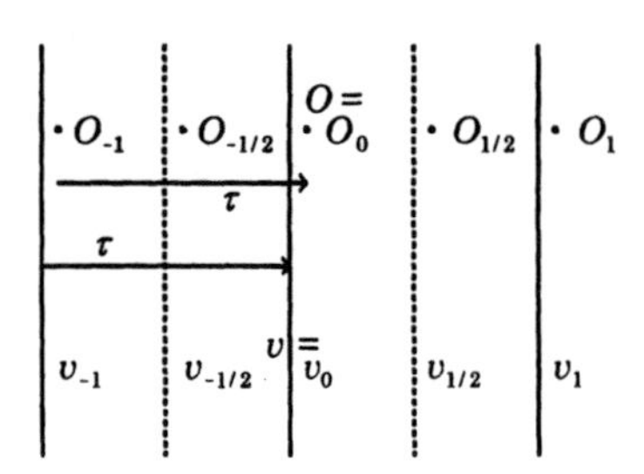

Figure 8.2.4 A lattice and a grid for a frieze group with fundamental translation τ

$\upsilon_n = \tau^n[\upsilon]$ for $n = 0, \pm1, \pm2, \ldots$; the line parallel to and midway between υ_n and υ_{n+1} is denoted by $\upsilon_{n+1/2}$. (See figure 8.2.4.) The following four theorems discuss which rotations, reflections, and glide reflections can belong to a frieze group. These isometry types are considered individually. Later in this section, the classification theorem describes how the various types interrelate.

Theorem 1. If a frieze group $\mathscr{G}$ with fundamental translation τ contains a nontrivial rotation ρ with center O, then $\rho = \sigma_O$. Moreover, $\mathscr{G}$ contains the half turns about the lattice points $O_{n/2}$ for $n = 0, \pm1, \pm2, \ldots,$ but no other rotations.

Proof. $\mathscr{G}$ will contain the translation $\upsilon = \rho\tau\rho^{-1}$ whenever it contains ρ, and by theorem 6.9.9, $\vec{\upsilon} \parallel \vec{\tau}$ only if ρ is the identity or a half turn. In that case, $\sigma_{O_{n/2}} = \sigma_O \tau^n$ by corollary 6.4.5, so $\mathscr{G}$ contains $\sigma_{O_{n/2}}$. Finally, suppose $\mathscr{G}$ contains another rotation φ. Then $\varphi = \sigma_P$ for some point P by the first sentence of this proof, and $\mathscr{G}$ must contain the translation $\sigma_P \sigma_O$ with vector $2(P - O)$, which then must equal $n\vec{\tau}$ for some integer n. It follows that $P = O_{n/2}$. ♦

Theorem 2. If a frieze group $\mathscr{G}$ with fundamental translation τ contains a reflection across a nonhorizontal line υ, then $\upsilon \perp \vec{\tau}$. Moreover, $\mathscr{G}$ contains the reflections across grid lines $\upsilon_{n/2}$ for $n = 0, \pm1, \pm2, \ldots,$ but no other reflections across nonhorizontal lines.

Proof. Imitate the proof of theorem 1, citing theorem 6.4.4 instead of corollary 6.4.5. ♦

Theorem 3. A frieze group $\mathscr{G}$ contains a reflection across at most one horizontal line.

Proof. If $\mathscr{G}$ contained reflections across distinct horizontal lines g and h, then it would contain the translation $\sigma_g \sigma_h$, whose vector is perpendicular to $\vec{\tau}$, hence not a multiple of $\vec{\tau}$. ♦

Theorem 4. If a frieze group $\mathscr{G}$ contains a glide reflection γ with nonzero vector V, then its axis g is horizontal and $V = \frac{1}{2} n\vec{\tau}$ for some integer n. Moreover, all glide reflections in $\mathscr{G}$ with nonzero vector have the same axis.

Proof. γ^2 is the translation with vector $2V$, hence $2V = n\vec{\tau}$ for some integer n. Suppose β is a glide reflection in $\mathscr{G}$ with nonzero vector $\frac{1}{2} m\vec{\tau}$ and axis f. Then $f \parallel g \parallel \vec{\tau}$, $\beta = \varphi\sigma_f$ and $\gamma = \sigma_g\psi$, where φ and ψ are the translations with vectors $\frac{1}{2} m\vec{\tau}$ and $\frac{1}{2} n\vec{\tau}$. Thus $\mathscr{G}$ contains $\gamma\beta = \varphi\sigma_f\sigma_g\psi = \varphi\chi\psi$, where $\chi = \sigma_f\sigma_g$ is a translation with vector W, which is nonzero and vertical unless $f = g$. Translation $\gamma\beta$ has vector $\frac{1}{2} m\vec{\tau} + \frac{1}{2} n\vec{\tau} + W$, which must be a multiple of $\vec{\tau}$. That can't happen if W is nonzero. ♦

Before beginning the classification theorem itself, try to predict its outcome. What are the symmetry groups of the example friezes you've seen so far? What others can you imagine? Use theorems 1 to 4 to guide you. To aid visualization and memory, use simple bounded friezes constructed from familiar patterns, such as letters of the alphabet. For example, the frieze

···AAAAAAAAAA···

has the same group as the pattern in figure 8.2.2; it consists of all powers of the fundamental translation τ, and reflections across all vertical lines separating or bisecting A cells. The symmetry group of the frieze

···BBBBBBBBBB···

contains the same translations, the reflection across the horizontal center line, and the glide reflections obtained from them by composition. The next three letters give nothing new, but the group of the frieze

···FFFFFFFFFF···

contains just the translations. So does the next in sequence, but

···HHHHHHHHHH···

has a much richer group: *all* the isometries mentioned so far, plus the half turns about the midpoints of the vertical lines bisecting or separating H cells. Then, nothing new occurs until

···NNNNNNNNNN···,

whose group contains the translations and those half turns, but none of the reflections or glide reflections. Try the rest of our alphabet, and any others you know. Are these five the only possible frieze types?

Why limit the search to friezes constructed from single letters? Try pairs! A schedule for a class that meets twice a week might display this pattern:

$$\cdots \mathsf{MWMWMWMWMW} \cdots.$$

Its group contains the translations and half turns mentioned earlier, reflections across vertical lines that quadrisect **MW** cells, and glide reflections whose vectors are odd multiples of $\frac{1}{2}\vec{\tau}$. Another pattern uses lowercase letters:

$$\cdots \mathsf{pbpbpbpbpb} \cdots.$$

Its group contains the same translations and glide reflections as the previous one, but none of the half turns or line reflections.

There are many more pairs, and triples, etc. Keep trying! How do you know when you're done? That's what the classification theorem will tell you.

Unguided, you may get lost in the complicated proof of the theorem. Its underlying structure, a *decision tree*, not only organizes the proof, but provides a practical way to classify friezes. Given a frieze group $\mathscr{G}$, we'll ask in succession four questions:

(O) Is there a rotation in $\mathscr{G}$?
(h) Is there a reflection across a horizontal line?
(v) Is there a reflection across a nonhorizontal line?
(γ) Is there a glide reflection?

Once those are answered, you'll be able to specify exactly what isometries belong to $\mathscr{G}$, and determine its conjugacy class. Thus the possible frieze classes can be identified as

(O & h & v & γ)
(O & h & v & *not* γ)
⋮
(O & *not* h & v & *not* γ)
⋮
(*not* O & *not* h & *not* v & *not* γ).

These are organized as a *tree* in figure 8.2.5; space permitted highlighting just one of the sixteen possible outcomes. This kind of tree grows upside down like a family tree; the *root*'s at the top, *branches* grow downward, and the *leaves* are at the bottom.

So far you've seen examples from only seven different frieze classes, but sixteen seem possible now. Are there really nine more? No. Proving the classification theorem, you'll find that questions (O), (h), (v), and (γ) are not independent; some listed outcomes are impossible for geometric reasons. In fact, you'll prune the tree severely, removing several large branches.

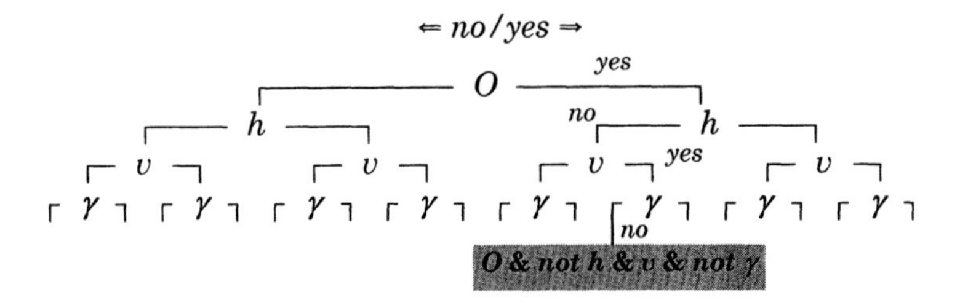

Figure 8.2.5 Frieze tree before pruning

Now, start the proof of the classification theorem. Consider a frieze group $\mathscr{G}$, and suppose it contains a rotation with center O. You're analyzing the right half of the frieze tree, branch (O), as indicated by figure 8.2.6. By theorem 1, you know exactly which rotations $\mathscr{G}$ contains: the half turns about lattice points $O_{n/2}$ for $n = 0, \pm 1, \pm 2, \ldots,$ and no others.

Next, suppose $\mathscr{G}$ also contains the reflection about a horizontal line h. You're analyzing branch $(O \,\&\, h)$ of the tree, as in figure 8.2.7. $\mathscr{G}$ must contain $\varphi = \sigma_O \sigma_h$. Then O must lie on h, else φ would be a glide reflection with axis perpendicular to $\vec{\tau}$. Therefore φ is the reflection across line $v \perp h$ through O. The answer to question (v) is affirmative; you can prune branch $(O \,\&\, h \,\&\, \textit{not } v)$. By theorems 2 and 3, you know exactly which line reflections $\mathscr{G}$ contains: σ_h, the reflections across grid lines $v_{n/2}$ for $n = 0, \pm 1, \pm 2, \ldots,$ and no others. Moreover, $\mathscr{G}$ contains glide reflections $\tau^n \sigma_h$ for $n = 0, \pm 1, \pm 2, \ldots,$ with axis h and vector $n\vec{\tau}$. Thus the answer to question (γ) is also affirmative; you can prune leaf $(O \,\&\, h \,\&\, v \,\&\, \textit{not } \gamma)$. If $\mathscr{G}$ contained a glide reflection φ with axis h and vector $(n + \frac{1}{2})\vec{\tau}$, then it would also contain the translation $\tau^{-n}\varphi$ with vector $\frac{1}{2}\vec{\tau}$, which is impossible. By theorem 4, therefore, $\mathscr{G}$ contains only the glide reflections $\tau^n \sigma_h$. This paragraph has completely described the members of $\mathscr{G}$ in terms of vector $\vec{\tau}$ and center O. It's the symmetry group of the frieze ···HHHH···.

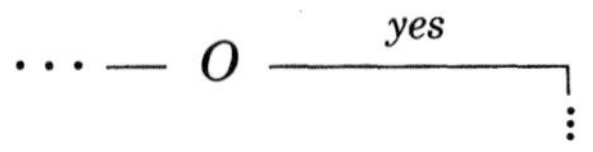

Figure 8.2.6
Branch (O)

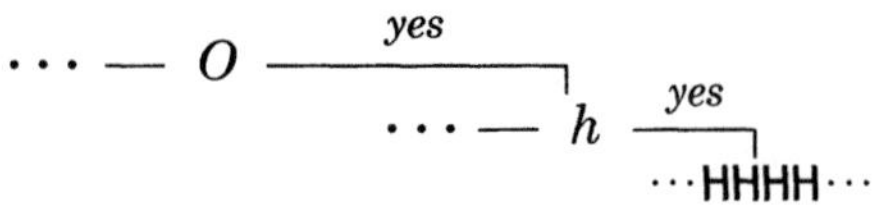

Figure 8.2.7
Branch $(O \,\&\, h)$

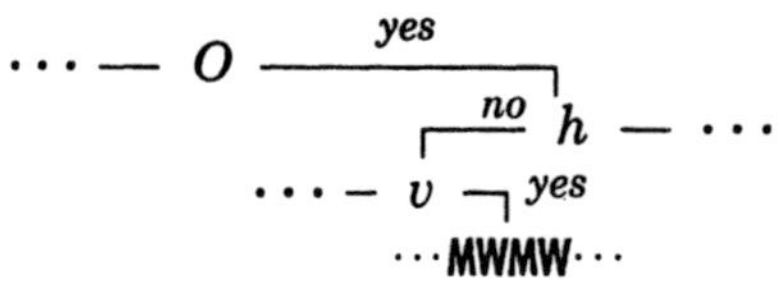

··· — O —yes— h — ···
no
··· — v —yes
···MWMW···

Figure 8.2.8 Branch
(O & *not* h & v)

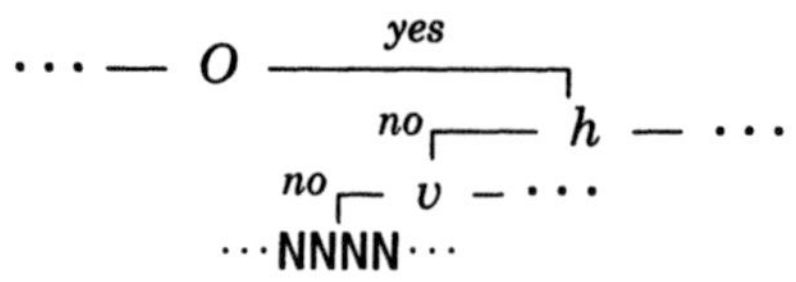

··· — O —yes— h — ···
no
no — v — ···
···NNNN···

Figure 8.2.9 Branch
(O & *not* h & *not* v)

Consider branch (O & *not* h & v), as indicated in figure 8.2.8. By theorem 4, $\mathcal{G}$ contains reflections across grid lines $v_{n/2}$ for $n = 0, \pm1, \pm2, \ldots$, but no others. $\mathcal{G}$ must contain $\varphi = \sigma_O \sigma_v$, a glide reflection with horizontal axis h through O and vector $2(O - P)$, where P is the foot of the perpendicular from O to v. Thus, the answer to question (γ) is affirmative; you can prune leaf (O & *not* h & v & *not* γ). Translation φ^2 also belongs to $\mathcal{G}$, so its vector $4(O - P) = m\vec{\tau}$ for some integer m. Were m even, $\mathcal{G}$ would contain $\sigma_h = \varphi\tau^{-m/2}$, contrary to the branch definition. Therefore m is odd: $m = 2k + 1$, and φ has vector $2(O - P) = \frac{1}{2} m\, \vec{\tau} = (k + \frac{1}{2})\vec{\tau}$. This equation implies that O, hence *all* lattice points, must lie midway between adjacent grid lines. Moreover, by this equation, $\mathcal{G}$ contains for each integer n the glide reflection $\varphi\tau^{n-k}$ with axis h and vector $(n + \frac{1}{2})\vec{\tau}$. It cannot contain glide reflection χ with axis h and vector $n\vec{\tau}$, else it would contain $\sigma_h = \chi\tau^{-n}$, which is impossible. By theorem 4, you've now described *all* members of $\mathcal{G}$, in terms of vector $\vec{\tau}$ and center O. It's the symmetry group of the frieze ···MWMW···.

The next branch to visit is (O & *not* h & *not* v), depicted in figure 8.2.9. By theorem 4, a glide reflection γ in $\mathcal{G}$ would have a horizontal axis, and $\mathcal{G}$ would contain $\gamma\sigma_O$, a glide reflection with a vertical axis, contrary to theorem 4 or the branch definition. The answer to question (γ) is negative; you can prune leaf (O & *not* h & *not* v & γ). You can now describe all members of $\mathcal{G}$ in terms of vector $\vec{\tau}$ and center O. It's the symmetry group of the frieze ···NNNN···.

Analyze branch (*not* O & h). It's easy to see—in figure 8.2.10—that the answers to questions (v) and (γ) must be *no* and *yes*. You can prune branch (*not* O & h & v) and leaf (*not* O & h & *not* v & *not* γ). $\mathcal{G}$ contains a glide reflection φ if and only if it contains the translation $\varphi\sigma_h$, hence it contains glide reflections $\sigma_h\tau^n$ for $n = 0, \pm1, \pm2, \ldots$, but no others. These sentences completely describe the members of $\mathcal{G}$ in terms of vector $\vec{\tau}$ and axis h. It's the symmetry group of the frieze ···BBBB···.

You can analyze the remaining branch (*not* O & *not* h), shown in figure 8.2.11, all at once. First, suppose $\mathcal{G}$ contains a reflection across a vertical line v. If $\mathcal{G}$ also contained a glide reflection γ with nonzero vector, then

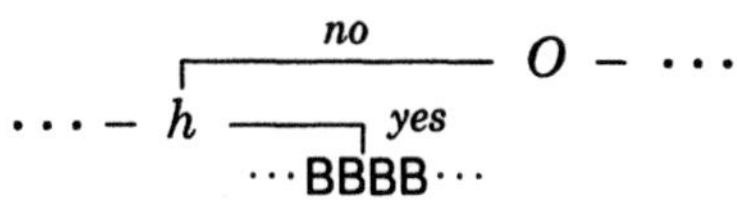

Figure 8.2.10
Branch (*not O* & *h*)

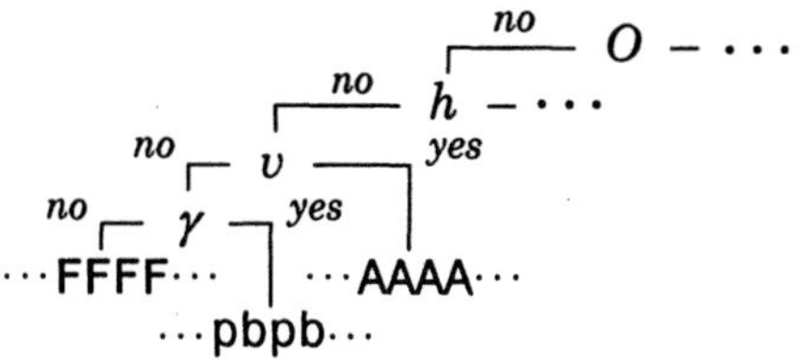

Figure 8.2.11
Branch (*not O* & *not h*)

γ would have a horizontal axis, and $\mathcal{G}$ would contain the half turn $\gamma\sigma_v$, contrary to the branch definition. You can prune leaf (*not O* & *not h* & v & γ), and use theorem 3 to describe all the reflections in $\mathcal{G}$ in terms of vector $\vec{\tau}$ and axis v. It's the symmetry group of the frieze ···AAAA···. Next, suppose $\mathcal{G}$ contains no reflection at all. If it contains no glide reflection, then it contains only the translations—it's the symmetry group of the frieze ···FFFF···. Finally, suppose $\mathcal{G}$ contains no reflection, but does contain a glide reflection γ with axis h and nonzero vector V. By theorem 4, h is horizontal and $V = \frac{1}{2}m\vec{\tau}$ for some integer m. Were m even, $\mathcal{G}$ would contain $\sigma_h = \gamma\tau^{-m/2}$, contrary to the branch definition. Therefore m is odd: $m = 2k + 1$, and γ has vector $\frac{1}{2}m\vec{\tau} = (k + \frac{1}{2})\vec{\tau}$. By this equation, $\mathcal{G}$ contains for each integer n the glide reflection $\gamma\tau^{n-k}$ with axis h and vector $(n + \frac{1}{2})\vec{\tau}$. It cannot contain glide reflection χ with axis h and vector $n\vec{\tau}$, else it would contain $\sigma_h = \chi\tau^{-n}$, which is impossible. By theorem 4, you've now described *all* members of $\mathcal{G}$ in terms of vector $\vec{\tau}$ and axis h. It's the symmetry group of the frieze ···pbpb···. Analysis of this last branch is complete, and the proof is finished! Here's the complete statement of the result:

Theorem 5 (Frieze classification theorem). Every frieze group is the symmetry group of a frieze similar to one of these:

···AAAAAAAAAA··· ···BBBBBBBBBB··· ···FFFFFFFFFF···
···HHHHHHHHHH··· ···NNNNNNNNNN··· ···MWMWMWMWMW···
···pbpbpbpbpb···

Moreover, you can classify any frieze by applying the decision tree in figure 8.2.12 to its symmetry group.

Two frieze groups classified alike by theorem 5 are symmetry groups of similar figures, hence they're conjugate, by theorem 8.1.1. On the other hand, the decisions required by theorem 5 to classify a frieze are all phrased

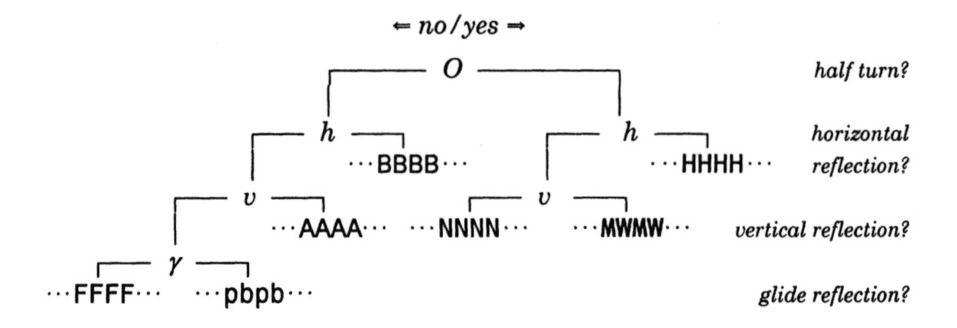

Figure 8.2.12 The frieze tree

in terms of geometric properties of its symmetries that are preserved under the conjugation $\psi \to \psi^\varphi$ defined by a similarity φ. For example, if φ maps frieze F to frieze F', and you ask question (O) of the figure 8.2.5 decision tree regarding F, you're asking whether the symmetry group $\mathscr{G}_F$ of F contains the half turn σ_O about some center O. You'll get the same answer if you ask whether symmetry group $\mathscr{G}_{F'}$ of F' contains the half turn $\sigma_{\varphi(O)} = \sigma_O^\varphi$ about some point $\varphi(O)$. Therefore, theorem 5 classifies conjugate frieze groups alike. In summary,

Theorem 6. Two frieze groups are classified alike by theorem 5 if and only if they're conjugate.

This text refers to the seven types of frieze groups by displaying alphabetic examples. Although effective, this is not standard notation. Two other notational systems have emerged as standard. One was established by the geometer L. Fejes Tóth.[20] Another has emerged from the crystallographic literature. Here are the corresponding group designations:

Frieze group notation

Example frieze	Fejes Tóth	Crystal
···AAAAAAAAAA···	$\mathfrak{F}_1^2$	*pm 1 1*
···BBBBBBBBBB···	$\mathfrak{F}_1^1$	*p 1 m1*
···FFFFFFFFFF···	$\mathfrak{F}_1$	*p1 1 1*
···HHHHHHHHHH···	$\mathfrak{F}_2^1$	*pmm2*
···NNNNNNNNNN···	$\mathfrak{F}_2$	*p1 1 2*
···MWMWMWMWMW···	$\mathfrak{F}_2^2$	*pma 2*
···pbpbpbpbpb···	$\mathfrak{F}_1^3$	*p1 a 1*

[20] Fejes Tóth 1964, section 3.

Fejes Tóth's notation stemmed from the order of his presentation. Groups $\mathfrak{F}_1$, $\mathfrak{F}_1^1$, $\mathfrak{F}_1^2$, and $\mathfrak{F}_1^3$ contain no half turn; the first contains no nontranslational symmetries, but the others contain horizontal, vertical, and glide reflections, respectively. Groups $\mathfrak{F}_2$, $\mathfrak{F}_2^1$, and $\mathfrak{F}_2^2$ all contain half turns; the first contains no other nontranslational symmetries, but the others contain horizontal and glide reflections, respectively. In the *pxyz* notation, x is m or 1 depending on whether the group contains a vertical reflection or not; y is m, a, or 1 depending on whether it contains a horizontal reflection, a glide but no horizontal reflection, or neither; and z is m or 1 depending on whether the group contains a half turn or not.

Artists have constructed friezes since prehistoric times. Owen Jones included examples of all seven frieze classes in the illustrations of ancient Egyptian ornaments in his classic text *The grammar of ornament*.[21] You can find hundreds of examples from other cultures in later studies of ornamentation.[22] The group-theoretic approach was developed to study a much harder problem: three-dimensional crystal symmetry. Its success in classifying the 230 crystal symmetry classes, and the route from that scientific work to its application in art is outlined in the introduction to this chapter.

Symmetry analysis of art work is where mathematics, art criticism, and anthropology intersect. It was born in Pólya's and Niggli's 1924 papers and Speiser's 1927 book. The world was soon engulfed in depression and war. In 1942 Niggli's niece Edith Müller wrote a famous Ph.D. thesis under Speiser's supervision.[23] She described the frieze and wallpaper groups in detail, then used them to classify the ornamentation of the splendid Moorish palace, the Alhambra, in Granada. Simultaneously, anthropologists were beginning to see the relevance of symmetry analysis for the study of native American cultures.[24] Thirty years passed, however, before that work was extended. In their elegant and profusely illustrated 1988 book *Symmetries of culture*, two pioneers in the application of symmetry analysis, anthropologist Dorothy K. Washburn and mathematician Donald W. Crowe, survey recent work. Anthropologists now infer differences in culture from differences in ornamental symmetry. Washburn has paid particular attention to the puzzling migratory patterns of the pre-European cultures of the American Southwest:

---

[21] Jones [1856] 1972.

[22] For example, all seven frieze patterns are found by Crowe and Washburn on San Ildefonso pottery (1985), by Gerdes and Bulafo in Inhambane basketry (1994a, 92), by the Hargittais in Hungarian needlework (1994, 135), and by McLeay in Australian cast ironwork (1994)

[23] Müller 1944: *Gruppentheoretische und strukturanalytische Untersuchungen der maurischen Ornamente aus der Alhambra in Granada.*

[24] Washburn and Crowe 1988, 12–14.

> Since a number of ... studies have shown that design structures within a population group are generally homogeneous and nonrandom, imported materials should be recognizable by their different structural layouts. In a study of black-on-white cylinder jar designs from Chaco Canyon, New Mexico, Washburn found that the locally made Pueblo II one-dimensional rotational (*p112*) design system was replaced in the Pueblo III period by two-dimensional layouts. Formal attributes of these atypical vessels suggested a nonlocal origin and, indeed, likely models for the form and design system can be found in the Mixtec and Toltec cultures of the Valley of Oaxaca.[25]

Artists of these cultures create designs that go beyond the limits of frieze group analysis. For example, Müller noted that close inspection of the Alhambra frieze patterns reveals that many are intertwined; such a pattern has a front and a back, like figure 8.2.13. These are really three-dimensional objects. The translations in their symmetry groups are all multiples of a single translation, but the groups may contain some of the three-dimensional isometries studied in chapter 7. Washburn and Crowe are deeply concerned with analysis of colored patterns such as those in the shaded figure 8.2.14. There you must determine whether an isometry that leaves fixed the outlines of various regions in a figure also preserves their colors, or perhaps permutes them. For references, consult Washburn and Crowe 1988 and Grünbaum and Shephard 1987.

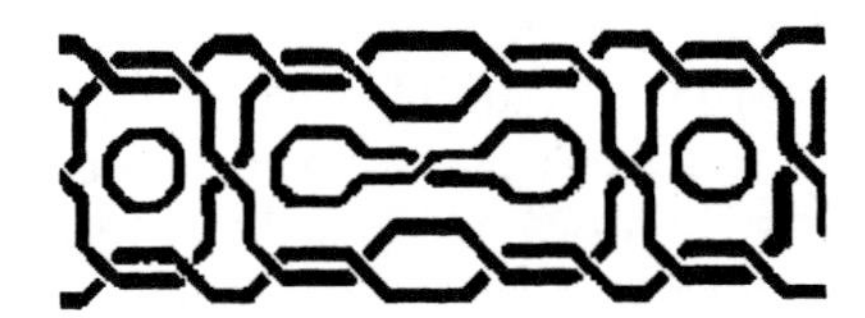

Figure 8.2.13
Intertwined
Alhambra frieze[26]

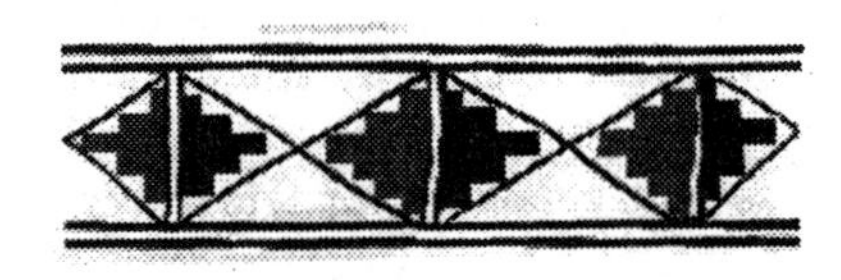

Figure 8.2.14
Bicolored
San Ildefonso frieze[27]

[25] Ibid. 27.

[26] Müller 1944, figure 19.

[27] Chapman 1970, plate 132r.

Paul NIGGLI was born in Zofingen, Switzerland, in 1888. His father was a Gymnasium teacher. During his school years, Niggli started work in mineralogy and geology by making a geological map of his home region. He entered the Eidgenössische Technische Hochschule (ETH) in Zürich in 1907, planning to become a secondary teacher. But he stayed on for the Ph.D., which he received in 1912. Niggli's first scientific position was at the Carnegie Geophysics Laboratory in Washington, D.C. There he began a series of fruitful collaborations that lasted throughout his career. From 1915 to 1920 he taught at the universities in Leipzig and Tübingen. During that period he wrote the first "user-friendly" text that explained to geoscience students the mathematical crystallography recently developed by Fedorov and Schönflies. In 1920 Niggli accepted concurrent full professorships in mineralogy and petrology at the ETH and the University of Zürich. He became a world leader in the field, editing the *Zeitschrift für Kristallographie* for twenty years, producing a long series of influential monographs and texts, and authoring over sixty research papers. He served as rector for several years at each of his universities. Niggli died in 1953.

8.3 Wallpaper ornaments

Concepts

- Wallpaper ornaments and wallpaper groups
- Fundamental translations
- Lattices and grids
- Rectangular, rhombic, square, centered rectangular, and hexagonal configurations
- Isomorphy
- Crystallographic restriction
- m-centers
- Classifying wallpaper ornaments
- Wallpaper group nomenclature

A *wallpaper ornament* is a symmetric plane figure that repeats indefinitely —or at least is always extensible—in two nonparallel directions. You can construct one by making a *basic* cell first, then juxtaposing duplicate cells as many times as you wish in two different, nonparallel directions and their opposites. Figures 8.3.1 and 8.3.2 present two examples, with basic cells identified. The symmetry group $\mathscr{W}$ of a wallpaper ornament, called a *wallpaper* group, consists of all plane isometries that leave the pattern fixed. It contains all compositions $\varphi^m\chi^n$ of powers of two nonparallel translations φ and χ (indicated in figures 8.3.1 and 8.3.2), but no other translations. Thus, $\mathcal{W}$ contains the translations with vectors $m\vec{\varphi} + n\vec{\chi}$ for $m, n = 0, \pm1, \pm2, \ldots$, but no others. Figure 8.2.3 is an example of a pattern whose symmetry group contains two nonparallel translations: vertical φ and horizontal χ as indicated. But it's *not* a wallpaper ornament, because its symmetry group contains *every* vertical translation, not just the powers of φ. This section presents an analysis and a classification of wallpaper groups, analogous to the study of frieze groups in section 8.2.

Lattice and grid

A glance at figures 8.3.1 and 8.3.2 shows that the translations φ and χ mentioned in the previous paragraph aren't unique. You could use other pairs $\varphi^k\chi^l$ and $\varphi^m\chi^n$ in place of φ and χ. (See exercise 8.5.19.) This ambiguity is inconvenient; theorems 1 and 2 provide a more definite way to specify the translations φ and χ.

Theorem 1. A wallpaper group $\mathcal{W}$ contains a shortest nontrivial translation τ.

Proof. Consider a point O as shown, for example, in figure 8.3.1. The points $P = \varphi^k\chi^l(O)$ for integers k and l are the lower ends of the ⌈ motifs. The distance condition $0 \neq OP \leq O\varphi(O)$ is satisfied by at least one but only finitely many of these points. Find such a point $P = \varphi^k\chi^l(O)$ with OP as small as possible, and set $\tau = \varphi^k\chi^l$. (In figure 8.3.1, $k = l = 1$ and $\tau = \varphi\chi$.) ♦

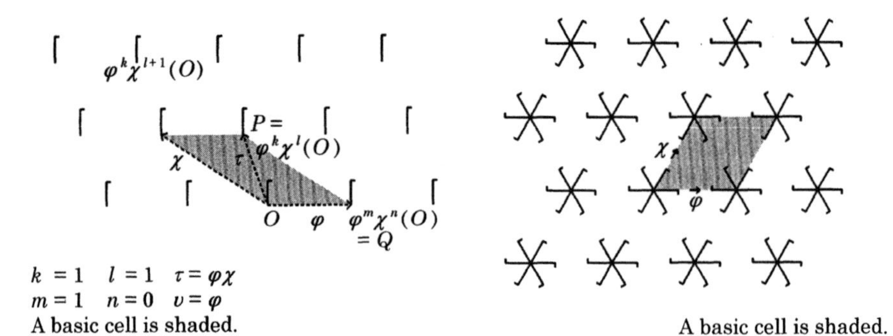

Figure 8.3.1
p1 wallpaper pattern

Figure 8.3.2
p6 wallpaper pattern

Translation τ in theorem 1 is still not uniquely determined; any translation φ in $\mathcal{W}$ with the same length as τ —for example, $\varphi = \tau^{-1}$ —is appropriate. Select one such translation τ, and call it the *first fundamental* translation for $\mathcal{W}$.

Theorem 2. Suppose the first fundamental translation τ has been selected for wallpaper group $\mathcal{W}$. Then $\mathcal{W}$ contains a translation υ not parallel to τ such that

- no translation in $\mathcal{W}$ not parallel to τ is shorter than υ,
- if $T = \tau(O)$ and $U = \upsilon(O)$ then $60° \le m\angle TOU \le 90°$,
- $m\angle OUT \le 60°$, and
- U doesn't lie on the same side of the perpendicular bisector k of $\overline{OT}$ as T.

Proof. Use the notation of the previous proof. Since $\varphi^k\chi^{l+1}$ isn't parallel to τ, the following condition is satisfied by at least one but only finitely many points $Q = \varphi^m\chi^n(O)$ for integers m and n: $\varphi^m\chi^n$ is not parallel to τ and $0 \ne OQ \le \varphi^k\chi^{l+1}(O)$. Find such a point Q with OQ as small as possible, and set $\upsilon = \varphi^m\chi^n$. (In figure 8.3.1, $m = 1$, $n = 0$, and $\upsilon = \varphi$.) Should $m\angle TOU \ge 90°$, replace υ by its inverse. If $m\angle TOU < 60°$, the hinge theorem would imply a contradiction—all angles of ΔOUT would measure less than $60°$ because $TU \ge OU \ge OT$. Similarly, $m\angle OUT > 60°$ would imply that all angles of ΔOUT would measure more than $60°$. Finally, U couldn't lie on the same side of k as T, for that would imply $OU > OT$. ♦

Select one such translation υ as described in theorem 2, and call it the *second fundamental* translation for $\mathcal{W}$. There's only one choice unless ΔOUT is equilateral. In that case, there are two—one point on each side of $\overleftrightarrow{OT}$ —and you may choose either one. According to the next result, the two fundamental translations determine *all* the translations in $\mathcal{W}$.

Theorem 3. For each translation ψ in $\mathcal{W}$ there exist unique integers m and n such that $\psi = \tau^m\upsilon^n$.

Proof. Use coordinate geometry with origin O so that points $T = \tau(O)$ and $U = \upsilon(O)$ have coordinates $<t,0>$ and $<0,u>$ with $t,u > 0$. Find real numbers x and y and integers m and n such that $\psi(O)$ has coordinates $<x,y>$, $mt \le x < mt + 1$, and $nu \le y < nu + 1$. Then $\mathcal{W}$ contains the translation $\omega = \psi\tau^{-m}\upsilon^{-n}$. Point $W = \omega(O)$ is distinct from T and U and lies on $\overline{OT}$ or $\overline{OU}$ or in the interior of parallelogram $OTVU$. $\mathcal{W}$ also contains translations $\omega\tau^{-1}$, $\omega\upsilon^{-1}$, and $\omega\tau^{-1}\upsilon^{-1}$, and $W = \omega\upsilon^{-1}(U) = \omega\tau^{-1}\upsilon^{-1}(V) = \omega\tau^{-1}(T)$. In the interior case, shown in figure 8.3.3, $\angle OWT$, $\angle TWV$, $\angle VWU$, and $\angle UWO$ would be smallest angles of

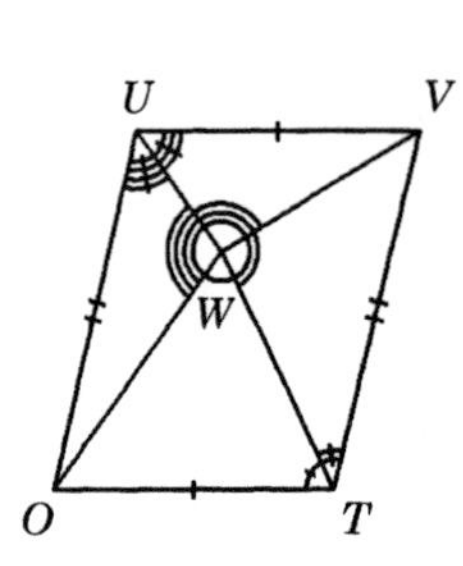

Figure 8.3.3
Proving theorem 3

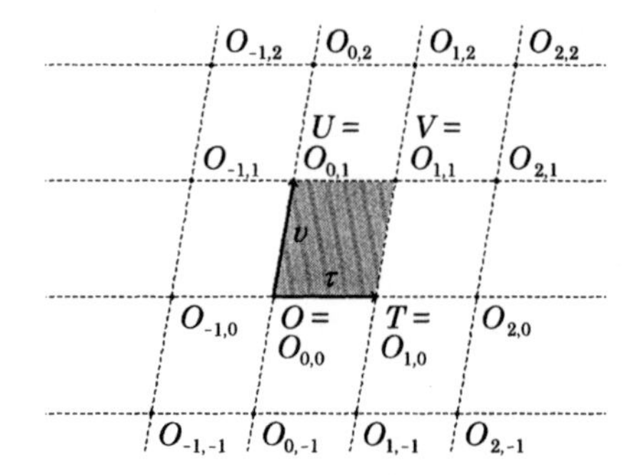

A fundamental cell is shaded.

Figure 8.3.4 Lattice and grid determined by a point O and fundamental translations τ and υ

ΔOWT, ΔTWV, ΔVWU, and ΔUWO, so each would measure at most $60°$, which is impossible because their total measure must be $360°$. Thus W must lie on $\overline{OT}$ or $\overline{OU}$. By theorem 1 or 2, $W = O$, hence $\psi = \tau^m\upsilon$ If $\tau^m\upsilon^n = \tau^{m'}\upsilon^{n'}$, then $\tau^{m-m'} = \upsilon^{n'-n}$, which is impossible unless $m - m' = n' - n = 0$, so m and n are unique. ♦

Given any point O, select a first fundamental translation τ as in theorem 1. Then theorem 2, with perhaps an arbitrary choice between two possibilities, yields a second fundamental translation υ. These define the *lattice* of points $O_{m,n} = \tau^m\upsilon^n(O)$ for all integers m and n, and the *grid* of lines parallel to τ or υ through lattice points. A parallelogram $O_{m,n}O_{m+1,n}O_{m+1,n+1}O_{m,n+1}$ is called a *fundamental cell.* No lattice points lie in its interior, and only these four lie on its edges. These notions are illustrated in figure 8.3.4. The rest of this section uses the notation

$$\begin{aligned} T &= O_{1,0} = \tau(O) & X &= \text{midpoint of } \overline{OT} \\ U &= O_{0,1} = \upsilon(O) & Y &= \text{midpoint of } \overline{OU} \\ V &= O_{1,1} = \tau\upsilon(O) & Z &= \text{midpoint of } \overline{OV}. \end{aligned}$$

Some lattice configurations require special considerations. When $\tau \perp \upsilon$, the configuration is called *rectangular,* after the shape of the fundamental cell. In the *rhombic* configuration, $|\tau| = |\upsilon|$ and the fundamental cells are rhombi. The *square* configuration is both rectangular and rhombic. In the *centered rectangular* configuration of figure 8.3.5, U lies on the perpendicular bisector of $\overline{OT}$. This terminology stems from a tendency to see U as the center of a rectangle in the lattice, rather than as the vertex of a parallelogram. The confusion extends further. Some authors don't impose minimality conditions on fundamental translations, but instead merely require that they satisfy the conclusion of theorem 3. They'd regard parallelogram *OSTU* in figure 8.3.5 as fundamental. This book calls *OSTU* a basic cell, but not fundamental, because $OS, OU > OT$. Some authors who take the opposing view note that *OSTU* is a rhombus, and call this configuration rhombic as well. One configuration is both rhombic and centered rectangular: when $OT = TU = UO$. In that case, regular hexagons like those in the figure 8.3.2 lattice stand out, and the configuration is called *hexagonal*.

Isomorphy

A wallpaper group might contain only translations—for example, consider the symmetry groups of the patterns in figures 8.3.1 and 8.3.6. Should all such groups be classified alike? It seems so; we tend to say that two such designs display only the translational symmetry required of all wallpaper patterns. A criterion that leads to this conclusion will be different from the one used in sections 8.1 and 8.2 to classify symmetry groups of polygons and friezes. The patterns in figures 8.3.1 and 8.3.6 are not similar; the latter was made from the former by changing the horizontal and vertical scaling independently. They're related by a transformation called an *affinity*, not one of the similarity transformations studied in this book. You can't expect the groups to be conjugate as defined in section 8.1.

We need a less restrictive criterion than conjugacy. Two wallpaper groups $\mathcal{W}$ and $\mathcal{W}'$ will be classified the same if their relationship has some but

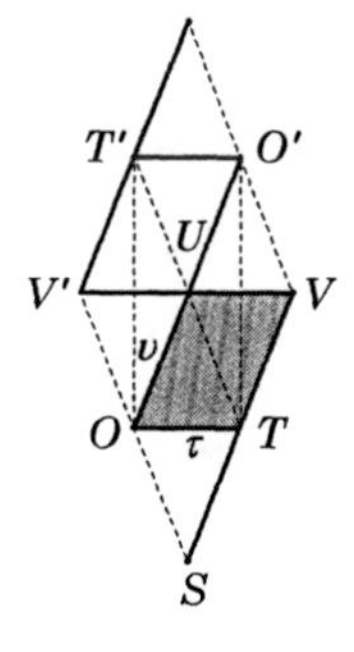

Grid lines are solid.
A fundamental cell is shaded.

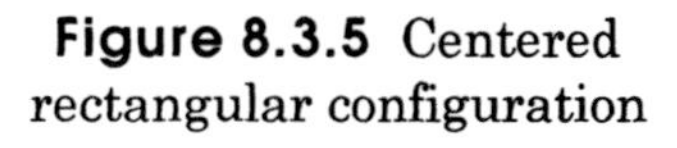

Figure 8.3.5 Centered rectangular configuration

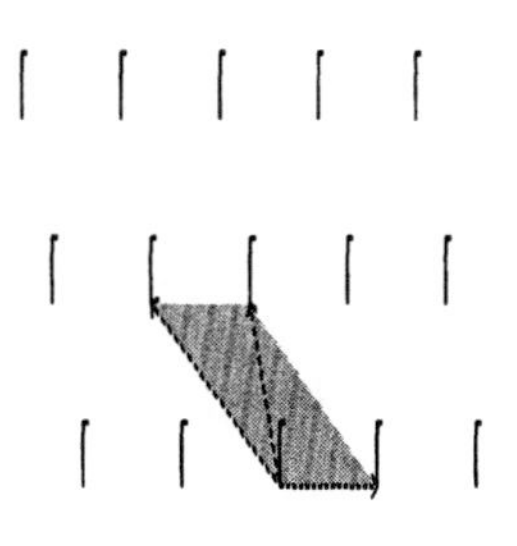

Figure 8.3.6 Figure 8.3.1 distorted

not all properties of conjugacy. The new criterion is *isomorphy*.[28] Transformation groups $\mathcal{G}$ and $\mathcal{G}'$ are called *isomorphic* if there's a bijection $\varphi \to \varphi'$ from $\mathcal{G}$ to $\mathcal{G}'$ —called an *isomorphism*—that preserves compositions. That is, for any members φ and χ of $\mathcal{G}$, $(\varphi\chi)' = \varphi'\chi'$. Conjugate groups are isomorphic. If ψ is an isometry and $\mathcal{G}'$ consists of the conjugates $\psi\varphi\psi^{-1}$ of members φ of $\mathcal{G}$, then the conjugation $\varphi \to \varphi' = \psi\varphi\psi^{-1}$ is an isomorphism, because

$$(\varphi\chi)' = \psi(\varphi\chi)\psi^{-1} = (\psi\varphi\psi^{-1})(\psi\chi\psi^{-1}) = \varphi'\chi'.$$

But as you'll see, isomorphic isometry groups are not necessarily conjugate. (Exercise 8.5.13 verifies that isomorphy is an inappropriate criterion for classifying symmetry groups of friezes.)

For example, the symmetry groups $\mathcal{W}$ and $\mathcal{W}'$ of figures 8.3.1 and 8.3.6 are isomorphic. If τ, υ and τ', υ' are pairs of fundamental translations for these groups, then by theorem 3, to any members φ and φ' of $\mathcal{W}$ and $\mathcal{W}'$ correspond unique integers k, l and k', l' such that $\varphi = \tau^k\upsilon^l$ and $\varphi' = (\tau')^{k'}(\upsilon')^{l'}$. Thus $\varphi \to \varphi' = (\tau')^k(\upsilon')^l$ is a bijection from $\mathcal{W}$ to $\mathcal{W}'$. In fact, it's an isomorphism—if $\chi = \tau^m\upsilon^n$, then

$$\begin{aligned}(\varphi\chi)' &= ((\tau^k\upsilon^l)(\tau^m\upsilon^n))' = (\tau^{k+m}\upsilon^{l+n})' = (\tau')^{k+m}(\upsilon')^{l+n} \\ &= ((\tau')^k(\upsilon')^l)((\tau')^m(\upsilon')^n) = \varphi'\chi'.\end{aligned}$$

This argument depended on few details of the figures—simply that they are wallpaper patterns whose symmetry groups consist solely of translations. It follows that any wallpaper group that consists solely of translations is isomorphic to the symmetry group of figure 8.3.1. The class of all such groups is called *p1*.

The following theorem lists several properties of isomorphisms. It clarifies the notion of isomorphy, and will be used to show that various wallpaper groups are *not* isomorphic.

Theorem 4. Suppose $\varphi \to \varphi'$ is an isomorphism between wallpaper groups $\mathcal{W}$ and $\mathcal{W}'$. Then $\iota' = \iota$ and $(\varphi^{-1})' = (\varphi')^{-1}$. If a member φ of $\mathcal{W}$ is a translation, glide reflection, reflection, or rotation through angle $360°/n$ for some integer $n > 0$, then so is the corresponding isometry φ' in $\mathcal{W}'$.

Proof. $\iota' = \iota$ because $\varphi'\iota' = (\varphi\iota)' = \varphi'$ for any φ. Moreover, $(\varphi^{-1})' = (\varphi')^{-1}$ because $\varphi'(\varphi^{-1})' = (\varphi\varphi^{-1})' = \iota' = \iota$.

If φ is a translation or a glide reflection, then its powers φ^n are all different, hence so are those of φ', which is therefore a translation or a glide reflection.

---

[28] This term stems from Greek words *isos* and *morphe* meaning *same* and *form*.

If φ were a translation but φ' a glide reflection, you could find a translation χ' in $\mathscr{W}'$ (one of its fundamental translations would do) such that $\chi'^2\varphi' \neq \varphi'\chi'^2$. By the previous paragraph, the corresponding element χ of $\mathscr{W}$ would be a translation or a glide reflection, hence χ^2 would be a translation and $\chi^2\varphi = \varphi\chi^2$. That would contradict the previous inequality. Thus φ' is a translation when φ is. This argument in reverse shows that φ' is a glide reflection when φ is.

If φ is a reflection, then φ' is a glide reflection by the previous paragraph. Further, $\varphi'^2 = \iota$ because $\varphi^2 = \iota$, so φ' is a reflection.

If φ is a rotation through angle $360°/n$ for some integer $n > 0$, then n is the first $k > 0$ for which $\varphi^k = \iota$. Since $(\varphi')^k = (\varphi^k)'$, it follows that n is the first $k > 0$ for which $(\varphi')^k = \iota$, hence φ' is a rotation through angle $360°/n$. ♦

Theorem 4 implies that any wallpaper group isomorphic to the symmetry group of figure 8.3.1 consists solely of translations. Lemma 5 summarizes the properties of such groups:

Lemma 5 (p1). Any wallpaper group that consists solely of translations is isomorphic to the symmetry group $\mathscr{W}$ of figure 8.3.1. The class of all such groups is called *p1*. No other wallpaper group is isomorphic to $\mathscr{W}$.

Lemma 5 is the first of a sequence of lemmas that together constitute the wallpaper classification theorem. Each lemma will define one or more classes of wallpaper groups $\mathscr{W}$ concisely in terms of the symmetries they include. From that definition will be derived a complete description of the membership of each $\mathscr{W}$ —the vectors, axes, centers, and angles of its symmetries—in terms of a lattice. Two wallpaper groups $\mathscr{W}$ and $\mathscr{W}'$ in the same class with different lattices are always isomorphic; you can define a bijection $\varphi \to \varphi'$ between them by describing φ uniquely in terms of its vector, axis, center, or angle and the lattice of $\mathscr{W}$, then specifying φ' as the isometry with the corresponding description in terms of the lattice of $\mathscr{W}'$. The bijection preserves composition because you can always describe the composition $\varphi\chi$ of symmetries in $\mathscr{W}$ in terms of the descriptions of φ and χ, and you can describe $\varphi'\chi'$ in terms of φ' and χ'. Your descriptions of $\varphi\chi$ and $\varphi'\chi'$ will differ only in their references to the lattices of $\mathscr{W}$ and $\mathscr{W}'$, so $(\varphi\chi)' = \varphi'\chi'$. This process is carried out in detail in the example preceding theorem 4. You can supply analogous details, if you wish, for the remaining classes of wallpaper groups.

Lemma 7, the next in the sequence, will classify the remaining wallpaper groups that contain no nontrivial rotations. One lemma per class would be cumbersome because there are seventeen different classes, so each lemma summarizes the results for several classes. The entire classification process will be outlined following lemma 7. Three more lemmas concerned with groups that contain various rotations will complete the classification theorem. The crystallographic notation for the groups—for example, the symbol *p1*—is discussed last in this section.

Reflectional symmetry

The next theorem begins to analyze the relationships of the symmetries that constitute wallpaper groups. What limitations result from the simultaneous presence of translations and glide reflections? The statement of this theorem is the most complicated one in the book. Suppose you're setting up a lattice to analyze a wallpaper group $\mathcal{W}$, and have selected the point O and first fundamental translation τ conveniently. Should $\mathcal{W}$ contain an odd isometry with axis g, then the second fundamental translation υ can be situated in a special way with regard to g, and $\mathcal{W}$ must also contain an odd isometry with axis parallel to g and vector closely related to the lattice as described in the theorem. The theorem will be used to provide explicit descriptions of the wallpaper groups that contain only trivial or $180°$ rotations.

Theorem 6. If wallpaper group $\mathcal{W}$ contains a glide reflection χ with axis l, then you can select the second fundamental translation υ and find a glide reflection γ in $\mathcal{W}$ with axis $g \parallel l$ and such that

(1) $|\tau| = |\upsilon|$, and — *rhombic configuration*
  (1a) $\tau\upsilon \parallel g$, and $\gamma = \sigma_g$ or $\gamma^2 = \tau\upsilon$, or
  (1b) $\tau\upsilon^{-1} \parallel g$, and $\gamma = \sigma_g$ or $\gamma^2 = \tau\upsilon^{-1}$; or

(2) U lies on the perpendicular bisector k of $\overline{OT}$, and — *centered rectangular configuration*
  (2a) $g \perp \tau$, and $\gamma = \sigma_g$ or $\gamma^2 = \tau^{-1}\upsilon^2$, or
  (2b) $g \parallel \tau$, and $\gamma = \sigma_g$ or $\gamma^2 = \tau$; or

(3) $\tau \perp \upsilon$, and — *rectangular configuration*
  (3a) $g \perp \tau$, and $\gamma = \sigma_g$ or $\gamma^2 = \upsilon$, or
  (3b) $g \parallel \tau$, and $\gamma = \sigma_g$ or $\gamma^2 = \tau$.

Proof. Suppose l is not parallel to τ. The parallel case is considered later. For each integer n, the glide reflection $\varphi = \tau^n\chi\tau^{-n}$ with axis $g = \tau^n[l] \parallel l$ belongs to $\mathcal{W}$; pick n so that g intersects grid line $\overleftrightarrow{OT}$ at a point $P \neq T$ between O and T.

Suppose g is not perpendicular to τ. The perpendicular case is considered later. Then $\sigma_g\tau\sigma_g$ and $\sigma_g\tau^{-1}\sigma_g$ are translations in $\mathcal{W}$ not parallel to τ but equally long. Exactly one of these, as shown in figure 8.3.7, must satisfy the second condition of theorem 2; call it υ. Depending on that selection, $g \parallel \tau\upsilon$ or $\tau\upsilon^{-1}$. (In the two situations shown, the acute angles α between τ and g are $40°$ and $55°$; they lie between $30°$ and $60°$, making $60° \leq$

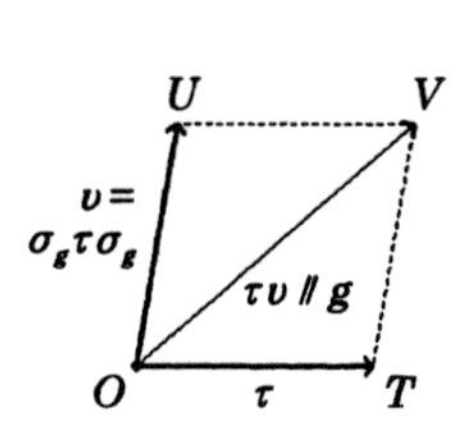

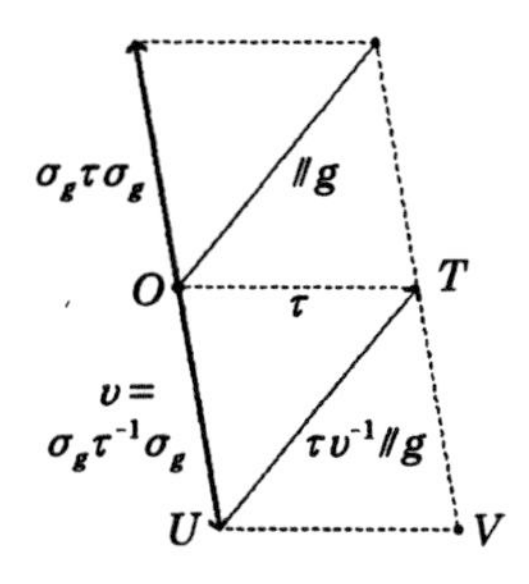

Figure 8.3.7
Proving theorem 6 when neither $g \parallel \tau$ nor $g \perp \tau$

m$\angle TOU \le 90°$ as required. Values of α outside that range would entail that $\mathscr{W}$ contain a nontrivial translation shorter than τ.) Consider first the case $g \parallel \tau\upsilon$. Since φ^2 is a translation in $\mathscr{W}$ parallel to g, you can find an integer n such that $\varphi^2(O)$ lies between $(\tau\upsilon)^n(O)$ and $(\tau\upsilon)^{n+1}(O)$ but $\varphi^2(O) \ne (\tau\upsilon)^{n+1}(O)$. That implies $\varphi^2(O) = (\tau\upsilon)^n(O)$, else $(\tau\upsilon)^{-n}\varphi^2(O)$ would be a lattice point inside the fundamental cell. If n is even, then $\gamma = (\tau\upsilon)^{-\frac{1}{2}n}\varphi$ is a reflection in $\mathscr{W}$ with axis parallel to g; if n is odd, then $\gamma = (\tau\upsilon)^{-\frac{1}{2}(n-1)}\varphi$ is a glide reflection in $\mathscr{W}$ with axis parallel to g and $\gamma^2 = \tau\upsilon$. The previous sentence is conclusion (1a) of the theorem. The case $g \parallel \tau\upsilon^{-1}$ is argued similarly, with $\tau\upsilon^{-1}$ and U in place of $\tau\upsilon$ and O; it leads to conclusion (1b).

Next suppose $g \perp \tau$. Use theorem 2 to find a second fundamental translation υ. Suppose, as in figure 8.3.8, that υ is not perpendicular to τ. The perpendicular case is considered later. $\mathscr{W}$ contains translations $\omega = \varphi\upsilon\varphi^{-1}$ and $\omega^{-1}\upsilon$; the latter is parallel to, hence not shorter than, τ. By theorem 2, U doesn't lie on the side of k opposite O. It doesn't lie on the same side either, because that would make $\omega^{-1}\upsilon$ shorter than τ. Therefore U lies on k. Since φ^2 is a translation in $\mathscr{W}$ parallel to $\tau^{-1}\upsilon^2$, the previous paragraph's argument with $\tau^{-1}\upsilon^2$ in place of $\tau\upsilon$ shows that $\mathscr{W}$ contains σ_g or a glide reflection γ such that $\gamma^2 = \tau^{-1}\upsilon^2$ —conclusion (2a) of the theorem.

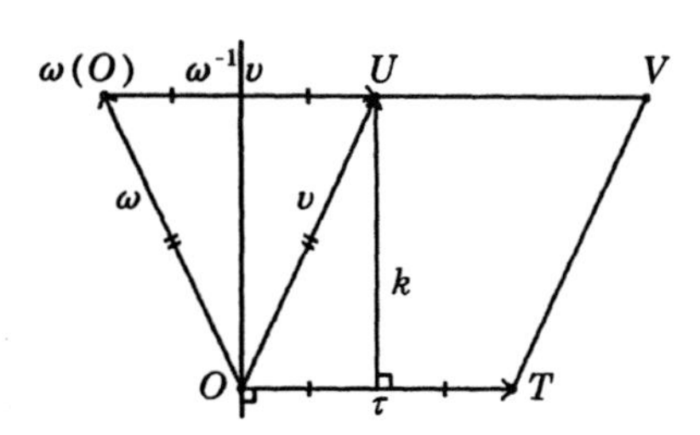

Figure 8.3.8
Proving theorem 6
when $g \perp \tau$ but not $\upsilon \perp \tau$

If $\upsilon \perp \tau$, the same argument with υ in place of $\tau^{-1}\upsilon^2$ leads to conclusion (3a) of the theorem.

Now suppose $g \parallel \tau$. Use theorem 2 to find a second fundamental translation υ. Then $\mathcal{W}$ contains translation $\omega = \varphi\upsilon\varphi^{-1}$, which has the same length as υ. If U didn't lie on k and υ weren't perpendicular to τ then $\omega\upsilon(O)$ would be a lattice point between but different from O and T, which is impossible. If U lies on k or $\upsilon \perp \tau$, the argument referred to in the previous paragraph, with τ in place of $\tau^{-1}\upsilon^2$, leads to conclusion (2b) or (3b) of the theorem. ♦

Now resume the classification of wallpaper groups begun in lemma 5. All wallpaper groups $\mathcal{W}$ contain a *p1* subgroup. What groups contain a glide reflection? Theorem 6 offers twelve possibilities for the relationship between a glide reflection and a lattice for $\mathcal{W}$. Place lattice point O on the axis of a glide reflection. In each of the (1) and (2) subcases with $\gamma \neq \sigma_g$, consider the glide reflection $\sigma = \upsilon^{-1}\gamma$. If $P = \sigma(O)$, then midpoint M of $\overline{OP}$ lies on the axis of σ, and $\sigma(M) = M$, hence σ is a reflection. Thus in cases (1) and (2), $\mathcal{W}$ contains a reflection.

Suppose $\mathcal{W}$ contains a glide reflection but no reflection or nontrivial rotation. For example, consider the symmetry group of figure 8.3.9. Theorem 6, subcase (3a) with $\gamma^2 = \upsilon$, or (3b) with $\gamma^2 = \tau$ must hold. Suppose the former, depicted in figure 8.3.10; subcase (3b) with $\gamma^2 = \tau$ leads to an analogous conclusion. Consider glide reflection $\tau\gamma = \gamma\tau^{-1}$. Find $P = \tau\gamma(O)$; midpoint M of $\overline{OP}$ lies on the axis of $\tau\gamma$. Find $M' = \tau\gamma(M)$; you can see that $\tau\gamma$ has the same vector $\tfrac{1}{2}\vec{\upsilon}$ as γ, but axis $\overleftrightarrow{MM'}$. Glide reflection $\upsilon\gamma = \gamma\upsilon$ has the same axis as γ but vector $\tfrac{3}{2}\vec{\upsilon}$. In fact, $\mathcal{W}$ contains glide reflection $\tau^m\upsilon^n\gamma$ for each m and n; its vector is $(n + \tfrac{1}{2})\vec{\upsilon}$. The axes of these glide reflections, determined by the value of m, constitute the lines perpendicular to τ midway between grid lines.

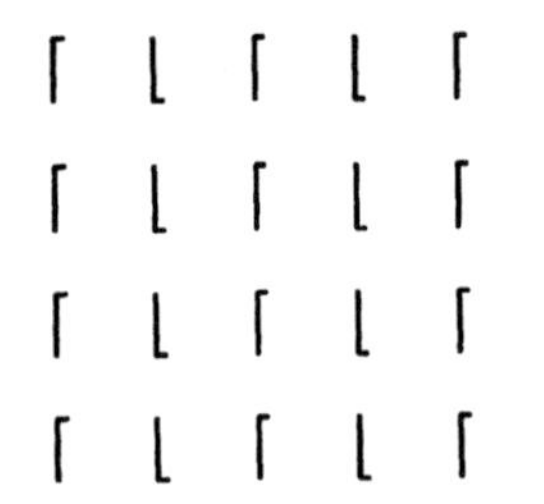

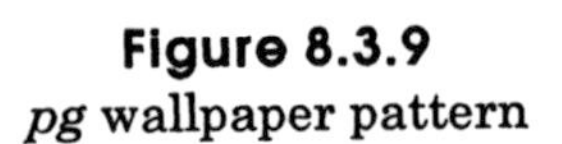

Figure 8.3.9
pg wallpaper pattern

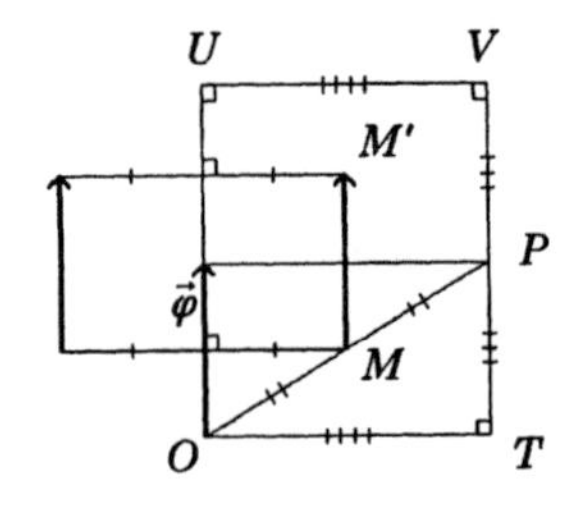

Figure 8.3.10
Analyzing case (3a)

Now suppose φ is *any* glide reflection in $\mathcal{W}$. If its axis h were not parallel to these axes, $\mathcal{W}$ would contain a nontrivial rotation $\varphi\gamma$. If it were parallel, but $\vec{\varphi} \neq (n + \frac{1}{2})\vec{v}$ for any integer n, then $\mathcal{W}$ would contain translation φ^2 parallel to v but not a power of v. Thus $h \perp \tau$ and $\vec{\varphi} = (n + \frac{1}{2})\vec{v}$ for some n. You can check that $v^n\gamma\varphi^{-1}$ is a translation parallel to τ, so $v^n\gamma\varphi^{-1} = \tau^{-m}$ for some m, hence $\varphi = \tau^m v^n \gamma$. Therefore $\mathcal{W}$ must consist precisely of the translations $\tau^m v^n$ and glide reflections $\tau^m v^n \gamma$ for all m and n. Moreover, since m and n identify the axis and vector of the glide reflection, each member of $\mathcal{W}$ has a *unique* description of this form. This paragraph, with the discussion under the previous heading, has shown that all wallpaper groups containing a glide reflection but no reflection or nontrivial rotation are isomorphic to the symmetry group of figure 8.3.9. These groups constitute a family called *pg*.

What groups contain a reflection, but no nontrivial rotation? Theorem 6 offers six possibilities for the relationship between a reflection's axis and the group's lattice. Place lattice point O on the axis of a reflection. In each subcase of theorem 6 with alternative $\gamma = \sigma_g$, the glide reflections $\varphi = \tau^m v^n \gamma$ belong to $\mathcal{W}$. Their axes are parallel to g and pass through lattice points or points midway between two lattice points. You can verify that vectors $\vec{\varphi}$ are multiples of the vector of the glide reflection in the other alternative of the subcase, and the correspondence between pairs m, n and the vector and axis is bijective. For example, in subcase (1a) the axis of φ passes through $\tau^{m-n}(O)$ or $\tau^{m-n-1}(X)$ depending on whether $m - n$ is even or odd, $\vec{\varphi} = \frac{1}{2}(\vec{m} + \vec{n})(\tau + v)$, and you can determine m and n from their sum and difference. In each subcase these axes are perpendicular to a translation in $\mathcal{W}$; you can use the argument of the previous paragraph to show that $\mathcal{W}$ can contain no further glide reflections. In the example subcase, displayed in figure 8.3.11, $\mathcal{W}$ contains no reflection with axis parallel to g through points midway between lattice points: $m - n$ and $m + n$ are both odd, so $\frac{1}{2}(m + n) \neq 0$. This happens in all four subcases of (1) and (2). But (3) is different; in subcase (3b), for example, displayed in figure 8.3.12, $\vec{\varphi} = n\vec{v}$, so $\mathcal{W}$ contains reflection $\tau^m\gamma$ across the axis through $\tau^{m-1}(X)$ if m is odd. According to the discussion under the previous heading, all case (1) and (2) groups are isomorphic, as are all case (3) groups. This paragraph has shown—or given instructions for showing—that all wallpaper groups containing a glide reflection but no reflection or nontrivial rotation are isomorphic to the symmetry group of figure 8.3.11 or that of 8.3.12. These groups constitute families called *cm* and *pm*. None of them is isomorphic to any *pg* group because they contain reflections and *pg* groups don't. No *cm* group $\mathcal{W}$ is isomorphic to any *pm* group $\mathcal{W}'$, because $\mathcal{W}$ contains glide reflections (with axes through points midway between lattice points) that commute with no reflections in $\mathcal{W}$, but every glide reflection in $\mathcal{W}'$ commutes with the reflection in $\mathcal{W}'$ across the same axis.

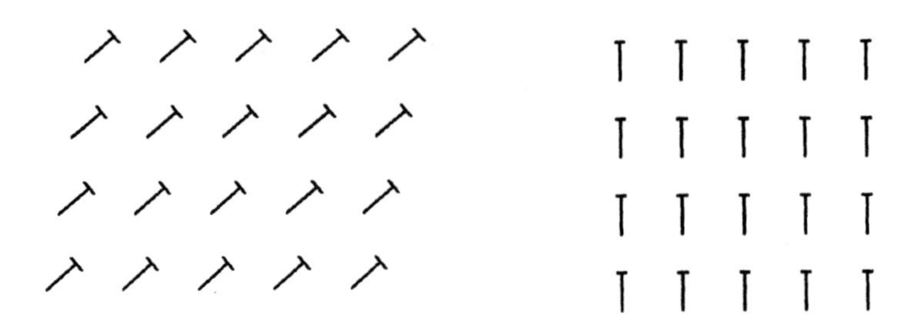

Figure 8.3.11
cm wallpaper pattern

Figure 8.3.12
pm wallpaper pattern

Lemma 7 summarizes the results of the previous three paragraphs:

Lemma 7 (pg, pm, and cm). Suppose wallpaper group $\mathcal{W}$ contains a glide reflection but no nontrivial rotation. If it contains no reflection, it's called a *pg* group and is isomorphic to the symmetry group of figure 8.3.9. If the axis of every glide reflection is the axis of some reflection, $\mathcal{W}$ is called a *pm* group, and is isomorphic to the group of figure 8.3.12. Otherwise, $\mathcal{W}$ is called a *cm* group, and is isomorphic to the group of figure 8.3.11. The isomorphism classes defined by lemmas 5 and 7 are disjoint.

Rotational symmetry

The next several theorems are technical steps that help narrow down the variety of rotations you must study to analyze a wallpaper group $\mathcal{W}$. Suppose you've used theorems 1 and 2 to set up a lattice and grid as in figure 8.3.4.

Theorem 8. If $\mathcal{W}$ contains rotations $\rho_{O,\theta}$ and $\rho_{P,\theta}$ then $O = P$ or $OP \geq \frac{1}{2}\,OT$.

Proof. $\mathcal{W}$ contains translation $\varphi = \rho_{P,\theta}\rho_{O,-\theta}$, so $\rho_{P,\theta}(O) = \varphi\rho_{O,-\theta}(O) = \varphi(O)$ and $2OP = OP + O\varphi(O) \geq O\varphi(O)$. By theorem 1, $O\varphi(O) = 0$ or $O\varphi(O) \geq OT$. The first case implies $O = P$. ♦

Theorem 9. If $\mathcal{W}$ contains a nontrivial rotation $\rho_{O,\theta}$, then there exists a point $P \neq O$ such that

- $\mathcal{W}$ contains $\rho_{P,\theta}$, and
- if $\mathcal{W}$ contains $\rho_{R,\theta}$, then $R = O$ or $OR \geq OP$.

Proof. $\mathscr{W}$ also contains $\rho_{T,\theta} = \tau\rho_{O,\theta}\tau^{-1}$. By theorem 8, $0 \neq OP \le OT$ for at least one but only finitely many points P such that $\mathscr{W}$ contains $\rho_{P,\theta}$. Choose one such that OP is smallest. ♦

***Theorem 10* (*Crystallographic restriction*).** If $\mathscr{W}$ contains a nontrivial rotation $\rho_{O,\theta}$, then the rotations about O in $\mathscr{W}$ form a cyclic group $\mathscr{C}$ with two, three, four, or six elements.

Proof. Replacing θ by $\theta \bmod 360°$ lets you assume $0° < \theta < 360°$. If $180° < \theta < 360°$, you could replace θ by $(-\theta) \bmod 360°$. Thus you can assume $0° < \theta < 180°$. Find a point P as described by theorem 8. Let $Q = \rho_{P,\theta}(O)$ and $R = \rho_{Q,\theta}(P)$, so that $\mathscr{W}$ also contains $\rho_{Q,\theta} = \rho_{P,\theta}\rho_{O,\theta}\rho_{P,\theta}^{-1}$ and $\rho_{R,\theta} = \rho_{Q,\theta}\rho_{P,\theta}\rho_{Q,\theta}^{-1}$. By theorem 9, $O = R$ or $OR \ge OP$. If $O = R$, then $\theta = 60°$. The next step is to show that $\overline{OP}$ and $\overline{QR}$ can *not* intersect at a point X interior to both as in figure 8.3.13. Were that the case, you could argue from congruence principles that $m\angle POR = \theta$, and that $m\angle QOP = m\angle OQP = \theta + m\angle OQR$ because ΔOPQ is isosceles; since $OR \ge QR$, applying the hinge theorem to ΔORQ would yield $m\angle OQR \ge m\angle QOR = m\angle QOP + m\angle POR = 2\theta + m\angle OQR$, hence $\theta = 0°$, contradiction! This shows that $\theta < 60°$ is impossible.

From $\theta \ge 60°$ it follows that the group $\mathscr{C}$ of all rotations about O in $\mathscr{W}$ is finite. Otherwise, $\mathscr{C}$ would contain rotations $\rho_{O,\theta}$ with arbitrarily small $\theta > 0°$. Suppose $\mathscr{C}$ has m elements. By theorem 8.1.4, $\mathscr{C}$ consists of rotations $\rho_{O,360°/m}^{k}$ for $k = 0$ to $m - 1$. From $\theta \ge 60°$ it follows that $m \le 6$. Figure 8.3.14, constructed like figure 8.3.13 with $\theta = \rho_{O,360°/5}$, shows that $m = 5$ is impossible, because $OR < OP$ in that case, too. (In fact, $OR \approx 0.38\,OP$.) ♦

When the cyclic group of rotations in $\mathscr{W}$ about a point O contains exactly m elements, O is called an *m-center* of $\mathscr{W}$. The symmetry group of the wallpaper pattern in figure 8.3.2 has 6-centers, 3-centers, and 2-centers

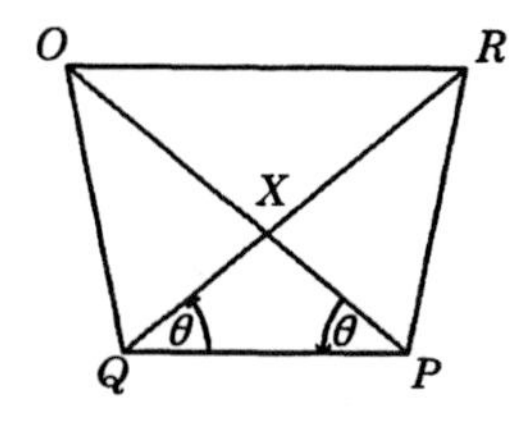

Figure 8.3.13 Proving theorem 10: $OR > OP$ and $\theta < 60°$ is impossible.

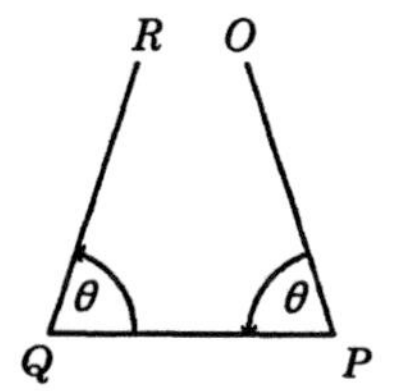

Figure 8.3.14 Proving theorem 10: $\theta = 360°/5$ is impossible.

(triangle vertices, incenters, and edge midpoints), but the figure 8.3.1 group has no centers at all.

Corollary 11. If $\mathcal{W}$ has a 4-center, then it can't have a 3- or 6-center.

Proof. In either case, $\mathcal{W}$ would contain a $30°$ rotation. ♦

Theorem 12. If O is an m-center of $\mathcal{W}$, so is $\varphi(O)$ for any φ in $\mathcal{W}$.

Proof. $\rho_{O,\theta}$ belongs to $\mathcal{W}$ if and only if $\rho_{\varphi(O),\theta} = \varphi\rho_{O,\theta}\varphi^{-1}$ belongs to $\mathcal{W}$. ♦

The classification theorem

A wallpaper group $\mathcal{W}$ is classified according to a decision tree much like the one used for frieze groups in section 8.2. The frieze tree is particularly elegant—you ask the same sequence of yes/no questions, no matter which branch the answers lead to. For the wallpaper groups, a tree like that would be too large and the questions too complicated to study conveniently. Instead, you select the first branch of the wallpaper tree by asking whether $\mathcal{W}$ has any m-centers, and if so, what's the largest m. Thus, five main branches extend from the root: no m-centers, 2-centers only, 3-centers only, 4-centers, and 6-centers. After determining the appropriate branch for $\mathcal{W}$, you ask several yes/no questions that depend on answers to previous questions. The complete tree is shown in figure 8.3.15.[29] Its leaves are labeled by the crystallographic designations of the seventeen different isomorphism classes of wallpaper groups. The four classes discussed so far—*p1*, *pg*, *cm*, and *pm*—constitute the rightmost four leaves of the tree. The crystallographic notation is discussed at the end of this section.

As noted under the heading **Isomorphy**, wallpaper groups falling into the same class are isomorphic. In fact, the seventeen classes are all disjoint. Most questions in figure 8.3.15 that lead you to classify two wallpaper groups differently are phrased to contradict plainly one of the properties of isomorphic wallpaper groups listed in theorem 4. Where the inference is not obvious, this section includes a short argument—for example, the one immediately before lemma 7.

The rest of the tree is discussed next in detail, branch by branch. The discussion is organized in lemmas 13, 14, 15, and 16, which summarize the results for groups with 6-centers, with 3-centers only, with 4-centers, and with 2-centers only. With lemmas 5 and 7, these constitute the proof of theorem 17, the wallpaper classification theorem.

---

[29] Adapted from Washburn and Crowe 1988, 128. Many details of this presentation of the classification theorem and its proof are adapted from Martin 1982, chapter 11, and Armstrong 1988, chapter 26.

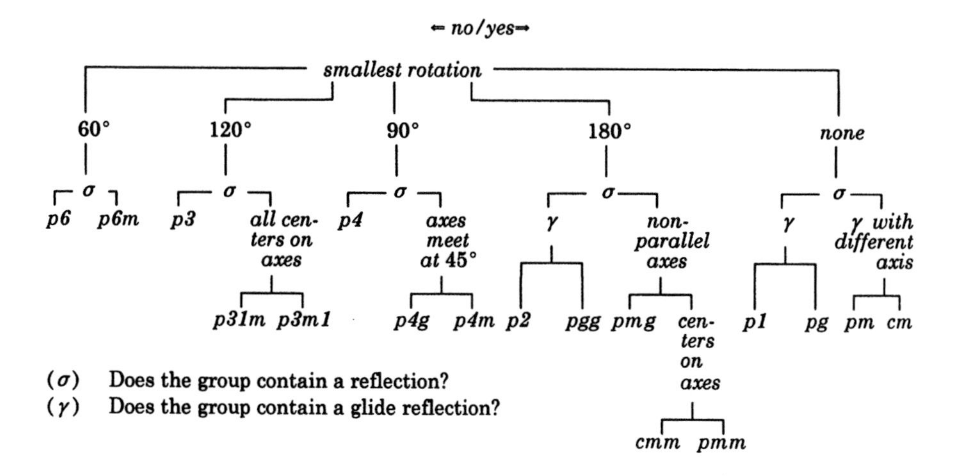

Figure 8.3.15 The wallpaper tree

Sixfold symmetry

Consider a wallpaper group $\mathcal{W}$ with a 6-center O. You're following the leftmost branch of the tree in figure 8.3.15. By corollary 11, $\mathcal{W}$ can't have a 4-center. Select a first fundamental translation τ for $\mathcal{W}$. Since translation $\upsilon = \rho_{O,60^\circ}\tau\rho_{O,-60^\circ}$ belongs to $\mathcal{W}$, is not parallel to τ, and has the same length, you may call it the second fundamental translation, so that $U = \rho_{O,60^\circ}(T)$, as in figure 8.3.16. By theorem 12, T and U are 6-centers. By theorem 8, any other center inside or on equilateral ΔOTU must lie outside or on the circles with centers O, T, and U with radius $r = \frac{1}{2}OT$. That is, it must lie in the shaded region Σ of figure 8.3.16. The incenter I of ΔOTU lies in Σ, and $\mathcal{W}$ contains $\rho_{O,60^\circ}\rho_{T,60^\circ} = \rho_{I,120^\circ}$. By theorem 8, no point $X \neq I$ in Σ can be the center of a 120° rotation in $\mathcal{W}$, because $IX < r$. Let O', T', and U' be the midpoints of the edges of ΔOTU as shown. Then $\mathcal{W}$ contains half turns

$$\rho_{I,120^\circ}\rho_{T,60^\circ} = \sigma_{O'} \qquad \rho_{I,-120^\circ}\rho_{O,60^\circ} = \sigma_{T'} \qquad \rho_{I,120^\circ}\rho_{O,60^\circ} = \sigma_{U'}.$$

No point $X \neq O', T', U'$ in Σ can be the center of a half turn in $\mathcal{W}$, because $O'X < r$. If $\mathcal{W}$ contained $\rho_{I,60^\circ}$, then it would contain σ_I, which is impossible because $IO' < r$; thus I is a 3-center. If $\mathcal{W}$ contained $\rho_{O',60^\circ}$, then it would contain $\rho_{O',120^\circ}$, which is impossible because $IO' < r$; thus O' is a 2-center. Similarly, T' and U' are 2-centers. A similar discussion with any lattice point in place of O shows that all lattice points are 6-centers; for all translations φ in $\mathcal{W}$, $\varphi(I)$ is a 3-center; $\varphi(O')$, $\varphi(T')$, and $\varphi(U')$ are 2-centers; and there are no other centers. The translations and rotations described in the previous sentence constitute the symmetry group $\mathcal{W}$ of figure 8.3.2. The class of all wallpaper groups isomorphic to $\mathcal{W}$ is called *p6*.

Suppose a wallpaper group $\mathcal{W}$ contains a *p6* subgroup $\mathcal{W}_6$ and another isometry φ. Continue using the lattice and grid described in the previous paragraph. Since $\mathcal{W}_6$ contains all possible translations and rotations, φ must be odd. By theorem 12, $\varphi(O)$ is a 6-center, hence a lattice point $O_{m,n}$. Odd isometry $\tau^{-m}\upsilon^{-n}\varphi$ belongs to $\mathcal{W}$ and fixes O, so it's the reflection σ_g across a line g through O. Further, $\sigma_g\tau\sigma_g$ is a translation in $\mathcal{W}$, so $\sigma_g(T) = \sigma_g\tau\sigma_g(O)$ is a lattice point, one of points $V_n = \rho_{O,n\cdot 60^\circ}(T)$ for $n = 0$ to 5. Note that $V_0 = T$ and $V_1 = U$. These points determine a regular hexagon with center at O, and g is one of its six lines of symmetry g_0 to g_5, shown in figure 8.1.10. Apparently, $\mathcal{W}$ contains

$$\sigma_{g_k}\sigma_{g_l} = \rho_{O,2(k-l)\cdot 30^\circ} = \rho_{O,(k-l)\cdot 60^\circ}$$

for each k and l, so $\mathcal{W}$ contains reflections across *all* these lines of symmetry. It follows that $\mathcal{W}$ must consist of $\mathcal{W}_6$ and the isometries $\varphi = \tau^m\upsilon^n\sigma_{g_k}$ for all m, n, and k. These isometries constitute the symmetry group $\mathcal{W}$ of figure 8.3.17. The class of all wallpaper groups isomorphic to $\mathcal{W}$ is called *p6m*.

Lemma 13 summarizes the results of the previous two paragraphs:

Lemma 13 (p6 and p6m). Suppose wallpaper group $\mathcal{W}$ contains a 60° rotation. If it contains no reflection, it's called a *p6* group and is isomorphic to the symmetry group of figure 8.3.2. Otherwise, $\mathcal{W}$ is called a *p6m* group and is isomorphic to the group of figure 8.3.17. The isomorphism classes defined by lemmas 5, 7, and 13 are all disjoint.

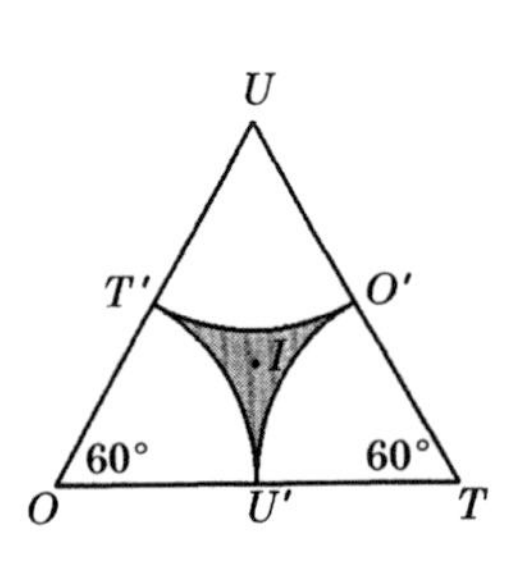

Figure 8.3.16
6-centers

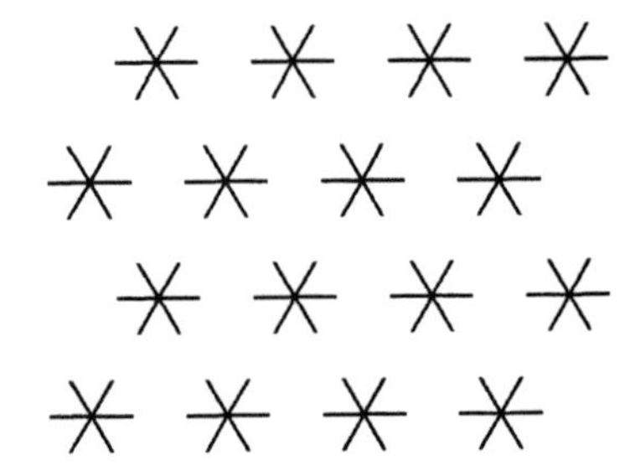

Figure 8.3.17
p6m wallpaper pattern

Threefold symmetry

Now suppose $\mathscr{W}$ is a wallpaper group with a 3-center O but no 6-center. It can have no 2- or 4-center, either. Consider any nontrivial translation φ in $\mathscr{W}$. Use theorems 9 and 12 to find a 3-center X at minimal nonzero distance from O. Then $\rho_{X,-120°}\rho_{O,-120°} = \rho_{Y,120°}$, where ΔOXY is equilateral, so Y is a 3-center. Repeating this argument shows that all heavy dots in figure 8.3.18 are 3-centers. (Ignore points I, A, and B.) They form the vertices of a family of equilateral triangles with edge length OX. No other points inside or on any of the six triangles with vertex O can be a center, because they're closer to O than X. Let T be the point such that $OXYT$ is a parallelogram. No other point A inside ΔXYT can be a center, because $\rho_{Y,-120°}\rho_{A,-120°} = \rho_{B,120°}$, where ΔAYB is equilateral and B lies inside or on ΔOXY. Repeating this argument shows that the heavy dots in figure 8.3.18 are the *only* centers. By theorem 12, $\varphi(O)$ is a 3-center. If $\varphi(O) = X$, then $\mathscr{W}$ would contain $\rho_{X,120°}\varphi = \rho_{I,120°}$, which is impossible. Thus $\varphi(O)$ can't be any of the six centers nearest O. But $\mathscr{W}$ contains translations $\tau = \rho_{X,-120°}\rho_{O,120°}$ and $\upsilon = \rho_{Y,-120°}\rho_{O,120°}$ with vectors $T - O$ and $U - O$. Therefore you can select τ and υ as basic translations for $\mathscr{W}$. The 120° rotations about the indicated centers and the translations $\tau^m\upsilon^n$ for all m and n constitute the symmetry group $\mathscr{W}$ of figure 8.3.19. The class of all wallpaper groups isomorphic to $\mathscr{W}$ is called *p3*.

Now suppose a wallpaper group $\mathscr{W}$, as in the previous paragraph, has a 3-center, no 6-center, and contains a *p3* subgroup $\mathscr{W}_3$ and another isometry φ. As before, it has no 2- or 4-center. This paragraph will show that $\mathscr{W}$ must contain the reflection across a line through one the centers. The following paragraph will determine that there are just two possibilities for $\mathscr{W}$. By theorem 12, $\varphi(O)$ is a center. Suppose $\varphi(O)$ is a lattice point. Then $\mathscr{W}$ contains a translation χ such that $\chi(\varphi(O)) = O$. Since $\mathscr{W}_3$ contains

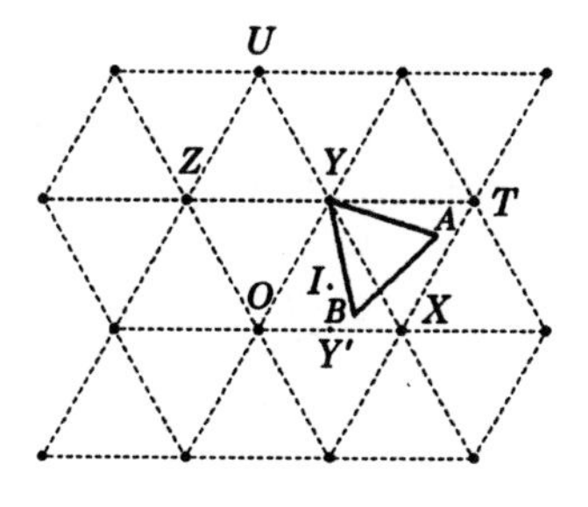

Figure 8.3.18 Analyzing groups with 3-centers only

Figure 8.3.19 *p3* wallpaper pattern

all possible translations and rotations, φ is odd, hence $\chi\varphi$ is odd. Since $\chi\varphi$ fixes O, it's the reflection across a line g through O, as required. Now suppose $\varphi(O)$ is not a lattice point. Then $\mathcal{W}$ contains a translation χ such that $\chi(\varphi(O))$ is one of the six centers nearest O. Suppose it's X; you can construct a similar discussion for the other possibilities. Let Y' be the midpoint of segment $\overline{OX}$. Then $\sigma_{Y'}\chi\varphi(O) = O$, so $\sigma_{Y'}\chi\varphi = \sigma_g$ for some line g through O. From figure 8.3.18 you can see that if C is a center, then so is $\sigma_{Y'}(C)$; since $\chi\varphi(X)$ is a center, so is $\sigma_g(X) = \sigma_{Y'}\chi\varphi(X)$. In fact, $\sigma_g(X)$ is one of the vertices V_i in figure 8.1.8 because $OX = O\sigma_g(X)$, and g is one of the lines g_i because it's the perpendicular bisector of $\overline{XV_i}$. The line $h \perp g_1 = O\vec{T}$ through Y' intersects g_1 at a point F. One of $\sigma_g\sigma_{g_\epsilon}$ and $\sigma_g\sigma_{g_\cap}$ is the $\pm 120°$ rotation ρ about O, so $\sigma_{Y'}\chi\varphi\rho = \sigma_{g_\epsilon}$ or $\sigma_{g_\cap}$. In the latter case, $\chi\varphi\rho = \sigma_{Y'}\sigma_{g_\cap}$, the glide reflection with axis h and vector $2(Y' - F)$, and $\mathcal{W}$ would contain the translation $(\chi\varphi)^2$ with nonzero vector $4(Y' - F) = X - Y$, which is shorter than $\vec{\tau}$, contradiction! In the former case, $\chi\varphi\rho = \sigma_{Y'}\sigma_{g_\epsilon} = \sigma_{Y'}\sigma_{O\breve{X}} = \sigma_{Y\breve{Y}''}$, as required.

The previous paragraph showed that if wallpaper group $\mathcal{W}$ contains a *p3* subgroup $\mathcal{W}_3$ and another isometry, it must contain σ_g for some line g through some center. Redefine O, and other features of figure 8.3.18 as necessary, to make O that center. Since X is a center, so is $\sigma_g(X)$. In fact, $\sigma_g(X)$ is one of the vertices V_i in figure 8.1.8 because $OX = O\sigma_g(X)$, and g is one of the lines g_i. Because $\mathcal{W}$ contains $\rho_{O,120°}$, it contains σ_{g_e} for all three even indices i or all three odd indices. But $\mathcal{W}$ can't contain reflections of *both* types, because their composition would be a rotation about O through an odd multiple of 60°, and O is not a 6-center. In either case, $\mathcal{W}$ consists of $\mathcal{W}_3$ and the compositions $\chi\sigma_{g_e}$ for all χ in $\mathcal{W}_3$ —as noted earlier, all even isometries in $\mathcal{W}$ are in $\mathcal{W}_3$, and if φ is an odd isometry in $\mathcal{W}$ then $\chi = \varphi\sigma_{g_e}$ is an even isometry in $\mathcal{W}$, hence in $\mathcal{W}_3$, and $\varphi = \chi\sigma_{g_e}$. The isometries χ and $\chi\sigma_{g_\epsilon}$ with χ in $\mathcal{W}_3$ constitute the symmetry group $\mathcal{W}$ of figure 8.3.20. The isometries χ and $\chi\sigma_{g_\cap}$ with χ in $\mathcal{W}_3$ constitute the symmetry group $\mathcal{W}'$ of figure 8.3.21. The classes of all wallpaper groups isomorphic to $\mathcal{W}$ and to $\mathcal{W}'$ are called *p31m* and *p3m1*. You may find it difficult to distinguish groups in these classes. The simplest criterion is that all *p3m1* centers lie on lines of symmetry, while some *p31m* centers do not. Thus, *p3m1* groups satisfy the following condition, but *p31m* groups do not: For every rotation ρ there's a reflection σ such that $\rho\sigma$ is a reflection. By theorem 4, *p3m1* and *p31m* groups cannot be isomorphic. Past authors have often reversed the designations *p3m1* and *p31m*—be careful! Exercise 8.5.20 explains in part why these groups are so easily confused.

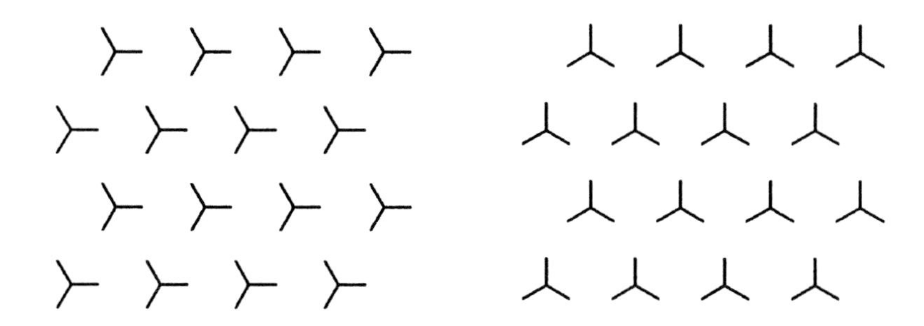

Figure 8.3.20
p31m wallpaper pattern

Figure 8.3.21
p3m1 wallpaper pattern

Lemma 14 summarizes the results of the previous three paragraphs:

Lemma 14 **(*p3, p31m, and p3m1*).** Suppose wallpaper group $\mathcal{W}$ contains a $120°$ rotation but no $60°$ rotation. If it contains no reflection, it's called a *p3* group and is isomorphic to the symmetry group of figure 8.3.19. If every center is on the axis of some reflection, $\mathcal{W}$ is called a *p3m1* group and is isomorphic to the group of figure 8.3.21. Otherwise, $\mathcal{W}$ is called a *p31m* group and is isomorphic to the group of figure 8.3.20. The classes defined by lemmas 5, 7, 13, and 14 are all disjoint.

Fourfold symmetry

Next, consider a wallpaper group $\mathcal{W}$ with a 4-center O. By corollary 11, it can have no 6- or 3-center. By theorem 12, $\varphi(O)$ is a 4-center for every isometry φ in $\mathcal{W}$. By theorem 9, there is a center $X \neq O$ closest to O. See figure 8.3.22. If $\mathcal{W}$ contained $\rho_{X,90°}$, then it would also contain a half turn $\sigma_N = \rho_{O,90°}\rho_{X,90°}$ with center N closer to O than X, contradiction! Thus X is a 2-center, and so is $Y = \rho_{O,90°}(X)$. Further, $\mathcal{W}$ contains $\rho_{Z,90°} = \sigma_X \rho_{O,-90°}$, where Z is the point such that $OXZY$ is a square. Therefore, Z is a 4-center. By theorem 12, $X' = \sigma_Z(X)$ and $Y' = \sigma_Z(Y)$ are 2-centers and $T = \sigma_X(O)$, $U = \rho_{O,90°}(T)$, and $R = \sigma_Z(O)$ are 4-centers. If φ is a translation in O then $O' = \varphi(O)$ must be a 4-center. If M is the midpoint of $\overline{OO'}$, so that $\varphi = \sigma_M \sigma_O$, then σ_M must belong to $\mathcal{W}$, hence $OM \geq OX$, hence $OO' \geq OT$. Therefore, $\tau = \sigma_X \sigma_O$ and $\upsilon = \sigma_Y \sigma_O$ are shortest nontrivial translations in $\mathcal{W}$, and $\tau(O) = T$, $\upsilon(O) = U$. Use them to set up the lattice as in figure 8.3.4. By theorem 8, if A and B are any centers, then $AB \geq OX$. Every point inside or on square $OPRQ$ lies within that distance of one of the centers already mentioned. Therefore there are no others inside or on it, and all centers are images of these under translations

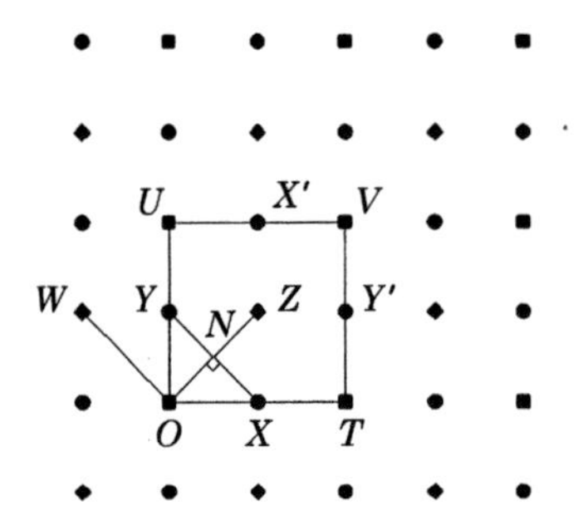

Figure 8.3.22 Analyzing groups with 4-centers

Figure 8.3.23 *p4* wallpaper pattern

in $\mathcal{W}$. Circles • in figure 8.3.22 indicate 2-centers, squares ▪ indicate lattice points, and diamonds ◆ indicate 4-centers that aren't lattice points. The translations and rotations catalogued in this paragraph constitute the symmetry group $\mathcal{W}$ of figure 8.3.23. The class of all wallpaper groups isomorphic to $\mathcal{W}$ is called *p4*.

Now suppose a wallpaper group $\mathcal{W}$, as in the previous paragraph, has a 4-center and contains a *p4* subgroup $\mathcal{W}_4$ and another isometry φ. As before, it has no 3- or 6-center. This paragraph will show that $\mathcal{W}$ must contain the reflection across a line through O, or across $\overleftrightarrow{XY}$. The following paragraph will determine that there are just two possibilities for $\mathcal{W}$. By theorem 12, $\varphi(O)$ is a center. Suppose $\varphi(O)$ is a lattice point. Then $\mathcal{W}$ contains a translation χ such that $\chi(\varphi(O)) = O$. Since $\mathcal{W}_4$ contains all possible translations and rotations, φ is odd, hence $\chi\varphi$ is odd. Since $\chi\varphi$ fixes O, it's the reflection across a line g through O, as required. Now suppose $\varphi(O)$ is not a lattice point. It must nevertheless be a 4-center since O is, so $\mathcal{W}$ contains a translation χ such that $\chi(\varphi(O)) = Z$. Then $\sigma_N\chi\varphi(O) = O$, so $\sigma_N\chi\varphi = \sigma_g$ for some line g through O. From figure 8.3.22 you can see that if C is a center, then $\sigma_N(C)$ is a center. Since $\chi\varphi(X)$ is a center, so is $\sigma_g(X) = \sigma_N\chi\varphi(X)$. In fact, $\sigma_g(X)$ is one of the four centers nearest O in figure 8.3.22 because $OX = O\sigma_g(X)$, and $g = \overleftrightarrow{OZ}$, $\overleftrightarrow{OU}$, $\overleftrightarrow{OW}$, or $\overleftrightarrow{OT}$, because it's the perpendicular bisector of $\overline{XX'}$. If $g = \overleftrightarrow{OZ}$, then $\chi\varphi = \sigma_N\sigma_{\overleftrightarrow{OZ}} = \sigma_{\overleftrightarrow{XY}}$, as required. Otherwise, one of the equations

$$\begin{aligned}
\sigma_N\sigma_{\overleftrightarrow{OU}}\rho_{O,\,90^\circ} &= \sigma_N\sigma_{\overleftrightarrow{OU}}\,\sigma_{\overleftrightarrow{OU}}\,\sigma_{\overleftrightarrow{OZ}} = \sigma_N\sigma_{\overleftrightarrow{OZ}} = \sigma_{\overleftrightarrow{XY}}\\
\sigma_N\sigma_{\overleftrightarrow{OW}}\sigma_O &= \sigma_N\sigma_{\overleftrightarrow{OW}}\,\sigma_{\overleftrightarrow{OW}}\,\sigma_{\overleftrightarrow{OZ}} = \sigma_N\sigma_{\overleftrightarrow{OZ}} = \sigma_{\overleftrightarrow{XY}}\\
\sigma_N\sigma_{\overleftrightarrow{OT}}\rho_{O,-90^\circ} &= \sigma_N\sigma_{\overleftrightarrow{OT}}\,\sigma_{\overleftrightarrow{OT}}\,\sigma_{\overleftrightarrow{OZ}} = \sigma_N\sigma_{\overleftrightarrow{OZ}} = \sigma_{\overleftrightarrow{XY}}
\end{aligned}$$

implies that $\sigma_{\overleftrightarrow{XY}}$ belongs to $\mathcal{W}$, as required.

The previous paragraph showed that if wallpaper group $\mathcal{W}$ contains a *p4* subgroup $\mathcal{W}_4$ and another isometry, it must contain $\sigma_{\overleftrightarrow{XY}}$ or σ_g for some line g through O. In the latter case, an argument in the previous paragraph shows that $g = \overleftrightarrow{OZ}$, $\overleftrightarrow{OU}$, $\overleftrightarrow{OW}$, or $\overleftrightarrow{OT}$. Since $\mathcal{W}$ contains the composition of the reflections across any two of these lines, it contains either all or none of them. Therefore $\mathcal{W}$ contains $\sigma_{\overleftrightarrow{OZ}}$ or $\sigma_{\overleftrightarrow{XY}}$. It can't contain both, because then $\mathcal{W}$ would contain $\sigma_{\overleftrightarrow{OZ}}\sigma_{\overleftrightarrow{XY}} = \sigma_N$, but N is no center. In the first case, $\mathcal{W}$ consists of $\mathcal{W}_4$ and the compositions $\chi\sigma_{\overleftrightarrow{OZ}}$ for all χ in $\mathcal{W}_4$ —as noted earlier, all even isometries in $\mathcal{W}$ are in $\mathcal{W}_4$, and if φ is an odd isometry in $\mathcal{W}$ then $\chi = \varphi\sigma_{\overleftrightarrow{OZ}}$ is an even isometry in $\mathcal{W}$, hence in $\mathcal{W}_4$, and $\varphi = \chi\sigma_{\overleftrightarrow{OZ}}$. The isometries χ and $\chi\sigma_{\overleftrightarrow{OZ}}$ with χ in $\mathcal{W}_4$ constitute the symmetry group $\mathcal{W}$ of figure 8.3.24. The isometries χ and $\chi\sigma_{\overleftrightarrow{XY}}$ with χ in $\mathcal{W}_4$ constitute the symmetry group $\mathcal{W}'$ of figure 8.3.25. The classes of all wallpaper groups isomorphic to $\mathcal{W}$ and to $\mathcal{W}'$ are called *p4m* and *p4g*. Note that *p4m* groups satisfy the following condition, but *p4g* groups do not: For every 90° rotation ρ there's a reflection σ such that $\rho\sigma$ is a reflection. By theorem 4, *p4m* and *p4g* groups cannot be isomorphic.

Lemma 15 summarizes the results of the previous three paragraphs:

***Lemma 15* (*p4, p4m, and p4g*).** Suppose wallpaper group $\mathcal{W}$ contains a 90° rotation. If it contains no reflection, it's called a *p4* group and is isomorphic to the symmetry group of figure 8.3.23. If every 4-center lies on the axis of some reflection, $\mathcal{W}$ is called a *p4m* group and is isomorphic to the group of figure 8.3.4. Otherwise, $\mathcal{W}$ is called a *p4g* group and is isomorphic to the group of figure 8.3.25. The classes defined by lemmas 5, 7, 13, 14, and 15 are all disjoint.

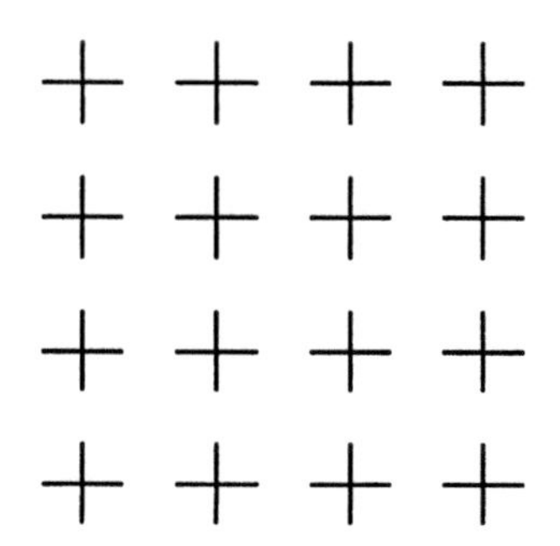

Figure 8.3.24
p4m wallpaper pattern

Figure 8.3.25
p4g wallpaper pattern

Twofold symmetry

Next, consider a wallpaper group $\mathscr{W}$ with a 2-center O but no 3-, 4-, or 6-center. Select basic translations τ and υ and let X, Y, and Z be the midpoints of segments between lattice points as in figure 8.3.26. Then $\mathscr{W}$ contains $\sigma_X = \tau\sigma_O$, $\sigma_Y = \upsilon\sigma_O$, and $\sigma_Z = \sigma_Y\sigma_O\sigma_X$, so X, Y, and Z are centers. If C is any center, then $\sigma_C\sigma_O$ is a translation in $\mathscr{W}$, so $\sigma_C(O) = \sigma_C\sigma_O(O)$ is a lattice point $O_{m,n}$, and $C = \frac{1}{2}mT + \frac{1}{2}nU$, using O as the origin of a coordinate system for this vector calculation. On the other hand, for any integers m and n there's a translation χ in $\mathscr{W}$ such that $\chi(\frac{1}{2}mT + \frac{1}{2}nU)$ is O, X, Y, or Z. Therefore all such points are centers. In short, the centers of $\mathscr{W}$ are these four points and their images under translations in $\mathscr{W}$. The translations and the half turns about these centers constitute the symmetry group $\mathscr{W}$ of figure 8.3.27. The class of all wallpaper groups isomorphic to $\mathscr{W}$ is called *p2*.

Now suppose a wallpaper group $\mathscr{W}$, as in the previous paragraph, has a 2-center O, no 3-, 4-, or 6-center, includes a *p2* subgroup $\mathscr{W}_2$, and contains σ_g for some line g. Select a first fundamental translation τ. By theorem 6 you can select the second fundamental translation υ and set up the lattice so that

(1) $|\tau| = |\upsilon|$ and
- (1a) $\tau\upsilon \parallel g$ or
- (1b) $\tau\upsilon^{-1} \parallel g$; or

(2) U lies on the perpendicular bisector k of $\overline{OT}$ and
- (2a) $g \perp \tau$ or
- (2b) $g \parallel \tau$; or

(3) $\tau \perp \upsilon$ and
- (3a) $g \perp \tau$ or
- (3b) $g \parallel \tau$.

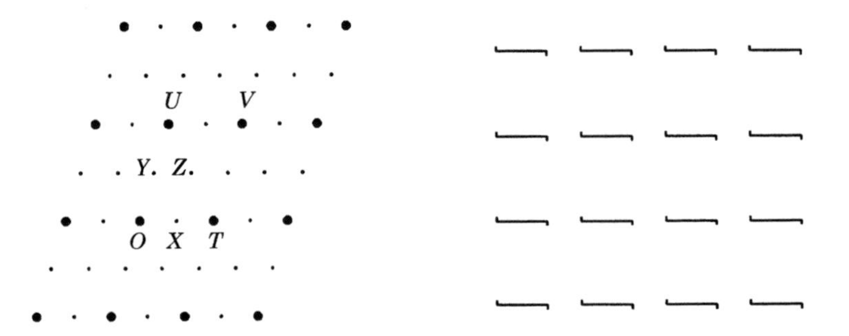

Figure 8.3.26 Analyzing groups with 2-centers

Figure 8.3.27 *p2* wallpaper pattern

The rest of this paragraph shows that in cases (1) and (2), g passes through a center. In subcases (1a) and (1b), by the argument in the first paragraph of the proof of theorem 6, you can assume that g intersects $\vec{OT}$ at a point $P \neq T$ between O and T, or it intersects $\vec{OU}$ at a point $P \neq U$ between O and U. In subcases (2a) and (2b), since $\sigma_O \sigma_X = \tau$ and $\sigma_O \sigma_Y = \upsilon$, you can assume $P \neq X, Y$ and P lies between O and X or Y. In subcase (1a), $\sigma_g(T)$ is a center, hence $\sigma_g(T) = Z$ or U, and X or O lies on g. In (1b), $\sigma_g(O)$ is a center, hence $\sigma_g(O) = Z$ or V, and X or T lies on g. In subcase (2a), $\sigma_g(O)$ is a center, hence $\sigma_g(O) = O$ or X, and O or Y lies on g. In (2b), $\sigma_g(X)$ is a center, hence $\sigma_g(X) = X$ or U, and O or Y lies on g.

In cases (1) and (2), redefine the lattice and grid if necessary, replacing O by a center on g, but using the same fundamental translations. $\mathcal{W}$ contains σ_g and the reflection $\sigma_g \sigma_O$ across the line $h \perp g$ through O.

In case (1) $\mathcal{W}$ contains reflections across all lines parallel or perpendicular to $\tau\upsilon$ through lattice points. More precisely, it consists of $\mathcal{W}_2$ and the compositions $\chi\sigma_{\vec{OV}}$ for all χ in $\mathcal{W}_2$ —all even isometries in $\mathcal{W}$ are in $\mathcal{W}_2$, and if φ is an odd isometry in $\mathcal{W}$ then $\chi = \varphi\sigma_{\vec{OV}}$ is an even isometry in $\mathcal{W}$, hence in $\mathcal{W}_2$, and $\varphi = \chi\sigma_{\vec{OV}}$. These isometries constitute the symmetry group $\mathcal{W}$ of figure 8.3.28. The class of all wallpaper groups isomorphic to $\mathcal{W}$ is called *cmm*.

In case (2) $\mathcal{W}$ contains reflections across all lines parallel or perpendicular to τ through lattice points. You can describe its members just like those of the *cmm* group in the previous paragraph if you refer to $\tau' = \tau\upsilon^{-1}$ and parallelogram $OSTU$ in figure 8.3.5 in place of τ and $OTVU$. It follows that $\mathcal{W}$ is isomorphic to the *cmm* group, hence it's a *cmm* group, too. (Parallelogram $OSTU$ is not a fundamental cell because neither τ' nor υ is a shortest nontrivial translation in $\mathcal{W}$. Nevertheless, equations $\tau' = \tau\upsilon^{-1}$ and $\tau = \tau'\upsilon$ imply that you can use τ' and υ to set up the isomorphism as discussed earlier under the heading **Isomorphy**.)

In subcase (3a) similar arguments show that $\mathcal{W}$ contains $\sigma_{\vec{OU}}$ or σ_h, where h is the perpendicular bisector of $\overline{OX}$. In the first situation,

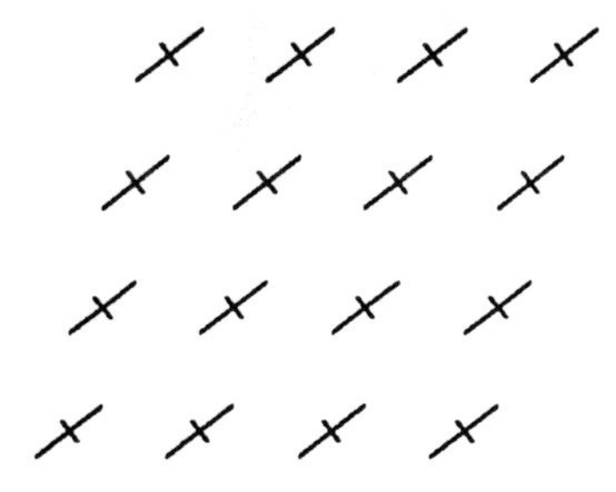

Figure 8.3.28
cmm wallpaper pattern

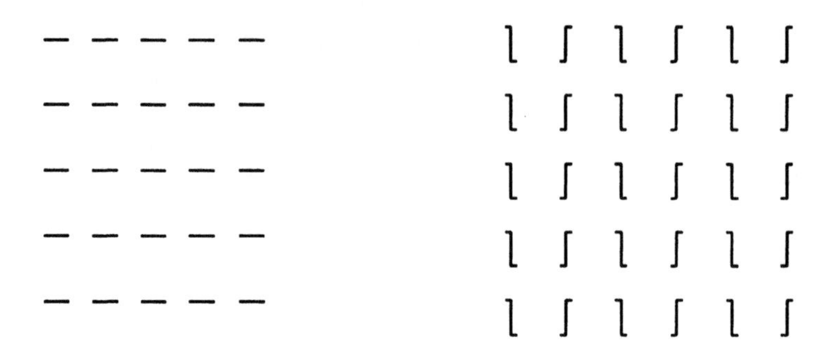

Figure 8.3.29
pmm wallpaper pattern

Figure 8.3.30
pmg wallpaper pattern

$\mathscr{W}$ contains reflections across all lines parallel to τ or υ through centers, and you can verify that $\mathscr{W}$ consists of $\mathscr{W}_2$ and the compositions $\chi\sigma_{O\bar{U}}$ for all χ in $\mathscr{W}_2$. These isometries constitute the symmetry group $\mathscr{W}$ of figure 8.3.29. In the second situation, $\mathscr{W}$ contains both σ_h and the reflection across $\sigma_X[h]$. More precisely, it consists of $\mathscr{W}_2$ and the compositions $\chi\sigma_h$ for all χ in $\mathscr{W}_2$. These isometries constitute the symmetry group $\mathscr{W}'$ of figure 8.3.30. The classes of all wallpaper groups isomorphic to $\mathscr{W}$ and to $\mathscr{W}'$ are called *pmm* and *pmg*. Note that *cmm* and *pmm* groups satisfy the following condition, but *pmg* groups do not: Every half turn commutes with some reflection. Moreover, *cmm* groups contain distinct commuting reflections, but *pmm* groups do not. By theorem 4, *cmm*, *pmm*, and *pmg* groups cannot be isomorphic. Subcase (3b) leads similarly to *pmm* and *pmg* groups.

Finally, suppose a wallpaper group $\mathscr{W}$ includes a *p2* subgroup $\mathscr{W}_2$, and contains a glide reflection but no reflection. Since the composition of a glide reflection with a half turn about a point on its axis is a reflection, no center can lie on the axis of any glide reflection in $\mathscr{W}$. Select a first fundamental translation τ. By theorem 6 you can select the second fundamental translation υ, set up the lattice, and find a glide reflection γ in $\mathscr{W}$ with axis g such that

(1) $|\tau| = |\upsilon|$ and
  (1a) $\tau\upsilon \parallel g$ and $\gamma^2 = \tau\upsilon$, or
  (1b) $\tau\upsilon^{-1} \parallel g$ and $\gamma^2 = \tau\upsilon^{-1}$; or

(2) U lies on the perpendicular bisector k of $\overline{OT}$ and
  (2a) $g \perp \tau$ and $\gamma^2 = \tau^{-1}\upsilon^2$, or
  (2b) $g \parallel \tau$ and $\gamma^2 = \tau$; or

(3) $\tau \perp \upsilon$ and
(3a) $g \perp \tau$ and $\gamma^2 = \upsilon$, or
(3b) $g \parallel \tau$ and $\gamma^2 = \tau$.

You can verify that cases (1) and (2) are impossible; the equations for γ^2 all require that $\gamma(O)$ be a center inside but not on a fundamental parallelogram. To visualize this, it helps to show first that you can assume in case (1) and subcase (2a) that g intersects $\overleftrightarrow{OT}$ at a point $P \neq O, X, T$ between O and T. Moreover, if γ is a glide reflection satisfying condition (3a), then $\gamma' = \sigma_O \gamma$ satisfies (3b), and vice versa. You can verify that $\mathcal{W}$ consists of $\mathcal{W}_2$ and the compositions $\chi\gamma$ for all χ in $\mathcal{W}_2$. These isometries constitute the symmetry group $\mathcal{W}$ of figure 8.3.31. The class of all wallpaper groups isomorphic to $\mathcal{W}$ is called *pgg*.

***Lemma 16* (*p2, cmm, pmm, pmg, and pgg*)**. Suppose wallpaper group $\mathcal{W}$ contains a half turn but no 90° or 60° rotation. If $\mathcal{W}$ contains no glide reflection, it's called a *p2* group and is isomorphic to the symmetry group of figure 8.3.27. If $\mathcal{W}$ contains a glide reflection but no reflection, it's called a *pgg* group and is isomorphic to the symmetry group of figure 8.3.31. Suppose $\mathcal{W}$ contains a reflection. If some half turn commutes with no reflection, $\mathcal{W}$ is called a *pmg* group and is isomorphic to the symmetry group of figure 8.3.30. If each half turn commutes with some reflection but distinct reflections never commute, $\mathcal{W}$ is called a *pmm* group and is isomorphic to the symmetry group of figure 8.3.29. Otherwise, $\mathcal{W}$ is called a *cmm* group and is isomorphic to the symmetry group of figure 8.3.28. The classes defined by lemmas 5, 7, and 13 to 16 are all disjoint.

Conclusion

The lemmas just listed constitute the proof of the following theorem:

***Theorem 17* (*Wallpaper classification theorem*)**. Each wallpaper group belongs to exactly one of the seventeen isomorphism classes defined by the tree in figure 8.3.15.

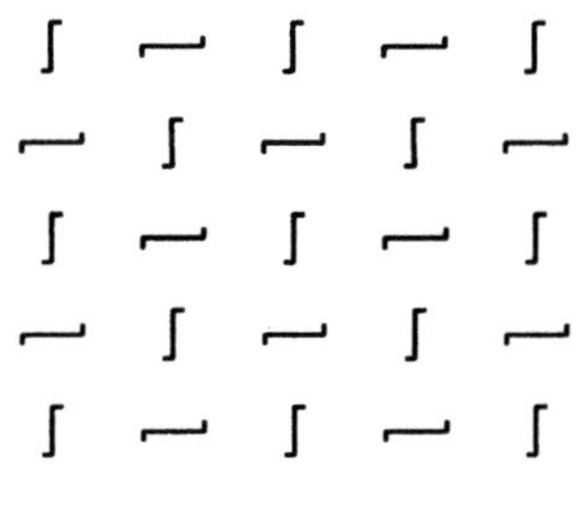

Figure 8.3.31
pgg wallpaper pattern

Authors have used several systems of notation for the isomorphism classes of wallpaper groups. Two systems have emerged as standard. One was established by the geometer L. Fejes Tóth.[30] Another arose from the crystallographic literature. *Methods of geometry* has used standard abbreviations for the four-symbol crystallographic designations. Table 8.3.1 shows the correspondence between this notation and Fejes Tóth's, indexed by the lemmas that summarize the discussion of the classes. For an explanation of the crystallographic symbols and correlation with five other systems, consult Schnattschneider 1978.[31]

Table 8.3.1

Crystallographic and Fejes Tóth symbols for wallpaper groups

Lemma	*Abbreviated crystallographic symbol*	*Fejes Tóth symbol*
5	*p1*	$\mathfrak{W}_1$
7	*cm*	$\mathfrak{W}_1^1$
	pm	$\mathfrak{W}_1^2$
	pg	$\mathfrak{W}_1^3$
13	*p6*	$\mathfrak{W}_6$
	p6m	$\mathfrak{W}_6^1$
14	*p3*	$\mathfrak{W}_3$
	p3m1	$\mathfrak{W}_3^1$
	p31m	$\mathfrak{W}_3^2$
15	*p4*	$\mathfrak{W}_4$
	p4m	$\mathfrak{W}_4^1$
	p4g	$\mathfrak{W}_4^2$
16	*p2*	$\mathfrak{W}_2$
	cmm	$\mathfrak{W}_2^1$
	pmm	$\mathfrak{W}_2^2$
	pmg	$\mathfrak{W}_2^3$
	pgg	$\mathfrak{W}_2^4$

[30] Fejes Tóth 1964, section 4.

[31] Schattschneider warns that several authors have confused the *p3m1* and *p31m* groups.

Evgraf Stepanovich FEDOROV was born in 1853 in Orenburg, Russia. His father was a military engineer. Fedorov grew up in St. Petersburg, where he attended a military gymnasium and engineering school. After graduation he alternated between military service and the study of medicine and several sciences. Eventually he became Director of Mines in the Urals. Fedorov started his major scientific work at age sixteen, a monograph on polyhedra and crystal structure that he finished twelve years later. It was published in 1885 in an obscure mineralogy journal. Although it received little attention, it was the basis of his major 1891 work that counted and classified the 230 types of symmetry groups of crystal structures. (This work included implicitly the seventeen wallpaper groups.) Fedorov corresponded with the better known German mathematician Arthur Schönflies, who was working in the same area. Each provided major parts of the work; they share credit. During his administrative career, Fedorov continued to contribute to mineralogy and crystallography as a sideline. Some of his works were translated, but he remained isolated. He died in 1919.

8.4 Polyhedra

Concepts

- Polyhedral surfaces
- Connected polyhedra
- Euler's theorem
- Regular polyhedra
- Dual polyhedra
- Cuboctahedron and rhombic dodecahedron
- Convex polyhedra
- Deltahedra
- Polyhedral symmetry groups
- Transitive symmetry groups

Polygonal regions and their boundaries were introduced in sections 3.8, 4.7, and 8.1, and used for describing portions of planes and reasoning about area. Some of them—for example, the regular polygons—are simpler than others. Simplicity is exploited in various computations and constructions: for example, calculating π in exercise 4.9.2, the Kin Kletso floor plan in figure 8.0.4, and the hexagonal pattern in figure 8.4.1, which could represent a tiled floor, chicken-wire fencing, or a cross section of a honeycomb. Such figures are beautiful and entertaining, as well; compare Kürschak's tile in figure 4.7.2 and the *p6* wallpaper pattern in figure 8.3.2. We study polygonal symmetry to explain the properties of these figures, and to enable us to use them effectively.

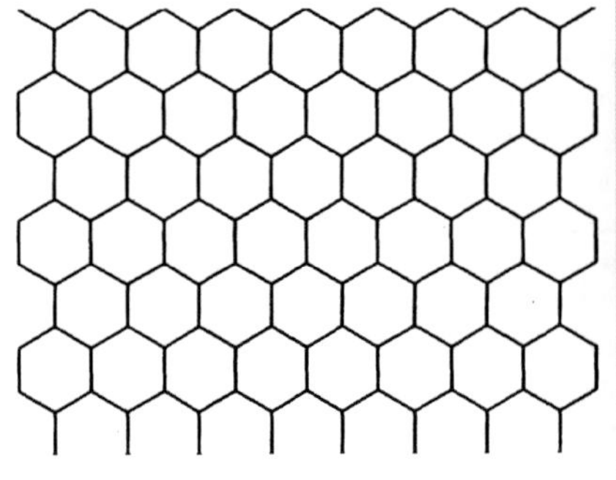

Figure 8.4.1
Hexagonal grid

Figure 8.4.2
Modular architecture[32]

Polyhedral regions were introduced in sections 3.10 and 4.8, and used for describing portions of space and reasoning about volume. Some of them, too, are simpler than others. This simplicity is exploited in various computations and constructions: for example, volume calculations in 3.10 and 4.8 and modular architecture in figure 8.4.2. And we find these figures beautiful and entertaining—see the ornamental lamp in figure 8.0.8.

This section develops a theory of polyhedra—*boundaries* of polyhedral regions and their symmetries—analogous to the theory of polygons. Unlike earlier sections of the chapter, this one doesn't attempt to cover its subject matter exhaustively. The theory of polyhedra is rich enough that entire books are devoted to various of its aspects. You'll find here only glimpses of them. Perhaps those will entice you to further study.

Polyhedra

Just as you built polygonal regions from triangles that intersect properly, you can construct polyhedral surfaces from polygonal regions—*faces*—that intersect properly. The faces don't need to be coplanar. For example, see figure 8.4.3, built from a quadrilateral and several triangular regions tilted this way and that according to the shading. (The χ values in the captions of figures 8.4.3 to 8.4.11 are explained under the heading **Euler's theorem**.)

You could flatten this surface into a plane, but not the open box surface in figure 8.4.4.

[32] Pearce 1978, plate 13.70.

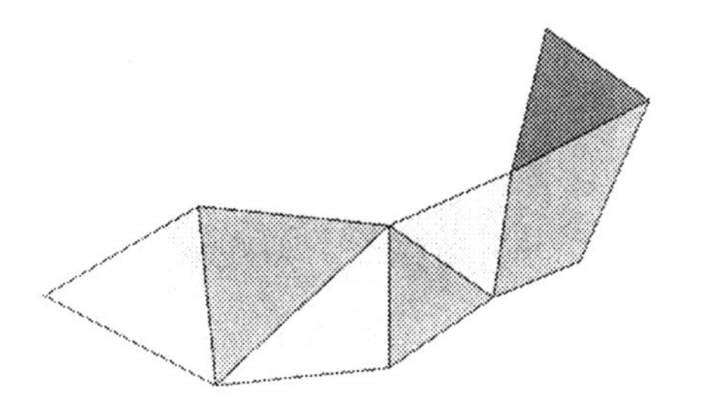

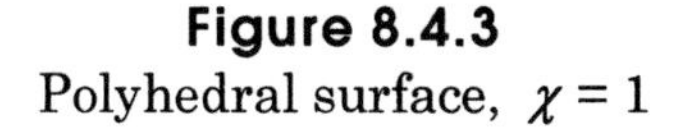

Figure 8.4.3
Polyhedral surface, $\chi = 1$

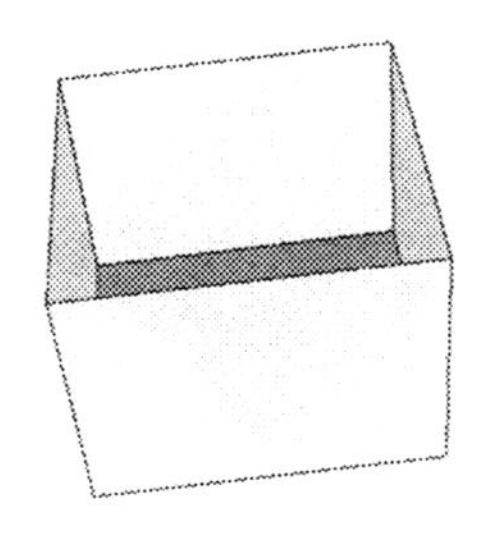

Figure 8.4.4
Open box, $\chi = 1$

Neither of the previous surfaces encloses a solid region, but the closed cube in figure 8.4.5 and the regular tetrahedron and octahedron in figures 8.4.6 and 8.4.7 certainly do. The faces of the closed surfaces in figures 8.4.8 to 8.4.10 meet in more complicated ways; each surface encloses two regions. Although the faces of the surface in figure 8.4.11 intersect as simply as possible, it's complicated in another way; the surface encloses a single region with a hole. For area and volume computations it's useful to consider unions of disjoint polygonal or polyhedral regions. Is it useful to consider unions of disjoint polyhedral surfaces, for example the union of figures 8.4.5 and 8.4.6? Complexity has arisen before the beginning of the theory of polyhedra! Should the theory comprehend *all* these examples, or just the simpler ones? Should it comprehend only boundaries of polyhedral regions, or should it also study surfaces like figures 8.4.3 and 8.4.4 that don't enclose a region?

The questions stem from the fact that each of these examples can be studied using some of the techniques that constitute the theory. And some methods of analysis—particularly Euler's theorem—*seem* to work with examples whose faces are interconnected in the most complicated ways. That is, they don't obviously fail. But it's difficult to apply them correctly to some figures, and

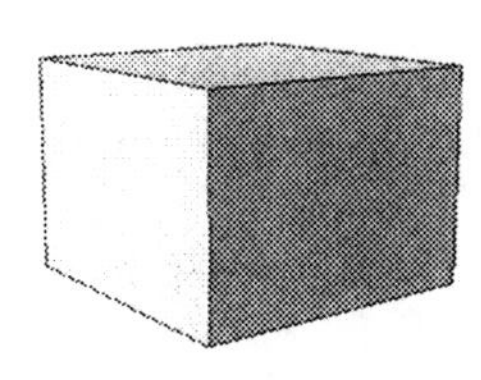

Figure 8.4.5
Closed cube, $\chi = 2$

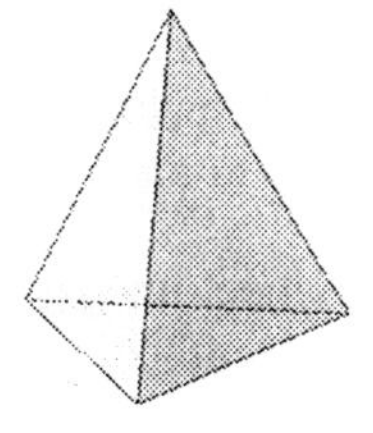

Figure 8.4.6 Regular tetrahedron, $\chi = 2$

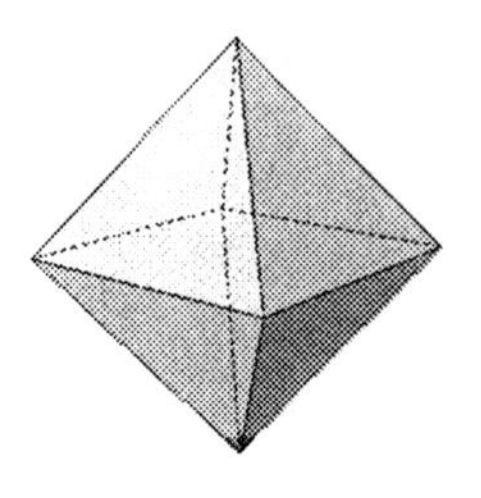

Figure 8.4.7 Regular octahedron, $\chi = 2$

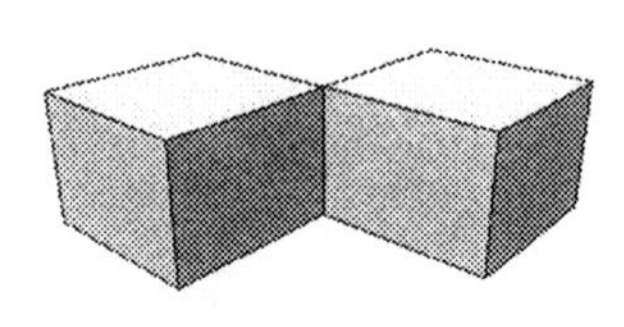

Figure 8.4.8 Two cubes joined along an edge, $\chi = 3$

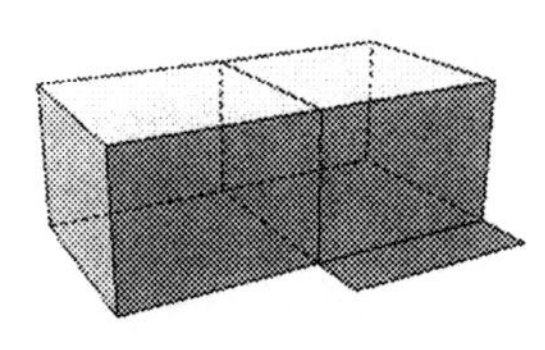

Figure 8.4.9 Two cubes joined along a face, with a flap, $\chi = 3$

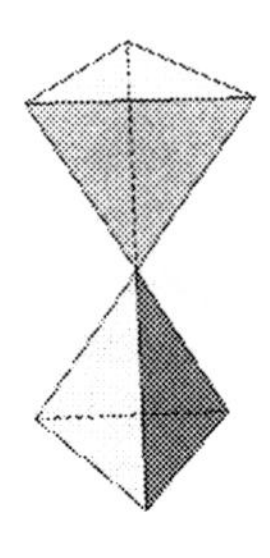

Figure 8.4.10 Two tetrahedra joined at a vertex, $\chi = 3$

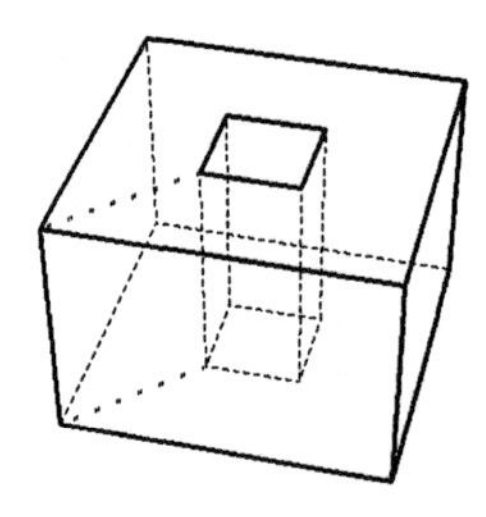

Figure 8.4.11 Cube with a hole, $\chi = 2$ or 0

in those cases they provide information that's hard to organize. Analysis and classification are simplified by restricting the class of examples under study.

The theory began with Euclid.[33] In book XIII of the *Elements*, he described the *regular* polyhedra in detail. Figures 8.4.5 to 8.4.7 show the cube and the regular tetrahedron and octahedron. The regular dodecahedron and

[33] Euclid [1908] 1956, book XIII, propositions 13 to 17.

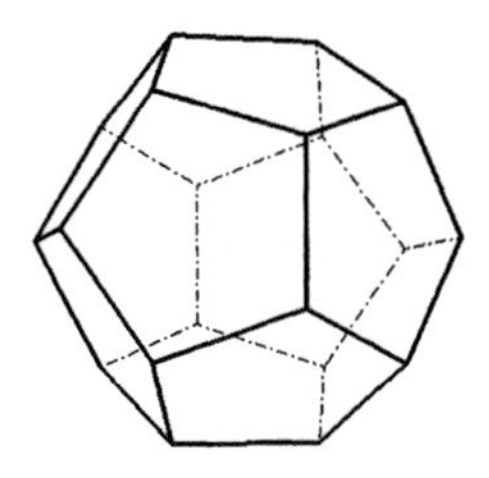

Figure 8.4.12 Regular dodecahedron, $\chi = 2$

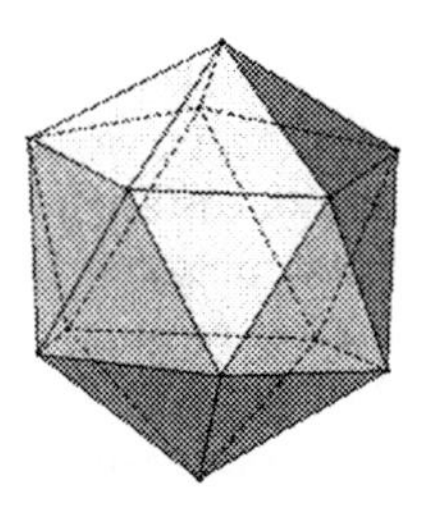

Figure 8.4.13 Regular icosahedron, $\chi = 2$

icosahedron[34] are displayed in figures 8.4.12 and 8.4.13. By Euclid's time, mathematicians had determined, presumably through investigation of models, how the various parts of these wonderful objects are related. Euclid didn't report that process. He provided instructions for constructing them with ruler and compasses (and presumably tools to cut and paste). But those constructions couldn't have been devised without prior knowledge of the figures. Euclid states that there are only five regular polyhedra, but he doesn't really specify criteria for being a polyhedron or for being regular. He didn't investigate the scope of his methods.

This subject came to be known as *solid* geometry, and mathematicians agreed that it's concerned with surfaces that enclose solid regions, not like figures 8.4.3 and 8.4.4. Are polyhedral regions, as introduced in section 3.10, the primary objects of study, or should their surfaces be regarded as more fundamental? That question wasn't important to Euclid, who was interested only in a few important examples. Evidently, you can't determine from a polyhedral region, considered as a point set, all details of the surface used to define it. For example, including or excluding the lightly dotted edges of the top and bottom faces of figure 8.4.11 has no effect on the enclosed region, but changes the character of the faces; with those edges included, no face has a hole. Euler was probably the first to consider the question and give a clear answer:

> Consideration of solids must be directed toward their surfaces. Indeed, once known from the surface by which it is included on all sides, a solid can be investigated routinely in a similar way, by which the types of its plane figures are defined in terms of its perimeter.[35]

In other words, you can study the region in terms of its surface structure, but not necessarily vice versa.

---

[34] *Tetra-*, *hexa-*, *octa-*, *dodeca-*, and *icosa-* are Greek prefixes meaning *four, six, eight, twelve,* and *twenty*. Some authors call a cube a regular *hexa*hedron!

[35] Euler [1758] 1953, 73, translated by the present author.

A theory of constructions of polyhedral regions, which would consider *all* structural details, would require attention to many inessential features not pertaining to the surface. Euler's suggestion has led instead to a theory of surface constructions. The objects under consideration include all their surface structure details, but no data directly describing regions they may enclose.

That can be a major problem. Think of a quilt, made of a million triangular patches, that covers a ball. Available to the theory is a database that tells which patches abut which along their various edges. Now consider a second database, identical to the first, but missing one patch. The first quilt encloses a region, the ball; but the second one doesn't. How do you determine that by inspecting the databases? After an exhausting search, you may find that the second database includes some patches—neighbors of the missing one —with edges not shared by others. Is that enough evidence to claim the second quilt doesn't enclose a region? No! Consider figure 8.4.9! It's difficult to describe precisely the class of objects studied by the theory under development.

It's generally agreed that the faces of these objects should be *simply connected polygonal regions*: polygonal regions whose boundaries form simple closed polygons.[36,37] To be sure, surfaces whose faces have disjoint components or holes do occur in practice, but you can always replace a troublesome face by one or more simply connected ones. For example, the top and bottom faces of figure 8.4.11 have holes. But if you cut them along the lightly dotted lines, you make them simply connected without altering the enclosed polyhedral region.

Consider a nonempty family $\mathcal{F}$ of simply connected polygonal regions, called *faces*, and the sets of their edges and vertices. (These are line segments and points; they may belong to more than one face.) There's at least one face, each face has at least three edges and three vertices, each edge has at least two vertices and borders at least one face, and each vertex is on at least two edges and one face. What assemblies of faces fall within the scope of the theory under development? $\mathcal{F}$ is said to *intersect properly* if the intersection of two distinct faces is always empty, an edge of each, or a vertex of each, and each edge is shared by at most two faces. Its union is called a *polyhedral surface*. Proper intersection requires that an edge have at most

[36] A *polygonal region* Π was defined in section 3.8, its *boundary* $\partial\Pi$ in 4.7, and *simple closed polygon* in 8.1.

[37] Points P and Q in the plane ε of Π are called *mutually accessible* if there's a polygon $P_0 \cdots P_{n-1}$ contained in $\partial\Pi$ or in $\varepsilon - \partial\Pi$ such that $P = P_0$ and $P_{n-1} = Q$. Mutual accessibility is clearly an equivalence relation on ε. Its structure is simplest when there are only three equivalence classes: $\partial\Pi$, the interior of Π, and its exterior. That happens just when Π is simply connected. Mutual accessibility thus affords a characterization of simple connectedness that seems more intuitive to the author than the definition just presented in the text. Proving they're equivalent, however, is very difficult. Since only the definition in the text is needed—for Euler's theorem—that proof is not included.

two incident vertices; that seems appropriate, because otherwise it would be hard to identify the edges and vertices from a drawing. Proper intersection also requires that an edge have at most two incident faces. The faces of the joined boxes in figures 8.4.8 and 8.4.9 do not intersect properly, so those objects lie outside the scope of the theory under development. It seems less appropriate to exclude those, because we use them commonly as containers. But including them would make the theory too complicated. The faces in figures 8.4.3 to 8.4.7 and 8.4.10 to 8.4.13 all intersect properly.

In the remainder of this section, whenever the term *polyhedral surface* is used, it refers to a nonempty properly intersecting family of faces as well as its union.

The *boundary* of a polyhedral surface is the union of its edges that border only one face. A polyhedral surface with finitely many faces but no boundary points is called a *polyhedron*. Each edge borders exactly two faces. The polyhedral surface in figure 8.4.3 that can be flattened, and the open box in figure 8.4.4, are not polyhedra according to this definition, but the regular polyhedra, the pierced cube, and the joined tetrahedra in figures 8.4.5 to 8.4.7 and 8.4.10 to 8.4.13 are. So is any union of finitely many examples such as these.

Euler's theorem

Consider the sets $\mathcal{V}$ of vertices and $\mathcal{E}$ of edges of a polyhedron $\mathcal{F}$. These sets have cardinalities $\#\mathcal{V}$, $\#\mathcal{E}$, and $\#\mathcal{F}$. The number $\chi = \#\mathcal{V} - \#\mathcal{E} + \#\mathcal{F}$ is called the *Euler characteristic* of $\mathcal{F}$. It's important to realize that $\mathcal{V}$ and $\mathcal{E}$ are *sets*—each member is counted only *once*, even if it's a vertex or edge of more than one face. The Euler characteristics of figures 8.4.3 to 8.4.13 are shown in their captions (including analogous χ values for the nonpolyhedral figures 8.4.3, 8.4.4, 8.4.8, and 8.4.9). What patterns appear? Figures 8.4.3 and 8.4.4, which enclose no region, have characteristic 1. Most of these examples have characteristic 2. Figures 8.4.8 to 8.4.10, which seem to enclose two regions each, have characteristic 3. Figure 8.4.11, with a hole, has characteristic 2 or 0, depending on whether you include the lightly dotted edges to make its top and bottom faces simply connected. Consider a polyhedral surface $\mathcal{F}$ that is the union of disjoint surfaces $\mathcal{F}_1$ and $\mathcal{F}_2$ with characteristics χ_1 and χ_2. Since the vertex and edge sets of $\mathcal{F}_1$ and $\mathcal{F}_2$ are disjoint, the characteristic of $\mathcal{F}$ is $\chi_1 + \chi_2$.

The Euler characteristic encodes some striking geometric properties of polyhedra, and provides a handy tool for reasoning about them. But it's difficult to state theorems about it that have both simple hypotheses and simple conclusions. Formulating a simple conclusion often requires stating a list of hypotheses to rule out exceptional cases. The next four paragraphs are aimed at a concise statement of one major result about the characteristic: Euler's theorem, which states very general conditions on polyhedra entailing $\chi = 2$. Figures 8.4.3, 8.4.4, 8.4.8, and 8.4.9 have been excluded because they're not polyhedra. The remaining examples with $\chi \neq 2$ are excluded by connectedness criteria.

If the following condition holds, a polyhedron $\mathscr{F}$ is called *vertex-connected*:

> for any distinct vertices V and W there exist sequences of edges E_0 to E_{n-1} and vertices V_1 to V_{n-1}, called a *path* of vertices and edges from V to W, such that if $V_0 = V$ and $V_n = W$, then $E_j = \overline{V_jV_{j+1}}$ for $j = 0$ to $n - 1$.

Unions of disjoint polyhedra, which can have arbitrarily large characteristic, will be excluded from Euler's theorem because they aren't vertex-connected. But that condition doesn't exclude figure 8.4.10, which is vertex-connected. If the following condition holds, a polyhedron $\mathscr{F}$ is called *face-connected*:

> for any distinct faces F and G there exist sequences of edges E_0 to E_{n-1} and faces F_1 to F_{n-1}, called a *path* of faces and edges from F to G, such that if $F_0 = F$ and $F_n = G$, then $E_j = F_j \cap F_{j+1}$ for $j = 0$ to $n - 1$.

In figure 8.4.14 the heavy edges indicate a path of vertices and edges from V_0 to V_4 and the shaded faces indicate a path of faces and edges from F_1 to F_5. Since each vertex of a polyhedron belongs to some face, all face-connected polyhedra are vertex-connected. But not vice versa; figure 8.4.10 isn't face-connected, because there's no path of edges and faces from the top face to the bottom face. Figures 8.4.5 to 8.4.7 and 8.4.11 to 8.4.13 are all face-connected.

Figure 8.4.11, the cube with a hole, shows that the Euler characteristic is sensitive to the connectivity of the faces. Without the lightly dotted edges, $\chi = 2$; adding one of them reduces χ to 1; adding the second makes $\chi = 0$. You can verify that adding further edges between existing vertices of the same face doesn't affect χ. This is a reason for requiring that the faces of a polyhedron be simply connected. The value of χ evidently has something to do with the number and nature of the holes in a polyhedron. Bore another hole from top to bottom alongside the first, adding additional edges to the

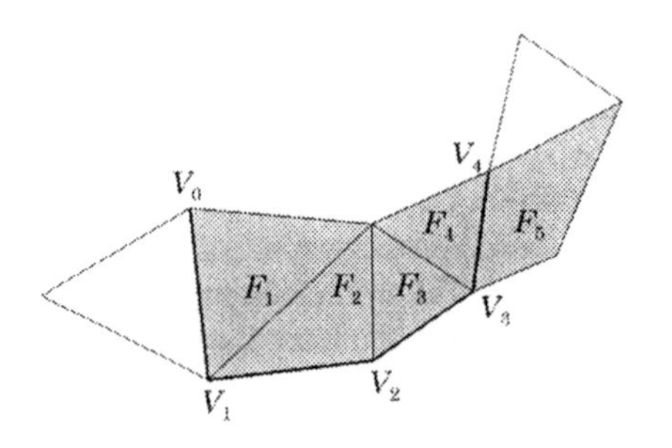

Figure 8.4.14

Path of vertices and edges from V_0 to V_4

Path of faces and edges from F_1 to F_5

top and bottom faces to keep them simply connected; that should result in $\chi = -2$. More holes, interconnected in various ways, should yield other values even more negative. The presence of holes yields values $\chi \neq 2$. In analogy to the terminology of polygonal regions, freedom from holes is called simple connectedness, defined in the next paragraph.

A *cycle* on a polyhedron $\mathcal{F}$ is a sequence of vertices V_0 to V_n and edges $E_j = \overline{V_j V_{j+1}}$ for $j = 0$ to $n - 1$ and $E_n = \overline{V_n V_0}$ such that the only edge intersections are $E_j \cap E_{j+1} = \{V_{j+1}\}$ for $j = 0$ to $n - 1$ and $E_n \cap E_0 = \{V_0\}$. $\mathcal{F}$ is called *simply connected* if for any cycle $\mathcal{Z}$ on $\mathcal{F}$, there exists a nonempty set $\mathcal{G}$ of faces, such that

- $\mathcal{Z}$ is the family of edges that border faces in both $\mathcal{G}$ and $\mathcal{F} - \mathcal{G}$,
- any path of faces and edges from a face in $\mathcal{G}$ to one in $\mathcal{F} - \mathcal{G}$ contains an edge in $\mathcal{Z}$.

The polyhedra in figures 8.4.5 to 8.4.7, 8.4.12, and 8.4.13 (the regular polyhedra) are all simply connected. But the cube with a hole in figure 8.4.11 is not; you can verify that the cycle consisting of the lightly dotted top and bottom edges and the vertical edges joining them does *not* divide the faces into disjoint sets $\mathcal{G}$ and $\mathcal{F} - \mathcal{G}$ as required.

How stable is the characteristic value $\chi = 2$ when you modify the polyhedron, maintaining connectivity properties? Experiment! For example, add a diagonal edge to a face of the cube or dodecahedron in figure 8.4.5 or 8.4.12; you add an edge *and* a face, so χ remains the same. Add a pyramid to another face; you add one vertex, n edges, and n new faces, but lose one old face, the base of the pyramid. Again, χ remains unchanged. These experiments provide the intuition that underlies the following famous theorem.

Theorem 1 (Euler's theorem). If the polyhedron $\mathcal{F}$ is face-connected and simply connected, then $\chi = 2$.

Proof. Define sequences of sets $\mathcal{B}_0 \subseteq \mathcal{B}_1 \subseteq \mathcal{B}_2 \subseteq \cdots$ of edges and $\mathcal{F}_0 \subseteq \mathcal{F}_1 \subseteq \mathcal{F}_2 \subseteq \cdots$ of faces as follows. First, select any face F_0; set $\mathcal{B}_0 = \phi$ and $\mathcal{F}_0 = \{F_0\}$. Given $\mathcal{B}_k$ and $\mathcal{F}_k$, form $\mathcal{B}_{k+1}$ by appending to $\mathcal{B}_k$ an edge B_{k+1} not in $\mathcal{B}_k$ but bordering exactly one face in $\mathcal{F}_k$. If there's none, stop the sequence. Otherwise, B_{k+1} borders exactly one other face F_{k+1}, which cannot belong to $\mathcal{F}_k$; append it to $\mathcal{F}_k$ to form $\mathcal{F}_{k+1}$. Notice that for each k a face belongs to $\mathcal{F}_k$ *if and only if* it has an edge in $\mathcal{B}_k$, and $\#\mathcal{B}_k = k = \#\mathcal{F}_k - 1$. Define $\mathcal{B} = \bigcup_k \mathcal{B}_k$ and $\mathcal{F}^* = \bigcup_k \mathcal{F}_k$.

Every face $F \neq F_0$ is linked to F_0 by a path of faces and edges in $\mathcal{B}$ and $\mathcal{F}^*$. To see this, suppose the sequences have just stopped with $\mathcal{B} = \mathcal{B}_k$ and $\mathcal{F}^* = \mathcal{F}_k$. Because $\mathcal{F}$ is face-connected, some path of edges and faces links F_0 to F. Proceed from F_0 crossing those edges. If one of them didn't belong to $\mathcal{B}_k$, the *first* that didn't would border exactly one face in $\mathcal{F}_k$, and the sequence would be able to progress. This paragraph implies $\mathscr{F} = \mathscr{F}^*$. When the sequences stop, $\mathscr{B} = \mathscr{B}_k$, $\mathscr{F} = \mathscr{F}_k$, and $\#\mathscr{B} = \#\mathscr{F} - 1$.

There's no cycle $\mathscr{Z}$ of vertices and edges in $\mathscr{E} - \mathscr{B}$. To see this, suppose there were such a cycle. Consider any edge in $\mathscr{Z}$ and the two faces F and F' it borders. By the previous paragraph they're linked by a path of faces and edges in $\mathscr{B}$. That would contradict the simple connectivity of $\mathscr{F}$ —any such path must contain an edge belonging to $\mathscr{Z}$.

Select any vertex V_0. For each j, every vertex $V \neq V_0$ is linked to V_0 by a path of vertices and edges not in $\mathscr{B}_j$. To verify this statement, proceed recursively. It's trivially true for $j = 0$. Suppose it's true for $j = k$. Then V is linked to V_0 by a path $\mathscr{P}$ of vertices and edges not in $\mathscr{B}_k$. If $\mathscr{P}$ doesn't contain B_{k+1}, then it's a path of vertices and edges not in $\mathscr{B}_{k+1}$, as desired. Otherwise, replace B_{k+1} in $\mathscr{P}$ by the concatenation of all remaining edges of the face F containing B_{k+1} that's not in $\mathscr{F}_{k+1}$. By the first paragraph of this proof, none of those are in $\mathscr{B}_{k+1}$, and because F is simply connected they form a path connecting the ends of B_{k+1}. The resulting path of vertices and edges not in $\mathscr{B}_{k+1}$ links V to V_0, as desired; the statement is true for $j = k + 1$.

By the previous paragraph, every vertex $V \neq V_0$ is linked to V_0 by a path of vertices and edges in $\mathscr{E} - \mathscr{B}$. In fact, this path is unique, for if two distinct paths of vertices and edges in $\mathscr{E} - \mathscr{B}$ linked V to V_0, there'd be a cycle $\mathscr{Z}$ of vertices and edges in $\mathscr{E} - \mathscr{B}$, contradicting the third paragraph of this proof.

The function λ that maps each vertex $V \neq V_0$ to the last edge $\lambda(V)$ of the path of vertices and edges in $\mathscr{E} - \mathscr{B}$ that links V to V_0 is a bijection from $\mathscr{V} - \{V_0\}$ to $\mathscr{E} - \mathscr{B}$, so $\#\mathscr{V} - 1 = \#(\mathscr{E} - \mathscr{B})$. It follows that $\#\mathscr{E} = \#(\mathscr{E} - \mathscr{B}) + \#\mathscr{B} = \#\mathscr{V} - 1 + \#\mathscr{F} - 1$, hence $\chi = \#\mathscr{V} - \#\mathscr{E} + \#\mathscr{F} = 2$. ♦

By 1630 Descartes knew the essence of theorem 1, but a more than a century elapsed before it was stated explicitly by Euler.[38] Euler and other mathematicians presented proofs, using various methods and with varying degrees of precision and rigor. That given here is due essentially to von Staudt.[39] Although the arguments in all these proofs are simple—von Staudt used only two sentences—several attempts were required before the intersection and connectivity hypotheses were stated precisely. Cromwell gives a detailed historical account.[40] Lakatos 1976 is an entire book in dialogue form devoted to this problem. For information on polyhedra with $\chi \neq 2$, consult the

[38] Euler [1758] 1953, number 33.

[39] Von Staudt 1847, section 50.

[40] 1997, chapter 5.

marvelous hand-lettered and illustrated book Stewart 1980, or a reference on combinatorial topology.[41]

Rademacher and Toeplitz present von Staudt's argument ominously.[42] All but one of the faces of the polyhedron represent fields in a low-lying island beset by the tides. The fields are walled by dikes along their edges; vertices represent villages situated at dike intersections. The remaining face— F_0 in the proof just given—is the surrounding cruel sea. One by one the fields are flooded by breaches in the dikes. $\mathcal{B}_k$ consists of the first k dikes that fail, and $\mathcal{F}_k$ is the set of fields flooded by those failures. Eventually all fields are flooded by a minimal set of breaches. Village V_0 in the proof is the seat of government. It can communicate with all other villages by sending messengers along unbroken dikes, because only a minimal set of breaches occurred. But the messengers have no choice of route, and any further breach will isolate at least one village.

Regular polyhedra

The *regular* polyhedra were described in detail by Euclid around 300 B.C.[43] Since Plato had discussed them about fifty years earlier, they're also known as *Platonic* figures.[44] Each has congruent regular polygonal faces, and the same number of faces meet at each vertex. For some $p \geq 3$ each face has p edges; since each edge lies on exactly two faces, $p\#\mathcal{F} = 2\#\mathcal{E}$. For some $q \geq 3$ each vertex lies on q edges; since each edge joins exactly two vertices, $q\#\mathcal{V} = 2\#\mathcal{E}$. These two equations with Euler's equation $\#\mathcal{V} - \#\mathcal{E} + \#\mathcal{F} = 2$ imply

$$\frac{2\#\mathcal{E}}{p} + \frac{2\#\mathcal{E}}{q} = 2 + \#\mathcal{E} \qquad \frac{1}{p} + \frac{1}{q} = \frac{1}{2} + \frac{1}{\#\mathcal{E}} > \frac{1}{2}.$$

This inequality prevents p and q from being very large. In fact, only five p, q pairs are possible. They're tabulated in table 8.4.1 with the corresponding values of $\#\mathcal{V}$, $\#\mathcal{E}$, and $\#\mathcal{F}$ and the name of the polyhedron.

According to the previous paragraph, table 8.4.1 lists the only *possibilities* for regular polyhedra—if they exist, they must have these properties. Do they really exist? You can convince yourself of that by building each one

---

41 For example, see Eves 1963–1965, chapter 15.

42 Rademacher and Toeplitz [1933] 1957, chapter 12.

43 Euclid [1908] 1956, book XIII.

44 Plato presented in *Timaeus* ([1965] 1971, section 21 ff.) a theory of the structure of matter, in which he assigned to the elements fire, air, water, and earth basic units having the shapes of the regular tetrahedron, octahedron, icosahedron, and the cube. He attributed their various chemical and physical properties to the way in which these units are constructed from 30° and 45° right triangles. The dodecahedron—the regular polyhedron most like a perfect sphere—he assigned to the universe as a whole.

Table 8.4.1 The regular polyhedra

p	q	$\frac{1}{p}+\frac{1}{q}$	$1/\#\mathscr{E} = \frac{1}{p}+\frac{1}{q}-\frac{1}{2}$	$\#\mathscr{V}$	$\#\mathscr{E}$	$\#\mathscr{F}$	*name*
3	3	$^2/_3$	$^1/_6$	4	6	4	*tetra*hedron
3	4	$^7/_{12}$	$^1/_{12}$	6	12	8	*octa*hedron
3	5	$^8/_{15}$	$^1/_{30}$	12	30	20	*icosa*hedron
4	3	$^7/_{12}$	$^1/_{12}$	8	12	6	*cube*
5	3	$^8/_{15}$	$^1/_{30}$	20	30	12	*dodeca*hedron

from the required number of equilateral triangles, squares, or regular pentagons. But how do you know they fit *exactly*? You can easily use a Cartesian coordinate system to find the vertices of a cube: all eight possibilities for $<\pm 1, \pm 1, \pm 1>$. Each edge has length 2. Those points with two positive coordinates and one negative form the vertices of a regular tetrahedron whose edges have length $2\sqrt{2}$. The centroids of the six faces of the cube constitute the vertices $<\pm 1, 0, 0>$, $<0, \pm 1, 0>$, $<0, 0, \pm 1>$ of a regular octahedron $\mathscr{O}$ whose edges have length $\sqrt{2}$. Any face T of $\mathscr{O}$ and the nearest vertex V of the cube also form a regular tetrahedron $\mathscr{T}$, and the faces of $\mathscr{T}$ that contain V are coplanar with the faces of $\mathscr{O}$ that intersect T. This shows that the dihedral angles of $\mathscr{T}$ and $\mathscr{O}$, hence of any[45] regular tetrahedron and octahedron, are supplementary.

Consider adjacent faces ΔVWX and $\Delta VWX'$ of that octahedron, and points P, Q, and Q' on edges $\overline{VW}$, $\overline{WX}$, and $\overline{WX'}$: for example,

$$\begin{aligned} V &= <1, 0, 0> & P &= tV + (1-t)W = <t, 1-t, 0> \\ W &= <0, 1, 0> & Q &= tW + (1-t)X = <0, t, 1-t> \\ X &= <0, 0, 1> & Q' &= tW + (1-t)X' = <0, t, -1+t>. \\ X' &= <0, 0, -1> \end{aligned}$$

Determine $t > 0$ so that $PQ = QQ'$ and therefore $\Delta PQQ'$ is equilateral; you get

$$\begin{aligned} (PQ)^2 &= 6t^2 - 6t + 2 & 0 &= (PQ)^2 - (QQ')^2 = 2(t^2 + t - 1) \\ (QQ')^2 &= 4t^2 - 8t + 4 & t &= -\tfrac{1}{2} + \tfrac{1}{2}\sqrt{5}. \end{aligned}$$

[45] This conclusion requires an argument that any two regular tetrahedra are similar, hence have the same dihedral angles, and an analogous argument for regular octahedra. You can supply those, using methods described in chapter 3.

Construct $R = tX + (1-t)V = \langle 1-t, 0, t \rangle$, so that ΔPQR is equilateral. Continue constructing points on all twelve edges, with coordinates

$$\langle \pm t, \pm(1-t), 0 \rangle, \quad \langle \pm(1-t), 0, \pm t \rangle, \quad \langle 0, \pm t, \pm(1-t) \rangle.$$

(In each case all four $\pm$ combinations occur.) Repeated application of the SSS congruence principle shows that these points form twenty congruent triangles: the faces of a regular icosahedron with edge $QQ' = 3 - \sqrt{5}$.

Eight of these faces lie within faces of the octagon; corresponding icosahedral and octahedral faces share the same centroid, with coordinates $\langle \pm\frac{1}{3}, \pm\frac{1}{3}, \pm\frac{1}{3} \rangle$. All eight $\pm$ combinations occur. In exercise 8.5.25 you'll calculate the coordinates of the centroids of the other twelve icosahedral faces, and show that altogether they form the twenty vertices of a regular dodecahedron with edge $-\frac{1}{3} + \frac{1}{3}\sqrt{5}$.

The programs that drew the regular polyhedra in figures 8.4.5 to 8.4.7, 8.4.12, and 8.4.13 actually used the coordinates derived in the previous three paragraphs.

Dual polyhedra

The construction of a regular dodecahedron in the previous paragraph suggests a more general method for analyzing polyhedra. On each of the n faces of a polyhedron $\mathscr{P}$ select a point C symmetrically situated with respect to its vertices; can you regard those n points as the vertices of a new polyhedron $\mathscr{P}'$? In the special case just considered, each face of $\mathscr{P}$ was a regular polygon and C its center. Applied to the regular polyhedra, the method is spectacularly successful:

n	*Polyhedron $\mathscr{P}$ with n faces*		*New polyhedron $\mathscr{P}'$ with n vertices*
4	Regular tetrahedron	→	Regular tetrahedron
6	Cube	⤨	Regular octahedron
8	Regular octahedron		Cube
12	Regular dodecahedron	⤨	Regular icosahedron
20	Regular icosahedron		Regular dodecahedron

In each case, carrying out the construction twice in succession produces a polyhedron similar to the original, related to it by a contraction about their common center. For this reason, $\mathscr{P}'$ is called the *dual* of $\mathscr{P}$. The cube and regular octahedron are dual, as are the regular dodeca- and icosahedra. The regular tetrahedron is *self-dual*; its dual is a similar tetrahedron related to the original by the reflection across their common center followed by a contraction about it. These relationships are illustrated by figures 8.4.15 to 8.4.19.

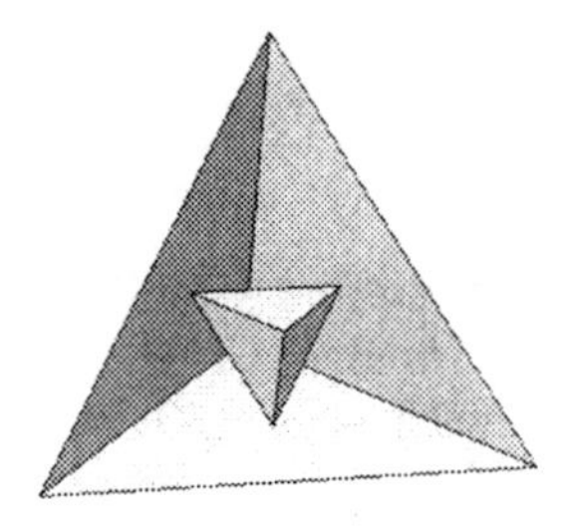

Figure 8.4.15 Regular tetrahedron and its dual

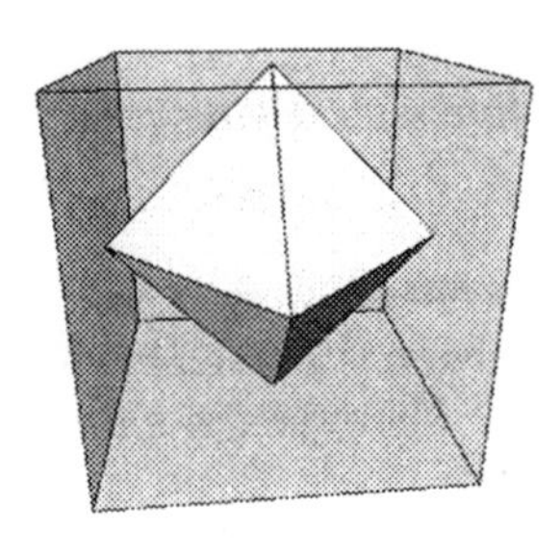

Figure 8.4.16 Cube and its dual

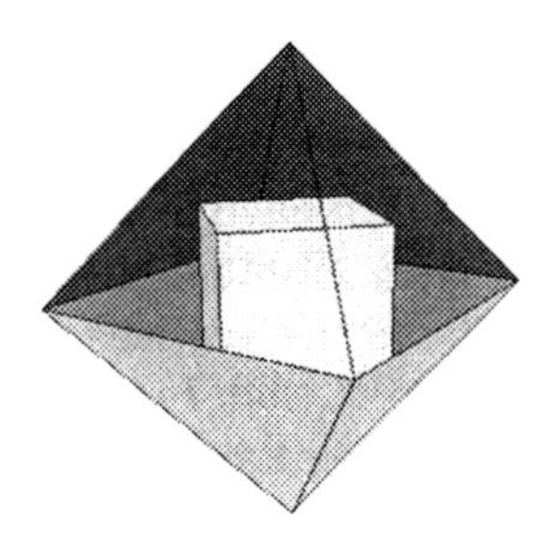

Figure 8.4.17 Regular octahedron and its dual

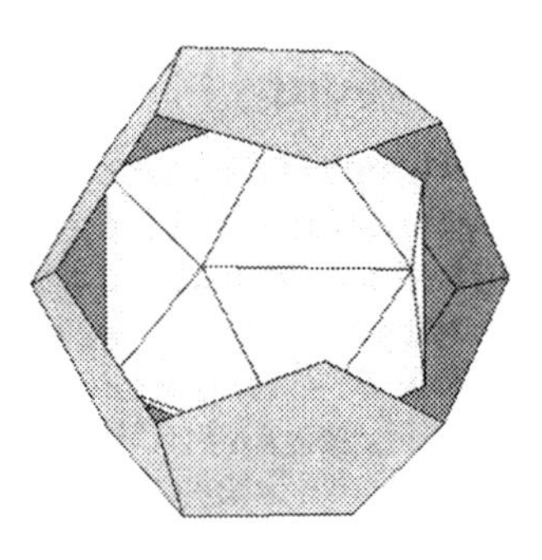

Figure 8.4.18 Regular dodecahedron and its dual

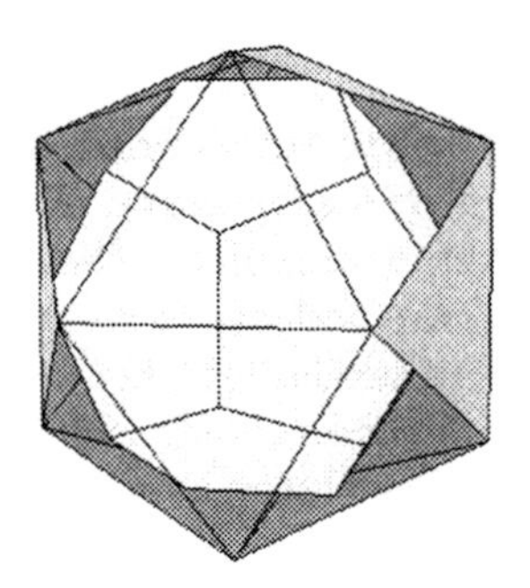

Figure 8.4.19 Regular icosahedron and its dual

How general is this dualization method? It seems clear that you can dualize any polyhedron $\mathcal{P}$ whose faces have centers of some sort.[46] Even if $\mathcal{P}$ is convex, though, you may have trouble identifying a dual polyhedron $\mathcal{P}'$. Knowledge of its vertices doesn't necessarily tell you how to join them with edges and faces. You'll gain more familiarity with dual polyhedra in the rest of this section and some exercises in section 8.5.

Cuboctahedron and rhombic dodecahedron

Each vertex of a regular polyhedron is surrounded by the same number of regular faces, all with the same number of edges. What possibilities arise when you relax one those constraints? For example, is there a polyhedron with each vertex surrounded by two triangles and two quadrilaterals? Is there a highly symmetric one? Euler's theorem is the theoretical tool for attacking such questions, so you need to rephrase the question, including its hypotheses: Is there such a face-connected and simply connected polyhedron? If so, suppose it has f_3 triangular faces and f_4 quadrilaterals. Let $\mathcal{V}$ and $\mathcal{E}$ denote its vertex and edge sets. Visiting all vertices and counting edges, triangles, and quadrilaterals you find

$$4\#\mathcal{V} = 2\#\mathcal{E} \qquad 2\#\mathcal{V} = 3f_3 \qquad 2\#\mathcal{V} = 4f_4$$

and by Euler's theorem,

$$\#\mathcal{V} - \#\mathcal{E} + (f_3 + f_4) = 2.$$

Using the first three equations to eliminate all unknowns but $\#\mathcal{V}$ from the last, you find $\#\mathcal{V} = 12$, which entails $\#\mathcal{E} = 24$, $f_3 = 8$, and $f_4 = 6$. You need eight triangles and six quadrilaterals, and you must connect them to get twenty-four edges and twelve vertices.

In fact, you can make a highly symmetric polyhedron from these components by truncating the vertices of a cube, obtaining the object in figure 8.4.20. Because the cube and regular octahedron are so closely related, you could just as well truncate an octahedron, as in figure 8.4.21. The resulting polyhedron $\mathcal{P}$ is called a *cuboctahedron.*

Can you visualize its dual polyhedron $\mathcal{P}'$? That must have fourteen vertices, one at the center of each face of $\mathcal{P}$. If the dual of $\mathcal{P}'$ is to be similar to $\mathcal{P}$, then $\mathcal{P}'$ must have twelve faces. It would be a dodecahedron, but not regular, because regular dodecahedra have twenty vertices, not fourteen. It couldn't be dual to the cuboctahedron as defined under the previous heading, because the centroids of the faces surrounding a vertex of a cuboctahedron aren't coplanar.

---

[46] You don't really have to use *centers.* In each of figures 8.4.15 to 8.4.19 you could move any vertex of the dual polyhedron slightly off center, and still have a polyhedron clearly related to the original.

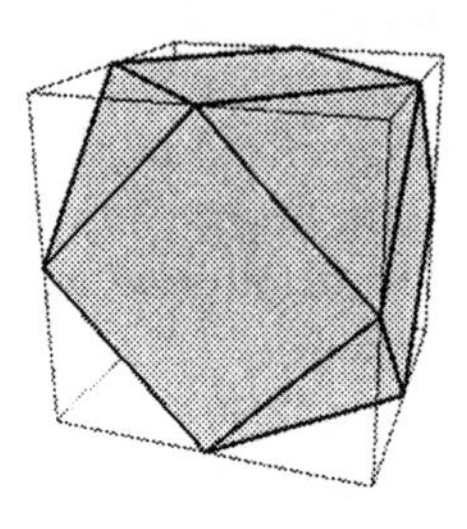

Figure 8.4.20 Truncating a cube to make a cuboctahedron

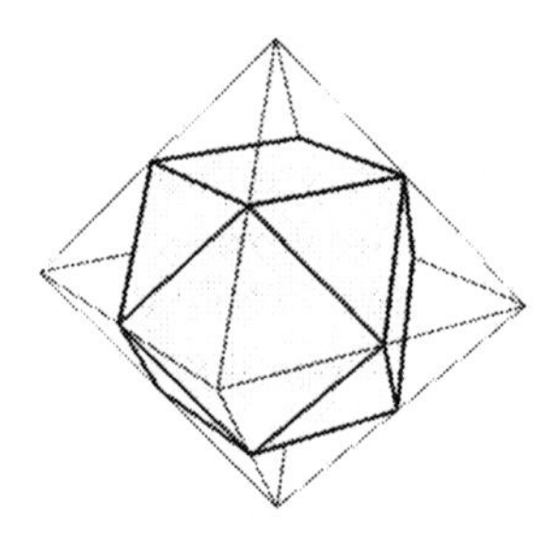

Figure 8.4.21 Truncating a regular octahedron to make a cuboctahedron

To construct $\mathscr{P}'$, use a Cartesian coordinate system to divide space into cubical cells separated by planes perpendicular to the axes, through "lattice" points with integral coordinates. Consider one cell $\mathscr{C}$. It has four pairs of opposite edges, which lie on four planes through its center. Planes parallel to these through other lattice points dissect every cell into six square pyramids with common apex at the cell center. Thus each edge $\overline{WY}$ of $\mathscr{C}$ is coplanar with the centers X and Z of two cells adjacent to $\mathscr{C}$, and $WXYZ$ is a rhombus. Each of the six faces of $\mathscr{C}$ forms the base of one of the pyramids constituting an adjacent cell. The union of $\mathscr{C}$ and these six pyramids is a polyhedron $\mathscr{P}'$ with a total of twelve congruent rhombic faces. (Tallying four rhombi at the apices of each of the six pyramids counts each rhombus twice.) $\mathscr{P}'$ is called a *rhombic dodecahedron*; it's displayed in figure 8.4.22. $\mathscr{P}'$ has fourteen vertices: the apices of the six pyramids and the eight vertices of $\mathscr{C}$. Each vertex of $\mathscr{P}'$ is shared by three obtuse or four acute face angles. Since the centers of the faces of $\mathscr{P}'$ are the midpoints of the edges of $\mathscr{C}$, the dual of $\mathscr{P}'$ (as defined under the previous heading) is a cuboctahedron. Although the cuboctahedron was reportedly known to Archimedes about 250 B.C., the rhombic dodecahedron was first discussed by Johannes Kepler around 1610.

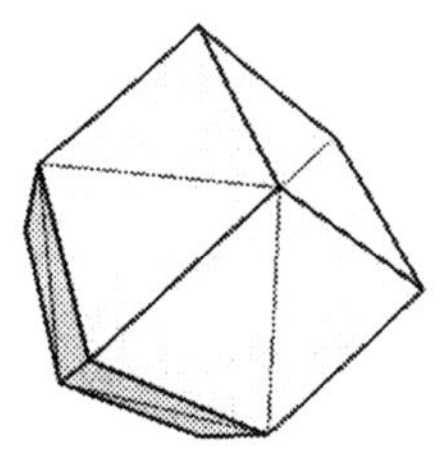

Figure 8.4.22
Rhombic docecahedron

Deltahedra

All faces of the cuboctahedron are regular polygons, but they don't all have the same number of edges. All faces of the rhombic dodecahedron are congruent, but they're not regular. What's possible if you require that all faces be congruent regular polygons?

First, no such polyhedron can have only regular faces with more than five sides, because then each angle at a vertex would be at least $120°$, and three of them would total more than $360°$. If all faces were regular pentagons, only three could meet at any vertex. Visiting all vertices and counting edges and faces, you'd find that $3\#\mathcal{V} = 2\#\mathcal{E} = 5\#\mathcal{F}$. Euler's theorem would yield $2 = \#\mathcal{V} - \#\mathcal{E} + \#\mathcal{F} = \frac{1}{10}\#\mathcal{V}$, hence $\#\mathcal{V} = 20$ and $\#\mathcal{F} = 12$: the polyhedron must be a regular dodecahedron. If all faces were squares, it would be a cube.

If all faces of a polyhedron are equilateral triangles, it's called a *deltahedron*. (Equilateral triangles resemble the Greek letter Δ.) There are three regular deltahedra: the tetrahedron, octahedron, and icosahedron. You can make a regular octahedron by joining two pyramids along their congruent square bases; any such polyhedron is called a *dipyramid*. Two more deltahedra fall into this category: the equilateral triangular and pentagonal dipyramids shown in figures 8.4.23 and 8.4.24.

You can join other polyhedra along faces so that only equilateral triangles remain. For example, start with a triangular prism with square faces or the square antiprism constructed in exercise 4.8.2, and erect equilateral pyramids on all square faces, to obtain the deltahedra in figures 8.4.25 and 8.4.26. Erecting pyramids on the faces of a polyhedron is called *stellation*, from the Latin word *stella* for *star*.

All deltahedra displayed so far have been convex. A polyhedron is *convex* if it has no vertices on different sides of any face plane. By adjoining polyhedra

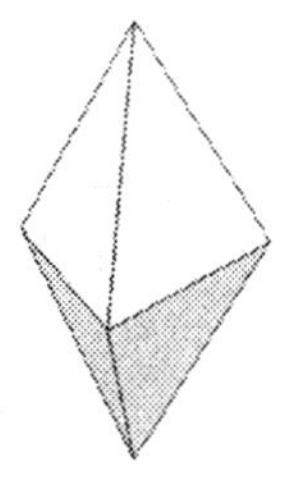

Figure 8.4.23 Equilateral triangular dipyramid

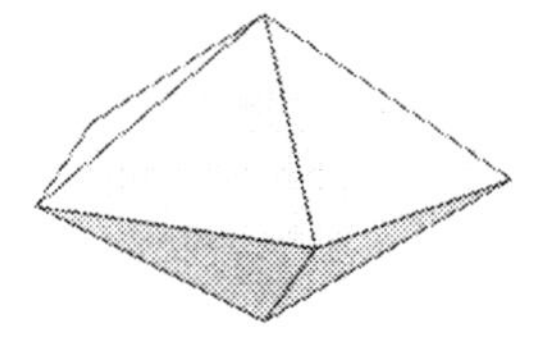

Figure 8.4.24 Equilateral pentagonal dipyramid

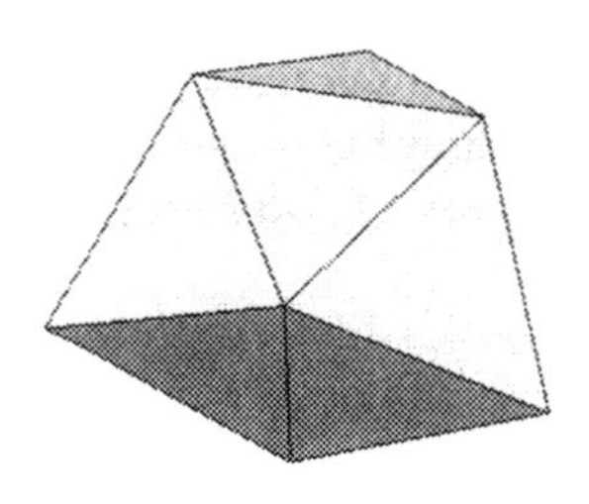

Figure 8.4.25 Deltahedral stellation of a triangular prism

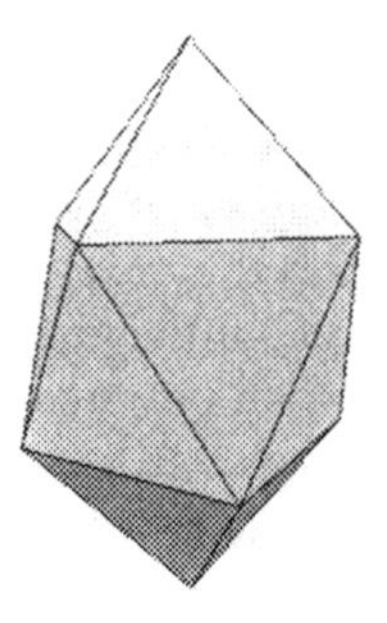

Figure 8.4.26 Deltahedral stellation of a square antiprism

you can construct *nonconvex* deltahedra freely, particularly if you use chains of regular tetrahedra, each adjoined to the next along a face.[47]

What other *convex* deltahedra are there? Suppose a deltahedron $\mathscr{P}$ has a vertex V belonging to just three faces—a *trihedral* vertex. Let X, Y, and Z denote the remaining vertices of those faces. $\mathscr{P}$ could be a regular tetrahedron. If not, edges $\overline{XY}$, $\overline{YZ}$, and $\overline{ZX}$ belong to three distinct faces that don't contain V. If they have a common vertex, then $\mathscr{P}$ is an equilateral triangular dipyramid. If not, two of those edges—for example $\overline{XY}$ and $\overline{YZ}$ —belong to faces ΔXYW and $\Delta YZW'$ with $W \neq W'$. If Y belonged to just one more face $\Delta YWW'$, then $ZXWW'$ would be a square, Y would be the apex of a pyramid with this square base, this pyramid would form half a regular octahedron, $VXYZ$ would be a regular tetrahedron erected on its face ΔXYZ, and ΔVXY and ΔXYW would be coplanar because the dihedral angles of a regular tetrahedron and a regular octahedron are supplementary. That's impossible; moreover, if Y belonged to *five* faces, $\mathscr{P}$ would fail to be convex. Therefore, the only convex deltahedra with trihedral vertices are the regular tetrahedron and the equilateral triangular dipyramid.

Now consider a convex deltahedron $\mathscr{P}$ with no trihedral vertex. Let v_4 and v_5 denote the numbers of vertices of $\mathscr{P}$ at which four and five vertices meet. Visiting all vertices and counting edges and faces, you'll find that

$$4v_4 + 5v_5 = 2\#\mathscr{E} = 3\#\mathscr{F}.$$

These equations and Euler's theorem yield

$$\begin{aligned}2 &= \#\mathscr{V} - \#\mathscr{E} + \#\mathscr{F} = (v_4 + v_5) - (2v_4 + \tfrac{5}{2}v_5) + (\tfrac{4}{3}v_4 + \tfrac{5}{3}v_5) \\ &= \tfrac{1}{3}v_4 + \tfrac{1}{6}v_5.\end{aligned}$$

[47] For an example study, see Trigg 1978.

Multiply by 6 to get

$$2v_4 + v_5 = 12.$$

Clearly $v_4 \le 6$. If $v_4 = 6$, then $v_5 = 0$ and $\mathscr{P}$ is a regular octahedron. If $v_4 = 5$, then $v_5 = 2$ and $\mathscr{P}$ is a pentagonal dipyramid. If $v_4 = 4$, then $v_5 = 4$; no such deltahedron has been displayed yet. If $v_4 = 3$, then $v_5 = 6$ and $\mathscr{P}$ is a stellated triangular prism as in figure 8.4.25. If $v_4 = 2$, then $v_5 = 8$, and $\mathscr{P}$ is a stellated square antiprism as in figure 8.4.26. If $v_4 = 1$, then $v_5 = 10$; no such deltahedron has been displayed yet. If $v_4 = 0$, then $v_5 = 12$ and $\mathscr{P}$ is a regular icosahedron.

In exercise 8.5.37 you'll show that a convex deltahedron with $v_4 = v_5 = 4$ must be a *Siamese* dodecahedron as shown in figure 8.5.12. In exercise 8.5.36 you'll show that the one remaining case in the previous paragraph, $v_4 = 1$ and $v_5 = 10$, is impossible. Therefore there are precisely eight types of convex deltahedra:

regular tetrahedron, octahedron, and icosahedron;
equilateral triangular and pentagonal dipyramids;
stellated equilateral triangular prism and square diprism;
Siamese dodecahedron.

Whereas the first five of these types were extremely familiar to ancient mathematicians, the last three were first discussed by Rausenberger (1915); the proof that there are only eight types is due to Freudenthal and van der Waerden (1947).

Polyhedral symmetry

Polygonal symmetry groups—the dihedral groups and their subgroups studied in section 8.1—are easy to visualize. The analogous three-dimensional groups—symmetry groups of polyhedra—are more complicated and harder to imagine. You'll benefit by handling models of the polyhedra as you read about their groups. Instructions for making models from railroad board and rubber bands are included in section 8.5, as well as references to other types of models.

A symmetry φ of a polyhedron $\mathscr{P}$ must map any quadruple $\langle V_1, \dots, V_4 \rangle$ of noncoplanar vertices of $\mathscr{P}$ to another such quadruple $\langle V_1', \dots, V_4' \rangle$; moreover, φ is the only isometry with $\varphi(V_i) = V_i'$ for all i. Given V_1 to V_4 there are only finitely many possibilities for V_1' to V_4', hence there are only finitely many possibilities for φ. That is, *the symmetry group* $\mathscr{G}$ *of a polyhedron is finite.* Therefore, it can contain no screw or glide reflection with nonzero vector; $\mathscr{G}$ must consist entirely of rotations and rotary reflections. To describe $\mathscr{G}$ for a given polyhedron, first determine its nontrivial rotations. The rotations about each axis a constitute a cyclic subgroup $\mathscr{G}_a$ of $\mathscr{G}$. If $\mathscr{G}_a$ has n elements, a is called an *n-fold* axis. It may be troublesome to find all the axes. Once you see an axis a, hold a model pointing a straight ahead of you; it's easy then to enumerate the elements of $\mathscr{G}_a$. Having found all rotational symmetries ρ, find any odd symmetry φ of $\mathscr{P}$. Each composition $\varphi\rho$ is again an odd symmetry; moreover, if χ is any odd symmetry of $\mathscr{P}$, then $\rho = \varphi^{-1}\chi$ is an even, hence rotational, symmetry and $\chi = \varphi\rho$. Thus *if* $\mathscr{G}$ *contains any odd symmetry* φ, *then exactly half its members are even symmetries* ρ; *its odd symmetries are the compositions* $\varphi\rho$.

As examples, consider right prisms $\mathscr{P}_n$ whose bases are regular polygons with n edges. When $n = 4$, adjust the distance between bases so that $\mathscr{P}_n$ is not a cube. Call the bases *horizontal* and the other faces *vertical*, and use similar terminology for lines parallel to them. Figure 8.4.27 shows examples for $n = 3$ and 6. (They're typical; some properties of $\mathscr{P}_n$ depend on whether n is odd or even.) Each prism has an n-fold vertical axis a of symmetry joining the centers of its bases. Midway between the bases are n twofold horizontal axes. For odd n each horizontal axis joins the center and midpoint of an opposite face and edge. For even n half the horizontal axes join centers of opposite faces; the other half join midpoints of opposite edges. Besides these $2n$ rotations, the symmetry group $\mathscr{G}$ of $\mathscr{P}_n$ contains reflection σ_ε across the plane midway between the bases, and the compositions of σ_ε with all $2n - 1$ nontrivial rotations. These compositions consist of $n - 1$ rotary reflections about a with mirror ε, and the reflections across n vertical planes through the horizontal axes. Thus $\mathscr{G}$ contains $4n$ symmetries in all: it's called a $\mathscr{D}_{nh}$ group.

How can you modify a prism $\mathscr{P}_n$ to maintain the vertical rotational symmetries but eliminate some or all of the others? First, the corresponding right pyramid $\mathscr{Q}_n$ in figure 8.4.28 displays the vertical rotations and reflec-

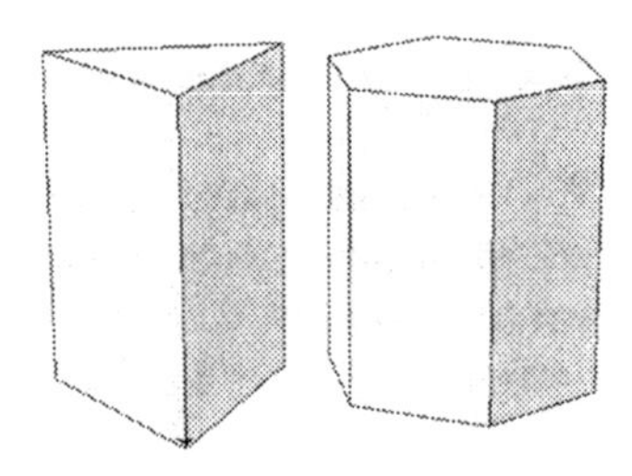

Figure 8.4.27
Prisms with $\mathscr{D}_{3h}$ and $\mathscr{D}_{6h}$ symmetry groups

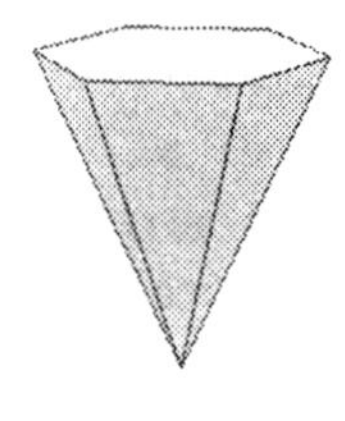

Figure 8.4.28
Pyramid with $\mathscr{C}_{6v}$ symmetry group

tions, but none of the others. Its symmetry group, with $2n$ elements, is called a $\mathscr{C}_{nv}$ group. In this notation, due to Schönflies (1891) and explained by Cromwell,[48] symbols $\mathscr{D}$ and $\mathscr{C}$ evoke analogies with the dihedral and cyclic groups studied in section 8.1; subscripts h and v indicate the presence of horizontal and vertical plane reflections.

Next, you could append a polyhedron $\mathscr{Y}$ with n-fold cyclic but not dihedral symmetry to one base of a prism $\mathscr{P}_n$ but not the other. Figure 8.4.29 is an example. The Y motif appears painted on the prism, but regard it as constructed from molding with rectangular cross section and glued on, so that the result is a polyhedron $\mathscr{Q}$. The symmetry group of $\mathscr{Q}$ consists only of the n vertical rotations. It's called a $\mathscr{C}_n$ group, and is isomorphic to the corresponding planar cyclic group studied in section 8.1.

You could also append copies of $\mathscr{Y}$ to *both* bases of $\mathscr{P}_n$. If you align them properly, as in figure 8.4.30, the symmetric group of the resulting polyhedron will consist of the n vertical rotations and the half turns about the n horizontal axes; the polyhedron displays no reflectional symmetry. This is called a $\mathscr{D}_n$ *dihedral* group. It's isomorphic to the corresponding planar dihedral group studied in section 8.1: the vertical rotations correspond to the rotations in the planar group; the half turns, to the planar line reflections.

Make a copy of $\mathscr{Y}$ with opposite orientation. (Apply glue to its opposite face.) Append $\mathscr{Y}$ to one base of a prism $\mathscr{P}_n$ and its opposite to the other, as in figure 8.4.31, where the motifs are Y and Y. (These opposite motifs appear alike in the figure because you're looking down at the outside of the top face and down at the inside of the bottom face.) The symmetry group of the resulting polyhedron consists of the vertical rotations and the corresponding rotary reflections with horizontal mirror. It's called a $\mathscr{C}_{nh}$ group.

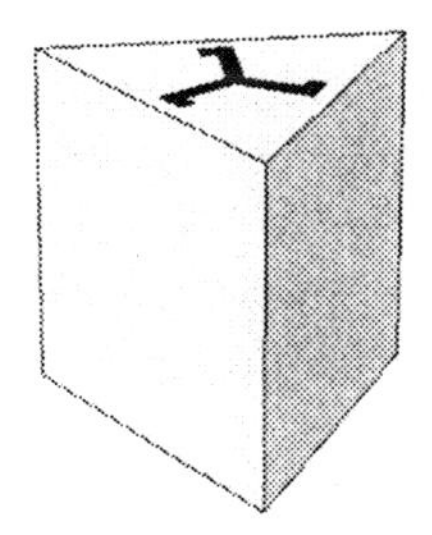

Figure 8.4.29
Polyhedron with cyclic
$\mathscr{C}_3$ symmetry group

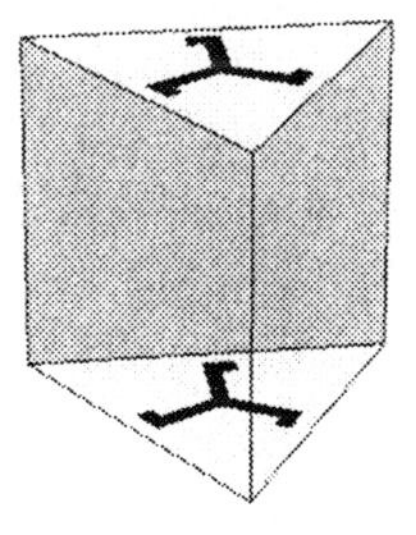

Figure 8.4.30
Polyhedron with dihedral
$\mathscr{D}_3$ symmetry group

[48] Cromwell 1997, chapter 8.

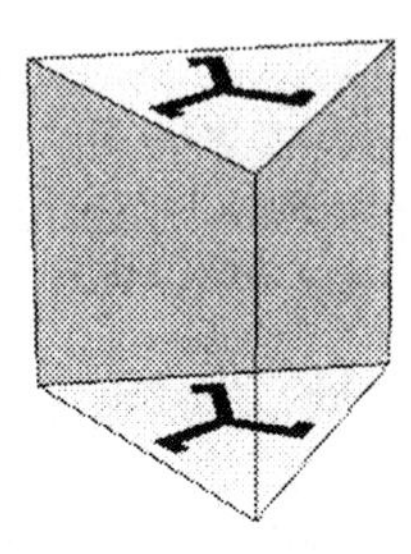

Figure 8.4.31
Polyhedron with
$\mathscr{C}_{3h}$ symmetry group

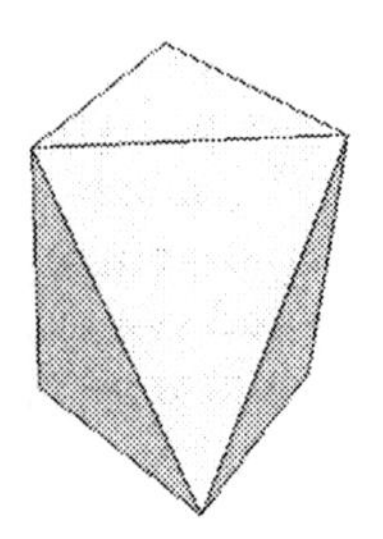

Figure 8.4.32
Antiprism with
$\mathscr{D}_{3v}$ symmetry group

Now consider a right antiprism $\mathscr{Q}_n$ whose regular polygonal bases have n edges. When $n = 3$, adjust the distance between bases to ensure that $\mathscr{Q}_n$ is not a regular octahedron—this example is shown in figure 8.4.32. Its symmetry group $\mathscr{G}$ is called a $\mathscr{D}_{nv}$ group. This contains the same vertical rotations and plane reflections as a $\mathscr{D}_{nh}$ group, and the same numbers of horizontal rotations and rotary reflections. But in these two types of symmetry group, the relationships of the horizontal rotations and rotary reflections to the vertical rotations and reflections are different. In particular, no $\mathscr{D}_{nv}$ group contains reflection σ_ε across horizontal plane ε, but it contains reflection σ_O across the center O of $\mathscr{Q}_n$ (midpoint of the segment joining the base centers) just when n is odd; all $\mathscr{D}_{nh}$ groups contain σ_ε, and they contain σ_O just when n is even.

In exercise 8.5.39 you'll append to the bases of a prism $\mathscr{P}_{2n}$ two opposite copies of a structure with $\mathscr{C}_n$ but not $\mathscr{D}_n$ symmetry, aligned so that the symmetry group $\mathscr{G}$ of the resulting polyhedron contains the same vertical rotations and rotary reflections as a $\mathscr{D}_{nv}$ group, but no vertical plane reflections or horizontal rotations. $\mathscr{G}$ is called an $\mathscr{S}_{2n}$ group.

In previous examples of polyhedral symmetry groups, all nonvertical axes were twofold. That's clearly not the case for regular polyhedra, which are considered next.

Each of the four altitudes of a regular tetrahedron is a threefold axis, and the three lines joining midpoints of its opposite edges are twofold axes. Thus its symmetry group $\mathscr{G}$ contains 12 rotations. (Count the identity just once!) Since $\mathscr{G}$ contains an odd symmetry—for example, the reflection across the perpendicular bisector of any of its six edges—it must have twenty-four members altogether. $\mathscr{G}$ is called a $\mathscr{T}_d$ group. Since there are only twenty-four permutations of its vertex set, *the symmetries of a regular tetrahedron correspond bijectively to the permutations of its vertices.*

The previous sentence explains why we regard regular tetrahedra as highly symmetric—their symmetries include all possible permutations of their vertex sets. Since a symmetry of any polyhedron preserves vertex adjacency, not every highly symmetric polyhedron can have this property; it holds in this case because any two vertices of a tetrahedron are adjacent. However, a regular tetrahedron shares a significant property with other highly symmetric polyhedra. Its symmetry group $\mathcal{G}$ is *transitive* in the following sense. Suppose each triple $<V, e, F>$ and $<V', e', F'>$ consists of an incident vertex, edge, and face; then $\mathcal{G}$ contains a symmetry φ such that $\varphi(V) = V'$, $\varphi(e) = e'$, and $\varphi(F) = F'$.

A cube has three fourfold axes joining centers of opposite faces, four threefold axes joining opposite vertices, and six twofold axes joining midpoints of opposite edges. Its symmetry group $\mathcal{G}$ thus contains twenty-four rotations. Since it contains plane reflections, $\mathcal{G}$ must have forty-eight members altogether. You can see it's transitive.

Any symmetry of a cube must preserve the set of face centers. Therefore it's a symmetry of the cube's dual octahedron. Similarly, every symmetry of a regular octahedron must be a symmetry of its dual cube. These two cubes are homothetic about their common center (see figures 8.4.16 and 8.4.17), so they have exactly the same symmetry group, which must also be the symmetry group of the octahedron. Since any two cubes or any two regular octahedra are similar, their symmetry groups are conjugate. That is, *all cubes and regular octahedra have conjugate symmetry groups.* These are called $\mathcal{O}_h$ groups.

The regular dodecahedron and icosahedron are related like the cube and regular octahedron; each is homothetic with the dual of its dual. Therefore, *all regular dodecahedra and icosahedra have conjugate symmetry groups.* They're called $\mathcal{I}_h$ groups.

A regular dodecahedron has six fivefold axes joining centers of opposite faces, ten threefold axes joining opposite vertices, and fifteen twofold axes joining midpoints of opposite edges. Its symmetry group $\mathcal{G}$ thus contains sixty rotations. Since it contains plane reflections, $\mathcal{G}$ must have one hundred twenty members altogether. You can see it's transitive.

Earlier in this section a regular tetrahedron was constructed by selecting half the vertices of a cube $\mathcal{C}$. From such a tetrahedron you can easily reconstruct $\mathcal{C}$, so every symmetry of the tetrahedron is also a symmetry of $\mathcal{C}$. Since all regular tetrahedra are similar, you can conclude that *every $\mathcal{T}_d$ group is conjugate to a subgroup of an $\mathcal{O}_h$ group.*

Figure 8.4.33 shows both tetrahedra constructed from a cube according to the previous paragraph. They interpenetrate; you can dissect their faces and edges to form a polyhedron that Kepler called a *stella octangula.* It has the same vertex set as the cube. Its symmetry group, called a $\mathcal{T}_h$ group,

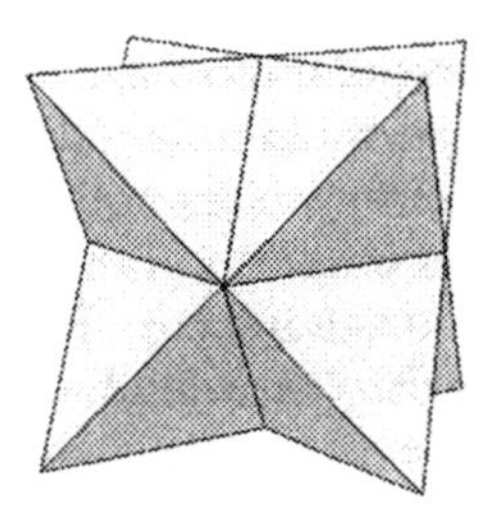

Figure 8.4.33
Stella octangula, with $\mathscr{T}_h$ symmetry group

consists of all symmetries of one of the tetrahedra, plus their compositions with the reflection about the center of the cube.

Euclid constructed a regular dodecahedron $\mathscr{D}$ from a cube, so that the vertices of the latter are a subset of those of $\mathscr{D}$. In (See exercise 8.5.25.) Therefore every symmetry of the cube is also a symmetry of the dodecahedron; you can conclude that *every $\mathscr{O}_h$ group is conjugate to a subgroup of an $\mathscr{I}_h$ group.*

By appropriately appending ϒ motifs to all faces of a regular tetrahedron, octahedron, or icosahedron $\mathscr{P}$, you can build a polyhedron with all the rotational symmetries of $\mathscr{P}$, but no others. Its symmetry group is called a *tetrahedral, octahedral,* or *icosahedral* group, abbreviated $\mathscr{T}$, $\mathscr{O}$, or $\mathscr{I}$ group. By the previous discussion *every tetrahedral group is conjugate to a subgroup of an octahedral group, and every octahedral group is conjugate to a subgroup of an icosahedral group.*

The previous paragraphs have described nearly all finite groups of three-dimensional isometries. Table 8.4.2 displays the complete list, in Schönflies' notation. A few of its entries represent classes of polyhedra not previously discussed. Exercise 8.5.40 requires you to supply example polyhedra for them.

The German and French mineralogists Johann Friedrich Christian Hessel and Auguste Bravais proved in 1830 and 1848 that this list includes all possible finite groups of three-dimensional isometries. Hessel's work was published obscurely, however, and not noticed until its republication in 1891. For complete proofs, consult the books by Martin, Senechal, and Cromwell.[49] Cromwell presents a decision tree for these groups in the same style as those in sections 8.2 and 8.3. Bravais' results played the same role in the classification of the crystallographic symmetry groups as the far simpler study of finite groups of plane isometries—cyclic and dihedral groups—plays in the section 8.3 classification of wallpaper patterns. Bravais' work led directly to the monumental crystal study by Schönflies and Fedorov forty years later.

---

[49] Martin 1982, section 17.2; Senechal 1990; and Cromwell 1997, chapter 8.

Table 8.4.2 The finite three-dimensional isometry groups

Type	*Members*	*Example*
$\mathscr{C}_1$	Just the identity ι	Tetrahedron with no equal edges
$\mathscr{C}_s$	ι and a plane reflection	Right pyramid with isosceles nonequilateral triangular base
$\mathscr{C}_i$	ι and a point reflection	See exercise 8.5.40.
$\mathscr{C}_2$	ι and a half turn	Right pyramid with nonrhombic nonrectangular parallelogram base
$\mathscr{C}_{2v}$	ι, a vertical half turn, and two vertical plane reflections	Right pyramid with nonsquare rectangular base
$\mathscr{C}_{2h}$	ι, a vertical half turn, a horizontal plane reflection, and a point reflection	See exercise 8.5.40.
$\mathscr{D}_2$	ι, a vertical half turn, and two horizontal half turns	See exercise 8.5.40.
$\mathscr{D}_{2v}$	ι, a vertical half turn, reflections across perpendicular planes containing its axis, half turns about horizontal axes midway between them, and $\pm 90°$ rotary reflections with horizontal mirror	See exercise 8.5.40.
$\mathscr{D}_{2h}$	ι, a vertical and two horizontal half turns with perpendicular axes, and reflections across the three planes they determine	Right rectangular prism with unequal dimensions
$\mathscr{C}_n$ $\mathscr{C}_{nv}$ $\mathscr{C}_{nh}$ $\mathscr{S}_{2n-2}$ $\mathscr{D}_n$ $\mathscr{D}_{nv}$ $\mathscr{D}_{nh}$ $\mathscr{T}$ $\mathscr{T}_d$ $\mathscr{T}_h$ $\mathscr{O}$ $\mathscr{O}_h$ $\mathscr{I}$ $\mathscr{I}_h$		Discscussed earlier for $n \geq 3$

Karl Georg Christian VON STAUDT was born in Rothenburg ob der Tauber in Bavaria in 1798. His parents' families had belonged to the ruling nobility of that free city of the Holy Roman Empire for many generations. Rothenburg was absorbed into the Electorate of Bavaria in 1802. Von Staudt was schooled first at home, then at the Gymnasium in nearby Ansbach. His spent his first university years at Göttingen, where he studied with and served as assistant to Gauss. Von Staudt was awarded the Ph.D. in astronomy from the university at Erlangen in 1822, for work already done under Gauss' direction. He worked first in secondary schools in Würzburg and Nürnberg, and achieved some note as a teacher. In 1832 von Staudt married Jeanette Drechsler, daughter of a Nürnberger judge. She died in 1848 after bearing a son and a daughter. In 1835 von Staudt was appointed Professor of Mathematics at Erlangen. He came to a faculty preoccupied with the immediately practical aspects of mathematics and science, but he confined his research to narrow areas of geometry and number theory. Von Staudt became one of the first in Germany to stress the abstract framework of geometry in order to achieve deep understanding and results of great generality. In particular, he showed how to develop real and complex projective geometry from synthetic axioms, independent of other areas of geometry. He was able to incorporate this work in his lectures only after 1842. Von Staudt published four books on geometry during 1847–1860. During that period he also devoted considerable effort to university governance and administration. His activity was curtailed during the 1860s by asthma. Von Staudt died in 1867 at Erlangen.

Auguste BRAVAIS was born in 1811 in Annonay, about seventy-five kilometers south of Lyons, the ninth of ten children of a physician. Although he won honors in mathematics in school, he had trouble with university entrance examinations, which delayed his matriculation for a year. But in 1828 he entered l'École Polytechnique in Paris and soon became a top student. In 1831, wanting to see the world, he joined the Navy and became an officer. During the next few years he studied plant oceanography, and with his brother published research of some note. Auguste earned the Ph.D. from the university at Lyons in 1837, for research on surveying and on the stability of ships. The next year he joined an expedition to explore northern Scandinavia; he soon published related research work in several areas of science. Back in France, he became a mountaineer, ascending peaks in the Alps to study minerals. In 1844 Bravais was appointed to teach astronomy at Lyons. The next year, he became Professor of Physics at l'École Polytechnique. In 1848–1849 he completed a classification of the finite groups of motions in three dimensions, as well as a catalog of the types of crystal lattices. This work led directly to the complete classification of crystals forty years later by Schönflies and Fedorov. Bravais was appointed to l'Académie des Sciences in 1854. Two years later, ill from overwork, he resigned from l'École Polytechnique and the Navy. Bravais died in 1863. His important book on crystallography was published in 1866.

8.5 Exercises

Concepts

Examples of functional and ornamental symmetry
Subgroup relationships among symmetry groups
Interior of a simple closed polygon
Triangulation
Convex polygons
Involutions
Generating sets of groups
Generators of translation groups
Polyhedral models
Schlegel diagrams
Inradii and circumradii of regular polyhedra
Interior of a polyhedron

This section contains forty-one exercises related to earlier parts of the chapter. Their chief goal is to introduce the detailed study of examples of the types of symmetry considered here. In eleven of these exercises you'll examine functional and ornamental designs incorporating polygonal symmetry, friezes, wallpaper patterns, and polyhedral symmetry. You'll determine their symmetry groups and see how they're related. In five more exercises you'll search for further examples of symmetric design. Perhaps you can challenge other readers to classify these and find even more interesting ones. A few exercises complete some discussions in earlier sections. But there's no effort here to introduce new topics. To extend the scope of the inquiries in this text, go to the literature. The references in earlier sections, particularly those in the chapter introduction, should point the way. *Bon voyage!*

General exercises

The first two exercises encourage you to become alert to examples of symmetry all around you. When you discover a new example, you should consider whether the symmetry is functional or ornamental, and what other types of symmetry could have served the design just as well.

Exercise 1. Figure 8.5.1 shows some household objects. Describe their symmetries. Is symmetry necessary for these designs? Could the designers have used other symmetry types?

Exercise 2. Find and describe more examples of household symmetry, comparable to those in figure 8.5.1.

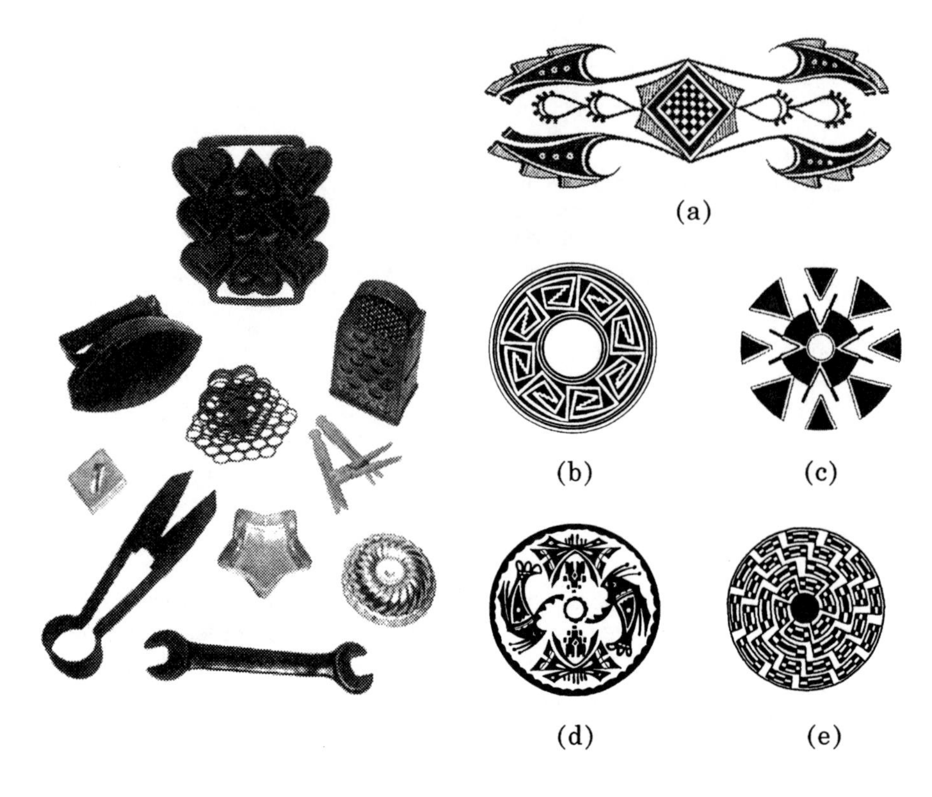

Figure 8.5.1
Household symmetry examples[50]

Figure 8.5.2
Five symmetric designs[51]

Exercises on polygons and polygonal symmetry

In exercises 3 to 5 you'll continue exploring examples of symmetry, finding and analyzing functional and ornamental polygonal designs. After that, two exercises remind you of the role of matrix algebra in this subject, and study the relationships between various polygonal symmetry groups. The last two exercises investigate the notion of regular polygon a little more deeply than section 8.1, and make the connection between the notion of simple closed polygon used here, and that of polygonal region introduced in chapter 3.

Exercise 3. Identify the symmetry groups of the ornaments in figure 8.5.2, from the San Ildefonso, Mimbres, Mescalero Apache, Cochiti, and Pima cultures of the southwestern United States.

---

[50] Photograph by the author.

[51] From Kate 1997.

Exercise 4. Classify uppercase Latin and Greek letters according to their symmetry groups. Use the most symmetric typefaces you can find. How about letters in other alphabets?

Exercise 5. Survey functional and ornamental symmetric polygons. Find examples of designs or ornaments with $\mathscr{C}_n$ or $\mathscr{D}_n$ symmetry groups, for $n = 2, 3, 4, \ldots$. Go up to as large an n as you can. Consider these categories or others you might think of:

(a) Household objects, as in figure 8.5.1.
(b) Coins, medals, medallions. Find familiar examples with seven or eleven vertices.
(c) Nuts, bolts, and other tools and machinery. What kind of nut has five edges?
(d) Logos, as in figure 8.5.3.
(e) Road signs. For example, the Kansas highway sign in figure 8.5.4.
(f) Architecture: rooms and buildings.
(g) Ornament. Analyze the plate by contemporary San Ildefonso artist Tse-pe in figure 8.5.5. (It's rich brown, with smooth and rough textures and a turquoise pebble.) Which symmetry groups are

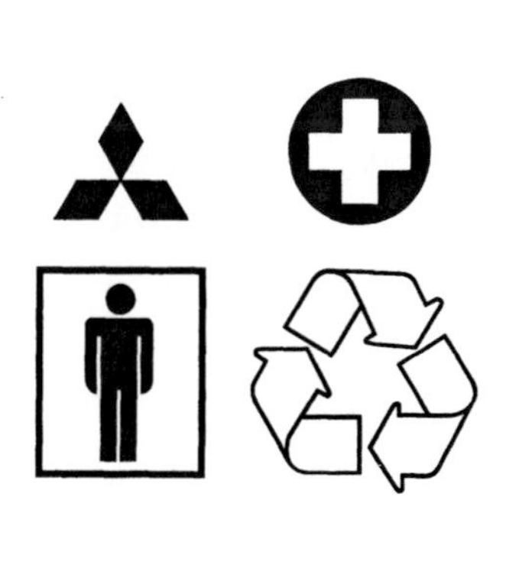

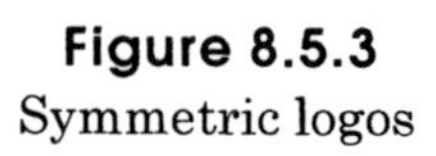

Figure 8.5.3
Symmetric logos

Figure 8.5.4
Kansas highway sign[52]

[52] Photograph by the author.

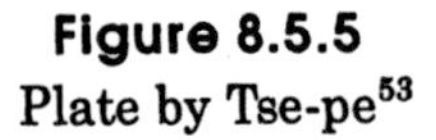

Figure 8.5.5
Plate by Tse-pe[53]

Figure 8.5.6
Tesuque jar[54]

employed in the two-century-old jar in figure 8.5.6 from nearby Tesuque Pueblo? Find other examples from various cultures.

(h) Flowers, fruits, vegetables.

Exercise 6. Find the matrices for the elements of a $\mathscr{D}_6$ group with center at the origin.

Exercise 7, Part 1. For which m and n is a $\mathscr{C}_m$ group a subgroup of a $\mathscr{C}_n$ group? In that case, how many subgroups are isomorphic to it? Are they all conjugate?

Part 2. Same as part 1 for $\mathscr{C}_m$ and $\mathscr{D}_n$.

Part 3. Same as part 1 for $\mathscr{D}_m$ and $\mathscr{D}_n$.

Part 4. How many subgroups has a $\mathscr{D}_6$ or $\mathscr{D}_{13}$ group?

Exercise 8. Suppose the n vertices of a simple closed polygon Π fall in order on a circle.

Part 1. Show that Π is regular if its edges are all congruent. What if the vertices are not in order, or don't fall on a circle?

Part 2. Show that Π is regular if its vertex angles are all congruent and n is odd. What if n is even?

---

[53] Dillingham 1994, 243.

[54] Frank and Harlow 1974, plate 39.

Exercise 9, Part 1. Define the notion *interior* of a simple closed polygon Π. *Suggestion*: This is harder than it seems. First, show how to find some point E that clearly should be called exterior. The interior points X are those for which the segment $\overline{EX}$ crosses Π an odd number of times. Now you need to define the word *crosses*.

Part 2. Let $\bar{\Pi}$ be the union of a simple closed polygon Π and its interior. Devise instructions—a *triangulation* algorithm—for dividing $\bar{\Pi}$ into triangular regions that intersect only along edges. This will show that $\bar{\Pi}$ is a polygonal region as defined in section 3.8. *Suggestion*: This, too, is harder than it seems, and you probably won't find a completely reliable algorithm. But you'll learn by proposing one and defending it against or accepting others' objections. If Π is convex—that is, two vertices never lie on different sides of any edge line—the problem is simpler. You'll find it discussed in books on computational geometry; for example, see Preparata and Shamos 1985.

Exercises on frieze groups

Exercises 10 to 12 continue your exploration of functional and ornamental symmetry examples. After those, two exercises clarify the relationship between conjugacy and isomorphy of frieze groups, and introduce the use of generators for describing isometry groups. Exercise 15 asks you to devise a frieze classification tree more compact than the one in section 8.2.

Exercise 10. Identify the symmetry groups of the friezes in figure 8.5.7 from various cultures of the southern and southwestern United States.

Exercise 11. Classify friezes constructed from repeated letters of the upper case Latin and Greek alphabets, using the typefaces you chose in exercise 4. Which friezes are classified the same, even though the letters are classified differently? Which frieze groups fail to occur? Can you construct examples of those using repeated *pairs* of letters? Which pairs?

Exercise 12. Survey functional and ornamental friezes. Consider these categories and others you may think of:

(a) border ornaments on clothing and flat art works;
(b) pottery decoration;
(c) architectural ornaments, fences, balusters, and gratings;
(d) tools.

Find friezes representing all seven classes.

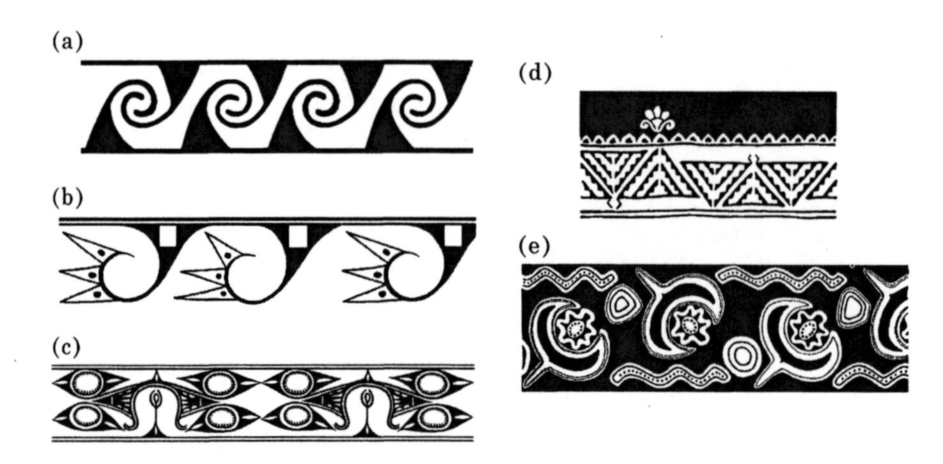

Figure 8.5.7
Five American friezes[55]

Exercise 13. In section 8.2, frieze groups are partitioned into conjugacy classes. But frieze groups may be isomorphic without being conjugate. Divide them into isomorphism classes. *Suggestion*: To prove that two of these groups are nonisomorphic, concentrate on their *involutions*—self-inverse elements different from the identity.

Exercise 14. A subset $\mathscr{S}$ of a group $\mathscr{G}$ *generates* it if no proper subgroup of $\mathscr{G}$ contains $\mathscr{S}$. For example, a finite cyclic isometry group is generated by a set consisting of a single rotation; the corresponding dihedral group is generated by a set consisting of that rotation and a single reflection. Find generating sets as small as possible for each of the conjugacy classes of frieze groups. Which have two-element generating sets? Which have generating sets consisting solely of involutions?

Exercise 15. Since there are seven conjugacy classes of friezes and $7 < 2^3$, it must be possible to classify a frieze with at most three successive questions. Design a tree with that property; try to keep the questions as simple as possible.[56]

---

[55] From Kate 1997. Figure 8.5.7(e) has been altered slightly to enhance its symmetry.

[56] Designing efficient decision trees is a challenge. For example, the key to central Californian marine shelled gastropods in Light et al. 1961, 243–259, organizes about ninety species. Unlike the frieze tree, that one doesn't ask the same question of each branch at each level, so it can be more efficient. But its depth is twelve, considerably greater than the theoretically optimal $7 = \lceil \log_2 90 \rceil$. You may have to ask as many as twelve questions to (continued...)

Exercises on wallpaper groups

These five exercises on wallpaper groups continue the emphasis of the previous exercises. You'll survey and analyze more examples of symmetry and clarify the use of generators for describing these groups. Exercise 20 investigates the subgroup relationships between wallpaper groups. It shows why you can't use this concept to describe why one wallpaper pattern might seem more symmetric than another, and may give you an idea why it's sometimes so hard to classify wallpaper patterns.

Exercise 16. Identify the symmetry groups of the ten wallpaper patterns in figure 8.5.8.

Exercise 17. Construct examples of all classes of wallpaper patterns from repeated letters of the upper- or lowercase Latin, Greek, or Cyrillic alphabets. Use the typefaces you chose in exercise 4. When possible, use single-letter motifs. You may vary their spacing to produce different symmetries. Use motifs based on two or more letters if required. But use letter O only as absolutely necessary. (You should have to resort to that in only three cases.)

Exercise 18. Use this book's references, your library, museums, and commercial shops to find more examples of wallpaper patterns. In particular, find examples of the classes not represented in exercise 16. Challenge other readers to classify them.

Exercise 19. Suppose φ and χ are the fundamental translations of a wallpaper group $\mathcal{W}$, and consider any integers m_1, m_2, n_1, and n_2. Show that $\varphi^{m_2}\chi^{n_2}$ and $\varphi^{m_1}\chi^{n_1}$ generate the group of all translations in $\mathcal{W}$ if and only if

$$\det\begin{bmatrix} m_1 & n_1 \\ m_2 & n_2 \end{bmatrix} \neq 0.$$

Exercise 20. Consider $p3m1$ and $p1m3$ groups $\mathcal{G}$ and $\mathcal{H}$. Show that each is isomorphic to a subgroup of the other. Find another pair of wallpaper group classes with this property. *Suggestion*: Find a $p3m1$ pattern inside a $p1m3$ pattern, and vice versa.

At first thought, it seems reasonable to call a figure G "at most as symmetric as" a figure H if the symmetry group $\mathcal{G}$ of G is isomorphic to a subgroup of the symmetry group $\mathcal{H}$ of H. But exercise 8.5.20 belies

---

[56] (...continued)
a particular specimen—for example, *Acmaea limatula.*

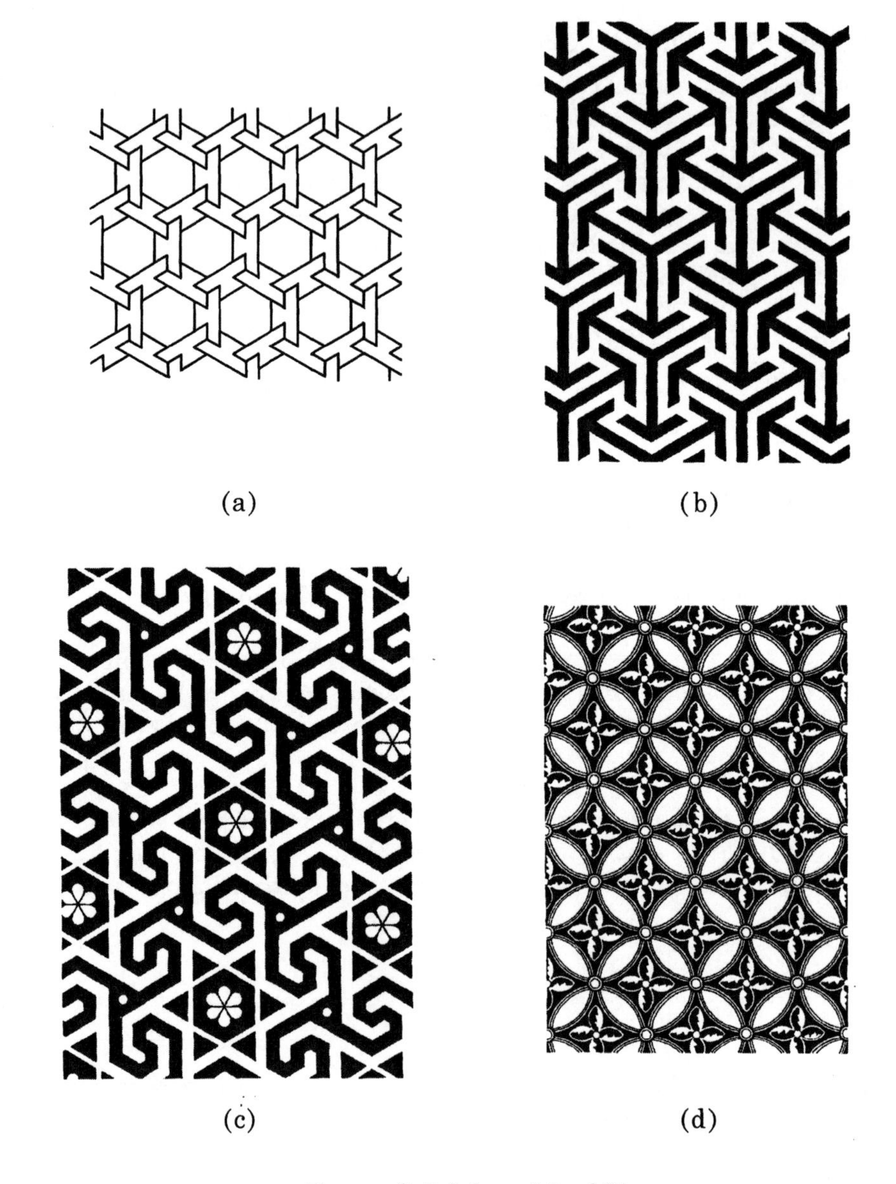

Figure 8.5.8 (part 1 of 3)
Example wallpaper patterns[57]

[57] (a) is an Iranian design from Wade 1982. (b) and (c) are Arab ornaments and (d) to (j), Japanese kimono and obi designs, from Audsley and Audsley [1882] 1968. The latter authors and their contemporaries termed these *diaper* ornaments, not because of their functionality, but because *diaper* stems from Greek words meaning *all white*. These patterns are designed to be printed on plain white cloth!

Figure 8.5.8 (part 2 of 3)
Example wallpaper patterns

that; it's possible that each group is isomorphic to a subgroup of the other, yet the two figures have different symmetry properties. You can show that this can't happen for cyclic and dihedral groups, or for the frieze groups. Perhaps this phenomenon accounts for the difficulty we sometimes have in distinguishing the symmetry properties of wallpaper patterns.

(i) (j)

Figure 8.5.8 (part 3 of 3)
Example wallpaper patterns

Exercises on polyhedra

Models are the best tools for studying polyhedra. Beautiful, expensive commercial products are available.[58] Since they're either ready-made or require assembly and fastening with glue, they don't offer much experience in construction. You don't get the insight that results from trying several times to fit the faces together. However, there's a simple way to construct models with inexpensive materials.

To make a regular tetrahedron, for example, first decide on the edge length. 3 inches works well.[59] You need a piece of stiff cardboard—*railroad board* is best—large enough that you can cut from it five equilateral triangles with edges somewhat longer than 3 inches. Cut out one equilateral triangle T as in figure 8.5.9 to use as a template. Use right triangle trigonometry to compute the T edge length $3 + \frac{1}{2}\sqrt{3} \approx 3.87$ inches, and to locate points A, B, and C near the vertices $\frac{1}{4}$ inch from the edges. Puncture T at

[58] See Symmetrics [no date], Cuisenaire Dale Seymour Publications 1998, and Key Curriculum Press 1997–1998, or more recent editions of these catalogs.

[59] Stewart (1980) recommended these specifications. The edge length, tab width, hole diameter, rubber-band dimension, and rigidity of the railroad board must be compatible. Different specifications could produce a model that sags or buckles.

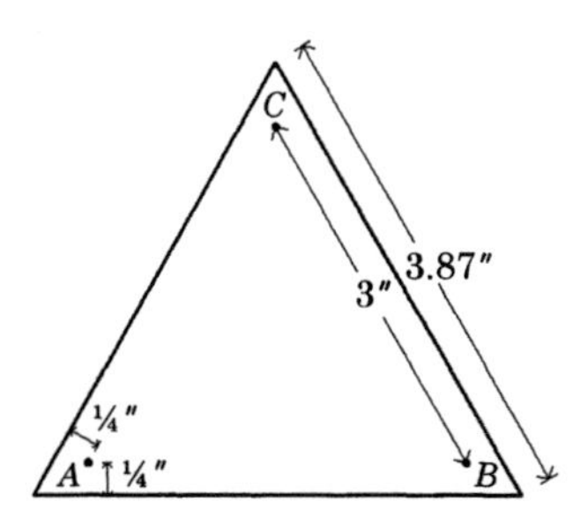

Figure 8.5.9 Equilateral triangular face template

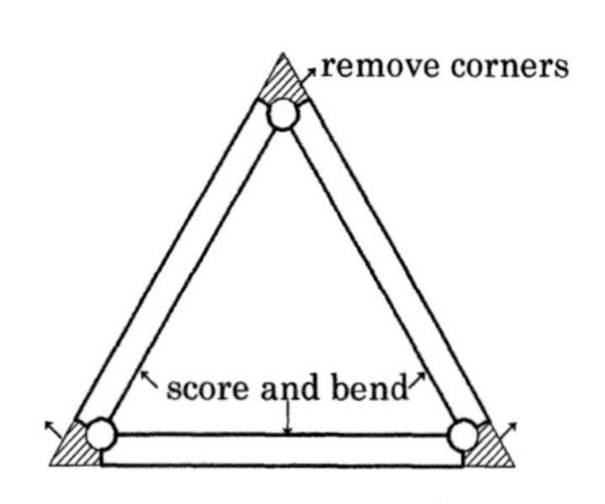

Figure 8.5.10 Equilateral triangular face for models

those spots with holes just large enough to accommodate a pencil point. Use the punctured T as a template to make four copies T' with pencil marks at A, B, and C. With straightedge and stylus score the edges of ΔABC. With pencil mark the segments S from those points perpendicular to the nearest edges. Use a hand punch to make circular holes with diameter ¼ inch centered at A, B, and C. Make scissor cuts along the remaining portions of segments S, removing the corners from triangles T'. The resulting face components resemble figure 8.5.10. Bend the rectangular flaps away from you along the scores. Use them as exterior tabs with 1¼-inch rubber bands to attach faces to each other. You've now assembled a regular tetrahedron with exterior tabs. The tabs make the model less than ideal in appearance, but that's an acceptable trade-off for ease of manufacture, disassembly, and reuse.

You should make a stock of triangular, square, pentagonal, hexagonal, …, tabbed faces, with the same edge length, and acquire a supply of rubber bands of the appropriate type. When you're finished with a particular model, you can disassemble it for easy storage, or for reuse of its components.[60]

Exercise 21. Make and thoroughly inspect models of the polyhedra considered in section 8.4.

Exercise 22. Describe the following intersections, including all possible cases:

(a) of a regular tetrahedron with a plane parallel to two opposite edges,
(b) of a cube with a plane perpendicular to the line joining two opposite vertices,
(c) of a regular octahedron with a plane parallel to two opposite faces,

[60] Stretched rubber bands may deteriorate if the model is stored assembled.

(d) of a regular dodecahedron with a plane parallel to two opposite faces,

(e) of a regular icosahedron with a plane perpendicular to the line joining two opposite vertices.

A *Schlegel* diagram for a convex polyhedron is a perspective image of its edges on the plane of one face F, with exterior focal point so close to F that the images of all the other edges fall within F. For example, figure 8.5.11 is a Schlegel diagram of a cube, with focus near the center of a face F. The outside square is the image of F; the inside one, that of the opposite face. Evidently, a Schlegel diagram for a polyhedron with n faces divides the plane into $n - 1$ polygonal regions and a surrounding unbounded region representing the face nearest the focus.

Exercise 23 Why is the use of Schlegel diagrams restricted to *convex* polyhedra? Make Schlegel diagrams for the regular polyhedra, cuboctahedron, rhombic dodecahedron, equilateral triangular and pentagonal prisms and dipyramids, square antiprism, and the deltahedrally stellated triangular prism.

Exercise 24. Consider a regular polygon P with n edges. Describe the duals of the right prism with base P and the corresponding equilateral pyramid, dipyramid, and antiprism.

Exercise 25, Part 1. Following the discussion under the section 8.4 heading **Regular polyhedra**, consider a regular octahedron $\mathscr{O}$ whose vertices are the points with coordinates ± 1 on the axes of a Cartesian coordinate system. Locate points on the edges of $\mathscr{O}$ that form the vertices of a regular icosahedron $\mathscr{I}$. Find the coordinates of the centroids of the faces of $\mathscr{I}$, and show that they form the vertices of a regular dodecahedron. *Suggestion*: You'll need the exact formula for the cosine of an interior angle of a regular pentagon —see exercise 4.9.1 and section 5.5.

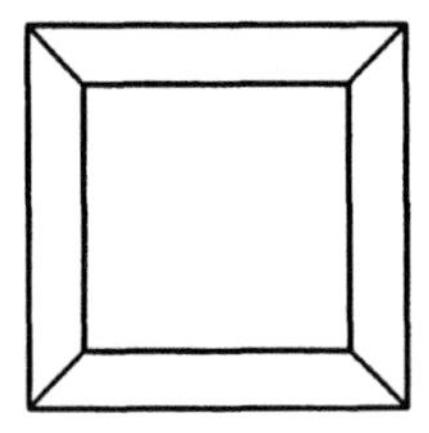

Figure 8.5.11 Schlegel diagram of a cube

Part 2. Compare this proof that a regular dodecahedron exists with Euclid's.[61]

Exercise 26. Define the notions *inradius* r and *circumradius* R of a regular polyhedron $\mathscr{P}$. The closer R/r is to 1, the closer $\mathscr{P}$ is to spherical. For each regular polyhedron derive a formula for this ratio. Do your results justify Plato's claim that the regular dodecahedron is most like a perfect sphere?

Exercise 27. For each regular polyhedron, compute the dihedral angle between adjacent faces. (You'll need this information for beveling should you plan to build finely joined models.)

Exercise 28. The paragraph in section 8.4 that enumerates the symmetry group $\mathscr{T}_d$ of the regular tetrahedron only mentions eighteen isometries. What are the other six?

Exercise 29, Part 1. Each symmetry of a cube permutes the four lines through opposite vertices. The twenty-four rotational symmetries correspond bijectively with those permutations. Label the axes and choose angle parameters for the rotations conveniently, then describe such a bijection explicitly.

Part 2. Describe in detail the odd symmetries of a cube.

Exercise 30. Define the notion *interior* of a convex polyhedron. Show that the interior is a convex point set. Can you extend this notion to apply to *any* polyhedron?

Stanislaus Ferdinand Victor SCHLEGEL was born in 1843 in Frankfurt am Main. He studied in Berlin during 1860–1863, and became a secondary-school teacher, serving from 1866 on in Breslau, Stettin, Mecklenburg, and Hagen. He earned the Ph.D. from the University at Leipzig in 1881, did some research on Grassmann's *Ausdehnungslehre* (a precursor of vector analysis) and other geometrical topics, and wrote many expository books and papers on mathematics for teachers and engineers. Schlegel turned to school administration in Hagen in 1893 and died there in 1905.

[61] Euclid [1908] 1956, book XIII, proposition 17.

Exercise 31. Prove that for any face-connected simply connected polyhedron with v vertices, e edges, and f faces,

(a) $2e \geq 3v$ (b) $2e \geq 3f$
(c) $3f \geq 6 + e$ (d) $3v \geq 6 + e$
(e) $6v \geq 12 + 2e \geq 12 + 3v$ (f) $2v \geq 4 + f$.

Exercise 32. Show that a face-connected simply connected polyhedron must have at least one face with fewer than six edges.

Exercise 33, Part 1. Show that no face-connected simply connected polyhedron has exactly seven edges.

Part 2. For what n is there such a polyhedron with exactly n edges?

Exercise 34. A European football is made of leather pieces in the shapes of regular pentagons and hexagons. They're sewn together so that each pentagon is surrounded by hexagons and each hexagon is surrounded alternately by pentagons and hexagons. From Euler's theorem, deduce how many pentagons and hexagons must be used. How many vertices and edges are there? How is the ball related to the regular polyhedra? What's the symmetry group of the ball? Draw its Schlegel diagram.

Exercise 35. Find and describe in detail all deltahedra with eight faces that aren't regular octahedra.

A polyhedral vertex incident with exactly n edges is called *n-valent.*

Exercise 36. Show that there's no convex deltahedron with one 4-valent and ten 3-valent vertices.

Exercise 37. Show that a convex deltahedron $\mathscr{D}$ with eight vertices, half 4-valent and half 5-valent, must be connected like the *Siamese dodecahedron* in figure 8.5.12. Show that there actually is a deltahedron with this structure.[62]

Exercise 38. Identify the symmetry groups of the cuboctahedron, rhombic dodecahedron, the deltahedra in figures 8.4.25 and 8.4.26, and the Siamese dodecahedron.

[62] This deltahedron was studied first—at least in print—by Rausenberger (1915), then again by Freudenthal and van der Waerden (1947). Their publications addressed audiences of German and Dutch secondary school teachers. The latter authors named the polyhedron to suggest a resemblance to Siamese twins.

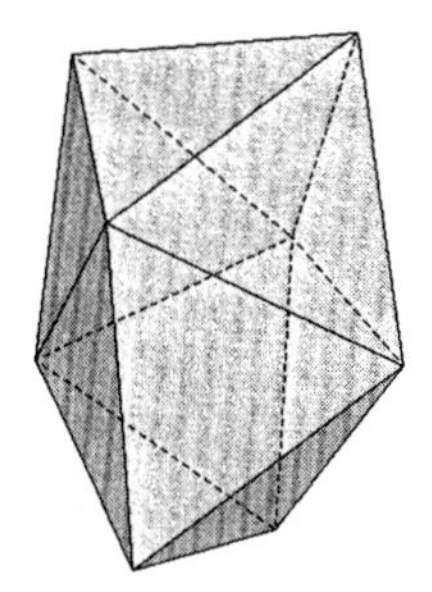

Figure 8.5.12
Siamese dodecahedron

PLATO was born about 427 B.C. in Athens, the son of noble, aristocratic parents. Through family connections he knew the philosopher Socrates, who was certainly the greatest influence on his development. The mathematician Theodorus of Cyrene was also in Athens during Plato's youth, and is reported to have been his teacher. After Socrates' execution in 399 B.C., Plato traveled for several years, then returned to Athens and founded there a school, the Academy, which he headed until his death about 347 B.C.

Under Plato's direction, the Academy became the center of studies in science, philosophy, and law; it was perhaps the first university. Its members included Theaetetus and Eudoxus, who made major discoveries in geometry essential to Euclid's later codification of that subject. A correspondent was the Pythagorean scientist and statesman Archytas of Tarentum, who was a major influence on Plato's own ideas.

Plato's writings, most in the form of dialogues, all survive, and have maintained a major influence on western philosophy ever since. They include inquiries into virtually all areas of the philosophy of his time. Plato was not himself a mathematician, so mathematics appears only occasionally in his works. The main example is in the dialogue *Timaeus*, in which Plato, through the words of the probably fictitious, vaguely Pythagorean title character, expounds his theory of the structure of matter based on properties of the regular polyhedra.

Timaeus was written during 368–348 B.C. (though its setting is some fifty years earlier). In 368 and 361 B.C., Plato visited cities in Sicily and in southern Italy, where Pythagoreans were influential, hoping to put into practice there his theory of the ideal political state. After meddling into local affairs, he was imprisoned, and apparently was extricated only through the efforts of his friend Archytas. Plato's political aspirations were never realized, but these escapades perhaps enhanced the Pythagorean influence on his late writings and on the mathematical work among the scholars at the Academy.

Plato's greatest student was Aristotle, who left the Academy to found his own school, the Lyceum. The Academy, however, continued in existence until A.D. 529.

Exercise 39. Construct a polyhedron with an $\mathscr{S}_{2n}$ symmetry group, in the style of figure 8.4.30. Show how to construct *convex* polyhedra with $\mathscr{C}_n$, $\mathscr{D}_n$, $\mathscr{S}_{2n}$, $\mathscr{C}_{nh}$, and $\mathscr{T}_h$ symmetry groups.

Exercise 40. Construct polyhedra with symmetry groups $\mathscr{C}_i$, $\mathscr{C}_{2h}$, $\mathscr{D}_2$, and $\mathscr{D}_{2v}$, as described in table 8.4.2.

Exercise 41. Use this book's references, your library, museums, commercial shops, and buildings in your neighborhood to find functional and ornamental examples of polyhedra with as many different symmetry groups as possible. Expand the selection in exercises 1 and 2.

Appendix

A

Equivalence relations

Concepts
Binary relation
Reflexivity, symmetry, and transitivity
Equivalence relation
Equivalence class

A *binary relation* on a set S is a set R of ordered pairs $<x, y>$ of elements $x, y \in S$. If $<x, y> \in R$, then x and y are said to *stand in the relation* R; this is written $x R y$. Certain binary relations are mathematical versions of the idea of "likeness." Precisely, R is an *equivalence relation* on S if it satisfies three conditions:

$x R x$ for each $x \in S$	*Reflexivity*
$x R y \Rightarrow y R x$ for any $x, y \in S$	*Symmetry*
$x R y$ & $y R z \Rightarrow x R z$ for any $x, y, z \in S$	*Transitivity*

Many equivalence relations appear unannounced in this book: for example, the congruence relations on the sets of all segments, of all angles, and of all triangles, and the similarity relation on the set of all triangles.

An equivalence relation always effects a classification of the elements of S into sets of like elements. The *equivalence class* of an element $x \in S$ is the set

$$x/R = \{ s \in S : x R s \}.$$

Reflexivity implies $x \in x/R$, so each $x \in S$ belongs to at least one equivalence class. By transitivity, and by symmetry and transitivity,

$$x \in y/R \ \& \ s \in x/R \Rightarrow y\,R\,x \ \& \ x\,R\,s \Rightarrow y\,R\,s \Rightarrow s \in y/R$$

$$\begin{aligned} x \in y/R \ \& \ s \in y/R &\Rightarrow y\,R\,x \ \& \ y\,R\,s \Rightarrow x\,R\,y \ \& \ y\,R\,s \\ &\Rightarrow x\,R\,s \Rightarrow s \in x/R. \end{aligned}$$

You can restate these two arguments:

$$\begin{aligned} x \in y/R &\Rightarrow x/R \subseteq y/R \\ x \in y/R &\Rightarrow y/R \subseteq x/R; \end{aligned}$$

that is,

$$x \in y/R \Rightarrow x/R = y/R.$$

It follows that

$$x \in (y/R) \cap (z/R) \Rightarrow x \in y/R \ \& \ x \in z/R \Rightarrow x/R = y/R = z/R,$$

hence each $x \in S$ belongs to *exactly one* equivalence class.

Appendix B

Least upper bound principle

Concepts

Upper bounds and least upper bounds
Lower bounds and greatest lower bounds
Least upper and greatest lower bound principles
Theoretical computation of the digits of the least upper bound

When parts of calculus are developed rigorously, based on fundamental properties of the real number system, the *least upper bound principle* is often used to justify propositions about limits of various kinds. This appendix derives it from detailed considerations of decimal expansions.[1] The principle is easy to state. An *upper bound* of a set S of real numbers is a number $b \geq$ each member of S. For example, 2 is an upper bound for the set S of values $f(x) = x^2/(1 + x^2)$, indicated by the dotted line in figure B.1. The least upper bound principle says that if a nonempty set S of real numbers has *any* upper bound, then it has a *least* upper bound.

A companion principle involves *lower* bounds: If a nonempty set S of real numbers has *any* lower bound, then it has a *greatest* lower bound. You can test your understanding of this discussion by creating an analogous one for the *greatest lower bound principle*.

---

[1] For further information, consult Mostow, Sampson, and Meyer 1963, chapter 4.

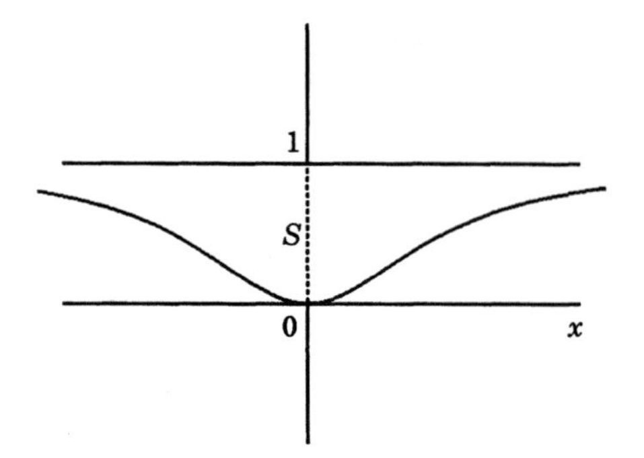

Figure B.1
The set S of values
$f(x) = x^2/(1+x^2)$

If a set S has a *maximum* element x —that is, $x \in S$ and $x \geq$ each member of S —then x must be the least upper bound. However, many bounded sets have no maximum element—for example, the set S in figure B.1. As you can see, any number ≥ 1 is an upper bound for S, and 1 is the least upper bound, but $1 \notin S$.

To prove the least upper bound principle, consider a nonempty set S with upper bound b. You'll see how to construct the decimal expansion

$$k_0.k_1k_2k_3\cdots = k_0 + \frac{k_1}{10} + \frac{k_2}{10^2} + \frac{k_3}{10^3} + \cdots$$

of the least upper bound K of S. Here, k_0 is an integer and $k_1, k_2, k_3, \ldots$ are selected from the digits 0 to 9. (If $k_0 \geq 0$, the expansion is usually written $k_0.k_1k_2k_3\cdots$; but if $k_0 < 0$, you need a subtraction to obtain the usual expansion.)

Step 0: *Determine* k_0. S has an integer upper bound: the first integer $\geq b$. Not every integer is an upper bound, however—select any $x \in S$ and consider the first integer $\leq x$. Therefore, you can find an integer k_0 such that

k_0 .... *is not* an upper bound for S,
$k_0 + 1$ .... *is* an upper bound for S.

Step 1: *Determine* k_1. Consider the eleven numbers

$$k_0, k_0 + 1/10, \ldots, k_0 + 9/10, k_0 + 1.$$

The first is not an upper bound for S, but the last is. Therefore, you can find k_1 among 0 to 9 such that

$k_0 + 0.k_1$ .... *is not* an upper bound for S,
$k_0 + 0.k_1 + 1/10$ .... *is* an upper bound for S.

Step $n+1$: *Determine* k_{n+1}, assuming you've already found $k_0, \ldots, k_n$ such that

$K_n = k_0 + 0.k_1 \cdots k_n$ ...... *is not* an upper bound for S,
$K_n + {}^1\!/\!_{10^n}$ .... *is* an upper bound for S.

Consider the eleven numbers

$$K_n, K_n + {}^1\!/\!_{10^{n+1}}, \ldots, K_n + {}^9\!/\!_{10^{n+1}}, K_n + {}^1\!/\!_{10^n}.$$

The first is not an upper bound for S, but the last is. Therefore, you can find k_{n+1} among 0 to 9 such that

$K_{n+1} = k_0 + 0.k_1 \cdots k_{n+1}$ ...... *is not* an upper bound for S,
$K_{n+1} + {}^1\!/\!_{10^{n+1}}$ .... *is* an upper bound for S.

Continue this process indefinitely, obtaining $k_2, k_3, k_4, \ldots$ in succesion, such that for each $n > 0$,

$K_n = k_0 + 0.k_1 \cdots k_n$ .... *is not* an upper bound for S,
$K_n + {}^1\!/\!_{10^n}$ .... *is* an upper bound for S.

The real number $K = k_0 + 0.k_1 k_2 k_3 \cdots$ constructed by step is (0), (1), (2),... is an upper bound for S. To see this, suppose, on the contrary, that $K < x \in S$. Then $x - K > 0$. Find a positive integer n such that ${}^1\!/\!_{10^n} < x - K$. Then $K_n + {}^1\!/\!_{10^n} \le K + {}^1\!/\!_{10^n} < x$, contradicting the result just established that $K_n + {}^1\!/\!_{10^n}$ is an upper bound for S.

Moreover, no number $L < K$ can be an upper bound for S. To see that, find a positive integer n such that ${}^1\!/\!_{10^n} < K - L$. Then

$$\begin{aligned} L < K - {}^1\!/\!_{10^n} &= [\,k_0 + 0.k_1 k_2 \cdots k_n k_{n+1} \cdots\,] - {}^1\!/\!_{10^n} \\ &= K_n + \left[\frac{k_{n+1}}{10^{n+1}} + \frac{k_{n+2}}{10^{n+2}} + \cdots \right] - {}^1\!/\!_{10}{}^n \\ &\le K_n + [\,{}^9\!/\!_{10^{n+1}} + {}^9\!/\!_{10^{n+2}} + \cdots\,] - {}^1\!/\!_{10^n} \\ &\le K_n + {}^1\!/\!_{10^n} - {}^1\!/\!_{10^n} = K_n. \end{aligned}$$

If L were an upper bound, this inequality would contradict the result established earlier that K_n is *not* an upper bound for S.

You've seen that K is an upper bound for S, but no number $< K$ is. Therefore, K is the *least* upper bound of S, as claimed.

Appendix C

Vector and matrix algebra

Concepts

Scalars
Vectors, rows and columns, matrices
Adding and subtracting vectors and matrices
Multiplying them by scalars
Products of vectors and matrices, scalar and dot products
Systems of linear equations, linear substitution
Transposition
Unit vectors and identity matrices
Gauss and Gauss–Jordan elimination
Invertible and singular matrices, inverses
Determinants

This appendix summarizes the elementary linear algebra used in this book. Much of it is simple vector and matrix algebra that you can learn from the summary itself, particularly if you devise and work through enough two- and three-dimensional examples as you read it. Some of the techniques summarized here require you to solve systems of linear equations by methods covered in school mathematics and commonly subsumed under the title *Gauss elimination*. There are no examples here to lead you through a full review of elimination, so if you need that, you should consult a standard linear algebra text.[1] Almost all the linear algebra used in the book is two- or three-

[1] For example, Larson and Edwards 1991, chapter 1.

dimensional, so there's little need for the full multidimensional apparatus, particularly for determinants. However, many simple techniques, even in three dimensions, are best explained by the general theory summarized here.

In the context of vector and matrix algebra, numbers are often called *scalars*. For the material in this appendix, the scalars could be any complex numbers, or you could restrict them to real numbers. Applications in this book only need real scalars.

Vectors

An n-tuple (pair, triple, quadruple, ...) of scalars can be written as a horizontal row or vertical column. A column is called a *vector*. In this book, a vector is denoted by an uppercase letter; in this appendix it's in the range O to Z. Its entries are identified by the corresponding lowercase letter, with subscripts. The row with the same entries is indicated by a superscript t. For example, consider

$$X = \begin{bmatrix} x_1 \\ \vdots \\ x_n \end{bmatrix} \qquad X^t = [x_1, \ldots, x_n].$$

You can also use a superscript t to convert a row back to the corresponding column, so that $X^{tt} = X$ for any vector X. Occasionally it's useful to consider a scalar as a column or row with a single entry.

In analytic geometry it's convenient to use columns of coordinates for points. Coefficients of linear equations are usually arranged in rows. For points, that convention tends to waste page space. This book uses the compact notation $<x_1, x_2, x_3>$ to stand for the column $[x_1, x_2, x_3]^t$.

You can *add* two vectors with the same number of entries:

$$X + Y = \begin{bmatrix} x_1 \\ \vdots \\ x_n \end{bmatrix} + \begin{bmatrix} y_1 \\ \vdots \\ y_n \end{bmatrix} = \begin{bmatrix} x_1 + y_1 \\ \vdots \\ x_n + y_n \end{bmatrix}.$$

Vectors satisfy *commutative* and *associative* laws for addition:

$$X + Y = Y + X \qquad X + (Y + Z) = (X + Y) + Z.$$

Therefore, as in scalar algebra, you can rearrange repeated sums at will and omit many parentheses.

The *zero* vector and the *negative* of a vector are defined by the equations

$$O = \begin{bmatrix} 0 \\ \vdots \\ 0 \end{bmatrix} \qquad -X = -\begin{bmatrix} x_1 \\ \vdots \\ x_n \end{bmatrix} = \begin{bmatrix} -x_1 \\ \vdots \\ -x_n \end{bmatrix}.$$

Clearly,

$$-O = O \qquad X + O = X$$
$$-(-X) = X \qquad X + (-X) = O.$$

You can regard vector *subtraction* as composition of negation and addition. For example, $X - Y = X + (-Y)$, and you can rewrite the last equation displayed above as $X - X = O$. You should state and verify appropriate manipulation rules.

You can *multiply* a vector by a scalar:

$$Xs = \begin{bmatrix} x_1 \\ \vdots \\ x_n \end{bmatrix} s = \begin{bmatrix} x_1 s \\ \vdots \\ x_n s \end{bmatrix}.$$

This product is also written sX.[2] You should verify these manipulation rules:

$$X1 = X \qquad X0 = O \qquad X(-s) = -(Xs) = (-X)s$$
$$X(-1) = -X \qquad Ot = O$$

$$(Xr)s = X(rs) \qquad \textit{(associative law)}$$

$$X(r + s) = Xr + Xs \qquad \textit{(distributive laws)}$$
$$(X + Y)s = Xs + Ys.$$

Similarly, you can *add* and *subtract* rows X^t and Y^t with the same number of entries, and define the *zero* row and the *negative* of a row. The *product* of a scalar and a row is

$$sX^t = s[x_1, \ldots, x_n] = [sx_1, \ldots, sx_n].$$

These rules are useful:

$$X^t \pm Y^t = (X \pm Y)^t \qquad -(X^t) = (-X)^t \qquad s(X^t) = (sX)^t.$$

Finally, you can multiply a row by a vector with the same number of entries to get their *scalar product*:

$$X^t Y = [x_1, \ldots, x_n] \begin{bmatrix} y_1 \\ \vdots \\ y_n \end{bmatrix} = x_1 y_1 + \cdots + x_n y_n.$$

A notational variant used in analytic geometry is the *dot* product: $X \cdot Y = X^t Y$. (Don't omit the dot. Vector entries are often point coordinates, and

---

[2] Xs is more closely compatible with matrix multiplication notation, discussed later. Each form has advantages, so this book uses both.

the juxtaposition XY usually signifies the *distance* between X and Y.) With a little algebra you can verify the following manipulation rules:

$$O^tX = 0 = X^tO \qquad (sX^t)Y = s(X^tY) = X^t(Ys)$$
$$X^tY = Y^tX \qquad (-X^t)Y = -(X^tY) = X^t(-Y)$$

$$(X^t + Y^t)Z = X^tZ + Y^tZ \qquad \textit{(distributive laws)}$$
$$X^t(Y + Z) = X^tY + X^tZ.$$

Matrices

An $m \times n$ *matrix* is a rectangular array of mn scalars in m rows and n columns. In this book, a matrix is denoted by an uppercase letter; in this appendix it's in the range A to O. Its entries are identified by the corresponding lowercase letter, with double subscripts:

$$A = \underbrace{\begin{bmatrix} a_{11} & \cdots & a_{1n} \\ \vdots & & \vdots \\ a_{m1} & \cdots & a_{mn} \end{bmatrix}}_{n \text{ columns}} \Bigg\} \; m \text{ rows.}$$

A is called *square* when $m = n$. The a_{ij} with $i = j$ are called *diagonal* entries. $m \times 1$ and $1 \times n$ matrices are columns and rows with m and n entries, and 1×1 matrices are handled like scalars.

You can *add* or *subtract* $m \times n$ matrices by adding or subtracting corresponding entries, just as you add or subtract columns and rows. A matrix whose entries are all zeros is called a *zero* matrix, and denoted by O. You can also define the *negative* of a matrix, and the *product* sA of a scalar s and a matrix A. Manipulation rules analogous to those mentioned earlier for vectors and rows hold for matrices as well; check them yourself.

You can *multiply* an $m \times n$ matrix A by a vector X with n entries; their product AX is the vector with m entries, the products of the rows of A by X:

$$AX = \begin{bmatrix} a_{11} & \cdots & a_{1n} \\ \vdots & & \vdots \\ a_{m1} & \cdots & a_{mn} \end{bmatrix} \begin{bmatrix} x_1 \\ \vdots \\ x_n \end{bmatrix} = \begin{bmatrix} a_{11}x_1 + \cdots + a_{1n}x_n \\ \vdots \\ a_{m1}x_1 + \cdots + a_{mn}x_n \end{bmatrix}.$$

You can verify the following manipulation rules:

$$OX = O = AO \qquad (sA)X = (AX)s = A(Xs)$$
$$(-A)X = -(AX) = A(-X)$$

$$(A+B)X = AX + BX \qquad \textit{(distributive laws)}$$
$$A(X+Y) = AX + AY.$$

The definition of the product of a matrix by a column was motivated by the notation for a system of m linear equations in n unknowns x_1 to x_n; you can write $AX = R$ as an abbreviation for the system

$$\begin{cases} a_{11}x_1 + \cdots + a_{1n}x_n = r_1 \\ \quad\vdots \qquad\qquad\quad \vdots \\ a_{m1}x_1 + \cdots + a_{mn}x_n = r_n \ . \end{cases}$$

Similarly, you can *multiply* a row X^t with m entries by an $m \times n$ matrix A; their product X^tA is the row with n entries, the products of X^t by the columns of A:

$$X^tA = [x_1, \ldots, x_m] \begin{bmatrix} a_{11} & \cdots & a_{1n} \\ \vdots & & \vdots \\ a_{m1} & \cdots & a_{mn} \end{bmatrix}$$
$$= [x_1a_{11} + \cdots + x_ma_{m1}, \ \ldots, \ x_1a_{1n} + \cdots + x_ma_{mn}].$$

Similar manipulation rules hold. Further, you can check the *associative law*

$$X^t(AY) = (X^tA)Y.$$

You can *multiply* an $l \times m$ matrix A by an $m \times n$ matrix B. Their product AB is an $l \times n$ matrix that you can describe two ways. Its columns are the products of A by the columns of B, and its rows are the products of the rows of A by B:

$$AB = \begin{bmatrix} a_{11} & \cdots & a_{1m} \\ \vdots & & \vdots \\ a_{l1} & \cdots & a_{lm} \end{bmatrix} \begin{bmatrix} b_{11} & \cdots & b_{1n} \\ \vdots & & \vdots \\ b_{m1} & \cdots & b_{mn} \end{bmatrix}$$
$$= \begin{bmatrix} a_{11}b_{11} + \cdots + a_{1m}b_{m1} & \cdots & a_{11}b_{1n} + \cdots + a_{1m}b_{mn} \\ \vdots & & \vdots \\ a_{l1}b_{11} + \cdots + a_{lm}b_{m1} & \cdots & a_{l1}b_{1n} + \cdots + a_{lm}b_{mn} \end{bmatrix}.$$

The i,kth entry of AB is thus $a_{i1}b_{1k} + \cdots + a_{im}b_{mk}$. You can check these manipulation rules:

$$AO = O = OB \qquad (sA)B = s(AB) = A(sB)$$
$$(-A)C = -(AC) = A(-C)$$

$$(A+B)C = AC + BC \qquad \textit{(distributive laws)}$$
$$A(C+D) = AC + AD.$$

The definition of the product of two matrices was motivated by the formulas for linear substitution; from

$$\begin{cases} z_1 = a_{11}y_1 + \cdots + a_{1m}y_m \\ \quad\vdots \qquad\qquad \vdots \\ z_l = a_{l1}y_1 + \cdots + a_{lm}y_m \end{cases} \qquad \begin{cases} y_1 = b_{11}x_1 + \cdots + b_{1n}x_n \\ \quad\vdots \qquad\qquad \vdots \\ y_m = b_{m1}x_1 + \cdots + b_{mn}x_n \end{cases}$$

you can derive

$$\begin{cases} z_1 = (a_{11}b_{11} + \cdots + a_{1m}b_{m1})x_1 + \cdots + (a_{11}b_{1n} + \cdots + a_{1m}b_{mn})x_n \\ \quad\vdots \qquad\qquad\qquad\qquad \vdots \\ z_l = (a_{l1}b_{11} + \cdots + a_{lm}b_{m1})x_1 + \cdots + (a_{l1}b_{1n} + \cdots + a_{lm}b_{mn})x_n . \end{cases}$$

That is, from $Z = AY$ and $Y = BX$ you can derive $Z = (AB)X$. In short, $A(BX) = (AB)X$. From this rule, you can deduce the *general associative law*:

$$A(BC) = (AB)C.$$

Proof: jth column of $A(BC) = A(j$th column of $BC)$
$= A(B(j$th column of $C))$
$= (AB)(j$th column of $C)$
$= j$th column of $(AB)C$. ♦

The *commutative* law $AB = BA$ *doesn't* generally hold. For example,

$$\begin{bmatrix} 0 & 0 \\ 0 & 1 \end{bmatrix}\begin{bmatrix} 0 & 0 \\ 1 & 0 \end{bmatrix} = \begin{bmatrix} 0 & 0 \\ 1 & 0 \end{bmatrix} \qquad \begin{bmatrix} 0 & 0 \\ 1 & 0 \end{bmatrix}\begin{bmatrix} 0 & 0 \\ 0 & 1 \end{bmatrix} = \begin{bmatrix} 0 & 0 \\ 0 & 0 \end{bmatrix}.$$

This example also shows that the product of nonzero matrices can be O.

Every $m \times n$ matrix A has a *transpose* A^t, the $n \times m$ matrix whose j,ith entry is the i,jth entry of A:

$$A^t = \begin{bmatrix} a_{11} & \cdots & a_{1n} \\ \vdots & & \vdots \\ a_{m1} & \cdots & a_{mn} \end{bmatrix}^t = \begin{bmatrix} a_{11} & \cdots & a_{m1} \\ \vdots & & \vdots \\ a_{1n} & \cdots & a_{mn} \end{bmatrix}.$$

The following manipulation rules hold:

$$A^{tt} = A \qquad O^t = O$$
$$(A + B)^t = A^t + B^t \qquad (sA)^t = s(A^t).$$

The transpose of a vector is a row, and vice-versa, so this notation is consistent with the earlier use of the superscript t. If A is an $l \times m$ matrix and B is an $m \times n$ matrix, then

$$(AB)^t = B^t A^t.$$

Proof: j,ith entry of $(AB)^t = i,j$th entry of AB
$= (\ i$th row of $A)(j$th column of $B)$
$= (\ j$th column of $B)^t(i$th row of $A)^t$
$= (\ j$th row of $B)(i$th column of $A)$
$= j,i$th entry of B^tA^t. ♦

Consider vectors with n entries. Those of the jth *unit* vector U^j are all 0 except the jth, which is 1. For any row X^t with n entries, X^tU^j is the jth entry of X^t. For any $m\times n$ matrix A, AU^j is the jth column of A. For example,

$$U^1 = \begin{bmatrix} 1 \\ 0 \\ \vdots \\ 0 \end{bmatrix} \qquad X^t U^1 = [x_1, \ldots, x_n]\begin{bmatrix} 1 \\ 0 \\ \vdots \\ 0 \end{bmatrix} = x_1.$$

$$AU^1 = \begin{bmatrix} a_{11} & a_{12} & \cdots & a_{1n} \\ a_{21} & a_{22} & \cdots & a_{2n} \\ \vdots & \vdots & & \vdots \\ a_{m1} & a_{m2} & \cdots & a_{mn} \end{bmatrix}\begin{bmatrix} 1 \\ 0 \\ \vdots \\ 0 \end{bmatrix} = \begin{bmatrix} a_{11} \\ a_{21} \\ \vdots \\ a_{m1} \end{bmatrix}.$$

The $n\times n$ matrix I whose jth column is the jth unit vector is called an *identity* matrix. Its only nonzero entries are the diagonal entries 1. Clearly, $I^t = I$. For any $m\times n$ matrix A, $AI = A$. *Proof*:

jth column of $AI = A(j$th column of $I)$
$= AU^j = j$th column of A. ♦

In particular, for any row X^t with n entries, $X^tI = X^t$.

Similarly, you may consider rows with m entries. The *unit* rows $(U^i)^t$ are the rows of the $m\times m$ identity matrix I. You can verify that for any column X with m entries, $(U^i)^tX$ is the ith entry of X. For any $m\times n$ matrix A, $(U^i)^tA$ is the ith row of A. This yields $IA = A$ for any $m\times n$ matrix A. In particular, $IX = X$ for any column X of length m.

Gauss elimination

The most common algorithm for solving a linear system $AX = R$ is called *Gauss elimination*. Its basic strategy is to replace the original system step by step with equivalent simpler ones until you can analyze the resulting system easily. Two systems are called *equivalent* if they have the same sets of solution vectors X. You need only two types of operations to produce the simpler systems:

I. interchange two equations,
II. subtract from one equation a scalar multiple of a different equation.

Obviously, type (I) operations don't change the set of solution vectors; they produce equivalent systems. If you perform a type (II) operation, subtracting s times the ith equation from the jth, then any solution of the original system clearly satisfies the modified one. On the other hand, you can reconstruct the original from the modified system by subtracting $(-s)$ times its ith row from its jth, so any solution of the modified system satisfies the original—the systems are equivalent.

The simpler systems ultimately produced by Gauss elimination have matrices A of special forms. A linear system and its matrix A are called *upper triangular* if $a_{ij} = 0$ whenever $i > j$, and *diagonal* if $a_{ij} = 0$ whenever $i \neq j$.

The first steps of Gauss elimination, called its *downward pass*, are type (I) and (II) operations that convert the original $m \times n$ system

$$\begin{cases} a_{11}x_1 + a_{12}x_2 + \cdots + a_{1n}x_n = r_1 \\ a_{21}x_1 + a_{22}x_2 + \cdots + a_{2n}x_n = r_2 \\ \quad\vdots \qquad\qquad\qquad\qquad \vdots \\ a_{m1}x_1 + a_{m2}x_2 + \cdots + a_{mn}x_n = r_m \end{cases}$$

into an equivalent upper triangular system:

$$\begin{cases} a_{11}x_1 + a_{12}x_2 + \cdots + a_{1m}x_m + \cdots + a_{1n}x_n = r_1 \\ \qquad\quad a_{22}x_2 + \cdots + a_{2m}x_m + \cdots + a_{2n}x_n = r_2 \\ \qquad\qquad\quad \ddots \qquad\qquad\qquad\qquad\quad \vdots \\ \qquad\qquad\qquad\quad a_{mm}x_m + \cdots + a_{mn}x_n = r_m \end{cases} \quad \text{if } m \le n, \text{ or}$$

$$\begin{cases} a_{11}x_1 + a_{12}x_2 + \cdots + a_{1n}x_n = r_1 \\ \qquad\quad a_{22}x_2 + \cdots + a_{2n}x_n = r_2 \\ \qquad\qquad\quad \ddots \qquad\qquad \vdots \\ \qquad\qquad\qquad\quad a_{nn}x_n = r_n \\ \qquad\qquad\qquad\qquad\; 0 = r_{n+1} \\ \qquad\qquad\qquad\qquad\quad \vdots \\ \qquad\qquad\qquad\qquad\; 0 = r_m \end{cases} \quad \text{if } m > n.$$

The algorithm considers in turn the diagonal coefficients a_{11} to $a_{m-1\,m-1}$, called *pivots*. If a pivot is zero, search downward for a nonzero coefficient; if you find one, interchange rows—a type (I) operation—to make it the new pivot. If not, then proceed to the next equation. Use a nonzero pivot a_{kk} with type (II) operations to eliminate the x_k terms from all equations after the kth. This process clearly produces an equivalent upper triangular system.

You can apply the downward pass to *any* linear system. In this book, it's used mostly with *square* systems, where $m = n$. Until the last heading of this appendix—*Gauss–Jordan elimination*—assume that's the case.

If the downward pass yields a square upper triangular matrix with no zero pivot, the original system and its matrix are called *nonsingular*. This property is independent of the right-hand sides of the equations; it depends only on the original matrix A. In the nonsingular case, you can perform more type (II) operations—constituting the *upward pass*—to convert *any* system $AX = R$ to an equivalent *diagonal* system:

$$\begin{cases} a_{11}x_1 & & & = r_1 \\ & a_{22}x_2 & & = r_2 \\ & & \ddots & \vdots \\ & & a_{nn}x_n & = r_n \,. \end{cases}$$

This system clearly has the unique solution

$$X = \langle r_1/a_{11}, r_2/a_{22}, \ldots, r_n/a_{nn} \rangle.$$

Given any $n \times p$ matrix C, you can repeat the process p times to solve equation $AB = C$ for the unknown $n \times p$ matrix B. If you solve the linear systems $AX = j$th column of C for $j = 1$ to p and assemble the solutions X as the corresponding columns of B, then $AB = C$. *Proof*:

jth column of $AB = A(j$th column of $B)$
$= A$(solution X of $AX = j$th column of C)
$= j$th column of C. ♦

On the other hand, if A is *singular*—the downward pass yields an upper triangular matrix with a zero pivot—then you can construct a nonzero solution of the *homogeneous* system $AX = O$. For example, the system

$$\begin{cases} 2x_1 + 3x_2 + 4x_3 = 0 \\ \qquad 0x_2 + 5x_3 = 0 \\ \qquad\qquad 6x_3 = 0 \end{cases}$$

has solution $X = \langle -1.5s, s, 0 \rangle$ for *any* values of s. In general, proceed back up the diagonal, solving the system as though you were performing the upward pass. When you encounter a zero pivot, give the corresponding X entry an arbitrary value—the parameter s in this example. Use a distinct parameter for each zero pivot.

The previous two paragraphs are crucial for the theory of matrix inverses, hence they're worth recapitulation. If an $n \times n$ matrix A is nonsingular—the downward pass yields an upper triangular matrix with no zero pivot—then for every $n \times p$ matrix C, the equation $AB = C$ has a unique solution B. But if A is singular, then at least one such equation—in particular, $AX = O$ —has multiple solutions.

Matrix inverses

A matrix A is called *invertible* if there's a matrix B such that $AB = I = BA$. Clearly, invertible matrices must be square. A zero matrix O isn't invertible, because $OB = O \neq I$ for any B. Also, some nonzero square matrices aren't invertible. For example, for every 2×2 matrix B,

$$\begin{bmatrix} 0 & 0 \\ 0 & 1 \end{bmatrix} B = \begin{bmatrix} 0 & 0 \\ 0 & 1 \end{bmatrix} \begin{bmatrix} b_{11} & b_{12} \\ b_{21} & b_{22} \end{bmatrix} = \begin{bmatrix} 0 & 0 \\ b_{21} & b_{22} \end{bmatrix} \neq I,$$

hence the leftmost matrix in this display isn't invertible. When there exists B such that $AB = I = BA$, it's unique; if also $AC = I = CA$, then $B = BI = B(AC) = (BA)C = IC = C$. Thus an invertible matrix A has a unique *inverse* A^{-1} such that

$$AA^{-1} = I = A^{-1}A.$$

Clearly, I is invertible and $I^{-1} = I$.

The inverse and transpose of an invertible matrix are invertible, and any product of invertible matrices is invertible:

$$(A^{-1})^{-1} = A \qquad (A^t)^{-1} = (A^{-1})^t \qquad (AB)^{-1} = B^{-1}A^{-1}.$$

Proof: The first result follows from the equations $AA^{-1} = I = A^{-1}A$; the second, from $A^t(A^{-1})^t = (A^{-1}A)^t = I^t = I$ and $(A^{-1})^t A^t = (AA^{-1})^t = I^t = I$. The third follows from $(AB)(B^{-1}A^{-1}) = ((AB)B^{-1})A^{-1} = (A(BB^{-1}))A^{-1} = (AI)A^{-1} = AA^{-1} = I$ and equation $(B^{-1}A^{-1})(AB) = I$, which you can check. ♦

The main result about inverses is that a square matrix A is invertible if and only if it's nonsingular. *Proof*: If A is nonsingular, use Gauss elimination to solve equation $AB = I$. To show that also $BA = I$, the first step is to verify that A^t is nonsingular. Were that not so, you could find $X \neq O$ such that $A^tX = O$, as mentioned under the previous heading. But then $X = I^tX = (AB)^tX = B^tA^tX = B^tO = O$ —contradiction! Thus A^t must be nonsingular, and you can solve equation $A^tC = I$. That entails

$$\begin{aligned} BA &= I^tBA^{tt} = (A^tC)^tBA^{tt} = C^tA^{tt}BA^{tt} = C^tABA^{tt} = C^tIA^{tt} \\ &= C^tA^{tt} = (A^tC)^t = I^t = I. \end{aligned}$$

Thus B is the inverse of A. Conversely, if A has an inverse B, then A must be nonsingular, for otherwise you could find $X \neq O$ with $AX = O$, which would imply $O = BAX = IX = X$ —contradiction! ♦

Determinants

The *determinant* of an $n \times n$ matrix A is

$$\det A = \textstyle\sum_\varphi a_{1,\varphi(1)} a_{2,\varphi(2)} \cdots a_{n,\varphi(n)} \operatorname{sign} \varphi,$$

where the sum ranges over all $n!$ permutations φ of $\{1,\ldots,n\}$, and $\operatorname{sign}\varphi = \pm 1$ depending on whether φ is even or odd. In each term of the sum there's one factor from each row and one from each column. For example, the permutation $1,2,3 \rightarrow 3,2,1$ is odd because it just transposes 1 and 3, so it corresponds to the term $a_{13}a_{22}a_{31}(-1)$ in the determinant sum for a 3×3 matrix A. For the theory of permutations, consult a standard algebra text.[3]

Usually you don't need the full apparatus of the theory of permutations. Most of the determinants you'll meet in this book are 2×2 or 3×3, and for them it's enough to write out the sums in full. For the 2×2 case there are two permutations of $\{1,2\}$, and

$$\det A = \det \begin{bmatrix} a_{11} & a_{12} \\ a_{21} & a_{22} \end{bmatrix} = a_{11}a_{22} - a_{12}a_{21}.$$

Clearly, the determinant of a 2×2 matrix is zero if and only if one row or column is a scalar multiple of the other.

For the 3×3 case, there are six permutations of $\{1,2,3\}$ and

$$\det A = \det \begin{bmatrix} a_{11} & a_{12} & a_{13} \\ a_{21} & a_{22} & a_{23} \\ a_{31} & a_{32} & a_{33} \end{bmatrix} = \begin{aligned} &a_{11}a_{22}a_{33} + a_{12}a_{23}a_{31} + a_{13}a_{21}a_{32} \\ &- a_{13}a_{22}a_{31} - a_{11}a_{23}a_{33} - a_{12}a_{21}a_{33}. \end{aligned}$$

Figure C.1 shows a handy scheme for remembering this equation. The indicated diagonals in the diagram, with their signs, contain the factors of the terms in the determinant sum.

The most important properties of determinants are closely tied to the linear system techniques summarized under the previous two headings.

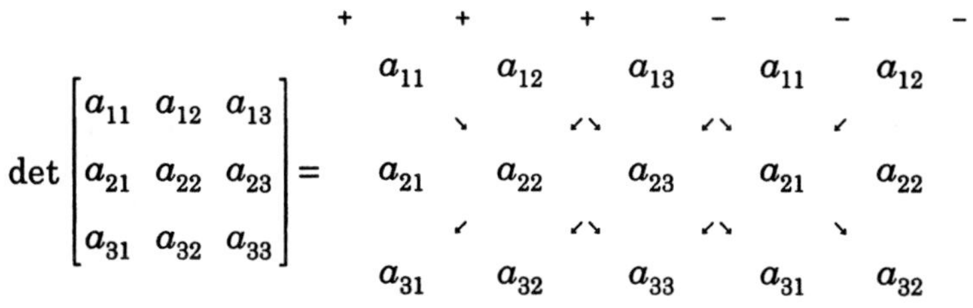

$$\det \begin{bmatrix} a_{11} & a_{12} & a_{13} \\ a_{21} & a_{22} & a_{23} \\ a_{31} & a_{32} & a_{33} \end{bmatrix} = \begin{array}{ccccccccc} + & & + & & + & & - & & - & & - \\ a_{11} & & a_{12} & & a_{13} & & a_{11} & & a_{12} \\ & \searrow & & \swarrow\searrow & & \swarrow\searrow & & \swarrow & \\ a_{21} & & a_{22} & & a_{23} & & a_{21} & & a_{22} \\ & \swarrow & & \swarrow\searrow & & \swarrow\searrow & & \searrow & \\ a_{31} & & a_{32} & & a_{33} & & a_{31} & & a_{32} \end{array}$$

Figure C.1 Evaluating a 3×3 determinant

---

[3] For example, Mostow, Sampson, and Meyer 1963, section 10.3.

First, the determinant of an upper triangular matrix is the product of its diagonal entries; in particular, identity matrices have determinant 1. *Proof*: Each term in the determinant sum except $a_{11}a_{22}\cdots a_{nn}$ contains at least one factor a_{ij} with $i > j$, and that factor must be zero. ♦

Next, if B results from a square matrix A by interchanging two rows, then $\det B = -\det A$. *Proof*: Each term in the sum for $\det B$ corresponds to a term with the opposite sign in the sum for $\det A$. ♦

A square matrix A with two equal rows has determinant zero. *Proof*: Interchanging them reverses the sign of the determinant but doesn't change the matrix. ♦

If all rows of square matrices A, B, and C are alike except that the ith row of A is the sum of the corresponding rows of B and C, then $\det A = \det B + \det C$. *Proof*: Each term in the sum for $\det A$ is the sum of the corresponding terms of $\det B$ and $\det C$. ♦

A square matrix A with a row of zeros has determinant zero. *Proof*: By the previous paragraph, $\det A = \det A + \det A$. ♦

If B results from a square matrix A by multiplying the ith row of A by a scalar s, then $\det B = s \det A$. *Proof*: Each term in the sum for $\det B$ is s times the corresponding term of $\det A$. ♦

If B results from a square matrix A by subtracting s times its ith row from its jth, then $\det B = \det A$. *Proof*: Construct C from A by replacing its jth row by $(-s)$ times its ith row, so that $\det B = \det A + \det C$. Construct D from A by replacing its jth row by its ith, so that $\det C = -s \det D$. Then D has two equal rows, so $\det D = 0$, hence $\det C = 0$, hence $\det B = \det A$. ♦

A square matrix A has determinant zero if and only if it's singular. *Proof*: By the previous discussion, $\det A$ is $(-1)^k$ times the product of the diagonal entries of the matrix that results from A through the downward pass of Gauss elimination; k is the number of row interchanges required in that process. ♦

An $n \times n$ matrix A has the same determinant as its transpose. *Proof*: $\det A^t$ is the sum of the terms $a_{\varphi(1),1}\cdots a_{\varphi(n),n} \operatorname{sign}\varphi$ for all the permutations φ of $\{1, \ldots, n\}$. You can rearrange each term's factors and write it in the form $a_{1,\chi(1)}\cdots a_{n,\chi(n)} \operatorname{sign}\varphi = a_{1,\chi(1)}\cdots a_{n,\chi(n)} \operatorname{sign}\chi$, where $\chi = \varphi^{-1}$. Since the permutations φ correspond one-to-one with their inverses χ, $\det A^t = \sum_\chi a_{1,\chi(1)}\cdots a_{n,\chi(n)} \operatorname{sign}\chi = \det A$. ♦

By the previous paragraph, you can formulate in terms of columns some of the earlier results that relate determinants to properties of their rows.

The next sequence of results leads slowly to the important equation $\det AB = \det A \det B$. Its proof uses some special matrices.

A *type (I) elementary matrix* E^{ij} results from the $n \times n$ identity matrix by interchanging its ith and jth rows, where $i \neq j$. Clearly, $\det E^{ij} = -1$. You can check that interchanging the ith and jth rows of any $n \times n$ matrix A yields the matrix $E^{ij}A$. Thus $\det E^{ij}A = \det E^{ij} \det A$ and $E^{ij}E^{ij} = I$, so E^{ij} is its own inverse.

A *type (II) elementary matrix* $E^{ij,c}$ results from the $n \times n$ identity matrix by subtracting c times its ith row from its jth, where $i \neq j$. Clearly, $\det E^{ij,c} = 1$. You can check that subtracting c times the ith row from the jth of any $n \times n$ matrix A yields the matrix $E^{ij,c}A$. Thus $\det E^{ij,c}A = \det E^{ij,c} \det A$ and $E^{ij,-c}E^{ij,c} = I$, so $(E^{ij,c})^{-1} = E^{ij,-c}$, another type (II) elementary matrix.

If D and A are $n \times n$ matrices and D is diagonal, then $\det DA = \det D \det A$. *Proof*: Each row of DA is the product of the corresponding row of A and diagonal entry of D. ♦

If A and B are $n \times n$ matrices, then $\det AB = \det A \det B$. *Proof*: If AB has an inverse X, then $A(BX) = (AB)X = I$, so A has an inverse. Thus if A is singular, so is AB, and $\det AB = 0 = \det A \det B$. Now suppose A is invertible. Execute the downward pass of Gauss elimination on A, performing type (I) and type (II) operations until you get an upper triangular matrix U. Each operation corresponds to left multiplication by an elementary matrix, so $E_k E_{k-1} \cdots E_2 E_1 A = U$ for some elementary matrices E_1 to E_k. The diagonal of U contains no zero, so you can perform more type (II) operations until you get a diagonal matrix D. Thus $E_l E_{l-1} \cdots E_{k+2} E_{k+1} U = D$ for some more elementary matrices E_{k+1} to E_l. This yields

$$E_l E_{l-1} \cdots E_k \cdots E_2 E_1 A = D \qquad A = E_1^{-1} E_2^{-1} \cdots E_k^{-1} \cdots E_{l-1}^{-1} E_l^{-1} D.$$

These inverses are all elementary matrices and

$$\begin{aligned} \det AB &= \det(E_1^{-1} E_2^{-1} \cdots E_k^{-1} \cdots E_{l-1}^{-1} E_l^{-1} DB) \\ &= \det E_1^{-1} \det E_2^{-1} \cdots \det E_k^{-1} \cdots \det E_{l-1}^{-1} \det E_l^{-1} \det D \det B \\ &= \det(E_1^{-1} E_2^{-1} \cdots E_k^{-1} \cdots E_{l-1}^{-1} E_l^{-1} D) \det B \\ &= \det A \det B. \end{aligned}$$ ♦

Determinants play important roles in analytic geometry. In three dimensions, these involve the *cross product* of two vectors. For a detailed description, see section 5.6.

Gauss-Jordan elimination

To solve a linear system $AX = B$ whose $m \times n$ matrix A isn't square, first perform the downward pass of Gauss elimination, converting it to an upper triangular system as described earlier. Instead of the upward pass, however, complete the process described next, called *Gauss–Jordan elimination.*

If $m > n$, then the last $m - n$ equations have the form $0 = r_k$ with $m < k \le n$. If any of these $r_k \ne 0$, the system has no solution. Otherwise, you can ignore those last equations, and the system is equivalent to an $m \times m$ upper triangular system. Proceed as described earlier.

If $m < n$, use further type (II) operations to eliminate all x_k terms above diagonal entries with coefficients $a_{kk} \ne 0$. If any equation in the resulting equivalent system has the form $0 = r_k$ with $r_k \ne 0$, the original system has no solution. Otherwise, ignoring equations of the form $0 = 0$, proceed backward through the system as in the upward pass of Gauss elimination. When you encounter a zero pivot, assign the corresponding X entry a parameter representing an arbitrary value. Use a different parameter for each zero pivot.

Bibliography

Most items in this bibliography are annotated to provide a guide through the rich literature related to *Methods of geometry*. Many have appeared in several editions. Often, information about the first has historical interest, while a later edition contains more material or is more accessible to you. In such cases both earlier and later publication dates are given, as in

Altshiller-Court, Nathan. [1935] 1964.

The remaining data for such items refers to the later edition.

Book citations include wherever possible an International Standard Book Number (ISBN), and a Library of Congress (LC) catalog number. *Caution!* A book may have similar editions with different ISBNs. Moreover, the LC number isn't standard, unless the book was cataloged at publication time and its number included on the copyright page. From one library to the next, the LC number may vary slightly—particularly its last digits. A number cited may be what the present author found on a book he used, or it may come from a catalog and thus correspond to a slightly different edition.

Alexanderson, Gerald L., and Kenneth Seydel. 1978. Kürschak's tile. *The mathematical gazette* 62: 192–196.

Altshiller-Court, Nathan. [1935] 1964. *Modern pure solid geometry.* Second edition. New York: Chelsea Publishing Company. ISBN: 0-828-40147-0. LC: QA475.C6. Perhaps the *only* advanced book on three-dimensional synthetic Euclidean geometry.

American Mathematical Society. 1991–. *1991 Mathematical subject classification.* A current hypertext version is accessible on the Internet: http://www.ams.org/msc.

Archimedes. [1912] no date. On the equilibrium of planes. In *The works of Archimedes*, edited by Thomas L. Heath, 189–202. New York: Dover Publications. LC: QA31.A692. Originally published by Cambridge University Press.

Aristotle. 1975. *Aristotle's Posterior Analytics.* Translated by Jonathan Barnes. Oxford: Clarendon Press. ISBN: 0-198-24089-9. LC: B441.A5.B371. Barnes' notes are clear and helpful, and there's a good bibliography.

Armstrong, M. A. 1988. *Groups and symmetry*. New York: Springer-Verlag New York. ISBN: 0-387-96675-7. LC: QA171.A76. A recommended text, somewhat above the level of *Methods of geometry*.

Artin, Emil. 1957. *Geometric algebra*. New York: Interscience Publishers. LC: QA251.A7. Chapter 2 is an elegant presentation of *plane affine geometry*: the consequences of the incidence and parallel axioms. Artin's approach, which dates from the 1920s, relies on algebraic notions related to groups, rings, fields, and vector spaces. It can be extended easily to three dimensions. Later chapters connect this with advanced linear algebra topics. Artin was one of the founders of modern algebra.

Audsley, W. and G. Audsley [1882] 1968. *Designs and patterns from historic ornament; Formerly titled, Outlines of ornament in the leading styles*. New York: Dover Publications. ISBN: 0-486-21931-3. LC: NK1539.A8. For students of art and art history.

Ball, W. W. Rouse, and H. S. M. Coxeter. [1892] 1987. *Mathematical recreations and essays*. Thirteenth edition. New York: Dover Publications. ISBN: 0-486-25357-0. LC: QA95.B2. The first edition, by Ball alone, is perhaps the most noted book on recreational mathematics.

Bartle, Robert G. 1964. *The elements of real analysis*. New York: John Wiley & Sons. ISBN: 0-471-05464-X. LC: QA300.B29. An excellent reference for the fundamental results of analysis you need in various mathematical studies. The use of vector and matrix algebra is particularly effective.

Beck, Anatole, Michael N. Bleicher, and Donald W. Crowe. 1969. *Excursions into mathematics*. New York: Worth Publishers. LC: QA39.2B42. Particularly reader-friendly, at a level comparable to that of *Methods of geometry*, are chapters

1. "Euler's formula for polyhedra, and related topics," by Crowe;
3. "What is area?" by Beck; and
4. "Some exotic geometries," by Crowe.

Behnke, Heinrich, Friedrich Bachmann, Kuno Fladt, Wilhelm Süss, et al. [1960] 1974. *Fundamentals of mathematics*. Volume 2, *Geometry*. Translated by Sydney H. Gould. Cambridge, MA: MIT Press. ISBN: 0-262-02069-6. LC: QA37.2B413. Top geometry researchers contributed to this survey, published originally as *Grundzüge der Mathematik*. This is a logical place for you to continue your study after completing *Methods of geometry*. Its target audience is prospective secondary-school teachers in Germany.

Berger, Marcel. 1987. *Geometry*. Two volumes. Translated by M. Cole and S. Levy. Berlin: Springer-Verlag. ISBN: 3-540-11658-3, 3-540-17015-4. LC: QA445.B413. Classical geometry from a modern viewpoint. This survey's breadth and depth are both remarkable, in spite of its terseness.

Birkhoff, George David. 1932. A set of postulates for plane geometry, based on scale and protractor. *Annals of mathematics*, second series. 33: 329–345. In this paper, Birkhoff, a Harvard professor and the leading American mathematician of his era, introduced the technique of using the real numbers explicitly in a geometric axiom system. Later, he incorporated his approach into the school text Birkhoff and Beatley [1941] 1959.

Birkhoff, George David, and Ralph Beatley. [1941] 1959. *Basic geometry*. Providence: American Mathematical Society / Chelsea Publishing Company. ISBN: 0-828-40120-9. LC: QA455.B45. Solutions and teachers' manuals are also available. The first attempt to introduce Birkhoff's axioms (Birkhoff 1932) into school mathematics. World War II detracted from its possible impact. Beatley was a professor of education at Harvard.

Blackwell, William. 1984. *Geometry in architecture*. Berkeley: Key Curriculum Press. ISBN: 1-559-53018-9. LC: NA2760.B53. The author, an architect, presents many geometrical aspects of his profession.

Boldt, H. 1934. Raumgeometrie und Spiegelungslehre. *Mathematische Zeitschrift* 38:104–134. This attempt to extend Thomsen's reflection calculus (1933a) to three dimensions is the source of exercises 7.7.31 and 7.7.32 in *Methods of geometry*. Boldt was Thomsen's Ph.D. student. His work didn't lead to much further development.

Bonola, Roberto. [1911] 1955. *Non-Euclidean geometry: A critical and historical study of its developments*. Translated with appendices by H. S. Carslaw and introduction by Federigo Enriques, containing the George Bruce Halsted translations of János Bolyai, "The science of absolute space," and Nikolai Lobachevski, "The theory of parallels." New York: Dover Publications. ISBN: 0-486-60027-0. LC: QA685.B6. The sources of hyperbolic geometry. Carslaw's work first appeared in 1911, Bonola's in 1906, Lobachevski's in 1840, and Bolyai's in 1832.

Bourbaki, Nicolas. 1939. *Éléments de mathématique, Première partie, livre 1, fascicule 1, Théorie des ensembles, Fascicule des résultats*. Actualités Scientifiques et Industrielles, 846. Paris: Hermann, 1939. LC: QA37.B646. *Bourbaki* is a pseudonym for a group of French mathematicians who began in the 1930s to reformulate all of higher mathematics in the abstract form pioneered in Göttingen in the 1920s. Their membership changes over the years. They succeeded to the extent that most of the present author's university mathematical education was presented in their style. This first fascicle of their multivolume encyclopedia summarizes the set theory volume Bourbaki 1968.

———. 1968. *Elements of mathematics: Theory of sets*. Reading, MA: Addison-Wesley Publishing Company. LC: QA37.B64613. See the preceding entry.

Bradley, A. Day. 1979. Prismatoid, prismoid, generalized prismoid. *American mathematical monthly* 86:486–490.

Brianchon, Charles-Julien, and Victor Poncelet. 1820. Brianchon and Poncelet on the nine point circle theorem. Translated by Morris Miller-Slotnick. In Smith [1929] 1959, 337–338. An excerpt from "Recherches sur la détermination d'une hyperbole équilatère, au moyen de quatre conditions données," *Annales de mathématiques* 11:205–220.

Brody, J. J., Catherine J. Scott, and Steven A. LeBlanc. 1983. *Mimbres pottery: Ancient art of the American Southwest.* Introduction by Tony Berlant. New York: Hudson Hills Press, in association with American Federation of Arts. ISBN: 0-933-92046-6. LC: E99.M76.B765.

Burckhardt, J. J. 1967/1968. Zur Geschichte der Entdeckung der 230 Raumgruppen. *Archive for history of exact sciences*, 4:235–246. A partial account of the classification of the 230 types of crystal symmetry groups reported in Fedorov [1891] 1971 and Schönflies 1891.

Cajori, Florian. 1890. *The teaching and history of mathematics in the United States.* Bureau of Education Circular of Information, number 3. Washington: Government Printing Office. LC: QA11.C3.

———. 1919. *A history of mathematics.* Second edition. New York: Macmillan. LC: QA21.C15.

Carnap, Rudolf. 1966. *Philosophical foundations of physics: An introduction to the philosophy of science.* Edited by Martin Gardner. New York: Basic Books. LC: QC6.C33. This book started as Gardner's notes from a course given by Carnap, one of the founders of the logical positivist philosophy of science. For many years, Gardner, a major popularizer of mathematics, wrote the *Scientific American* column on recreational mathematics. Good bibliography.

Chapman, Kenneth M. 1970. *The pottery of San Ildefonso Pueblo.* Albuquerque: University of New Mexico Press. ISBN: 0-862-30157-6. LC: E99.S213.

Cohen, I. Bernard. 1995. *Science and the founding fathers: Science in the political thought of Jefferson, Franklin, Adams, and Madison.* New York: W. W. Norton. ISBN: 0-393-03501-8. LC: E302.5.C62. The dean of American historians of science, Cohen is professor emeritus at Harvard.

Coxeter, Harold Scott Macdonald. 1969. *Introduction to geometry.* Second edition. New York: John Wiley & Sons. ISBN: 0-471-50458-0. LC: QA445.C67. *The* classic geometry text of the 1950–1999 era. Coxeter's presentation is much broader than that of *Methods of geometry*, but also more terse.

———, and Samuel L. Greitzer. 1967. *Geometry revisited.* New York: Mathematical Association of America. ISBN: 0-883-85619-0. LC: QA473.C6. Originally published by Random House. A wonderful little book aimed at advanced secondary-school students and their teachers. Coxeter is the leading geometer of our era; Greitzer, a secondary-school teacher.

———, and William O. J. Moser. 1965. *Generators and relations for discrete groups.* Second edition. New York: Springer-Verlag New York. LC:

QA171.C7. An advanced text that prepares mathematicians for research involving symmetry groups.

Cromwell, Peter. 1997. *Polyhedra*. Cambridge, UK: Cambridge University Press. ISBN: 9-521-55432-2. LC: QA491.C76. Beautiful and thorough survey of the subject, including its history.

Crowe, Donald W. 1986. *Symmetry, rigid motions, and patterns*. (HiMap Module 4.) Arlington, MA: Consortium for Mathematics and its Applications. Crowe, a geometer, is one of the pioneers in the use of symmetry analysis to study cultures. This high-school curriculum module introduces symmetry groups, and summarizes Dorothy K. Washburn's use of them to study communication patterns in the prehistoric Chaco culture of the southwestern United States. Short, focused bibliography.

———, and Dorothy K. Washburn. 1985. Groups and geometry in the ceramic art of San Ildefonso. *Algebras, groups and geometries* 3: 263–277.

Cuisenaire Dale Seymour Publications. 1998. *Secondary math and science*. White Plains, NY: Cuisenaire Dale Seymour Publications. ISBN: 7-690-01335-X. This catalog of mathematics materials for secondary-school students and teachers lists items from various publishers. It includes many geometry books and manipulatives.

Daniels, Norman. 1975. Lobachevsky: Some anticipations of later views on the relation between geometry and physics. *Isis* 66:75–85. A historical research paper with a good bibliography.

Davis, David Roy. 1949. *Modern college geometry*. Reading, MA: Addison-Wesley Publishing Co. LC: Q474.D29. A typical advanced Euclidean geometry text—the first one the present author ever used for teaching geometry.

Dehn, Max. 1900. Über raumgleiche Polyeder. *Nachrichten von der Königlichen Gesellschaft der Wissenschaften zu Göttingen*, 345–354. The first published solution to any problem in Hilbert's famous list (Hilbert 1900). Dehn, his Ph.D. student, showed that any axiom system for polyhedral volume must include a principle that permits use of some integral calculus methods.

Derive: A mathematical assistant for your personal computer. 1989–. Soft Warehouse, Honolulu. The least elaborate mathematical desktop software package that performs symbolic algebra computations.

Desargues, Gérard. 1648. Desargues on perspective triangles. Translated by Leo G. Simons. In Smith [1929] 1959, 307–310. The source of Desargues' famous theorem was this note from a 1648 book on perspective by his student A. Bosse. You can find the original text in *Oeuvres de Desargues*, volume 1, edited by M. Poudra, Paris: Leiber, 1864, 413–415. LC: QA33.D4.

Descartes, René. [1637] 1954. *The geometry of René Descartes*. Translated by David Eugene Smith and Marcia L. Latham. New York: Dover

Publications. ISBN: 0-486-60068-8. LC: QA33.D5. The first book on analytic geometry. This translation was first published in 1925.

Dillingham, Rick. 1994. *Fourteen families in Pueblo pottery*. Foreword by J. J. Brody. Albuquerque: University of New Mexico Press. ISBN: 0-826-31499-6. LC: E99.P9.D54. The art of the Pueblo cultures has always exploited symmetry. The work of some contemporary artists, particularly the Acoma potters featured in this book, is wondrously intricate. Dillingham was an artist and art scholar. Brody is an authority on the ancient and current art of the southwestern United States.

Duncan, Alastair. 1988. *Art deco*. London: Thames and Hudson. ISBN: 0-500-20230-3. LC: NX600.A6.D86. Style with bibliography.

Ellers, Erich W. 1979. Factorization of affinities. *Canadian journal of mathematics* 31:354–362. This research paper, and other sources cited there, show how you can express arbitrary linear transformations as compositions of simpler ones, analogous to the representation of isometries as compositions of reflections in chapter 6 of *Methods of geometry*.

Euclid. [1908] 1956. *The thirteen books of Euclid's Elements*. Translated from the text of Heiberg, with introduction and commentary by Sir Thomas L. Heath. Second edition. Three volumes. New York: Dover Publications. ISBN: 0-486-60088-2. LC: QA31.E83. This translation was first published by Cambridge University Press in 1908.

Euler, Leonhard. [1758] 1953. Elementa doctrinae solidorum. *Novi commentarii Academiae Scientiarum Petropolitanae* 4:14–17, 109–140. Reprinted in *Opera omnia*, volume 1: *Commentationes geometricae*, edited by Andreas Speiser, 71–93, Basel: Orell Füssli Turici, 1953. LC: Q113.E8. This paper states Euler's theorem on polyhedra.

———. [1775] 1968. Nova methodus motum corporum rigidorum determinandi. *Novi commentarii Academiae Scientiarum Petropolitanae* 20:29–33, 208–238. Reprinted in *Opera omnia*, series 2: *Opera mechanica*, volume 9: *Commentationes mechanicae ad theoriam corporum rigidorum pertinentes*, edited by Charles Blanc, 98–125, Basel: Orell Füssli Turici, 1968. LC: Q113E8. Euler derives equations for isometries without using matrix theory.

Eves, Howard. 1963–1965. *A survey of geometry*. Two volumes. Boston: Allyn and Bacon. LC: QA445.E9.

Fano, Gino. 1892. Sui postulati fondamentali della geometria proiettiva. *Giornale di matematiche di Battaglini* 30:106–132. An axiomatization of higher-dimensional projective geometry in a style similar to Hilbert 1899.

Farmer, David W. 1996. *Groups and symmetry: A guide to discovering mathematics*. Providence: American Mathematical Society. ISBN: 0-821-80450-2. LC: QA174.2.F37. This do-it-yourself introduction to symmetry analysis is more leisurely and less deep than *Methods of geometry*. Brief, focused bibliography.

Fedorov, Evgraf Stepanovich. 1891. Symmetry in the plane. *Proceedings of the Imperial St. Petersburg Mineralogical Society*, series 2, 28:345–390. The original paper that classified the seventeen wallpaper groups.

———. [1891] 1971. The symmetry of regular systems of figures. Ibid., 1–146. Monograph 3 in *Symmetry of crystals*, translated by David and Katherine Harker, ACA Monograph number 7, Buffalo: American Crystallographic Association, 50–131. ISBN: 0-686-60371-0. LC: QD901.A68.no.7. One of the original papers that classified the 230 crystal groups. See also Schönflies 1891, Burckhardt's historical account (1967/1968), and Senechal and Galiulin 1984.

Fejes Tóth, L. 1964. *Regular figures*. New York: Macmillan. ISBN: 0-080-10058-9. LC: QA601.F38. A classic and authoritative account of symmetry analysis.

Feuerbach, Karl Wilhelm. 1822. Feuerbach: On the theorem which bears his name. Translated by Roger A. Johnson. In Smith [1929] 1959, 339–345. An abridgement of *Eigenschaften einiger merkwürdiger Punkte des geradlinigen Dreiecks, und mehrerer durch sie bestimmten Linien und Figuren*, Nürnberg: Riegel und Wiessner.

Field, Judith Veronica. 1997. *The invention of infinity: Mathematics and art in the Renaissance*. Oxford: Oxford University Press. ISBN: 0-198-52394-7. LC: N7430.5.F52.

Forder, Henry George. [1928] 1958. *The foundations of Euclidean geometry*. New York: Dover Publications. LC: QA681.F71.

———. [1950] 1962. *Geometry: An introduction*. New York: Harper & Brothers. LC: QA445.F6. A brief, inviting survey of several fields of geometry.

Foster, Lorraine. No date. *The Alhambra, past and present: A geometer's odyssey*. Parts 1 and 2. Northridge: Department of Mathematics, California State University, Northridge. Videotapes. Foster finds at the Alhambra examples of all the different wallpaper designs, and outlines the underlying theory.

Frank, Larry, and Francis H. Harlow. 1974. *Historic pottery of the Pueblo Indians 1600–1800*. Photographs by Bernard Lopez. Boston: New York Graphic Society. ISBN: 0-887-40227-5. LC: E99.P9.F69.

Frege, Gottlob. 1980. *Philosophical and mathematical correspondence*. Edited by Gottfried Gabriel, Hans Hermes, Friedrich Kambartel, Christian Thiel, and Albert Veraart, abridged from the German edition by Brian McGuinness, translated by Hans Kaal. Chicago: University of Chicago Press. ISBN: 0-226-26197-2. LC: QA3.F74213. Includes the Frege–Hilbert correspondence.

Freudenthal, Hans. 1962. The main trends in the foundations of geometry in the 19th century. In *Logic, methodology and philosophy of science*:

Proceedings of the 1960 International Congress, edited by Ernest Nagel, Patrick Suppes, and Alfred Tarski, 613–621. Stanford: Stanford University Press. ISBN: 0-804-70096-6. LC: BC135.I55. This survey has a good bibliography.

———, and Bartel Leendert van der Waerden. 1947. Over een bewering van Euclides. *Simon Stevin* 25:115–128. This paper, targeted at Dutch secondary-school teachers, analyzes the convex deltahedra.

Gallian, Joseph A. 1994. *Contemporary abstract algebra*. Third edition. Lexington, MA: D. C. Heath & Co. ISBN: 0-669-33907-5. LC: QA162.G34. A reader-friendly introduction to algebra for undergraduates. It includes many considerations of symmetry groups.

Gerdes, Paulus, and Gildo Bulafo. 1994a. *Sipatsi: Technology, art and geometry in Inhambane*. Translated by Arthur B. Powell. Maputo: Instituto Superior Pedagógico Moçambique. This booklet and the next item are prime examples of the application of symmetry analysis in anthropology. Sipatsi are woven handbags. The authors are also promoting the use of cultural context in mathematical instruction.

———. 1994b. *Explorations in ethnomathematics and ethnoscience in Mozambique*. Maputo: Instituto Superior Pedagógico Moçambique. See the previous item.

Gillispie, Charles Coulston, editor. 1970. *Dictionary of scientific biography*. Sixteen volumes. New York: Charles Scribner's Sons. ISBN: 0-684-6962-2. LC: Q141.D5. The source of much material in the biographical notes in *Methods of geometry*.

Grünbaum, Branko, and Geoffrey C. Shephard. 1987. *Tilings and patterns*. New York: W. H. Freeman and Co. ISBN: 0-716-71193-1. LC: QA166.8.G78. This huge advanced reference provides encyclopedic coverage of all aspects of chapter 8 of *Methods of geometry*, and their many extensions.

Gupta, Haragauri Narayan. 1967. The theorem of three or four perpendiculars. *The mathematical gazette* 51:43–46.

Halsted, George Bruce. [1904] 1907. *Rational geometry: A text-book for the science of space*. Second edition. New York: John Wiley & Sons. LC: QA453.H28. Perhaps the *only* school geometry text in English that's based on Hilbert's axioms (Hilbert 1899). It was apparently unsuccessful.

Hargittai, István, editor. 1986. *Symmetry: Unifying human understanding*. New York: Pergamon Press. ISBN: 0-080-33986-7. LC: Q172.5.S95S94. Also published as volume 12B, numbers 1–4, of the journal *Computers and mathematics with applications*. A huge collection of papers on varied applications of symmetry.

———, and Magdolna Hargittai. 1994. *Symmetry: A unifying concept*. Bolinas, CA: Shelter Publications. ISBN: 0-898-15590-8. LC:

Q172.5.S95H37. An instructive and entertaining collection of examples of many kinds of symmetry.

Heil, Erhard. 1984. 3-circle theorem made visible. *Mathematics magazine* 57:56.

Hick, John. 1967. Ontological argument for the existence of God. In *The encyclopedia of philosophy*, volume 5, edited by Paul Edwards, 538–542. New York: Macmillan Publishing Company. ISBN: 0-028-94950-1. LC: B41.E5.

Hilbert, David. 1899. *Grundlagen der Geometrie, Festschrift zur Feier der Enthüllung des Gauß-Weber Denkmals in Göttingen.* Leipzig: Verlag von B. G. Teubner. LC: QA681.H5. Quotations in *Methods of geometry* are the present author's translations from that edition. Also available as *Foundations of geometry* (*Grundlagen der Geometrie*), second edition, translated by Leo Unger from the tenth German edition, revised and enlarged by Paul Bernays, LaSalle, IL: Open Court, 1971. ISBN: 0-875-48163-9. LC: QA681.H5813.

———. 1900. Mathematische Probleme. *Nachrichten von der Königlichen Gesellschaft der Wissenschaften zu Göttingen*, Mathematisch-Physikalisch Klasse, 253–297. A translation by Mary Winston Newson appeared as "Mathematical problems," *Bulletin of the American Mathematical Society*, 8 (1902):437–479. In this August 1900 address to the International Congress of Mathematicians in Paris, Hilbert suggested a list of open problems that might guide mathematics into the twentieth century. History has demonstrated the accuracy of his judgement.

Hilbert, David, and Stefan Cohn-Vosson. [1932] 1952. *Geometry and the imagination.* Translated by P. Nemenyi. New York: Chelsea Publishing Company. ISBN: 0-828-40087-3. LC: QA681.H6. The original Springer book *Anschauliche Geometrie* was based on Hilbert's l920–1921 lectures on visual geometry, written up by his assistant W. Rosemann. Its chapter II analyzes the frieze, wallpaper and crystal groups. The present author's love for geometry stems from this book.

Hilton, Peter, Derek Holton, and Jean Pedersen. 1997. *Mathematical reflections in a room with many mirrors.* New York: Springer-Verlag New York. ISBN: 0-387-94770-1. LC: QA93.H53. Chapter 5 of this lively book uses quilting as a route to the study of symmetry.

Hull, Lewis, Hazel Perfect, and J. Heading. 1978. Pythagoras in higher dimensions: three approaches. *The mathematical gazette* 62:206–211.

Huntington, Edward V. 1902. A complete set of postulates for the theory of absolute continuous magnitude. *Transactions of the American Mathematical Society* 3:264–279. One of the first axiomatizations of the real number system. Good bibliography.

International Congress of Philosophy. 1900. *Bibliothèque du Congrès International de Philosophie.* Volume 3, *Logique et histoire des sciences.* Paris: Librairie Armand Colin. LC: B20.I6.

Jeger, Max. [1964] 1966. *Transformation geometry.* Translated by A. W. Deicke and A. G. Howson. New York: John Wiley & Sons. LC: QA601.J4. Published originally by Räber Verlag as *Konstruktive Abbildungsgeometrie,* this little book roughly parallels chapter 6 of *Methods of geometry.* It pursues plane similarities and affine transformations more deeply, but pays little attention to analytic methods. Jeger's construction problems are ingenious and intriguing. A Luzerne secondary-school teacher, he died in 1991.

Johnson, Roger A. [1929] 1960. *Advanced Euclidean geometry (formerly titled 'Modern geometry'): An elementary treatise on the geometry of the triangle and the circle.* Edited by John Wesley Young. New York: Dover Publications. LC: QA474.J6. Probably the most comprehensive book of its genre.

Jones, Owen. [1856] 1972. *The grammar of ornament: Illustrated by examples of various styles of ornament.* New York: Van Nostrand Reinhold. ISBN: 0-442-24175-5. LC: NK1510.J7. The classic text on ornamentation. Look for the first edition in a rare-book library; its 112 color plates are considerably more vivid than those of the reprint.

Jones, William. 1706. *Synopsis palmarium matheseos, or, a new introduction to the mathematics.* London: J. Matthews. LC: QA35.J6. Cajori (1919, 158) claims this is the first publication to use the letter π (page 263) for the length of a semicircle with radius 1. The title word *palmarium* apparently refers to its small size; it's a *hand*book.

Kant, Immanuel. [1781] 1965. *Critique of pure reason.* Translated by Norman Kemp Smith. New York: St. Martin's Press. ISBN: 0-312-45010-9. LC: B2778.E5.S6.

Kaplan, Elliott D., editor. 1996. *Understanding GPS: Principles and applications.* Boston: Artech House Publishers. ISBN: 0-890-06793-7. LC: G109.5K36. A technical overview of the satellite-based Global Positioning System.

Kate, Maggie. 1997. *North American Indian motifs: CD-ROM and book.* Mineola, NY: Dover Publications. ISBN: 0-486-99945-9.

Key Curriculum Press. 1997–1998. *Mathematics catalog, grades 6–12.* Berkeley: Key Curriculum Press.

Kim, Scott. 1981. *Inversions: A catalog of calligraphic cartwheels.* Peterborough, NH: Byte Books. ISBN: 0-070-34546-5. LC: N7430.5K6. An ingenious collection of calligraphic designs that can be read upside down, backward, in a mirror, or over and over. For example, a 180° rotation turns Kim's rendition of his own name *Scott Kim* into that of the title *Inversions*.

Klein, Felix. [1908] 1939. *Elementary mathematics from an advanced standpoint: Geometry*. Translated from the third German edition by E. R. Hedrick and C. A. Noble. LC: QA461.W45. New York: Dover Publications. A broad survey of what the leading geometer of his era thought mathematics teachers in German academic secondary schools should know.

Klemm, Michael. 1982. *Symmetrien von Ornamenten und Kristallen*. Berlin: Springer-Verlag. ISBN: 0-387-11644-3. Advanced, reader-unfriendly treatment of the crystal groups in all dimensions.

Kneser, Hellmuth. 1931. Solution of problem 121. *Jahresbericht der Deutschen Mathematiker-Vereinigung* 41:69–72. In an earlier issue of the same journal, Gerhard Thomsen had asked whether the equation in exercise 6.11.39 of *Methods of geometry* is the shortest that's true for the reflections across *all* lines *a*, *b*, and *c*. Kneser showed that it is, except for some trivial cases. His argument involves paths in a regular hexagonal grid.

Kuhn, Thomas. [1962] 1970. *The structure of scientific revolutions*. Second edition. Chicago: University of Chicago Press. ISBN: 0-226-45808-3. LC: Q175.K95. This edition appeared in 1962 as volume 2, number 2, of the *International encyclopedia of unified science*.

Lakatos, Imre. 1976. *Proofs and refutations*. Edited by John Worrall and Elie Zahar. Cambridge, UK: Cambridge University Press. LC: QA8.4.L32. Dialogues concerning the philosophy and practice of mathematics, centered on the formulation of Euler's theorem about the characteristic of a polyhedron.

Larson, Roland E., and Bruce H. Edwards. 1991. *Elementary linear algebra*. Second edition. Lexington, MA: D. C. Heath & Company. ISBN: 0-669-39641-9. LC: QA184.L39. This clearly written standard text emphasizes applications more than most.

Legendre, Adrien-Marie. [1834] 1848. *Elements of geometry and trigonometry*. Translated by David Brewster, adapted by Charles Davies. New York A. S. Barnes & Company. LC: QA429.L42. A popular text by a mathematician of world stature.

Lekson, Stephen H. [1984] 1986. *Great Pueblo architecture of Chaco Canyon, New Mexico*. Albuquerque: University of New Mexico Press. ISBN: 0-826-30843-0. LC: E99.P9.L45. Originally published by the United States, Department of the Interior, National Park Service.

Levy, Lawrence S. 1970. *Geometry: Modern mathematics via the Euclidean plane*. Boston: Prindle, Weber & Schmidt. LC: QA447.L48. Two chapters of this elegant little text are roughly equivalent to chapter 6 of *Methods of geometry*, but include more material on similarities. The other two cover inversive geometry, using complex numbers. An appendix contains

Bolyai's theorem on polygons with equal area, the tacit but fundamental background for the axiomatic area theory in section 3.8 of *Methods of geometry*. Another gives an elementary proof of Gauss' fundamental theorem of algebra!

Light, Sol Felty, et al. 1961. *Intertidal invertebrates of the central California coast*. Second edition. Berkeley: University of California Press. ISBN: 0-520-02113-4. LC: QL164.L53.

Marchisotto, Anna Elena. 1990. *Mario Pieri: His contribution to mathematics and mathematics education*. Ph.D. dissertation, New York University. Includes a comprehensive bibliography of works by and about Pieri, and a translation of Pieri 1899.

Martin, George E. 1982. *Transformation geometry: An introduction to symmetry*. New York: Springer-Verlag. ISBN: 0-387-90636-3. LC: QA601.M36. The present author's favorite book on his favorite subject.

Mathematics magazine. 1984. Problem 1181. 57:304–305.

———. 1989. Problem 1333. 62:352.

McCleary, John. 1982. An application of Desargues' theorem. *Mathematics magazine* 55:233–235.

McLeay, Heather. 1994. A closer look at the cast ironwork of Australia. *Mathematical intelligencer* 16:61–65. Application of frieze group analysis.

Melzak, Zdzisław Alexander. 1983. *Invitation to geometry*. New York: John Wiley & Sons. ISBN: 0-471-09209-6. LC: QA445.M43. Brief, elegant survey.

Meredith, David. 1993. *X(Plore): Version 4.0 for DOS compatible computers*. Englewood Cliffs, NJ: Prentice-Hall. ISBN: 0-13-014226-3. This mathematical desktop software is adequate for the computations in *Methods of geometry*, except for symbolic algebra. It's by far the easiest to use, and the least expensive. *All* figures in *Methods of geometry* were produced by *X(Plore)*, except photographs and scans from other publications. Text was added to figures via *WordPerfect*.

Miller, Arthur I. 1972. The myth of Gauss' experiment on the Euclidean nature of physics. *Isis* 63:345–348. Miller's argument is attacked in the associated comments of George Goe and B. L. van der Waerden, ibid., 65(1974):83–87.

Modenov, P. S., and A. S. Parkhomenko. [1961] 1965. *Geometric transformations*. Volume 1, *Euclidean and affine transformations*. Translated and adapted from the first Russian edition by Michael B. P. Slater. New York: Academic Press. LC: QA601.M613. One of the earliest books on transformational geometry, this survey of basic results has no exercises. It was intended for teacher training and as supplementary material for various elementary courses. Volume 2 is titled *Projective transformations*.

Moise, Edwin E. [1963] 1990. *Elementary geometry from an advanced standpoint*. Third edition. Reading, MA: Addison-Wesley Publishing Company.

ISBN: 0-201-50867-2. LC: QA445.M58. A marvelous introduction to foundations of geometry, this was originally written for prospective teachers of the SMSG approach to geometry (Moise and Downs 1964), but can serve you very well, too.

———, and Floyd L. Downs Jr. 1964. *Geometry*. Teacher's edition. Reading, MA: Addison-Wesley Publishing Company. LC: QA453.M85. The first secondary-school text produced by the SMSG, or New Math, movement (Wooton 1965). Moise was a Harvard research mathematician who had specialized in topology. Downs was a teacher at Newton, MA, High School. Their mathematics is impeccable, and their overall content fairly traditional, except for their inclusion of solid geometry. The book has great appeal for professional mathematicians, who often don't understand why this approach to school geometry was so widely misunderstood and so often failed in practice. Moise [1963] 1990 considers this approach from an advanced viewpoint.

Mostow, George D., Joseph H. Sampson, and Jean-Pierre Meyer. 1963. *Fundamental structures of algebra*. New York: McGraw-Hill Book Company. LC: QA154.M68. An exceptionally comprehensive and clear algebra text written for a freshman (!) course at Yale.

Mueller, Ian. 1981. *Philosophy of mathematics and deductive structure in Euclid's Elements*. Cambridge, MA: MIT Press. ISBN: 0-262-13163-3. LC: QA31.E9M83. Good bibliography.

Muir, Thomas, and James Thomson. 1910. Letters to the editor. *Nature* 83:156, 217, 459–460. These establish that Muir and Thomson's father, professors at St. Andrews and Glasgow, coined the word *radian* around 1870. The second writer's brother was William Thomson, Lord Kelvin—a renowned physicist.

Müller, Edith. 1944. *Gruppentheoretische und strukturanalytische Untersuchungen der maurischen Ornamente aus der Alhambra in Granada* (Inaugural-Dissertation zur Erlangung der philosophischen Doktur-würde, vorgelegt der Philosophischen Fakultät II der Universität Zürich). Rüschlikon: Buchdruckerei Baublatt. In this doctoral thesis supervised by Andreas Speiser, Paul Niggli's niece presents the frieze and wallpaper groups in detail, then uses them to classify the ornamentation of the famous Moorish palace in Granada.

Niggli, Paul. 1924. Die Flächensymmetrien homogener Diskontinuen. *Zeitschrift für Kristallographie*, 60:283–298. The first clear presentation of the analysis and classification of the wallpaper groups.

———. 1926. Die regelmäßige Punktverteilung längs einer Geraden in einer Ebene. *Zeitschrift für Kristallographie*, 63:255–274. The first explicit presentation of the analysis and classification of the frieze groups.

Niven, Ivan M. 1956. *Irrational numbers*. Buffalo: Mathematical Association of America. ISBN: 0-88385-011-7. LC: QA247.5.N57.

Padoa, Alessandro. [1901] 1967. Logical introduction to any deductive theory. In *From Frege to Gödel: A source book in mathematical logic, 1879–1931*, edited by Jean van Heijenoort, Cambridge, MA: Harvard University Press, 1967, 118–123. ISBN: 0-674-32450-1. LC: QA9.V3. The editor's translation of "Essai d'une théorie algébrique des nombres entiers, précédé d'une introduction logique à une théorie déductive quelconque," in International Congress of Philosophy 1900, 309–365. The first clear, complete description of the axiomatic method as now used in higher mathematics.

Pascal, Blaise. [1658] 1948. The mind of the geometrician. The art of persuasion. In *Great shorter works of Pascal*, translated by Emile Cailliet and John C. Blankenagel, 189–211. Philadelphia: The Westminster Press. ISBN: 0-837-16072-3. LC: B1900.E5.C3. These papers, the first complete expositions of the axiomatic method, weren't published in entirety until long after Pascal's 1662 death.

Pasch, Moritz. [1882] 1926. *Vorlesungen über die neuere Geometrie*. Second edition. With an appendix, "Die Grundlegung der Geometrie in historischer Entwicklung," by Max Dehn. Berlin: Verlag von Julius Springer. Pasch was the first to apply really precise and rigorous techniques in foundations of geometry, although he hadn't adopted the abstract view of the axiomatic method. Dehn was a student of Hilbert and a major researcher in foundations of geometry.

Pearce, Peter. 1978. *Structure in nature is a strategy for design*. Cambridge, MA: MIT Press. ISBN: 0-262-16064-1. LC: NA2750.P4. Geometric design in architecture.

Peckhaus, Volker. 1994. Hilbert's axiomatic programme and philosophy. In *The history of modern mathematics*, volume 3, *Images, ideas, and communities*, edited by Eberhard Knobloch and David E. Rowe, Boston: Academic Press, 91–111. ISBN 0-12-599663-2. LC: QA21.S98. A technical study of Hilbert's program for establishing the consistency of mathematics.

Pedoe, Dan. 1976. *Geometry and the liberal arts*. New York: St. Martin's Press. ISBN: 0-312-32370-0. LC: QA445.P44.

Pieri, Mario. 1899. Della geometria elementare come sistema ipotetico deduttivo. Monografia del punto e del moto. *Memorie della Reale Accademia delle Scienze di Torino*, second series 49:173–222. This is translated and published as "Of the elementary geometry as a hypothetical-deductive system: Monograph of point and motion," Appendix C of Marchisotto 1990.

———. 1901. Sur la géométrie envisagée comme un système purement logique. In International Congress of Philosophy 1900, 367–404. Quotations in *Methods of geometry* from this paper are the present author's translations.

Plato. [1965] 1971. *Timaeus and Critias*. Translated with an introduction and an appendix on *Atlantis* by Desmond Lee. Harmondsworth, UK:

Penguin Books. ISBN: 0-14-044261-8. LC: B387.A5.L3. Contains Plato's theory of the structure of matter, based on properties of the regular polyhedra. *Warning*: At least one other edition of Plato—a Jowett translation—purges this material from *Timaeus* with no comment!

Playfair, John. [1795] 1860. *Elements of geometry*: *Containing the first six books of Euclid, with a supplement on the quadrature of the circle, and the geometry of solids, to which are added, elements of plane and sphericale trigonometry*. Philadelphia: J. D. Lippincott & Company. The source of the form of the parallel axiom used in *Methods of geometry*. Playfair is more noted as geologist than mathematician.

Pólya, Georg. 1924. Über die Analogie der Kristallsymmetrie in der Ebene. *Zeitschrift für Kristallographie* 60:278–82. This letter to the editor by one of the most influential and revered mathematical figures of this century is one of the first reports of the group-theoretic analysis of plane ornaments. It's quite readable, because Pólya only outlines the work. The immediately following paper by the editor, Niggli 1924, provided the details.

Preparata, Franco P., and Michael Ian Shamos. 1985. *Computational geometry*: *An introduction*. New York: Springer-Verlag. ISBN: 0-387-96131-3. LC: QA447.P735. Stemming from Shamos' 1978 Yale Ph.D. thesis, this book organized a new field: the geometry that underlies software for graphics, geometric design, and robot control.

Rademacher, Hans, and Otto Toeplitz. [1933] 1957. *The enjoyment of mathematics*: *Selections from mathematics for the amateur*. Translated and expanded by Herbert Zuckerman. Princeton: Princeton University Press. ISBN: 0-691-02351-4. LC: QA95.R2. Originally published as *Von Zahlen und Figuren*: *Proben mathematischen Denkens für Liebhaber der Mathematik*. Toeplitz worked with Hilbert in Göttingen during the 1920s.

Rausenberger, O. 1915. Konvexe pseudoreguläre Polyeder. *Zeitschrift für mathematische und naturwissenschaftliche Unterricht,* (1915): 35–142. This paper, targeted at German secondary-school teachers, analyzes the convex deltahedra.

Reid, Constance. 1970. *Hilbert*. With an appreciation of Hilbert's mathematical work by Hermann Weyl. New York: Springer-Verlag. ISBN: 0-387-94674-8. LC: QA29.H5.R4. Reid is a writer, not a mathematician. Her sister Julia Robinson made a major contribution to the solution of the problem in Hilbert 1900 on Diophantine equations.

Rich, Barnett. 1989. *Schaum's outline of theory and problems of geometry*. Second edition. Revised by Philip A. Schmidt. New York: McGraw-Hill. ISBN: 0-07-052246-4. LC: QA445.R53. A good source of elementary exercises.

Richards, Joan L. 1988. *Mathematical visions*: *The pursuit of geometry in Victorian England*. Boston: Academic Press. ISBN: 0-12-587445-6.

LC: QA443.5.R53. Nineteenth-century English elite had to demonstrate mastery of Euclidean geometry before studying other subjects. The author assesses the effect of that tradition.

Russell, Bertrand A. W. [1897] 1956. *An essay on the foundations of geometry.* With a new foreword by Morris Kline. New York: Dover Publications. ISBN: 0-415-14146-X. LC: QA681.R96. This philosophical study arose from Russell's Cambridge Ph.D. dissertation, written under Alfred North Whitehead's supervision. It's a particularly good survey of Kant's ideas and the nineteenth-century reaction to the discovery of non-Euclidean geometry. It doesn't necessarily represent Russell's later views.

———. [1951] 1967. *The autobiography of Bertrand Russell, 1872–1914.* Boston: Little, Brown & Company. ISBN: 0-049-21003-3. LC: B1649.R94. A32. Russell was a major figure in mathematics, logic, pacifism, analytical philosophy, and history of philosophy.

Russo, Lucio. 1998. The definitions of fundamental geometric entities contained in Book I of Euclid's *Elements. Archive for history of exact sciences,* 52:195–217.

Schattschneider, Doris. 1978. The plane symmetry groups: Their recognition and notation. *American mathematical monthly* 85:439–450. Schattschneider does us a service by discussing methods for distinguishing the wallpaper groups and comparing seven systems of notation for them.

———. 1990. *Visions of symmetry: Notebooks, periodic drawings, and related work of M. C. Escher.* New York: W. H. Freeman and Company. ISBN: 0-7167-2126-0. LC: NC263.E83A4. Analysis of Escher's art is a full subdiscipline of geometry. This is the definitive work. Its bibliography will lead you to the rest of the Escher literature.

Schneer, Cecil J. 1983. The Renaissance background to crystallography. *American scientist* 71:254–263.

Schönflies, Arthur. 1891. *Kristallsysteme und Kristallstruktur.* Leipzig: B. G. Teubner. LC: QD911.S65. One of the original works that classified the 230 crystal groups. See also Fedorov [1891] 1971 and Burckhardt's historical account (1967/1968).

Schwartzman, Steven. 1994. *The words of mathematics: An etymological dictionary of mathematical terms used in English.* Washington: Mathematical Association of America. ISBN: 0-88385-511-9. LC: QA5.S375.

Seidenberg, Abraham. 1974. Did Euclid's *Elements,* Book I, develop geometry axiomatically? *Archive for history of the exact sciences* 14:263–295. Seidenberg, a researcher in algebraic geometry, pursued as a sideline the history of ancient mathematics. Good bibliography.

Senechal, Marjorie. 1990. Finding the finite groups of symmetries of the sphere. *American mathematical monthly* 97:329–335.

———. 1995. *Quasicrystals and geometry.* Cambridge, UK: Cambridge University Press. ISBN: 0-521-57541-9. LC: QD926.S46. The recent discovery of aperiodic crystal structures surprised crystallographers and geometers alike. This monograph studies them with techniques presented in chapter 8 of *Methods of geometry*.

———, and George Fleck, editors. 1977. *Patterns of symmetry*. Amherst: University of Massachusetts Press. ISBN: 0-87023-232-0. LC: NX650.P65. P37. This collection of papers presented at the world's first Symmetry Festival includes studies in art, biology, crystallography, literature, and music.

———, and George Fleck, editors. 1988. *Shaping space: A polyhedral approach*. Boston: Birkhäuser Boston, Inc. ISBN: 0-8176-3351-0. LC: N7430.5S52. This collection of papers presented at or inspired by the 1984 Shaping Space conference addresses a general audience, includes material directed at secondary schools, and has an excellent bibliography.

———, and R. V. Galiulin. 1984. An introduction to the theory of figures: The geometry of E. S. Fedorov. *Topologie structurale/Structural topology* 10:5–22. A synopsis of Fedorov's first monograph, which led to his major crystallographic work. The article appears in both French and English, and includes a biographical sketch and portrait.

Shubnikov, Aleksei Vasilevich, and Vladimir Aleksandrovich Koptsik. 1974. *Symmetry in science and art*. Translated by G. D. Archard, edited by David Harker. New York: Plenum Press. ISBN: 0-306-30759-6. LC: Q172.5.S95.S4913. Comprehensive treatment of symmetry analysis and its applications, including many extensions of the techniques of *Methods of geometry*.

Sivaramamurti, C. 1966. *Our national emblem*. LC: CR115.I4.S57. New Delhi: Government of India, Ministry of Information & Broadcasting, Publications Division.

Smith, David Eugene. 1924. *Essentials of solid geometry*. Boston: Ginn and Company. LC: QA457.S58. The present author studied this text in secondary school. It was adapted from the earlier, more comprehensive Wentworth and Smith [1888] 1913.

———, editor. [1929] 1959. *A source book in mathematics*. New York: Dover Publications. LC: QA3.S63. Includes translations of the original sources of several results in *Methods of geometry*.

Speiser, Andreas. 1927. *Theorie der Gruppen von endlicher Ordnung: Mit Anwendung auf algebraische Zahlen und Gleichungen sowie auf die Kristallographie*. Second edition. Berlin: Springer-Verlag. LC: QA171.S6. This book's 1924 edition was one of the first modern higher-algebra texts. This edition was the first book to present the group-theoretic analysis of ornaments.

Staudt, Georg Karl Christian von. 1847. *Geometrie der Lage*. Nürnberg: Verlag der Fr. Korn'schen Buchhandlung. LC: QA471.S77. Pioneering development of synthetic geometry: concise, elegant, and readable. Section

4, number 50, is the source of the proof of Euler's theorem presented in *Methods of geometry*.

Steiner, Hans Georg. 1964. Frege und die Grundlagen der Geometrie. *Mathematisch-physikalische Semesterberichte* 10:175–186, 11:35–47. This expository paper discusses the Hilbert–Frege dispute, emphasizing technical points of Frege's philosophy.

Stevens, Peter S. 1974. *Patterns in nature*. Boston: Little, Brown and Company. ISBN: 0-316-81328-1. LC: QH81.S859. Beautifully illustrated study of symmetry in physics, biology, and geology.

Stewart, B. M. 1980. *Adventures among the toroids*: *A study of quasi-convex, aplanar, tunneled orientable polyhedra of positive genus having regular faces with disjoint interiors*; *being an elaborate description and instructions for the construction of an enormous number of new and fascinating mathematical models of interest to students of Euclidean geometry and topology, both secondary and collegiate, to designers, engineers and architects, to the scientific audience concerned with molecular and other structural problems, and to mathematicians, both professional and dilettante*; *with hundreds of exercises and search projects, many completely outlined for self-instruction*. Revised second edition. Privately published by its author. ISBN: 0-686-11936-3. LC: QA491.S75. Lettered and illustrated by hand. A professor at Michigan State University, Stewart died in 1994.

Stewart, Ian, and Martin Golubitsky. 1992. *Fearful symmetry*: *Is God a geometer?* London: Penguin Books. ISBN: 0-14-013047-0. LC: Q172.5.S95.S74. A study of symmetry in physics and biology. Its treatment of *breaking* symmetry is particularly interesting.

Symmetrics. No date. Polyhedron kits. Elkins Park, PA: Symmetrics. These classy kits consist of accurately beveled white plastic faces for various regular and semi-regular polyhedra. You must glue them together.

Synthese. 1977. Volume 34. This entire volume of the philosophical journal is dedicated to Hans Reichenbach. It contains several studies emphasizing conventionalism.

Thomas, George B., Jr., and Ross L. Finney. [1951] 1979. *Calculus and analytic geometry*. Fifth edition. Reading, MA: Addison-Wesley Publishing Company. ISBN: 0-130-14226-3. LC: QA303.T342. Originally published in 1951, this is still a standard text.

Thomsen, Gerhard. 1933a. *Grundlagen der Elementargeometrie in gruppenalgebraischer Behandlung*. Leipzig: B. G. Teubner. LC: QA681.T4. During the 1920s Thomsen discovered how to formulate many familiar properties of plane figures in terms of equations involving compositions of half turns and line reflections. His book contains that work and the start of its applications to the axiomatic foundations of geometry, projects pursued later by a school of geometers including the present author. Thomsen 1933b is an English summary.

———. 1933b. The treatment of elementary geometry by a group-calculus. *Mathematical gazette* 17:230–242. A translated excerpt from Thomsen 1933a, this is the only paper in English that describes Thomsen's work in any detail.

Toepell, Michael-Markus. 1986. *Über die Entstehung von David Hilberts „Grundlagen der Geometrie"*. Göttingen: Vandenhoeck & Ruprecht. ISBN: 3-525-40309-7. LC: QA681.H582. A study of the genesis of Hilbert 1899.

Trigg, Charles W. 1978. An infinite class of deltahedra. *Mathematics magazine* 51:55–57.

Trudeau, Richard J. 1987. *The non-Euclidean revolution*. Introduction by H. S. M. Coxeter. Boston: Birkhäuser Boston. ISBN: 0-8176-3311-1. LC: QA685.T75. Lively text with a good bibliography.

Tuller, Annita. 1967. *A modern introduction to geometries*. New York: Van Nostrand Reinhold Company. LC: QA473.T8. A survey of Euclidean and non-Euclidean geometries written for prospective and in-service secondary-school teachers.

W. H. H. H. 1883. Letters to the editor. *Nature* 28:87,104. The origin of the terms *circumcircle, excircle, incircle*, etc.

Wade, David. 1982. *Geometric patterns & borders*. New York: Van Nostrand Reinhold Company. ISBN: 0-442-29241-4. LC: NK1570.W28. A handbook for designers.

Waerden, Bartel Leendert van der. 1963. *Science awakening*. Translated by Arnold Dresden. New York: John Wiley & Sons. ISBN: 0-195-19076-9. LC: QA22.W214. A fascinating history of ancient mathematics by one of the founders of modern algebra, who turned historian later in his career.

Washburn, Dorothy K., and Donald W. Crowe. 1988. *Symmetries of culture: Theory and practice of plane pattern analysis*. Seattle: University of Washington Press, 1988. ISBN: 0-295-96586-X. LC: NK1570.W34. A wonderful account of symmetry analysis as an anthropological and archaeological technique. Washburn is the anthropologist; Crowe, the geometer. The book is profusely illustrated. Its extensive bibliography will lead you deeper into this field.

Wentworth, George, and David Eugene Smith. [1888] 1913. *Plane and solid geometry*. Boston: Ginn and Company. LC: QA453.W28. A popular secondary-school text. Its latter half, on solid geometry, was adapted and published separately as Smith 1924.

Wetzel, John E. 1978. On the division of the plane by lines. *American mathematical monthly* 75:647–656.

Weyl, Hermann. 1952. *Symmetry*. Princeton: Princeton University Press. ISBN: 0-691-02374-3. LC: N76.W4. This beautifully illustrated little book, by a renowned mathematician, is the best introduction to the group-theoretic analysis of symmetry and its application in art.

Whitehead, Alfred North. [1907] 1971. *The axioms of descriptive geometry.* New York: Hafner Publishing. In 1907, *descriptive geometry* meant the consequences of the incidence axioms.

Wiener, Hermann. 1890–1893. Über geometrische Analysen. *Berichte über die Verhandlungen der Königlichen Sächsischen Gesellschaft der Wissenschaften zu Leipzig, Mathematisch-Physikalische Klasse*, 52:13–23, 71–87, 245–267; 53:424–447, 644–673; 55:555–598. This series' title varies from installment to installment, but the table of contents in its last part shows how all it's all organized. Although isolated results date back to Euler, this is the first major work *devoted* to transformational geometry in the style of *Methods of geometry.* Wiener starts with the problem of describing the composition of two screws, then builds a general two- and three-dimensional theory. His writing—particularly in the third installment—is hampered by a lack of convenient terminology and notation for bijections.

Wilder, Raymond L. 1968. *Evolution of mathematical concepts: An elementary study.* New York: John Wiley & Sons. LC: QA9.W57. A researcher in topology, Wilder worked in history of mathematics as a sideline, pioneering the application of rigorous historiographic techniques.

Wills, Gary. 1978. *Inventing America: Jefferson's Declaration of Independence.* New York: Vintage Books. ISBN: 0-394-72735-5. LC: E221.W64. Word-by-word textual analysis by a journalist trained in philosophy.

Winternitz, Arthur. 1940. Zur Begründung der projektiven Geometrie: Einführung idealer Elemente unabhängig von der Anordnung. *Annals of mathematics*, second series, 51:365–390.

Wooton, William. 1965. *SMSG: The making of a curriculum.* New Haven: Yale University Press. LC: QA11.W74. A history of the beginning of the New Math movement by one of its founders. SMSG produced the text Moise and Downs 1964, on which the axiomatization in *Methods of geometry* is based.

Yale, Paul B. [1968] 1988. *Geometry and symmetry.* New York: Dover Publications. ISBN: 0-816-29964-1. LC: QA477.Y3. Originally published by Holden-Day. Yale's book is probably the most readable introduction to material beyond that in chapter 8 of *Methods of geometry.*

Index